MOIRE ANALYSIS OF STRAIN

PRENTICE-HALL INTERNATIONAL, INC., *London*
PRENTICE-HALL OF AUSTRALIA, PTY. LTD., *Sydney*
PRENTICE-HALL OF CANADA LTD., *Toronto*
PRENTICE-HALL OF INDIA PRIVATE LTD., *New Delhi*
PRENTICE-HALL OF JAPAN, INC., *Tokyo*

MOIRE ANALYSIS OF STRAIN

A. J. Durelli
V. J. Parks

Mechanics Division
The Catholic University of America

Prentice-Hall, Inc., Englewood Cliffs, New Jersey

Other Books on Stress Analysis by A.J. DURELLI and Associates

A.J. Durelli, E.A. Phillips, and C.H. Tsao, *Introduction to the Theoretical and Experimental Analysis of Stress and Strain.* New York: McGraw-Hill Co., 1958.

A.J. Durelli and W.F.Riley, *Introduction to Photomechanics.* Englewood Cliffs, N.J.: Prentice-Hall, Inc., 1965.

A.J. Durelli, *Applied Stress Analysis.* Englewood Cliffs, N.J.: Prentice-Hall, Inc., 1967.

Current printing (last digit):
10 9 8 7 6 5 4 3 2 1

13-599605-8
Library of Congress Catalog Card Number: 79–106149
Printed in the United States of America

ACKNOWLEDGMENTS

This book has developed from several research programs conducted by the authors with the support of the National Science Foundation and the Army Research Office (Durham). Without this help the writing of the book would not have been possible. The authors are particularly grateful to Dr. M. Gaus of NSF and Dr. S. Kumar of ARO(D) for their continuous understanding of the work and recommendations of financial support.

Several chapters of the book follow closely the text of papers published by the authors with several of their former associates, in particular C. Sciammarella, I. Daniel, S. Morse, W. F. Riley, and Fu-pen Chiang. The authors are indebted to J. A. Clark for his special contribution to Chapter 20, the section in Chapter 1 devoted to diffraction properties, and for his many very valuable suggestions.

Two courses have been based on the manuscript of this book, one organized by the Mechanical Engineering Department of the University of Strathclyde, in Glasgow, Scotland, and the other organized by the Instituto de Ingenieria of the Universidad Nacional Autonoma de Mexico. To Dr. A. S. Tooth and Dr. Juan Casillas, the authors are grateful for the opportunity of delivering their ideas to faculties and students.

A. J. Durelli
V. J. Parks

Washington, D.C.

CONTENTS

ESTHETIC CONSIDERATIONS: A PREFACE

Moire, the last addition to the experimental stress analyst's battery of tools, and one of the most refined and precise means of measurement, is one of the simplest, most primitive natural phenomena. The early man must have noticed it in fences made of several layers of bamboo canes and in the superposition of textiles of regular and coarse pitch. The modern man cannot miss it in springtime when he removes mosquito screens from the attic and puts them on his windows. The observation of this simple phenomenon of shadows, as the observation of everything that is simple in nature, can be the opportunity for some philosophical thinking, and the deepening of our knowledge of nature. Moire offers in particular the opportunity of esthetic experiences which deserve comments and careful analysis.

The eye is powerfully attracted by moire patterns. The attraction is due, to a large extent, to the changing character of these patterns. This variability is not found only in moire, and to clarify the point and increase understanding it is appropriate to start by reviewing changing patterns offered by other experimental stress analysis techniques. It will be seen later that moire presents distinctive effects and that in some cases may present effects more striking than those produced by the other techniques.

Two other features of moire patterns will also be described and illustrated. Moire patterns can easily be associated by the observer with images of the physical world and, in certain cases, moire patterns may be quite beautiful. These two features can also be found in other experimental stress analysis techniques.

1. Changing Patterns Obtained Using Whole-Field Experimental Stress Analysis Methods, Other than Moire

Photoelasticity, brittle coatings, and grids can present to the educated eye the effect of moving patterns. Figure P.1 shows what happens to a transparent

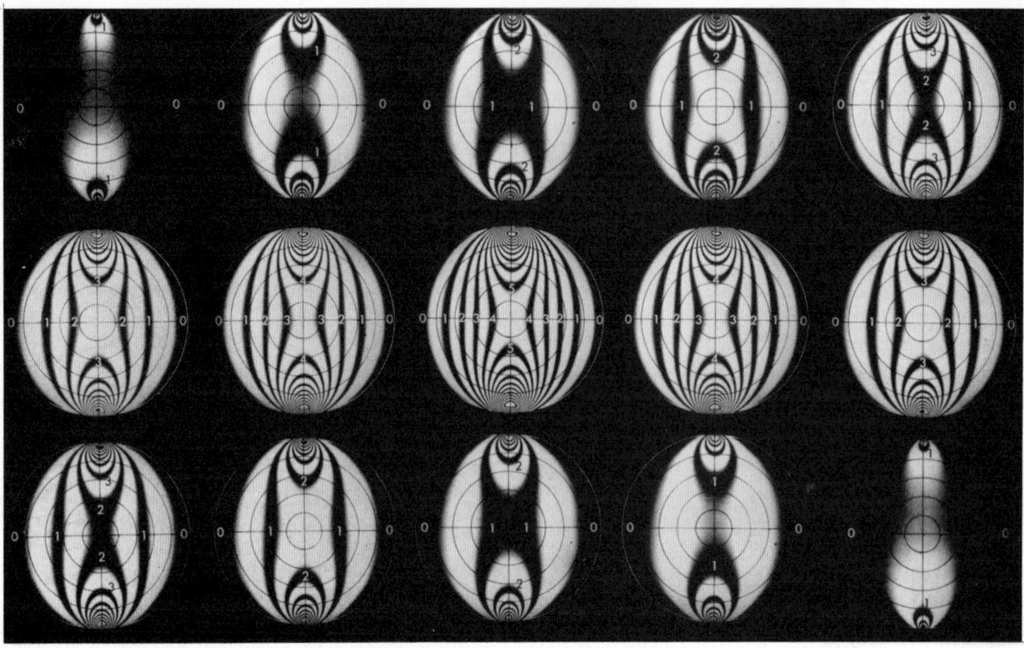

Figure P.1. Loading and Unloading. Isochromatics in a circular disk subject to a progressively increasing, and then decreasing, load.

disk in a field of polarized light when the disk is subjected to a progressively increasing diametral load and later the load is progressively decreased back to zero. The isochromatic-fringe orders increase and decrease with the increasing or decreasing of the load and at high levels of load cover the whole field of the disk. The effect can be obtained in soft materials like polyurethane rubbers by the application of small loads to these materials, when they are in a field of polarized light. (The necessary polariscopes and photographic cameras are relatively inexpensive and easy to use.) These patterns, however, are all proportional to the load.

More striking is the wave-propagation phenomenon observed with polarized light in this kind of material. Figure P.2 shows one of the earliest dynamic photoelasticity isochromatics. The eye is impressed by the transmission of stresses inside the body, from the point of load application to the point of support of the disk, and by the free disk bouncing in the air after the end of the impact. The evidence that the support of the disk does not know that the disk has been hit until much after the impact is particularly striking. Here the pattern is no longer proportional to the load, as it was in Fig. P.1, but changes continuously as time passes. Waves in the material used for this test travel at 2,500 inches per second, and high-speed motion-picture cameras are usually necessary to stop the phenomenon. The eye reconstructs it by looking at the photographs successively. (The instrumentation necessary to produce the impact and record the pictures is relatively expensive and requires appreciable skill in the operator.)

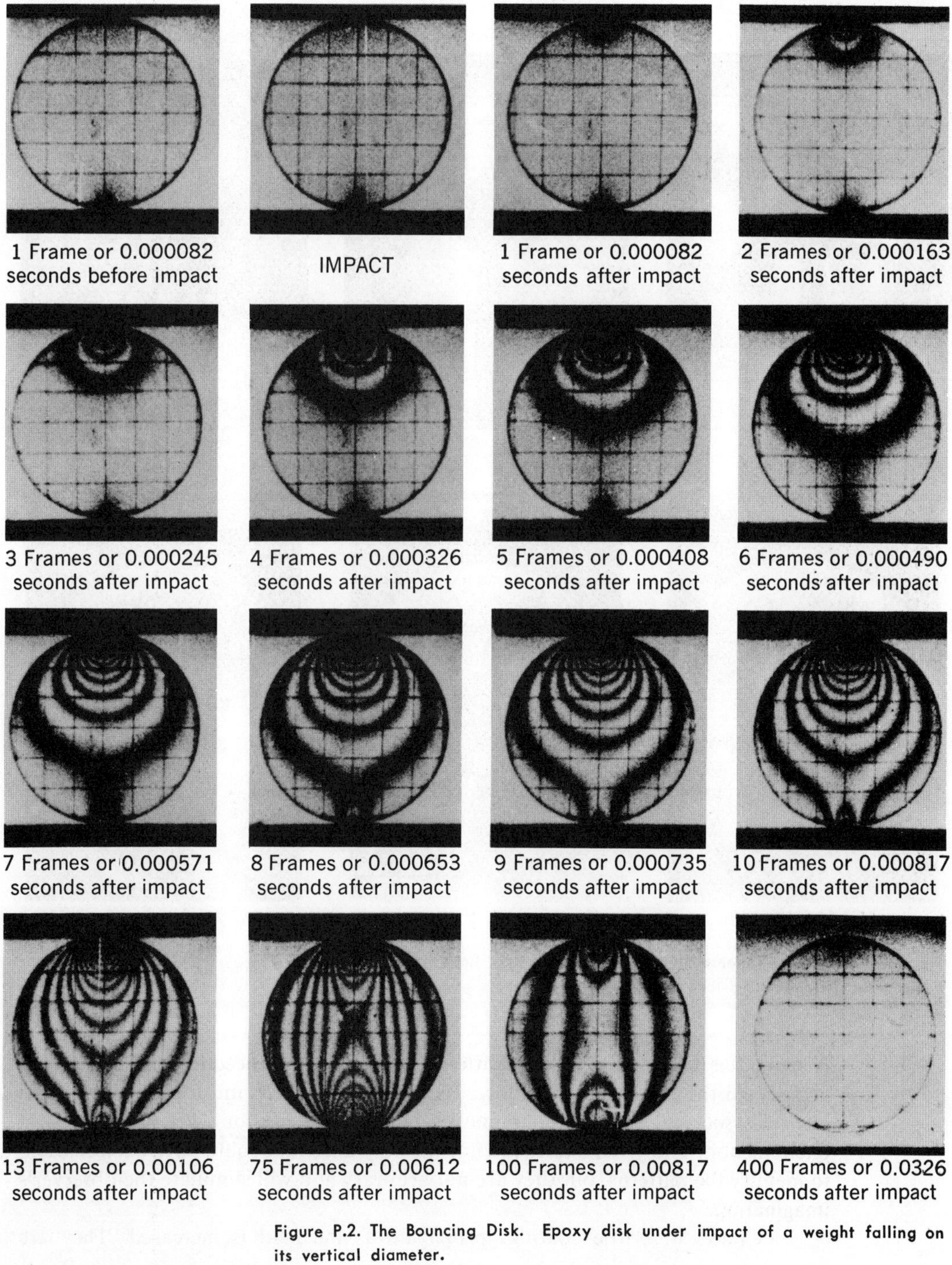

Figure P.2. The Bouncing Disk. Epoxy disk under impact of a weight falling on its vertical diameter.

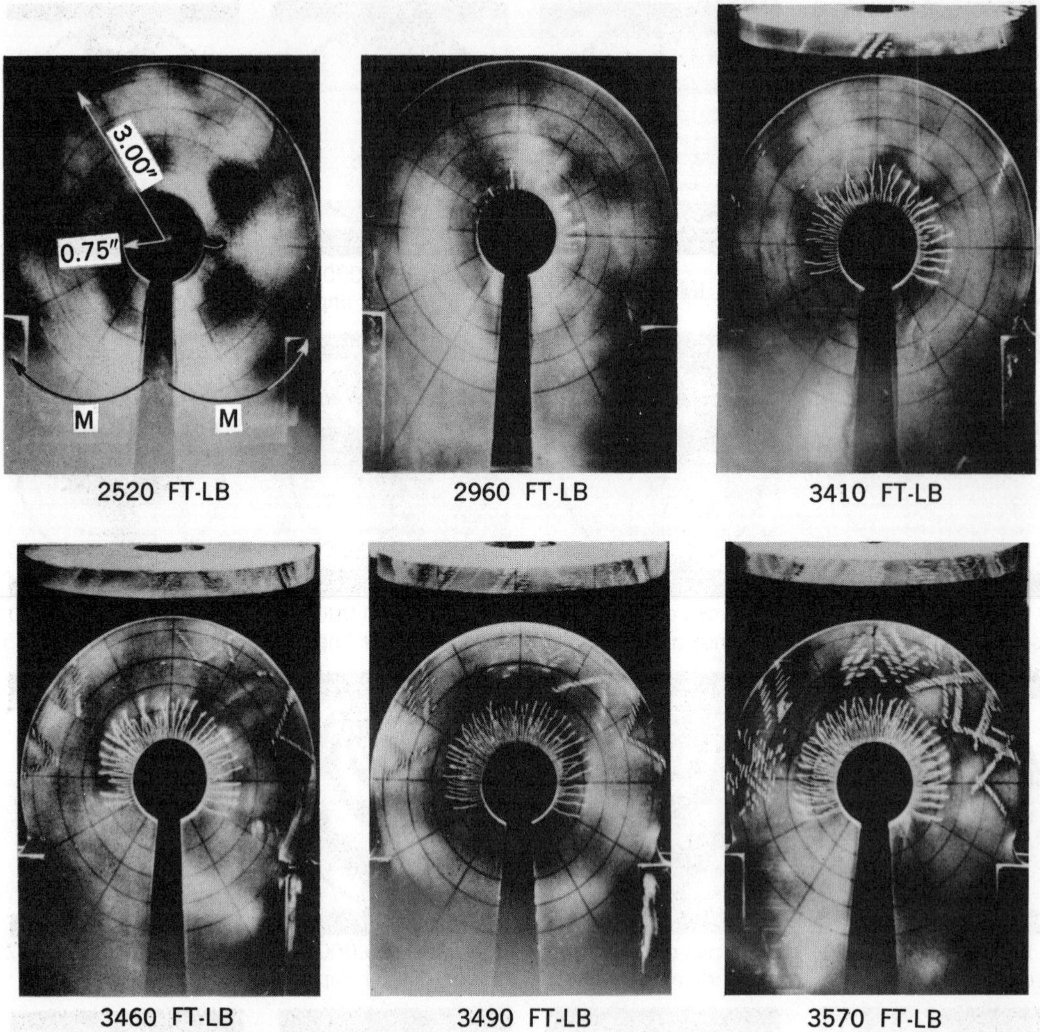

Figure P.3. Attack and Defense. Propagation of Lueders' lines in a mild-steel curved beam under pure bending (brittle-coating patterns).

A name has been given to this pattern and the same procedure will be used to identify all the patterns that follow. As will be seen later, most of these patterns can be associated with familiar images of everyday life, or with some physical phenomenon. The given names are not the only ones that could have been selected to identify the patterns, but they are not arbitrary and will stimulate the observer's imagination.

Cracks in brittle coatings propagate as the load is increased. They are usually difficult to see and to photograph but penetrants and fluorescent powders permit the recording of the crack extension. Figure P.3 shows the propagation of Lueders' lines in a curved mild-steel beam subjected to successively increasing levels of bending moment. The eye is attracted by the locally erratic behavior of the coating, within a general geometric regularity. The first cracks to appear are radial and penetrate deep in the field. The cracks appearing later are tangential and seem to build barriers that will stop the further penetration of the first cracks. (The recording of these photographs requires great skill in the operator.)

Grids also permit the observation of changing shapes. Figure P.4 shows the distortion of a Cartesian grid drawn on the surface of a urethane-rubber ring subjected to diametral compression. The eye, however, is mainly impressed when all these patterns are superimposed, as in Fig. P.5. The superposition strikes the eye more deeply and produces the impression of change. (These patterns are not

Figure P.4. The Squeezed Ring. Cartesian grid on a circular ring subjected to several levels of diametral compression.

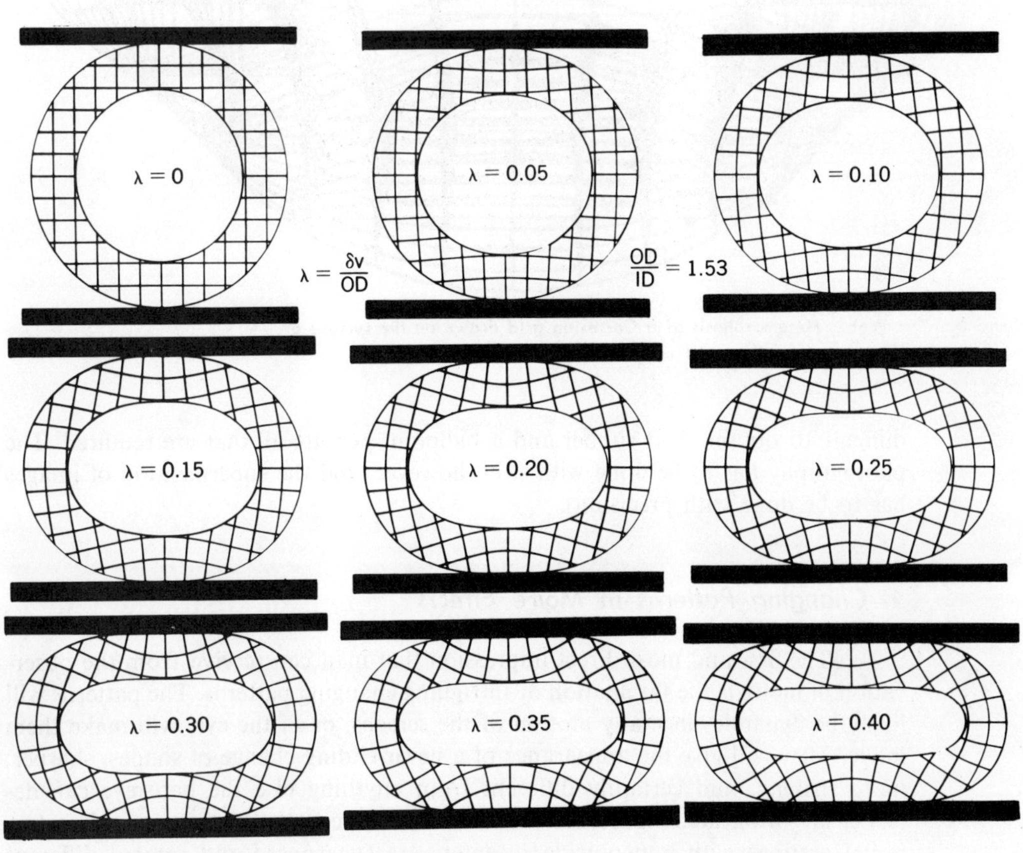

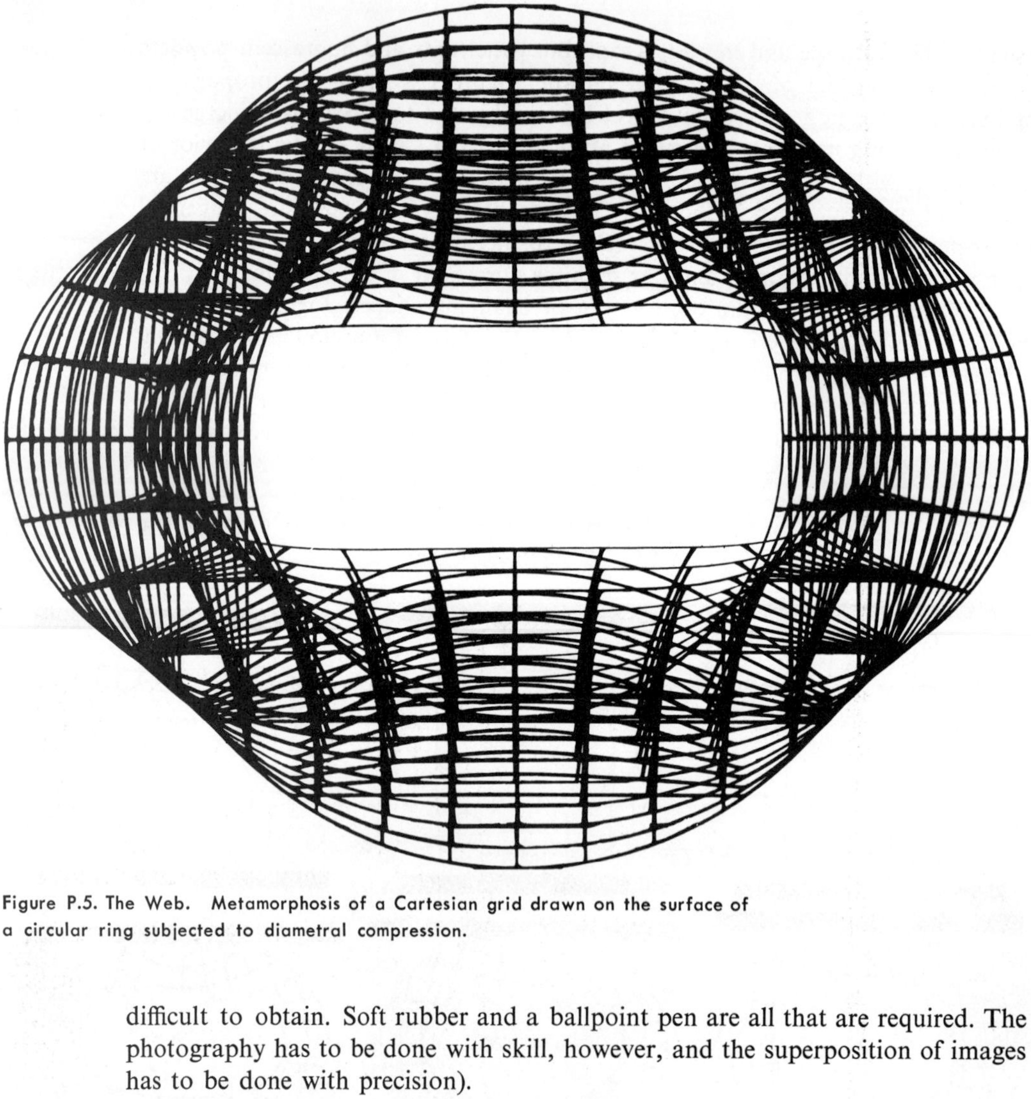

Figure P.5. The Web. Metamorphosis of a Cartesian grid drawn on the surface of a circular ring subjected to diametral compression.

difficult to obtain. Soft rubber and a ballpoint pen are all that are required. The photography has to be done with skill, however, and the superposition of images has to be done with precision).

2. Changing Patterns in Moire Effects

The first and most direct impression that man can receive from the observation of moire is the fascination of intriguing changing patterns. The patterns will likely be dynamic since any motion of the screens, or of the eye, will make them move. They will give the appearance of a never-ending change of shapes, sharper, more striking, and certainly different from anything else the bare eye can detect easily in nature. Figure P.6 shows some of the patterns obtained when two radial gratings with noncoincident centers are superposed and rotated different amounts one with respect to the other. The photographs stop the motion at certain times, but the rotation by hand, or by means of a motor, gives the impression of never-ending change.

When an elliptical grating is superposed on a grating made of parallel straight lines, and the elliptical grating is rotated, a large variety of changing patterns appear. Those shown in Figs. P.7 and P.8 are selected examples. If the ratio of the pitches of the gratings is changed, other kinds of changing patterns can be obtained. The feeling of change is even more pronounced when two elliptical gratings are used, one superposed on the other, as shown in Figs. P.9 and P.10. It would be much easier to appreciate this variability if the images were projected using a movie camera.

Some painters have tried to produce similar effects, for instance B.Cunningham in his "Equivocation"[1] and B. Riley in his "Current."[2] The main difference

[1] East Hampton Gallery, New York.
[2] The Museum of Modern Art, New York.

Figure P.6. Aurora Borealis. Isothetics associated with two radial gratings with noncoincident centers, rotating one in respect with the other.

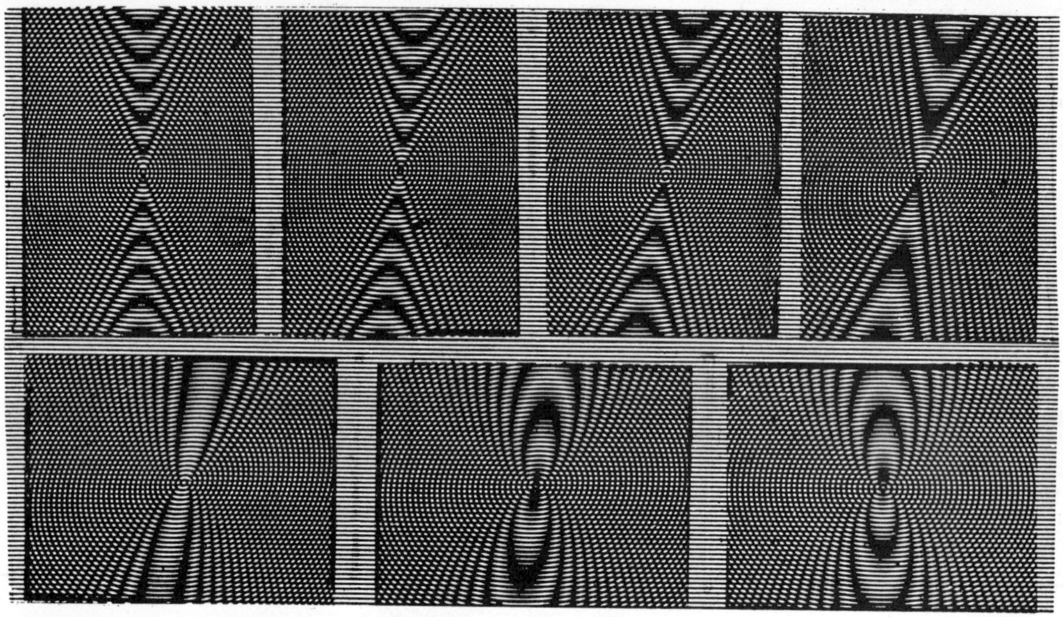

Figure P.7. Steady. . . . Unsteady. . . . Get-up. Moire associated with linear and elliptical gratings superposed. (Pitch of parallel lines in between those of major and minor axes of the ellipses.)

Figure P.8. From Propellers to Jets. Moire associated with linear and elliptical gratings superposed. (Pitch of parallel lines equals the pitch along the major axis of the ellipses.)

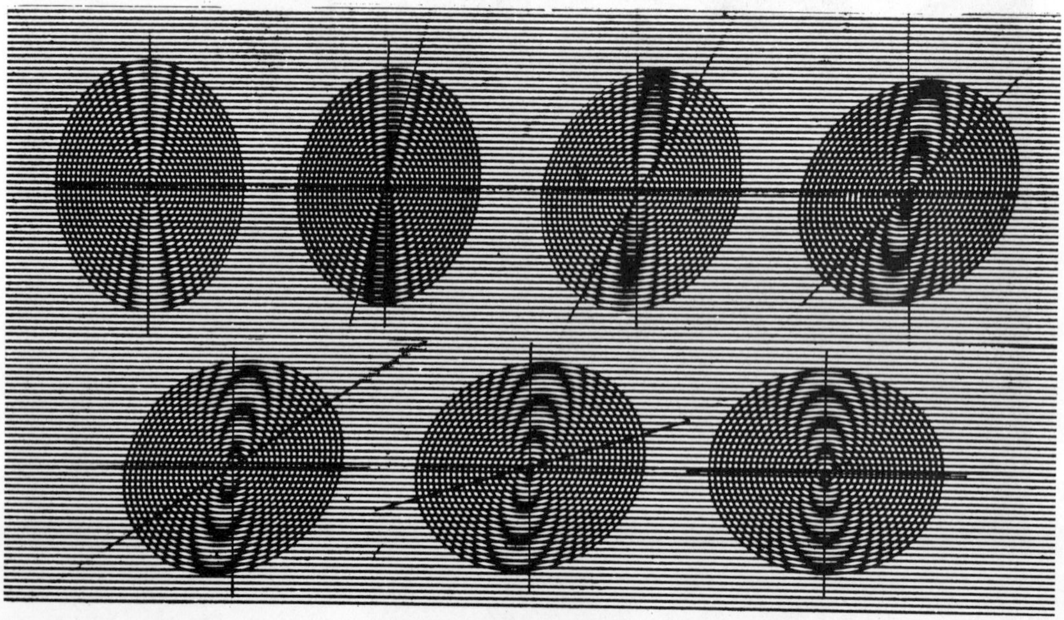

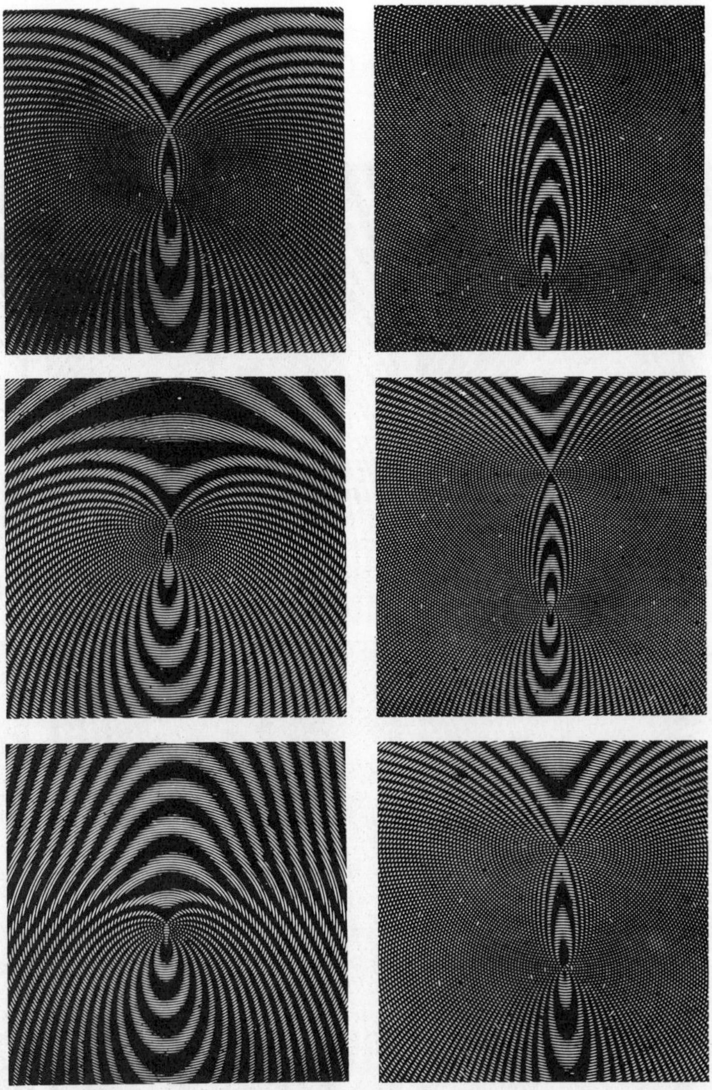

Figure P.9. Penetration and Dissolution. Moire associated with two elliptical gratings displaced along their minor axes.

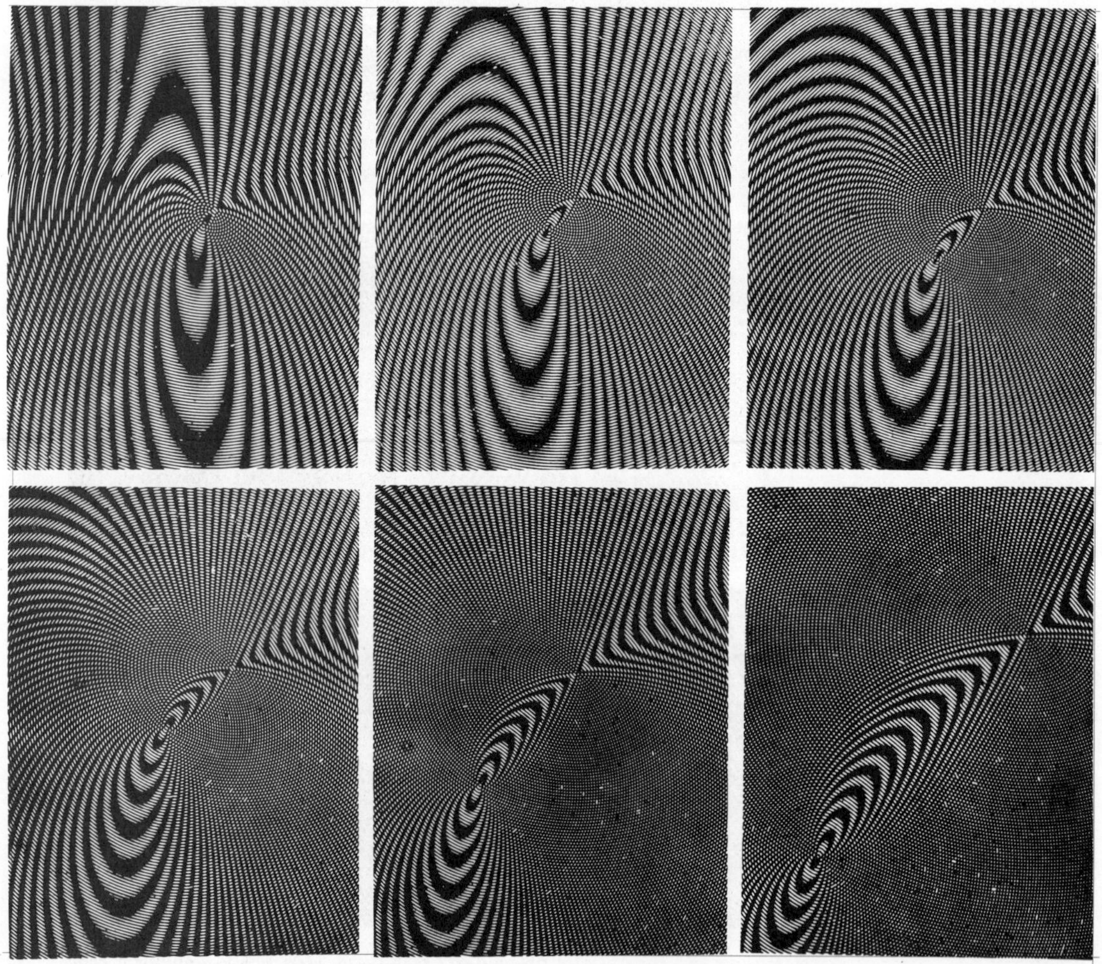

Figure P.10. The Hummingbird and the Flower. Moire associated with two elliptical gratings displaced along a line making 45° to their axes.

seems to be that frequently the work of the optical artist produces dizziness. In the statically produced moire patterns, the observer is the master of his senses.

To obtain moire patterns such as the ones shown above, two gratings on transparent film are required. The superposition of both gratings is easy and the photographic recording does not require special equipment. Coarse gratings are relatively easy to make, and inexpensive.

3. *Similarity of Patterns with Natural Phenomena and Artistic Expressions*

Photoelasticity and brittle coatings as well as moire patterns become frequently associated in the mind of the observer with completely unrelated patterns that can be found in nature, or with patterns developed using artistic techniques. Every one of the previous illustrations has been given a name that expresses some association of this kind. The following patterns are particularly suitable to these associations.

Figure P.11. The Totem Pole. Isochromatics in a rubber strip bonded along its sides to rigid bars and shrunk.

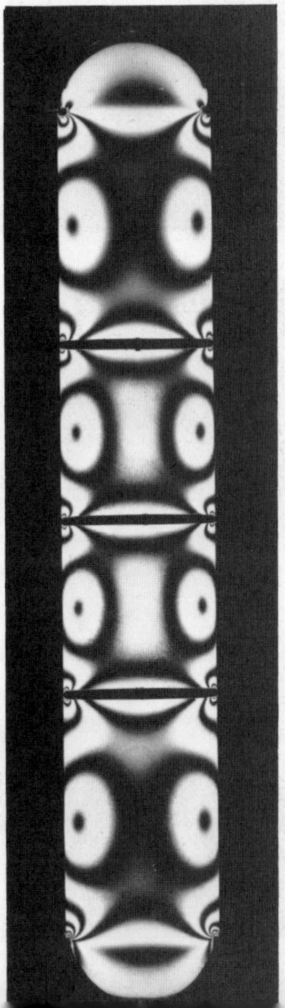

Figure P.11, as well as Figs. P.1 and P.2, have been obtained using a circular polariscope. These figures show only the family of fringes called isochromatics. If a plane polariscope is used, another family of fringes can be seen, the one called isoclinics. Isochromatics never intersect. The isoclinics can intersect each other and they intersect the isochromatics. This occurrence produces new optical effects, some extremely attractive. Figure P.12 is an example of the patterns obtained in rubber sheets cast around rigid inserts. (The materials necessary to produce these patterns are inexpensive and the casting techniques are easy to control.)

Isostatics (fine cracks) and isoentatics of a brittle-coating test on a curved beam (Fig. P.13) are sharply defined and give the stress distribution when the beam is subjected to pure bending. The fine lines produced by the cracks seem to continue the eyebrows of an observer looking through a keyhole.

Moire also presents patterns with which association can be easily found. Figure P.14 is the combination of a u- and v-isothetic pattern on the meridian slice of a rod subjected to uniform pressure on part of its surface. The smile of the face contrasts with the surprised expression of the eyes. Figure P.15 also reminds the observer of an eye, an eye that with curiosity and anxiety looks

Figure P.12. Arabesque. Isoclinics and isochromatics in a matrix shrinking about four inserts.

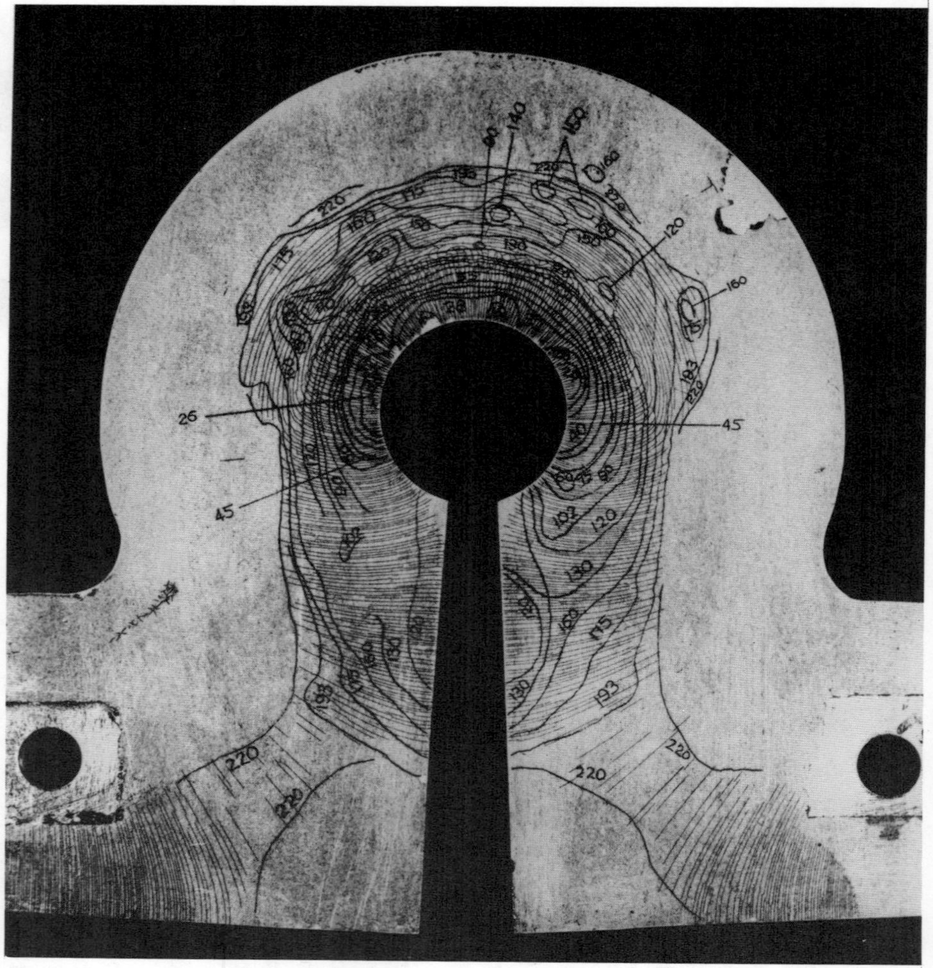

Figure P.13. The Key Hole. Isostatics and isoentatics in a brittle coating on a curved beam subjected to pure bending.

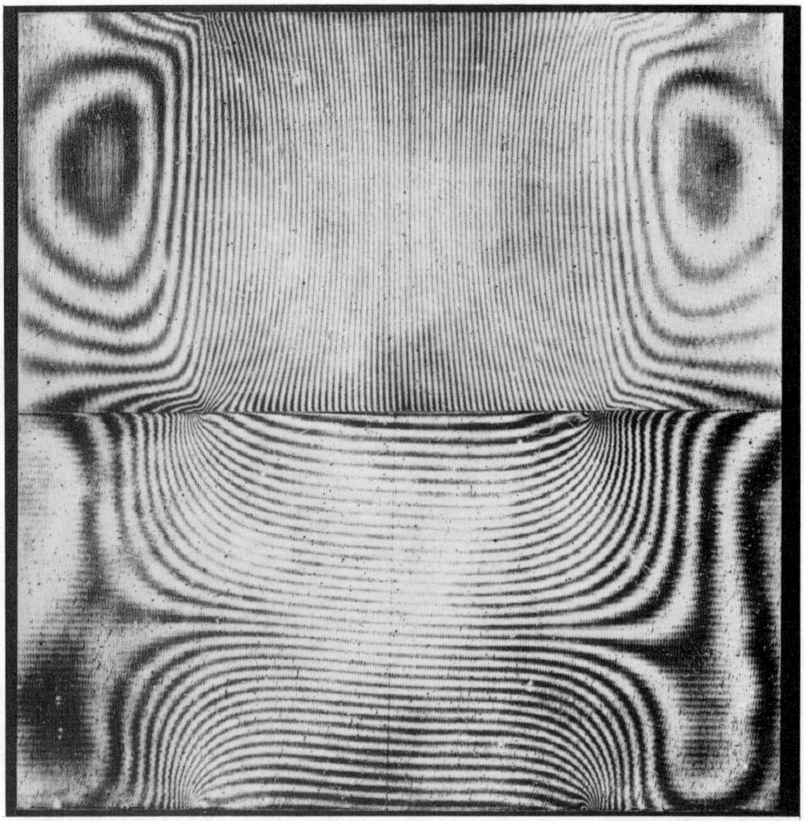

Figure P.14. The Man Who Could Not Make Up His Mind. Isothetics of *u* and *v* families in the meridian slice of a rod subjected to uniform pressure on part of its surface.

Figure P.15. The All-Seeing Eye. Moire associated with two elliptical gratings superposed.

Figure P.16. Bows and Arrows. Velocities of displacement components in a ring subjected to diametral load, obtained by moire of moire.

Figure P.17. Eye Behind Bars. Velocities of displacement components in a ring subjected to diametral load, obtained by moire of moire.

everywhere. Moire of moire, the method that permits the determination of velocities of displacement components, exhibits even more puzzling effects because of the superposition of fringes on the dense background of other fringes. Figures P.16 and P.17 are enlargements of parts of a circular ring subjected to diametral compression. The patterns in these figures are closely related to many abstract paintings.

4. *Beauty in the Patterns*

After some reflection over the observation of these patterns, a second thought comes to mind. Some of these fascinating patterns may actually be beautiful.

It is an open question whether artistic beauty can be produced mechanically or automatically, without deliberation, or whether at least the selective process of the human mind is necessary to artistic expression. There is little doubt, however, that some of the moire patterns, as well as some of the photoelastic and brittle-coating patterns, evoke an esthetic response in the observer. Prof.

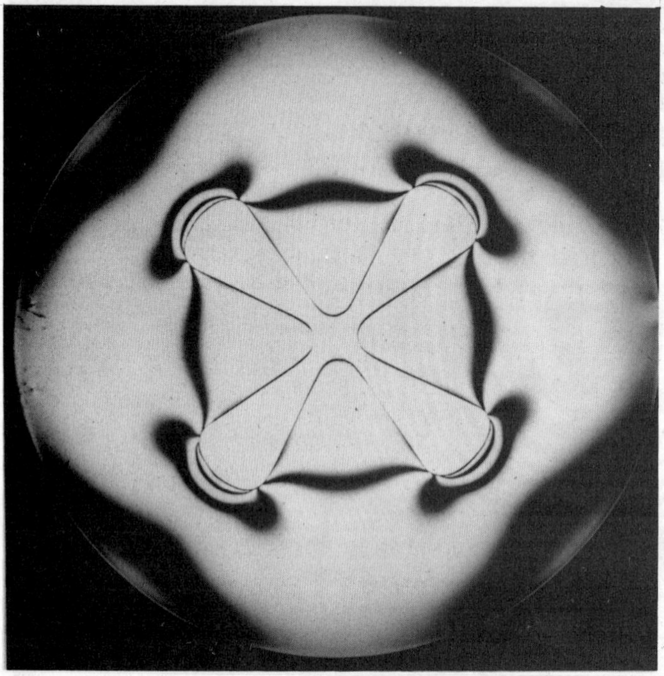

Figure P.18. The Morning Star. Light field isochromatics in a model of a solid propellant grain, subjected to uniform pressure.

Figure P.19. The Evening Star. Dark field isochromatics in a model of a solid propellant grain, subjected to uniform pressure.

Figure P.20. The Flower. Isochromatics around a circular hole in a plate subjected to equal biaxial load.

Oster has given a great deal of thought to this aspect of moire.[3] Like photography, and certainly better than many expressions of so-called "pop art," moire effects show attractive, fascinating, and in some cases beautiful patterns. It should be emphasized, however, that a definite difference is made here between a pattern that is strange, or unusual, or striking, or ambiguous, or dizziness-producing, or illusion-producing, and a pattern that is beautiful. Beauty may not be always a well-defined concept, and it may depend to a large extent on the developed esthetic capacity of the observer, but it exists, and cannot be identified with the unusual, with shock, with strangeness, nor with optical illusion.

Consider again, to start with, patterns obtained using polarized light. Figures P.18 and P.19 are isochromatics in a model of the transverse cross-section of a solid-propellant grain, when the model is loaded by uniform pressure applied at its outside boundary. Both pictures develop a feeling of smooth harmony to which the lack of concentration at the fillet certainly contributes. Figure P.20 is much sharper and the gradients near the discontinuity are steeper. The four

---

[3] G. Oster and Y. Nishijima, "Moire Patterns," *Scientific American*, Vol. 208, No. 5 (May 1963), pp. 54–63.

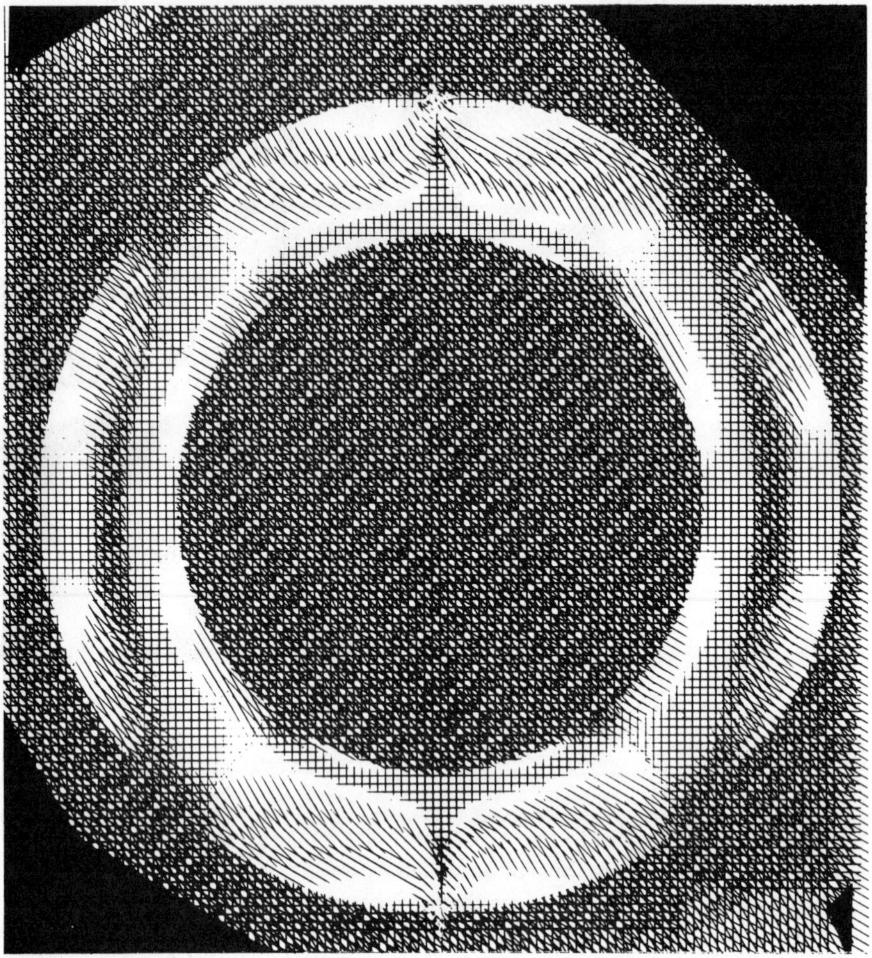

Figure P.21. Ring on a Rug. Isostatics obtained photographically from isoclinics in a ring subjected to diametral compression.

axes of symmetry and the continuous ring around the hole make it particularly attractive. The sharpness near the hole becomes a soft shade near the outside boundaries. The fringes are the isochromatics around a circular hole in a plate subjected to equal biaxial loading.

The previous effects were obtained using isochromatics. Figure P.21 shows the strange appearance of a film on which isostatics are obtained from isoclinics that have been printed, after each being filtered through a screen the orientation of which is the parameter of the corresponding isoclinic. Isoclinic and isochromatic patterns combined produce still stranger effects. Figure P.22 shows the superposition of fringes that cross each other different ways. The feeling is similar to the one associated with viewing a very elaborate tapestry or on Arabian ceramic.

Figure P.22. Mosaic. Isoclinics and isochromatics in a matrix shrinking about inserts.

Figure P.23. The Rising Sun through the Cracked Window. Isostatics near the transverse axis in a circular ring subjected to diametral compression.

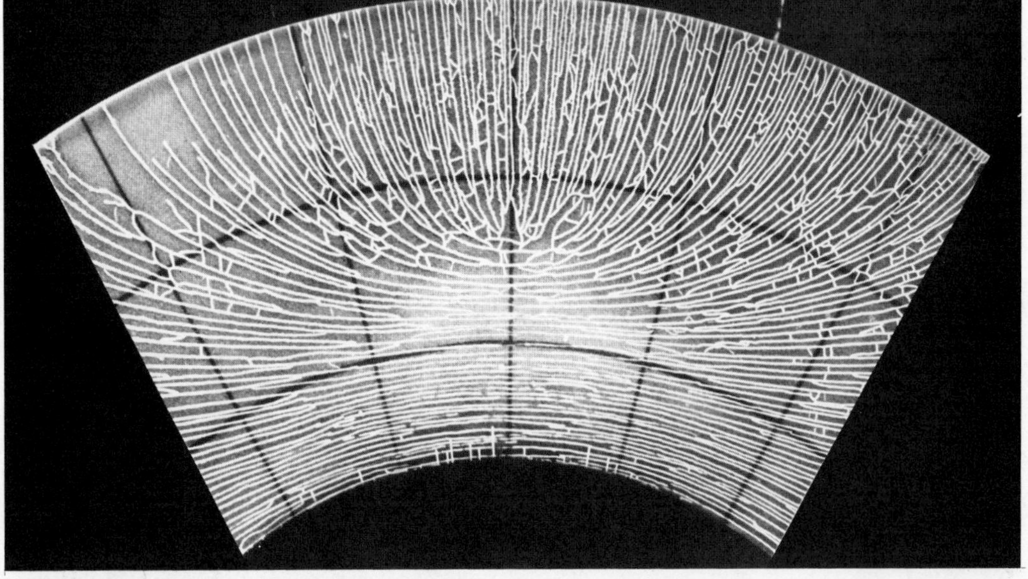

Isostatic patterns obtained using brittle coatings can exhibit a quite mysterious beauty. Since they are usually very dense, it is necessary to enlarge them appreciably to appreciate them. Figure P.23 shows parts of a ring diametrically loaded. The appearance of the cracks in the coating has been emphasized by using a penetrant and a fluorescent powder. Isostatics are usually obtained by graphical transformation of the isoclinics. The two families are everywhere perpendicular to each other; to avoid confusing them one is usually represented by solid lines and the other by dotted lines. The resultant drawing sometimes presents very attractive patterns. In Fig. P.24, the white space corresponding to the insert position is beautifully balanced by the "dynamic" concentration of lines around the singular point.

Coarse gratings can produce fascinating patterns of strange unusual beauty. The patterns shown in Figs. P.15, P.25, P.26, P.27, and P.28 are easily produced by superposing two slightly elliptical gratings and displacing one with respect to the other. Whereas photoelastic isochromatics are well-defined fringes with smooth shady edges, these fringes are saw-toothed, producing a stranger impression. Sometimes, as in Figs. P.25, P.26, and P.27, they die out in a series of white diamond shapes. It is not necessary to associate these patterns with any natural phenomenon; however, the impression of ripples in the water hit by a falling stone, in Fig. P.25, is unavoidable.

Figure P.24. Embracing Love. Isostatics in matrix shrunk around an elbow.

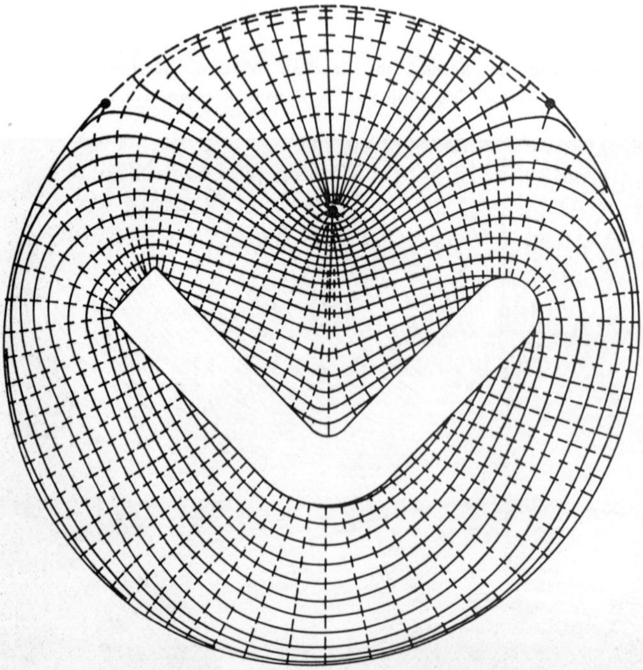

Figure P.25. Stones Falling on Water. Moire associated with two elliptical gratings, displaced along an axis.

Figure P.26. Disturbance of Security. Moire associated with two elliptical gratings displaced along an axis.

Figure P.27. Take-Off. Moire associated with two elliptical gratings, superposed eccentrically and rotated.

All the moire patterns shown so far have been produced by the superposition of two gratings. Very attractive patterns can also be produced by the moire-of-moire effect produced by the superposition of four gratings. Those shown in Figs. P.29 and P.30 are the result of the superposition of four slightly elliptical gratings displaced along a common axis.

There are many schools of art, many theories of art, and many types of artistic expression. They seldom agree on the vocabulary to be used or in the object of their activities. The last word in the development of artistic movements seems to be that the best art is the one that is so pure that it is not related to the representation of any physical phenomenon. All conceptual association should be absent from the perceptual abstraction. From this point of view, the sculptures of Praxiteles, the paintings of Michaelangelo, and the work of the Impressionists may easily be dismissed. It may even be claimed that Beethoven's Sixth Symphony is less of a masterpiece because its author called it the "Pastoral."

All the patterns shown here have been given names, and their names suggest concepts of the physical world. The suggestion of a free association cannot detract from their beauty. They are pure art in the sense that they do not represent any physical phenomenon. They *are* a physical phenomenon: an optical interference, a shadow, or a fracture. And they may have a mathematical interpretation in terms of the theory of continua.

Figure P.28. The Bride. Moire associated with two elliptical gratings displaced along their minor axis.

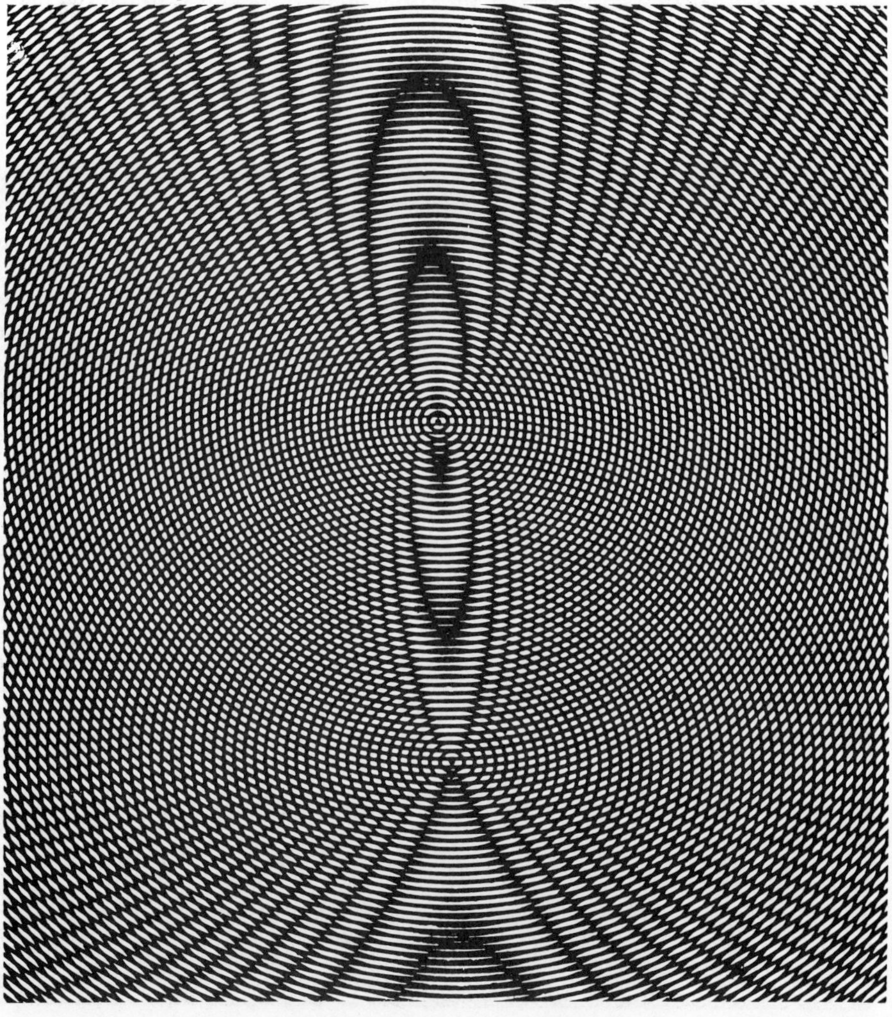

Since many of these patterns have not been arbitrarily produced, but deliberately selected, it would be rather difficult to develop a theory to explain their beauty or to serve as a guide for the future selection of other patterns. They have been selected because they create an esthetic feeling. Some are symmetrical, some asymmetrical. In some of them the edges are sharp, in others soft. Some exhibit fine lines, others broad bands. And although they are reproduced in black-and-white, most of them could as well have been reproduced in color. The link common to all of them is the pleasure they produce to the eye.

Since these patterns all have a scientific meaning and are obtained using techniques developed for scientific purposes, it may be appropriate to call this kind of art "log art," the art directly associated with science. It may also be appropriate to trace its origin to Plato, who wrote in *Philebus:* "By beauty of form I do not mean such beauty as that of animals or pictures, which the many would sup-

Figure P.29. Labyrinth. Moire of moire produced by the superposition of four elliptical gratings.

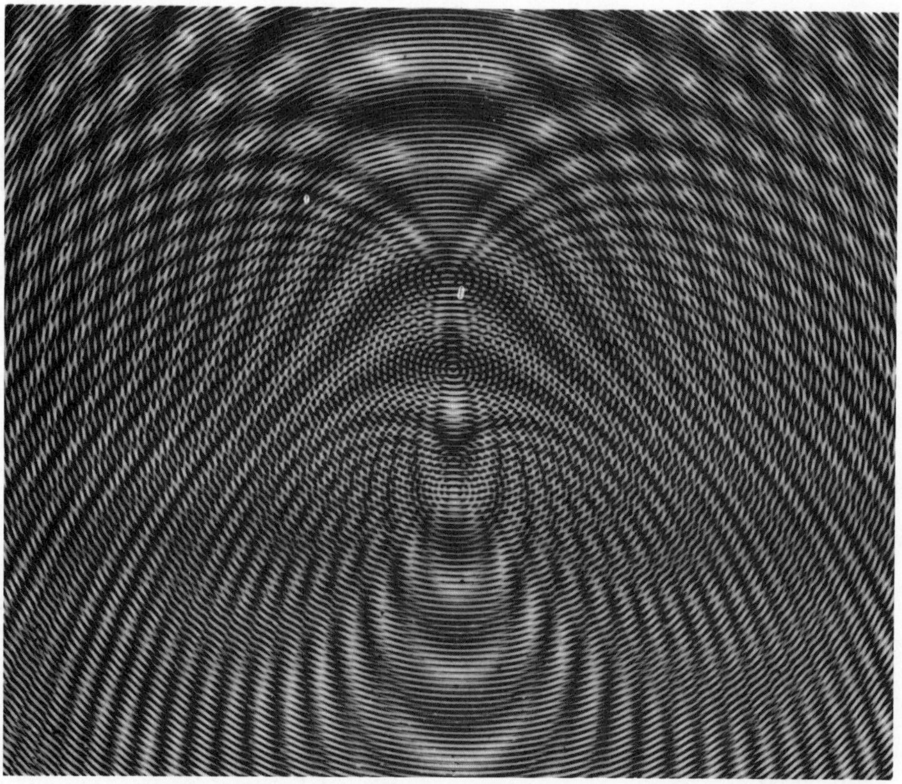

Figure P.30. Lights and Shadows. Moire of moire produced by the superposition of four elliptical gratings.

pose to be my meaning; but . . . understand me to mean straight lines and circles, and the plane or solid figures which are formed out of them by turning-lathes and rulers and measurers of angles; for these I affirm to be not only relatively beautiful, like other things, but they are eternally and absolutely beautiful. . . ."[4]

5. Scientific Meaning

These beautiful patterns have a measurable meaning in nature, and scientists in the last decade have given considerable thought to the problem of interpretation of moire fringes. Several chapters of this book will deal with that problem. One aspect should be emphasized here.

Shapes and changes in shapes are directly apprehended by our senses.

---

[4] *Dialogues of Plato*, 3rd ed. (tr. B. Jowett), Vol. IV, p. 625 (Oxford: Oxford University Press, 1892).

It is sufficient to open our eyes and see trees grow and clouds shrink or expand. A very low level of abstraction is necessary to grasp the concept of displacement. After it has been established that moire patterns of deformed bodies can be looked at as isothetics, or loci of points having the same value of component of displacement, the apparently complicated patterns are understandable in terms of geometry. The method can be used efficiently to measure rigid-body motions, but for the stress analyst moire is above all the means of determining the metamorphosis or change in shape of loaded bodies. This emphasis on direct geometry has two implications. The first point is important technically: no intermediate physical phenomenon is used to measure the change in shape. Whereas birefringence or ohmic resistance may be affected by the environment and their relationship with displacement may vary, moire records directly whatever geometric changes take place in the plane of the grating.

The second point is of psychological importance. Engineers, in particular civil engineers, have been trained to emphasize the concept of stress. Experimentalists in mechanics are used to determining strains, which they convert into stresses. The use of moire forces the mind to go back to the basic, intuitive, and uniquely defined concept of displacement. And the fact that derivatives of displacements can be obtained by moire-of-moire techniques helps also in emphasizing the importance of the knowledge of displacement and the realization that strains and stresses are further abstractions, much more difficult to define and to conceive physically.

It should be possible to present all the theory of elasticity in a systematic way, physically, starting from the concept of displacement, made visible using grids and moire, and following with spatial and time derivatives of displacements obtained using moire-of-moire techniques. A beginning in that direction was attempted by one of the authors in another publication.[5] The mathematical elaborations would then be related to physics and this relationship with physics would follow step-by-step the mathematical developments.

[5] A. J. Durelli, "Visual Representation of the Kinematics of the Continuum," *Exp. Mech.*, Vol. 6, No. 3 (March 1966), pp. 113–139.

LIST OF SYMBOLS

a	fraction of a fringe order related to fringe shift
A	amplitude of light
b	width of specimen
B	distance from edge of hole to edge of a specimen
BL	base length
d	distance from surface of a reflective plate to master grating
d	distance between dots in diffraction plane
d, d_1	density of grating lines in master and specimen gratings, respectively
d_1	distance between two master-grating lines along the specimen grating line
D	fringe density
DS	deformation sensitivity
e	distance on the specimen plane between a master grating line and its shadow
e	motion of the reflected image of a master grating due to the inclination of a reflective plate
e_x, e_y, e_{xy}	components of the strain tensor
e^L, e^E	see ϵ^L, ϵ^E, etc. for meaning of superscripts
e'_1, e'_2, e'_3	principal components of deviatoric strain tensor
E	modulus of elasticity
f	focal length of lens
f_σ	photoelastic material-stress-fringe value
F_σ, F_ε	photoelastic model-stress-fringe and model-strain-fringe values, respectively
g	gap between master and specimen gratings
G	shear modulus
H	beam height
i	angle between the light beam and the normal to the grating plane
I	intensity of light at a point in a moire pattern

I_{max}	maximum intensity of light between moire fringes
ID	inner diameter
k	constant
k_0, k_1	specimen-grating orders at two nearby points which coincide when deformed grating is shifted on itself
K	pitch-to-shift ratio
K	bulk modulus
l	difference between the path length of two parallel rays of light
l_i, l_f	initial (before deformation) and final (after deformation) lengths of a line segment in a body subjected to load
l, m	integral grating orders (where the pitches are about the same, l is the order of the master grating; where the pitches differ greatly, l is the order at the denser grating)
L	length
m	number of master-grating lines between two fringes
n	moire-fringe order
n	photoelastic-fringe order
n	series of integers (0, 1, 2, 3, . . .)
N	moire-of-moire-fringe order
N_s	number of master-grating lines in a shift to obtain moire of moire
OD	outer diameter
p	pressure
p	master-grating pitch
p_1	specimen-grating pitch
p_r	angular pitch, angle between two radial grating lines
P	load
P_{act}, P_{max}	actual load and maximum load, respectively
$P(x, y, z)$	point in specimen described by Cartesian coordinates
$P'(x', y', z')$	position of P after deformation
$\left.\begin{matrix} P_1(x_1, y_1, z_1) \\ P_0(x_0, y_0, z_0) \end{matrix}\right\}$	two nearby points in a specimen before deformation
$\left.\begin{matrix} P'_1(x'_1, y'_1, z'_1) \\ P'_0(x'_0, y'_0, z'_0) \end{matrix}\right\}$	position of points P_0 and P_1 after deformation
q	order of a grating line in master and specimen grating
r	a small whole number, or fraction greater than unity composed of two whole numbers
r	radius of dot in diffraction plane
r, s	Cartesian coordinates normal and parallel, respectively, to master grating lines direction
s	width of diffracted moire grating
s	shift of deformed grating or moire pattern on itself
S_1, S_2, S_3	principal components of deviatoric-stress tensor
$\dot{S}_1, \dot{S}_2, \dot{S}_3$	rates of S_1, S_2, S_3
SS	strain sensitivity

t	transmission coefficient, the ratio of output intensity to input intensity of light
t	model thickness
u_i	displacement component
u_r, u_θ	radial and circumferential displacements, respectively
$u_{45°}$	displacement component at 45° to the x- and y-directions
u_α	rotation of radial lines about a center in angular units
u, v, w	displacement components in the x-, y-, and z-directions, respectively (subscripts 1 or 0 relate displacements to points P_1 or P_0)
$u_n, u_{n-1}, v_n, v_{n-1}$	u and v displacements components at consecutive levels of load
$u_0, u_1, u_2, v_0, v_1, v_2$	displacement components of points P_0, P_1, P_2
$\dot{u}, \dot{v}$	displacement component velocities
$U_i(x, y)$	combined function representing $u(x, y)$ $(i = 1)$, and $v(x, y)$ $(i = 2)$
U_T	total displacement of a point
$\frac{\partial u}{\partial n}, \frac{\partial v}{\partial n}$	gradients at a point in the u- and v-fields, respectively (these are normal to the isothetic lines)
$\frac{\partial u}{\partial x}, \frac{\partial v}{\partial y}, \frac{\partial u}{\partial y}, \frac{\partial v}{\partial x}$	partial derivatives of the $u(x, y)$ and $v(x, y)$ fields
$\Delta u, \Delta v$	change in the Cartesian components of u *and* v between points P_1 and P_0 when deformation takes place
$\Delta x, \Delta y, \Delta x', \Delta y'$	Cartesian components of distance between points P_1 and P_0 (primes denote values after deformation)
$\frac{\Delta u}{\Delta x'}, \frac{\Delta v}{\Delta y'}, \frac{\Delta u}{\Delta y'}, \frac{\Delta v}{\Delta x}$ $\frac{\Delta u}{\Delta x''}, \frac{\Delta v}{\Delta y''}, \frac{\Delta u}{\Delta y''}, \frac{\Delta v}{\Delta x'}$	incremental ratios (correspond to partial derivatives under certain conditions)
v_f	fictitious displacement due to space between specimen and master grating
$\dot{W}$	rate at which stress does work in connection with the change in shape
α	angle between incident and reflected rays on a hologram
α	angle between direction of displacement at a point and coordinate axis
α	angle of diffracted ray of light measured from the direction of the undiffracted ray
α	angle of ray from a point on a moire pattern to the center of the camera lens, measured from plane of pattern
$\alpha_\delta, \alpha_r, \alpha_s$	parametric variables to determine rotation, from master pitch p, fringe angle ϕ, and spacing δ, δ_r, or δ_s, respectively
β	Angle of reflection of plate before rotation
$\beta_\delta, \beta_r, \beta_s$	parametric variables to determine specimen pitch from master pitch p, fringe angle ϕ, and spacing δ, δ_r, or δ_s, respectively
γ	angle between grating lines and a line of symmetry (that is, the angle between the principal section and a normal to the line of symmetry)

γ_{xy}	shear-strain (change in angle between the x- and y-directions)
γ_{max}	maximum shear strain
δ	distance between moire fringes
δ_f	initial fringe spacing due to mismatch
δ_r, δ_s	distance between fringes in the r- and s-directions
δ_v, δ_h	vertical and horizontal displacement, respectively
ϵ	direct or "normal" strain in the direction indicated by the subscript
ϵ_1, ϵ_2	principal strains
ϵ_{avg}	average strain
ϵ_m	initial fictitious strain corresponding to a mismatch
ϵ_t	tangential strain
ϵ_x, ϵ_y	direct or "normal" strains in the x- and y-directions
$\epsilon^L, \epsilon^E, \gamma^L, \gamma^E$	strains referred to Lagrangian (L) or Eulerian (E) definition. If no superscript is shown, it is assumed that both are the same
$\bar{\epsilon}$	natural strain
ζ_f	size of angle after deformation which was a right angle before deformation
ζ_i	size of angle before deformation which was a right angle after deformation
η	angle between direction of motion and normal to grating
θ	isoclinic angle, angle between direction of principal stress or strain and coordinate axis. A subscript indicates the coordinate axis
θ	angle of out-of-plane rotation of reflective plate. Subscripts represent the direction that, when associated with the plate normal, defines the plane of rotation.
θ	angle between grating line and co-ordinate axis
θ, θ'	angles between master and specimen gratings associated with the r- and s-axes, respectively. If gratings were initially aligned with r- or s-axes, these angles give rotation of r- or s-axes.
θ_x, θ_y	inclination after deformation of lines that were initially parallel to x- and y-axes.
λ	parametric ratio identifying load or displacement level. Subscripts n and $n-1$ indicate consecutive levels
λ	wavelength of light
ν	Poisson's ratio
ξ	angle on the master grating plane at the point where the optical axis of the camera intersects the grating plane, formed by a grating line and a line to some point in the grating plane
ρ	distance from a point in moire field to optical axis of camera
σ	normal stress
σ_1, σ_2	principal stresses
σ_x, σ_y	normal stresses in x- and y-directions
$\bar{\sigma}$	natural stress
σ^*	yield stress

σ_{avg}	average stress
σ_θ, σ_m	stresses in hoop and meridian plane of body, respectively
τ_{xy}	shear stress in x- and y-directions
τ_{max}	maximum shear stress
ϕ	angle between fringe and master grating
ϕ_u, ϕ_v	angles between the isothetic lines and the x-axis measured counterclockwise (the subscript designates the displacement component field of the isothetic line)
ψ_u, ψ_v	angles between the maximum gradients $\partial u/\partial n$ and $\partial v/\partial n$ and the x-axis
φ	angular phase shift between two parallel rays of light
ω	rigid rotation

DEFINITIONS

ISOBAR	Locus of points exhibiting the same value of normal stress (equal stress).
ISOBATHIC	Locus of points exhibiting the same value of thickness in a plate (equal thickness).
ISOCHROMATIC	Locus of points exhibiting the same value of maximum shear stress or maximum shear strain (equal shear stress, or strain).
ISOCLINIC	Locus of points at which the principal stress or strain exhibits the same inclination (equal inclination).
ISOCLINIC OF DISPLACEMENT	Locus of points at which the displacement exhibits the same inclination.
ISOENTATIC	Locus of cracked ends in a brittle-coating test.
ISOGONIC	Locus of points exhibiting the same value of angular displacement (equal angle).
ISOGYRO	Locus of points exhibiting the same value of rotation of a line that goes through the point (equal line rotation).
ISOKINETIC	Locus of points exhibiting the same value of displacement (equal displacement).
ISOPACHIC	Locus of points exhibiting the same value of transverse strain in a plate (equal change in thickness).
ISOPARAGOGIC	Locus of points exhibiting the same value of partial derivative of displacement components.
ISOSTATIC	Line the tangent to which coincides with the direction of the principal stress or strain.
ISOSTROPHIC	Locus of points exhibiting the same value of rigid rotation (equal rigid rotation).
ISOTACHIC	Locus of points exhibiting the same value of velocity.
ISOTENIC	Locus of points exhibiting the same value of direct strain (equal strain).
ISOTHETIC	Locus of points exhibiting the same value of displacement component (equal displacement component).
METAMORPHOSIS	A deformed shape.
TRAJECTORY OF DISPLACEMENT	Line the tangent to which coincides with the direction of the displacement.

INTRODUCTION

The word *moire* comes from the name of a silk fabric which, when folded, exhibits patterns of light and dark bands. The *moire effect*, as understood in this book, is the similar effect of light and dark bands, or *fringes*, produced by the superposition of two sets of gratings when certain required circumstances are satisfied.

The superposition of two similar motifs to produce a third, coarser pattern is a common phenomenon in mechanical vibrations, optics, acoustics, and electricity. The phenomenon is called *interference*. It is often found in the overlapping of waves. More generally, superposing motifs of all sorts may produce interference. In this book the effects of superposing geometric motifs (*gratings*) which lie in a plane will be considered. The interference produced by superposing gratings has come to be known as *moire* (or *moire patterns*).

The study of the moire effect can be conducted from several points of view. It is possible to start by studying the geometric relationships existing between the pitches and the relative position of two gratings, on the one hand, and the distance and the inclination of the moire fringes, on the other. It is also possible to look at the moire pattern as the result of adding or subtracting two functions expressed parametrically. Finally, it is possible to see the moire as a locus of points undergoing the same displacement. Depending on the application and the preference of the investigator, any of these approaches can be followed. In this book the three approaches will be reviewed, and the applications to strain and stress analysis will be emphasized.

In metrology, moire is essentially a means of measuring displacements. To take advantage of this physical phenomenon, some previous understanding of a few basic concepts in the kinematics of displacement and strain is necessary. Several sections will be included in this book to familiarize the reader with those concepts. The fundamental concept of displacement will be handled first. The applications of moire in the field of displacement measurement are direct, simple, and require only elementary knowledge of kinematics. The concept of strain is more complicated and will be handled later.

Historical Background

In order to obtain strains from moire patterns, two approaches have been followed: one that can be called "geometrical" and the other which consists of relating the fringes to the displacement field.

The geometrical interpretation seems to have been originated in a paper published in Dutch in 1945 by D. Tollenaar,[1] and was applied for the first time to the subject of strain determination in 1952 by J. K. Kaczer and F. Kroupa[2] of the Physics Institute of the Charles University in Prague.

Using the geometrical approach, one can study moire-pattern formation as the result of the intersections of the two systems of lines of the master and specimen gratings. Knowing the distance between master-grating lines and measuring the distance between fringes, it is possible by geometric analysis of the intersections of the two systems of lines to compute the distance between specimen-grating lines at a point and the corresponding change in direction. With these two data, direct (or "normal") and shear strains can be computed. The geometrical approach gives values of the strains that are the average values in a region limited by two fringes.

Similar approaches are presented in other papers.[3,4,5] R. Bromley[6] obtains results similar to those previously referred to by using tensor notation.

A different point of view in the analysis of moire patterns was presented in 1948 by R. Weller and B. M. Shepard.[7] These authors described the application of moire fringes to measuring displacements. In 1954, M. Dantu,[8] following the same lines, introduced the interpretation of moire patterns in terms of the components of the displacements. Although Dantu's presentation was limited to the field of small strains and deformations, it has the great advantage of relating the moire directly to all the previously developed knowledge of the theory of strain. Dantu's contributions[9] are fundamental in the field of the application of

---

[1] D. Tollenaar, "Moire—Interferentieverschijnselen bij rasterdruk," Amsterdam Instituut voor Grafische Technick, 1945.

[2] J. Kaczer and F. Kroupa, "The Determination of Strains by Mechanical Interference," *Czechoslovak Journal of Physics*, Vol. 1 (1952), p. 80.

[3] R. Lehman, and A. Wiemer, "Untersuchungen zur Theorie der Doppelraster als Mittel zur Messanzeige," *Feingerätetechnik Heft* (1953), pp. 5–199.

[4] J. D. C. Crisp, "The Measurement of Plane Strains by a Photoscreen Method," *Proc. SESA*, Vol. 15, No. 1 (1957).

[5] A. Vinckier and R. Dechaene, "Use of Moire Effect to Measure Plastic Strains," *Transactions of the ASME*, Paper No. 59-Met. 7.

[6] R. Bromley, "Two-Dimensional Strain Measurement by Moire," *Proc. Phys. Soc.*, Vol. 69, Part 3B (1956).

[7] R. Weller and B. M. Shepard, "Displacement Measurement by Mechanical Interferometry," *Proc. SESA*, Vol. 6, No. 1 (1948).

[8] M. Dantu, "Recherches Diverses d'Extensometrie et de Détermination des Contraintes," Conference faite au GAMAC, Feb. 22, 1954.

[9] M. Dantu, "Utilisation des Réseaux pour l'Étude des Déformations," Laborat. Central des Ponts et Chaussées, Publ. 57–6 (1957).

the moire method to strain analysis. Many of the ideas developed in this book were originated by him. An extensive although incomplete review of papers dealing with moire methods has been published by P. Theocaris.[10]

On the Theory of Strain

The determination of strain in the general case, using moire fringes, requires an appreciably more elaborated theory than the one with which experimental strain analysts are usually familiar.

Most strain-analysis methods are little influenced by rotations taking place at the point at which the strain is to be determined. This is not the case with moire. Rotation with no strain may produce appreciable moire response. The general case of strain analysis using moire requires a detailed study of strains and rotations. The chapter on finite strains was written to clarify this question.

On the Book's Plan

The relation of the moire effect to strains and displacements and the manner in which these relationships are treated in this book may be explained in terms of several variables. These variables and the book plan are illustrated in the table below.

Variables	Characteristic Symbols	Relationships Discussed in			
Master Grating	p (pitch)				
Specimen Grating	p_1 (pitch)		(Chap. 5)	} Chaps. 2 and 3	
Moire Fringes	u or v; or δ and ϕ	} Chap. 1			
Displacement Field	vector $(u + v)$				
Strain Field	$\epsilon_x, \epsilon_y, \gamma_{xy}, \epsilon_1, \epsilon_2$				} Chap. 4
Stress Field	$\sigma_x, \sigma_y, \tau_{xy}, \sigma_1, \sigma_2$				

The basic relationship of moire fringes (isothetics) to displacements is discussed in Chapter 1. The analytic relationship between gratings and fringes is presented in Chapter 2. This same relationship is presented in Chapter 3 in a geometric way and developed so as to be applicable to strain analysis. Chapter 4 presents the classic displacement–strain relationships. Chapter 5 discusses various properties of moire fringes. Either the approach in Chapter 3 or the approach given in Chapters 1 and 4 can be used to analyze strain, depending on the problem and the analyst's preference.

---

[10] P. Theocaris, "Moire Fringes: A Powerful Measuring Device," *Applied Mechanics Reviews*, Vol. 15 (May 1962), pp. 333–339. Updated in *Applied Mechanics Surveys*, eds. Abramson, Liebowitz, Crowley, and Juhasz (Spartan, 1966), pp. 613–628.

The first five chapters present the fundamentals of the moire effect. The following three chapters show different methods of evaluating the data. In Chapter 6, the method of analysis of displacements using the techniques discussed in Chapter 1 is presented. Chapter 7 gives examples of strain analysis using the displacement–derivative approach presented in Chapters 1 and 4. Chapter 8 gives examples of strain analysis using the geometric approach presented in Chapter 3. Chapters 10 through 13 deal with special procedures in the uses of moire. Chapters 14 through 16 deal with the experimental techniques, and the last five chapters deal with the solution of specific problems, including wave-propagation problems and large elastic and plastic strains.

The applications described have been conducted by the authors and their associates during the course of more than ten years. Since the moire techniques have changed appreciably in this length of time, some of the early applications have not been conducted with the benefit of the experience which could be used in the solution of problems conducted recently. It would have been practically impossible to repeat all those applications using the newly developed techniques.

It should also be pointed out that although an overall picture of the field of moire in the analysis of strain will be given, no attempt is made to write an all-inclusive treatise on moire. Some subjects, such as the use of diffraction effects and the application to the analysis of moments in the bending of plates, will only be summarized.

PART 1

FUNDAMENTAL PROPERTIES

1

THE MOIRE PHENOMENON

Since the concepts of displacement, its components along axes of reference, and their derivatives are fundamental to the understanding of the moire phenomenon, it is appropriate to start this chapter with a short comment on what displacements are.

1.1. Displacements

Displacement of a point or particle, by definition, is its change in position. Suppose that point P, in space, has a position given by its coordinates (x, y, z) in a Cartesian system of reference. Suppose that point P moves in space and reaches the position P' given by its coordinates (x', y', z'). The displacement of P to P' is a vector. It can be specified by its projections on the three coordinate axes (as the Cartesian components of the vector), as follows:

$$u = x' - x$$
$$v = y' - y$$
$$w = z' - z$$

Figure 1.1 shows the displacement of point P to its new position P' and the projection of the displacement on the x-y and x-z planes. Similar projection could be obtained on the y-z plane. (To avoid confusion, it has not been included in the figure.) In general, the moire effect is associated with movements of points, or displacements, taking place in a plane. To fix the idea, assume that this plane is the x-y plane. In what follows, only x- and y-coordinates and u- and v-displace-

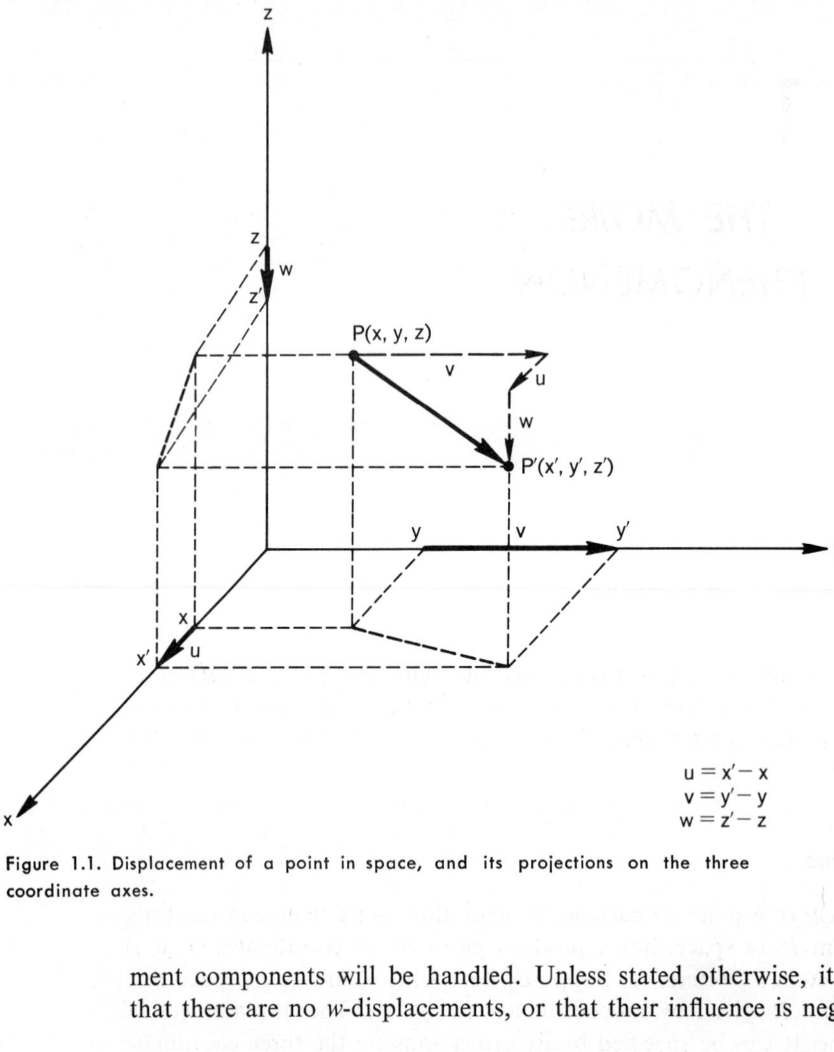

Figure 1.1. Displacement of a point in space, and its projections on the three coordinate axes.

ment components will be handled. Unless stated otherwise, it will be assumed that there are no *w*-displacements, or that their influence is negligible.

1.2. The Physical Phenomenon

Screens, grids, grilles, and *gratings* are arrays of alternately opaque and transparent (or dark and light) lines or dots. They are usually straight lines parallel to each other and at the same distance from each other (Fig. 1.2). Sometimes they are circles (Fig. 1.3) or radial lines (Fig. 1.4). Cross-gratings or dots (Fig. 1.5) are usually made of two systems of lines perpendicular to each other. The width of the opaque line and the width of the interline or the transparent line are usually the same, but in some cases they may be different. Unless otherwise stated, it will be assumed in what follows that the gratings are made of a system of parallel opaque

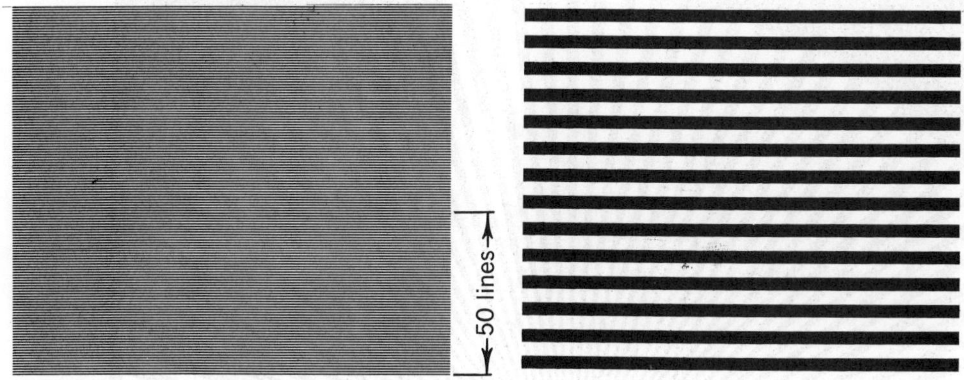

Figure 1.2. Typical gratings of straight parallel lines.

Figure 1.3. Grating of concentric circles with constant pitch and line width equal to interline width.

Figure 1.4. Grating of radial lines with variable pitch and line width equal to interline width.

and transparent, or dark and light, lines of equal width. In this book the word *grating* will be used exclusively to identify the array of lines or dots mentioned above. The word *grid* will refer to coarse systems of lines which produce no moire effect. When the grid method is used, the changes in distances between the lines of the grid are determined by individual measurements using microscopes, rulers, or comparators.

Two gratings are necessary to produce the moire effect; the *model* or *specimen grating*, and the *master grating*. Sometimes one may be the image of the other.

The distance between the master-grating lines is called the *pitch* and is represented by p. The number of master-grating lines per unit length is called the density of the master and is represented by d. The original pitch in both the specimen and the master grating may be different, but in general it is the same. The distance between the specimen-grating lines may have changed from the original distance, because of a deformation of the specimen. This distance will be called p_1. The corresponding density will be called d_1.

Figure 1.5. Grating of crossed lines (top) and grating of dots in a square array (bottom).

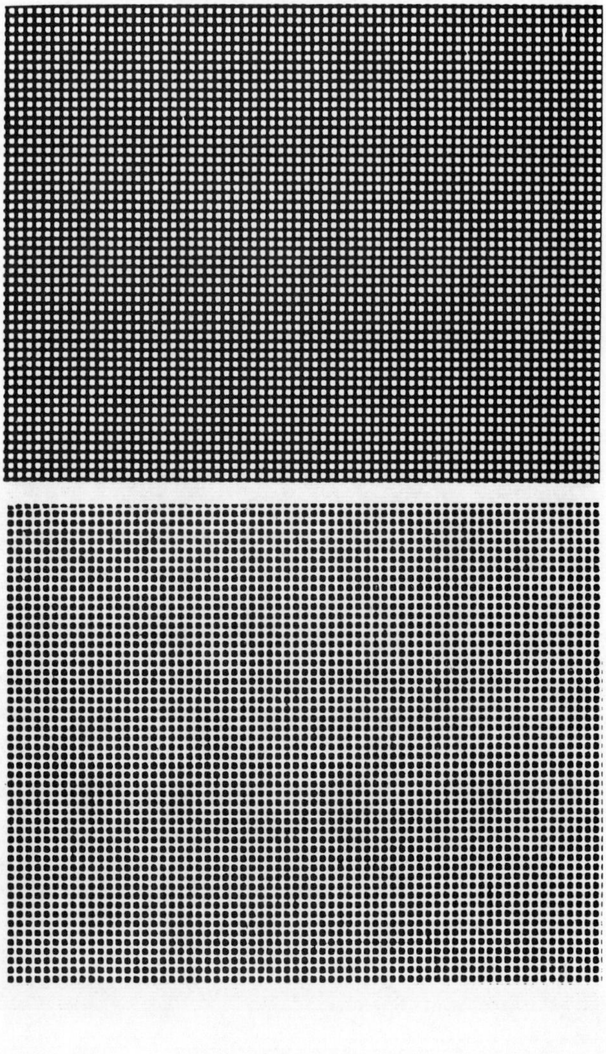

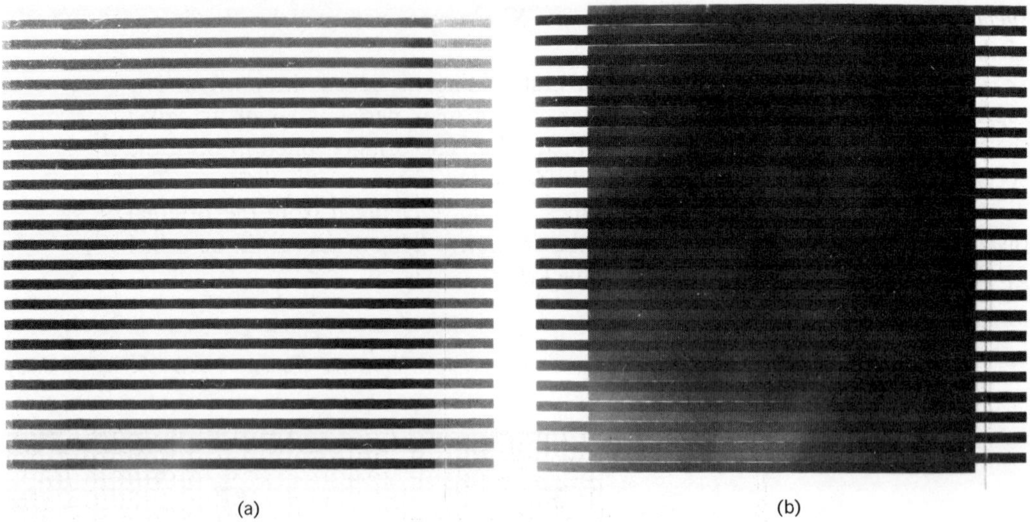

Figure 1.6. Gratings of parallel lines with same pitch and with line width equal to interline width. (a) Coincidence of lines. (b) Lines falling in between interlines.

Figure 1.7. Gratings of parallel lines with same pitch and with thinner line width than interline width. (a) Matching of lines. (b) Lines falling in between interlines.

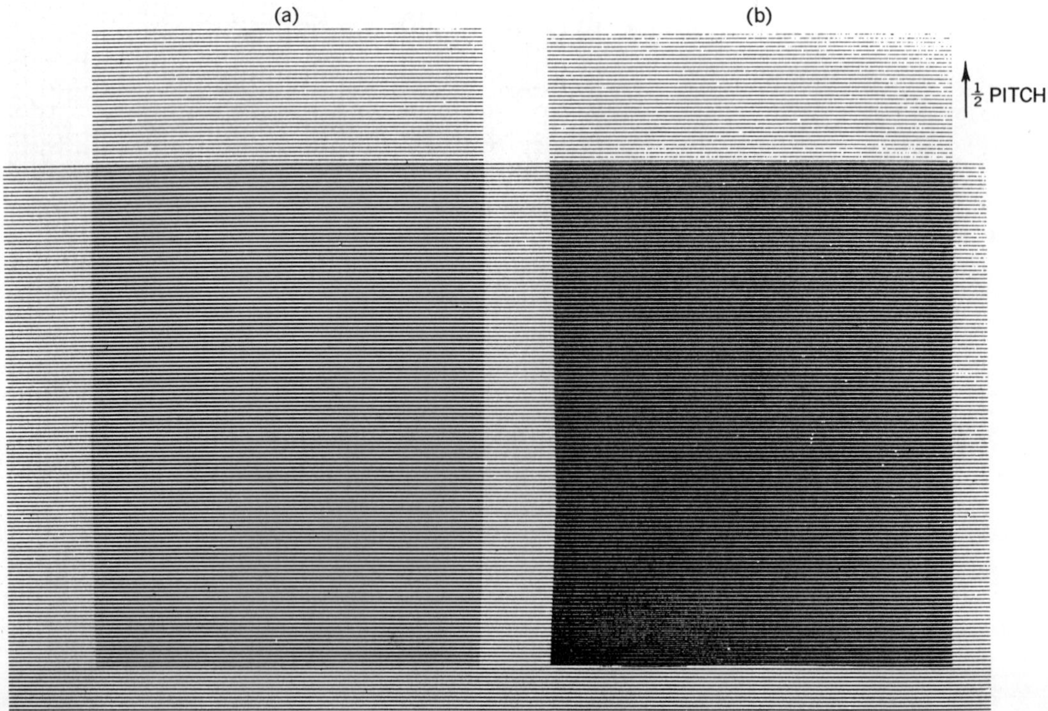

Suppose these gratings are printed on transparent sheets, one placed on top of the other, and light is allowed to pass through both of them. If the pitch of the lines is the same in both gratings and if the direction of their lines is the same, uniform light or uniform darkness will be observed, depending on whether the lines of one grating coincide exactly with the lines of the other grating, or whether the lines of one grating fall somewhere between the lines of the other grating (Figs. 1.6 and 1.7). In Fig. 1.6b the alternate dark and light bands are visible; however, if the density of fringes were sufficiently high, the eye would blend the effect and see uniform light.

The same phenomenon will be observed if, after the light goes through the first grating, it is reflected by either the front or the back surface of the second grating.

1.2.1. Relative Displacement of Two Gratings of the Same Pitch Along Their Common Perpendicular

Suppose that one of the gratings shown in Figs. 1.6 or 1.7 is displaced in its plane in the direction perpendicular to the grating-lines direction, and that the other grating is left in place. Darkness will be observed every time the lines of one grating fall between the lines of the other grating. Brightness will be observed every time the lines of one grating fall on top of the lines of the other grating, as in Fig. 1.6. Every complete cycle will correspond to the passage of a grating line through a specific point; in other words, every cycle (darkness or brightness) will indicate a displacement of one grating line equal to the pitch of the gratings.

Under these conditions, the phenomenon is rather difficult to observe because it is influenced very much by the small imperfections of the gratings. It could be recorded, however, either by eye or by a photocell. If, as explained above, the two gratings originally coincide and are displaced, being kept parallel to each other, along their common perpendicular in the plane, the relative motion of the gratings is given by

$$v = np \tag{1.1}$$

where n is the number of darkness or brightness cycles and p is the pitch of the gratings.

1.2.2. Displacement of One Grating Obliquely to the Direction of the Lines of Another of the Same Pitch

It is also obvious that if the movement of one grating with respect to the other is such that the lines of both gratings are always parallel, but the direction of the movement is inclined at an angle η with respect to the direction of their common perpendicular, then the relative displacement in the direction of the motion will be given by

$$\frac{v}{\cos \eta} = \frac{np}{\cos \eta} \tag{1.2}$$

and the component of the displacement in the direction perpendicular to the gratings direction is still

$$v = np$$

Obviously, a movement parallel to the direction of the gratings will produce no effect.

1.2.3. Mismatch-of-Gratings Interference

If two gratings have the same direction but different pitch, or if they have the same pitch but one is rotated with respect to the other, an interference is

Figure 1.8. Mechanism of formation of moire fringes using parallel straight line gratings. (a) Fringes due to rotation alone. (b) Fringes due to difference in pitch alone. (c) Fringes due to combination of rotation and difference in pitch.

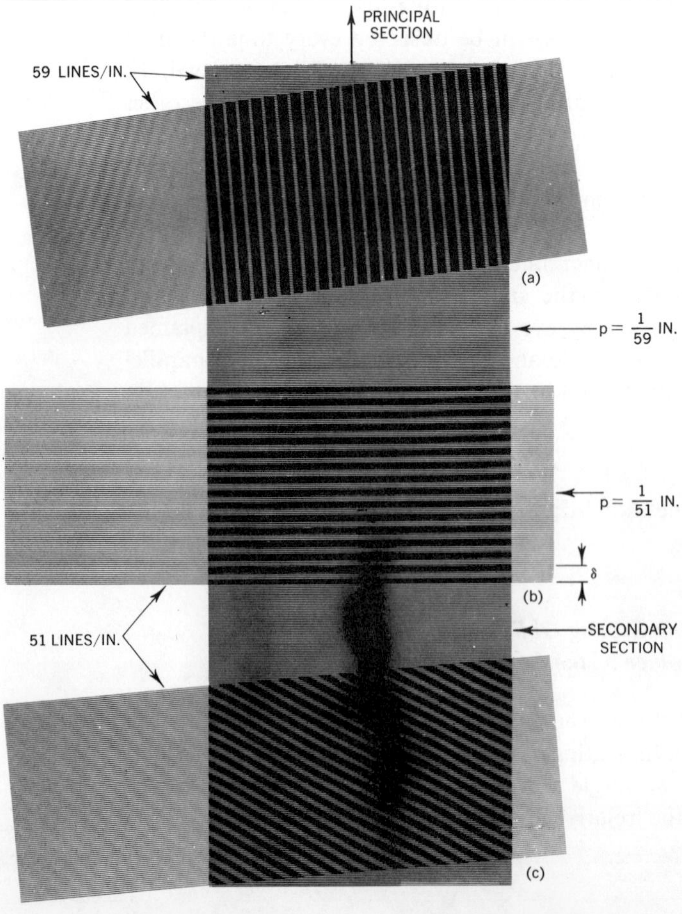

Figure 1.9. Moire fringes resulting from the superposition of two circular concentric gratings. The pitch of one is one-third greater than the pitch of the other.

produced (Fig. 1.8), shown as black and white fringes. It will be seen in Chapter 2 that each moire fringe can be defined as a locus of interference points of two superposed families of lines.

The distance between fringes is called *fringe spacing* and has been assigned the symbol δ. The number of fringes per unit length is called *fringe density* and is represented by $D = 1/\delta$. Any line perpendicular to the master-grating lines will be called a *principal section*, and a line parallel to the master-grating lines, a *secondary section*. Model and master gratings are assumed to be in the same plane.

The example of interference shown in Fig. 1.8 (moire effect is sometimes called *mechanical* interference) was obtained from sheets of cellophane with printed gratings of parallel straight lines with different pitches, superposed one on the other. Some exhibit relative rotation. The interference effect is obvious. The interference resulting from the superposition of two circular concentric gratings—the pitch of one being one-third larger than the pitch of the other—is shown in Fig. 1.9. Figure 1.10 illustrates the patterns produced by various amounts

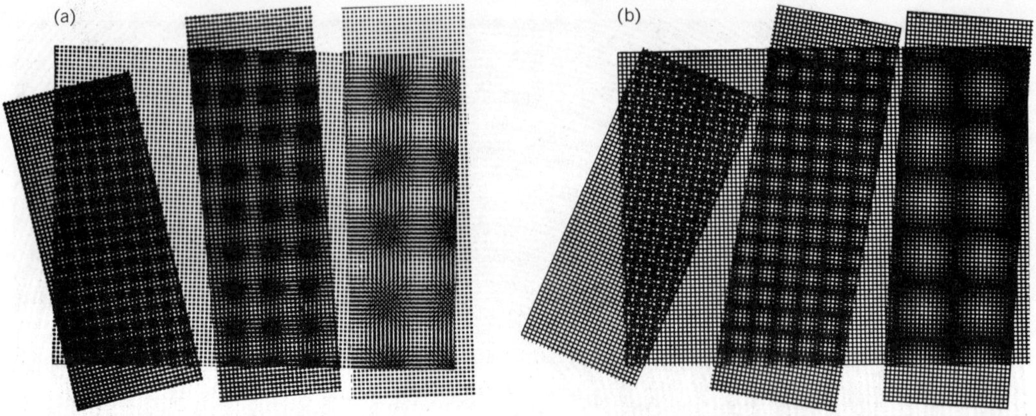

Figure 1.10. Moire patterns produced by rotation of a grating with respect to another of the same pitch. Top set: square array of dots. Bottom set: cross-grating.

of rotation of a cross-grating with respect to another cross-grating and of a grating of square dots with respect to another grating of square dots.

If the pitch of the gratings is appreciably smaller than 0.001 in., diffraction effects will be apparent and the phenomena will be much more complicated. In this book the gratings are considered to be sufficiently coarse to be studied with sufficient accuracy by means of rectilinear optics. A few exceptional situations will be considered in Sec. 1.8.

To illustrate better the moire phenomenon, the simple case where there is a relative displacement of the gratings in a direction perpendicular to their common grating direction, but where no rotation of the gratings is involved, will be considered in Sec. 1.3. In Sec. 1.4 the case of gratings subjected to pure rotation will be considered. In Sec. 1.5 the general situation, which includes both rotation and displacement perpendicular to the grating, will be shown.

1.3. Fundamental Property of Moire Fringes (Nonrotational Case)

The simple case of the relative displacement of one grating with respect to another of the same pitch, where the movement keeps the lines parallel to each other, has been considered in Secs. 1.2.1 and 1.2.2. It can be visualized as corresponding to a rigid-body translation.

Consider now the interference of two gratings with lines parallel to each other, but of different pitches p and p_1. The interference can be visualized as being obtained from uniform contraction or elongation of one grating with respect to the other. This mechanism of fringe formation is shown in Fig. 1.11. A dark fringe will appear at the points where an opaque strip falls over a transparent strip. When two opaque strips coincide, there is a maximum of light intensity and a light fringe appears. Assume that the master- and specimen-grating pitches were originally equal but that the specimen grating is now the one with the larger pitch p_1, due to specimen elongation. Assume also that the grating lines of specimen and master at the left edge of the figure coincide before and after deformation. A point P in the undeformed state has undergone a displacement equal to the pitch p of the master grating due to the elongation and is now at P'. Then the center line of the light fringe which goes through P' indicates a displacement in the direction of the principal section equal to p. At the center of the light fringe to the right of P' the displacement is equal to $2p$, and for the nth fringe the displacement is np. It can be concluded that the moire pattern gives the relative displacement in the direction of the principal section and the final position of the point that has moved (since the fringe coincides with the final position of the point) $u = np$. The same expression was obtained in Sec. 1.2. The pattern indicates that point P' in the deformed state of the model has undergone a displacement p.

The following relationship is also obvious: Calling m the number of lines of the master grating that fall between two fringes,

$$\delta = mp = (m \pm 1)p_1 \tag{1.3}$$

This permits the determination of the specimen pitch:

$$p_1 = p\frac{m}{m \pm 1} = \frac{\delta}{\dfrac{\delta}{p} \pm 1} \tag{1.4}$$

The plus-or-minus sign ($\pm$) indicates that the specimen pitch either increases (minus sign) or decreases (plus sign) with respect to the master pitch.

Notice that for this nonrotational case, the fringe direction coincides with the grating direction, secondary section of the system.

Equations (1.1) and (1.2) had been obtained by equating displacements.

Figure 1.11. Formation of moire fringes without relative rotation of the gratings. (It can be conceived as if the original specimen pitch p became p_1 by elongation of the specimen): $\epsilon^E = p/\delta$.

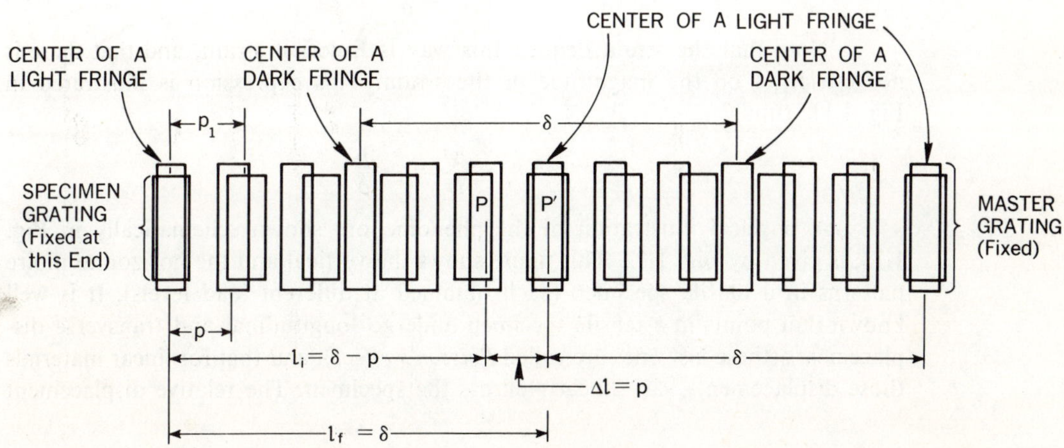

The same relations can also be described by equating the number of grating lines and fringes which essentially is equivalent to treating the reciprocals of the displacements, as shown below.

If between two fringes there are m master-grating lines, there will be $(m \pm 1)$ specimen-grating lines. And over a span of n fringes there will be nm master-grating lines and $n(m + 1)$ specimen-grating lines. Then, on any span the number of fringes n is the difference between the number of master-grating lines $n(m \pm 1)$ and specimen-grating lines nm over that span. In equation form:

Number of Fringes = (Specimen-Grating Lines) − (Master-Grating Lines)

$$n = \pm[n(m \pm 1) - nm] \tag{1.5}$$

If a unit span is taken, then the number of fringes is the fringe density D, the number of grating lines is the grating density, and the following relationship for densities holds:

$$D = \pm(d - d_1)$$

or

$$\frac{1}{\delta} = \pm\left(\frac{1}{p} - \frac{1}{p_1}\right) = \pm\frac{p_1 - p}{pp_1} \tag{1.6}$$

Further,

$$\frac{p}{\delta} = \pm\frac{p_1 - p}{p_1} \tag{1.6a}$$

This same relation could also be obtained directly from Eq. (1.4).

It will help the understanding of the phenomenon if a reference is made to the concept of strain that will be introduced later. For the case of little or no rotation, if the pitch p_1 is considered to be a distance that originally had a length p, then by definition the term $\pm(p_1 - p)/p_1$ gives the magnitude of the strain on the grating with p_1 pitch, normal to the grating lines. (See Chapter 4.) For this case the strain ϵ is expressed simply as

$$\epsilon^E = \pm\frac{p}{\delta} \tag{1.7}$$

Note that the strain defined this way is Eulerian strain, and that there is no restriction on the magnitude of the strain. This expression is illustrated in Fig. 1.11; thus

$$\epsilon^E = \pm\frac{\Delta l}{l_f} = \pm\frac{p}{\delta}$$

A physical illustration of the phenomenon, shown schematically in Fig. 1.11, is given by Fig. 1.12. This figure shows the vertical and the horizontal moire patterns in a tensile specimen (each obtained at different load levels). It is well known that points in a tensile specimen undergo longitudinal and transverse displacements (these last ones are called *Poisson's effect*), and that for linear materials those displacements vary linearly across the specimen. The relative displacement

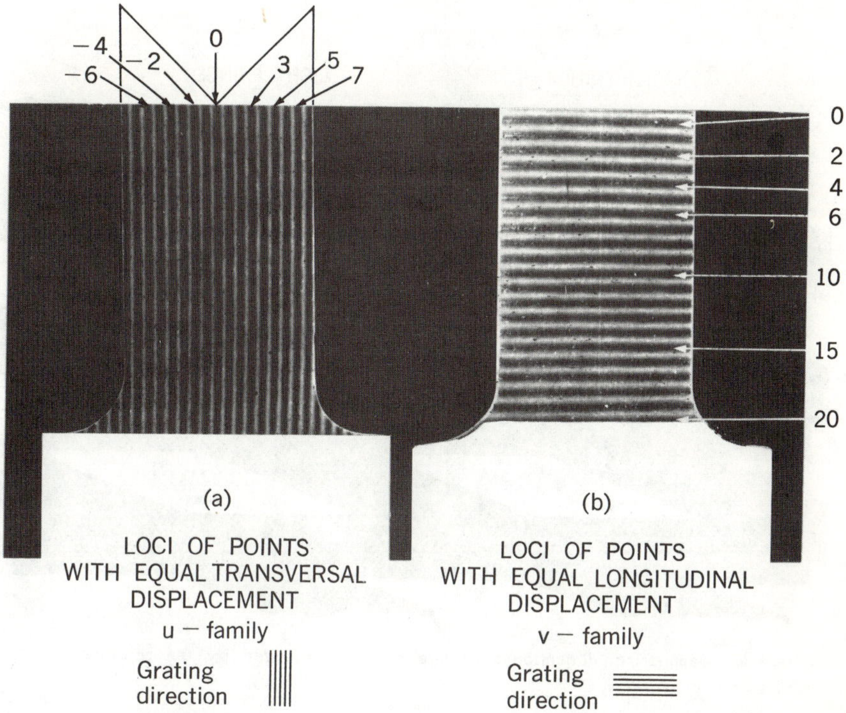

Figure 1.12. Moire patterns on a tensile specimen.

component of points in one fringe with respect to the points belonging to the immediate fringe is equal to the grating pitch. To determine the total displacement components at any point, it would be necessary to count the number of bright–dark cycles as the displacement occurred. Since this is not always practical, some considerations independent of the moire pattern must be made. In the u-family (Fig. 1.12a) it is obvious, for reasons of symmetry, that the fringe coinciding with the longitudinal axis of the specimen belongs to order zero (there is no transverse displacement at the axis). On the other hand, the zero order in the v-family can be determined if the movement of the platens of the testing machine is known. If both platens move the same distance in opposite directions, then the horizontal axis of the specimen has not moved, and the fringe coinciding with the axis exhibits order zero.

1.4. Fundamental Property of Moire Fringes (Pure Rotational Case)

If the model grating is not subjected to elongations or contractions in the direction perpendicular to the grating, but to rotation only, then the phenomenon

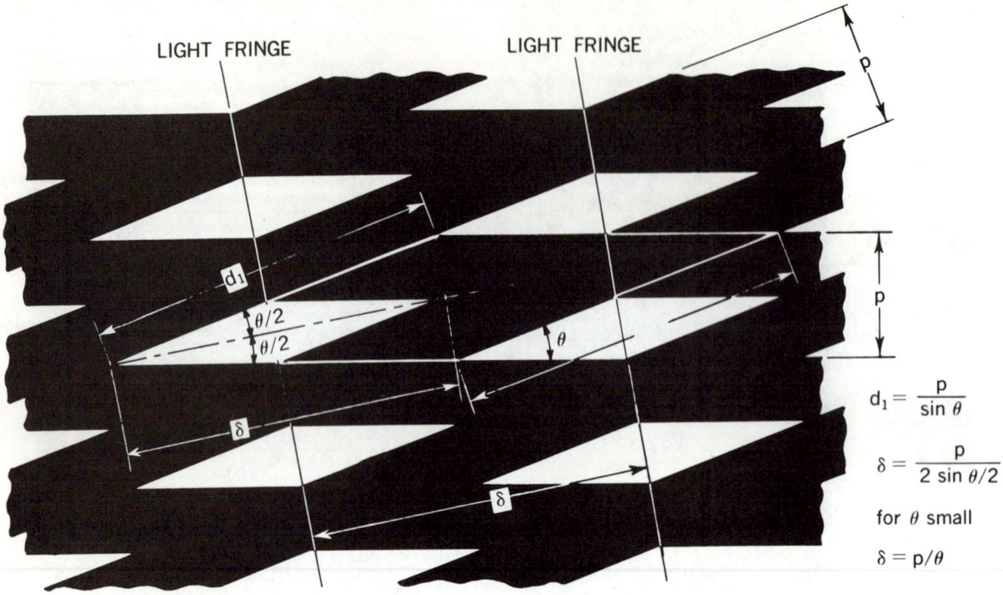

Figure 1.13. Geometrical deduction of fringe spacing produced by the relative rotation of two gratings.

is one of pure rotation of two like gratings with respect to each other. Such a case is shown in Fig. 1.13.

From the geometry of the intersecting lines shown in Fig. 1.13, the length of the side d_1 of the rhombus is given by

$$d_1 = \frac{p}{\sin\theta}$$

and that of the semidiagonal δ of the rhombus is given by

$$\delta = d_1 \cos\frac{\theta}{2}$$

$$\delta = \frac{p\cos\frac{\theta}{2}}{\sin\theta} \tag{1.8}$$

$$\delta = \frac{p}{2\sin\frac{\theta}{2}}$$

and, for small angles θ, by

$$\theta = \frac{p}{\delta} \tag{1.9}$$

This is exactly the same expression [Eq. (1.7)] obtained for strain in Sec. 1.3;

however, the two cannot be used simultaneously. To specify this they can be written

$$\epsilon\Big|_{\theta=0} = \frac{p}{\delta}$$

$$\theta\Big|_{\substack{u=0\\ \theta\ll 1}} = \frac{p}{\delta}$$

A simple means has been obtained, therefore, of determining the rotation of two equal gratings when the pitch and the distance between fringes are known.

From symmetry, it can be seen that the fringes bisect the angle formed by any master- and specimen-grating lines. Thus, if θ is the angle of rotation and ϕ is the angle between the master grating and the fringe (Fig. 1.14), then

$$\begin{aligned} \phi &= \frac{\pi}{2} \pm \frac{\theta}{2} \\ \theta &= \pm(2\phi - \pi) \end{aligned} \qquad (1.10)$$

The plus sign will be for the case where both ϕ and θ are measured in the same direction from the master grating.

Suppose the axis of rotation is chosen to coincide with the intersection of two grating lines, point O, shown in Fig. 1.14. The light fringe passes through this axis. It can be seen in the figure that this fringe connects intersections of specimen- and master-grating lines which coincided before rotation. These points, therefore, have not moved perpendicularly to the direction of the master-grating lines.

In other words, the specimen-grating lines at the intersections with the corresponding master-grating lines (moire fringe) have had no displacement in the y-direction. Therefore, the light fringe through the axis of rotation defines a line of zero y-displacement.

The fringe to the left of this zero displacement can be seen as the locus of intersections of specimen-grating lines with master-grating lines which were separated from them by a distance of 1 pitch width before rotation. This fringe represents a y-displacement of 1 pitch.

If the fringes are ordered about the zero fringe defined above as $n = -2, -1, 0, +1, +2$, etc., from right to left, then the y-displacement v can be written as

$$v = np \qquad (1.11)$$

It can be concluded here also, as in the case of pure elongation, that the moire pattern gives the relative displacement, in the direction of the principal section, of points the final position of which falls on the fringe.

It is important to point out here the meaning of the statement, "The displacement is referred to the final position of points."

The fringe n at any point gives a value of v (Eq. 1.11). This value of v is associated with the point on the deformed body at which n was read, and as such it is defined in terms of its deformed position.

If coordinates had been marked on the body before deformation, then it

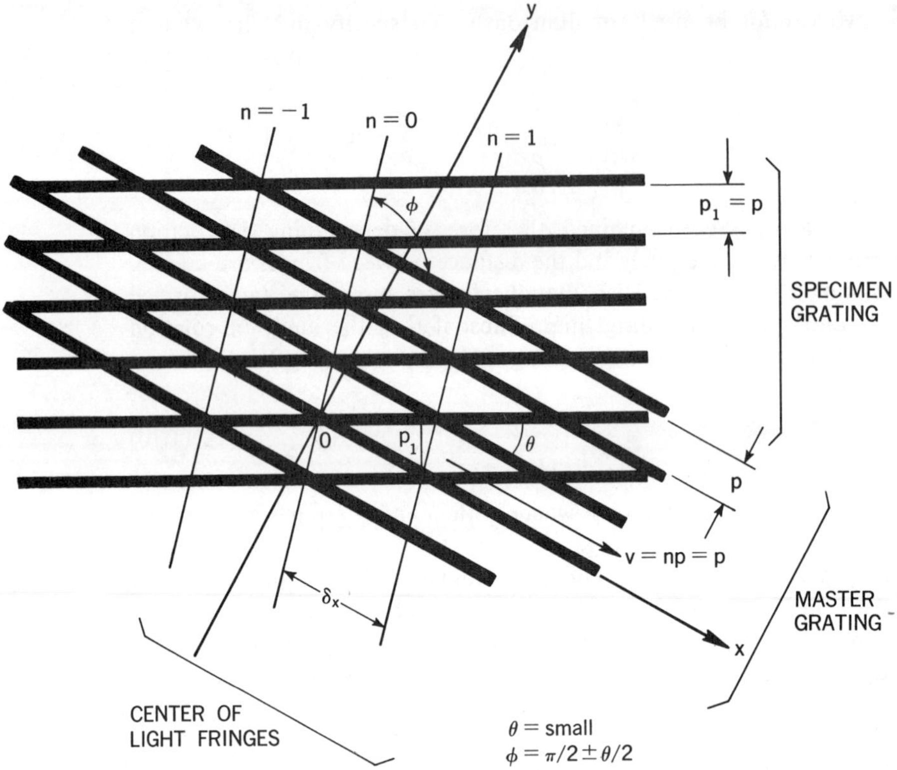

Figure 1.14. Moire fringes produced by pure rotation of one grating, relative to the other of the same pitch.

would be possible to refer the value of v to the coordinates of the point in the undeformed body. This is seldom found to be practical.

The importance of this will be appreciated when the derivatives of the displacements are obtained from the moire pattern.

A physical illustration of the phenomenon shown schematically in Fig. 1.14 is shown in Fig. 1.8a.

Notice that according to Eq. (1.10) the angle between the fringe and the normal to the grating is one-half the angle between the two gratings (rotation). For small angles of rotation, the fringes produced by rotation coincide practically with the direction perpendicular to the grating, or principal section of the system.

1.5. Fundamental Property of Moire Fringes (General Case, with Elongation and Rotation of Gratings)

Figure 1.15 illustrates the moire phenomenon in the general case when one of the gratings not only elongates but also undergoes rotations. Suppose that before the relative displacement of one grating with respect to the other, the lines of corresponding orders in both gratings coincide. After their relative displacement, there is a point at which the two lines of order q intersect and show no relative displacement in the direction perpendicular to the grating lines. The order of the fringe going through that point is zero. The same reasoning applies to the lines of the gratings immediate to the lines q, and which can be called $(q + 1)$. After the relative displacement, they have not moved. Their intersection will also belong to the fringe of order zero. On the other hand, the fringe of order 1 is produced by the displacement p of the points of the specimen grating in the direction of the principal section; the fringe of order 2 is produced by the displacement $2p$, and the nth fringe by the displacement np. Then the moire fringes are the loci of points with a relative displacement in the direction of the principal section (displacement component), which is equal to an integer number times the pitch of the master grating. These displacement components refer to points that are given in the deformed or final shape of the model. (The original system of reference has moved with the model, and its final position in general is not known.)

Each fringe is characterized by a parameter. This parameter is arbitrary, since we are considering relative displacements. This parameter is called *order of the fringes* and is assigned the letter n.

In the following, as in Fig. 1.1, a Cartesian system of axes x and y is used as a reference system. The component of the displacement parallel to the x-axis

Figure 1.15. Moire fringes are the loci of the points having the same value of component of displacement in the direction normal to the master grating (general case including relative rotation of the gratings). For the purpose of the illustration, the lines have been made thinner than the interlines and the deformation is very large.

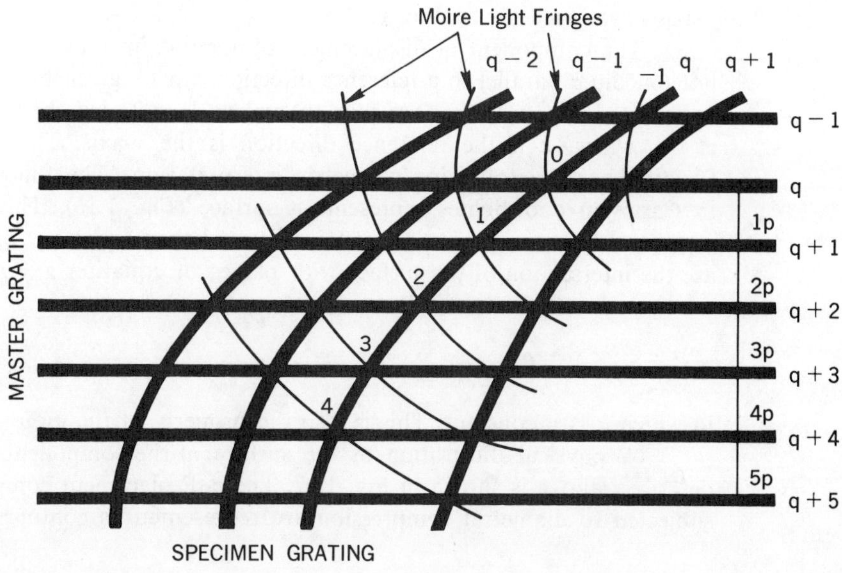

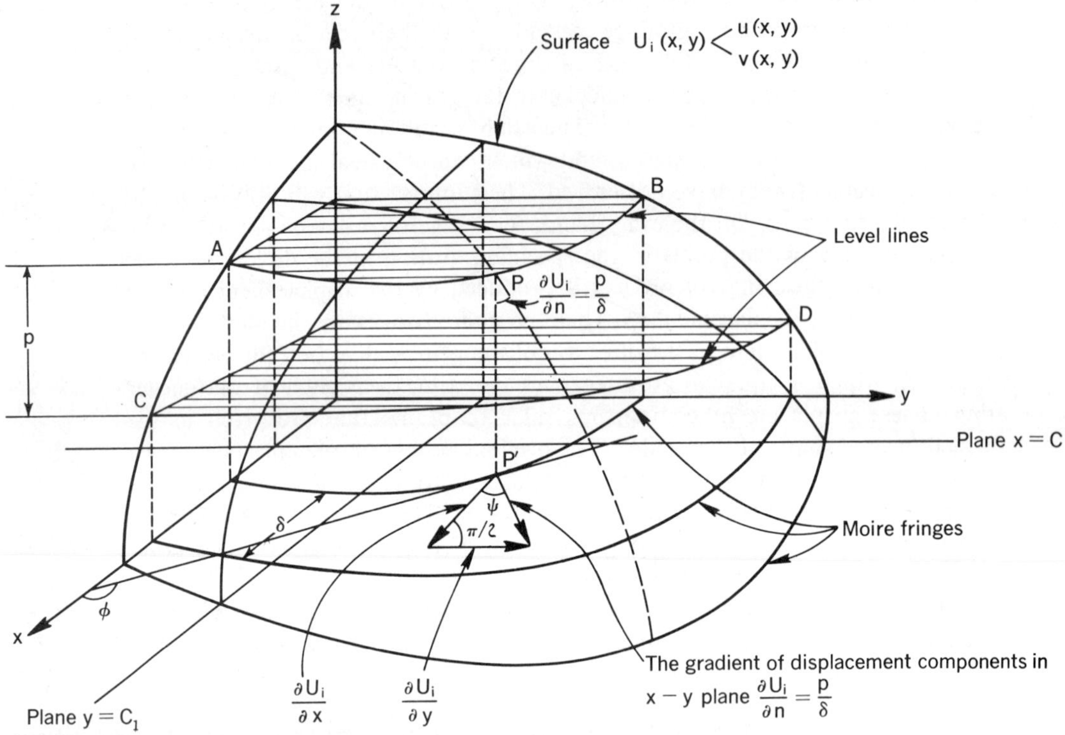

Figure 1.16. Surface of the components of displacements $u(x, y)$ or $v(x, y)$.

is given the symbol u and the component of the displacement parallel to the y-axis is assigned the symbol v.

The component of displacement of a point in a two-dimensional continuous medium parallel to a reference direction can be given by a function of two variables, $U_i(x, y)$. Here i can take the values 1 or 2, 1 if the reference direction is the x-axis, 2 if the reference direction is the y-axis. The above-mentioned function has the following geometric interpretation: The function $z = U_i(x, y)$ in Cartesian coordinates represents a surface (Fig. 1.16). This surface can be represented by the projection of its contour lines on the x-y plane. These lines are the intersection of the surface with planes of equation $z = np$. The resulting lines are given by

$$U_i(x, y) = k \tag{1.12}$$

in which k is a constant. This is also the property of the moire fringes.

A physical illustration of the surface of the component of displacements described above is shown in Fig. 1.17. The v-displacement components in a disk subjected to diametral compression are represented as contour lines on a relief

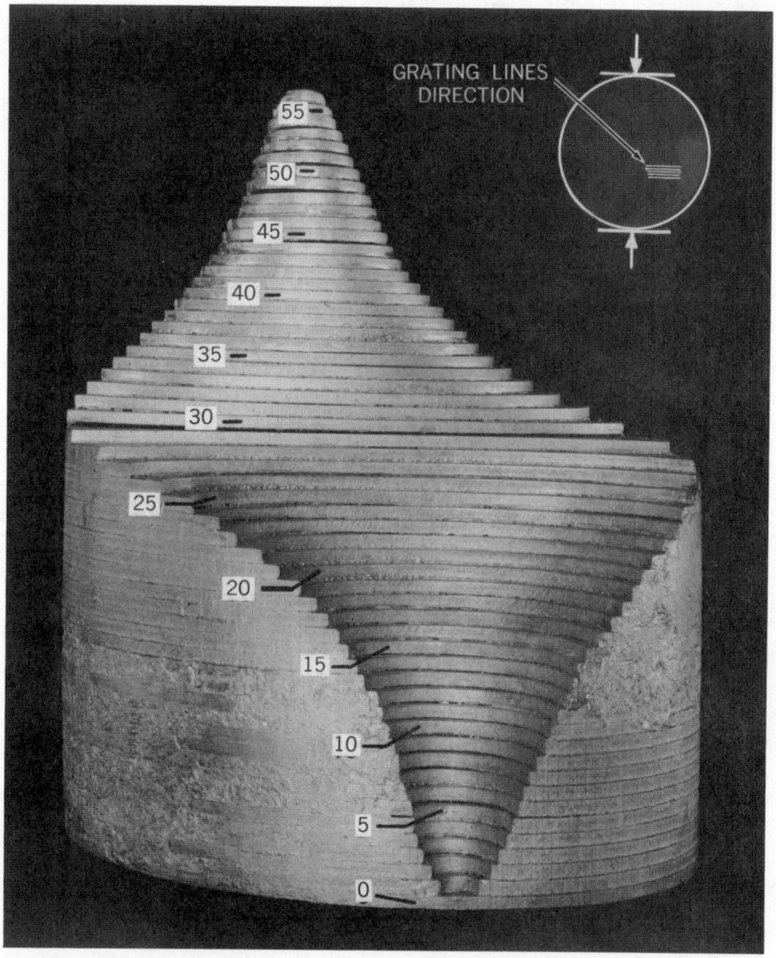

Figure 1.17. Moire fringes in a circular disk under diametral compression. They are shown as lines of constant level on a surface representing the components of displacement in the direction perpendicular to the master grating.

map. People in the field of topographic surveying are familiar with this kind of representation.

The projections of the contour lines on the x-y plane are the moire fringes, and are shown in Fig. 1.18.

The disk shown in Fig. 1.18 is made of rubber and the bottom of its vertical diameter has undergone no displacement in the process of loading. Therefore, the order of the moire fringe at that point is zero. Since the pitch of the grating is $\frac{1}{300}$ in., the fringe value of the moire fringes is $\frac{1}{300}$ in. To determine the vertical displacement of a point on a fringe, it is sufficient to multiply its order by the

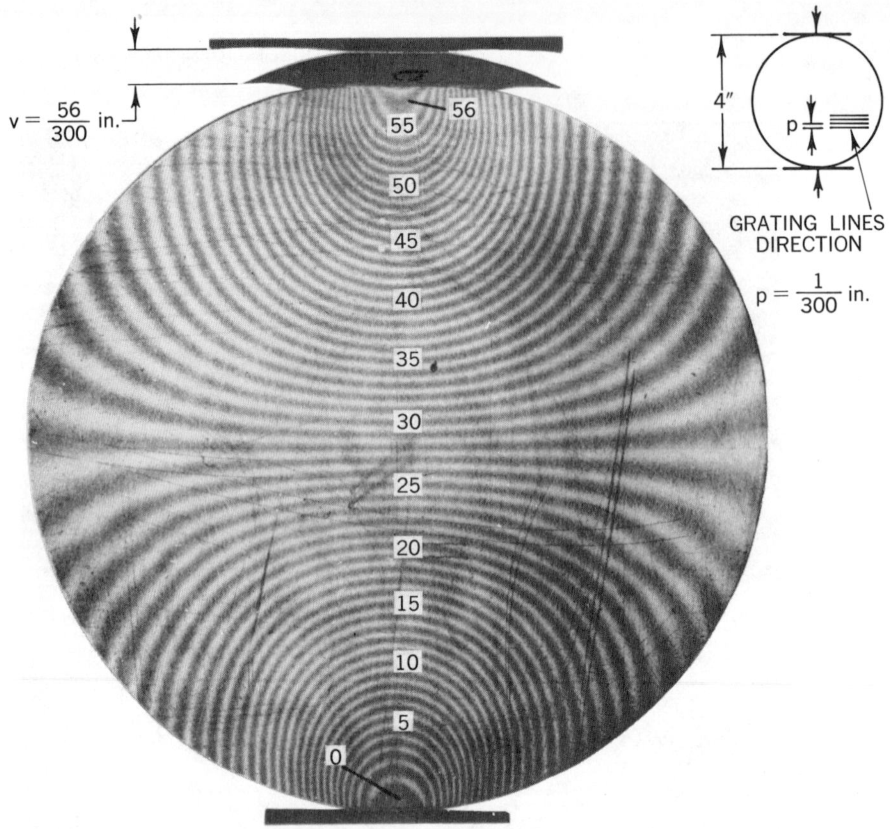

Figure 1.18. Moire fringes in a circular disk under diametral compression. Fringes are loci of points havinng the same value of vertical displacement. (Isothetics.)

fringe value $\frac{1}{300}$ in. The top of the vertical diameter has been displaced downward by

$$56 \times \tfrac{1}{300} \text{ in.}$$

This displacement can be verified by the displacement of the upper platen of the machine used to load the disk, the shadow of which is visible in Fig. 1.18.

The locus of points that have the same value of displacement component (u or v) will be called *isothetic* (of the u or v family).

The general meaning of the expression p/δ [Eqs. (1.7) and (1.9)] is also illustrated in Fig. 1.16. The gradient of the surface $z = U_i$ will be the slope of the surface in the direction normal to the constant-height lines (moire fringes). Since the height between moire fringes in the z-direction is p and the distance between moire fringes in the x-y plane is δ, then their ratio is the gradient

$$\frac{\partial U_i}{\partial n} = \frac{p}{\delta} \tag{1.13}$$

and is seen to resolve into Cartesian components in Fig. 1.16. These concepts are useful in strain analysis.

Figure 1.19 shows gradient components of the isothetic u-family. Call δ the distance between fringes and δ_x and δ_y the distance between fringes along the Cartesian directions. The change in displacement component u for points belonging to two adjacent fringes is p. The incremental values of the displacement components and of the change in position can be written as

$$\Delta u = p$$

$$\Delta x = \delta_x$$

$$\Delta y = \delta_y$$

Then the gradient components are

$$\frac{\Delta u}{\Delta x} = \frac{p}{\delta_x}$$

$$\frac{\Delta u}{\Delta y} = \frac{p}{\delta_y}$$

and in the limit

$$\frac{\partial u}{\partial x} = \frac{p}{\delta_x} \tag{1.14}$$

$$\frac{\partial u}{\partial y} = \frac{p}{\delta_y} \tag{1.15}$$

Figure 1.19. Gradient components of the isothetic u-family along coordinate axes.

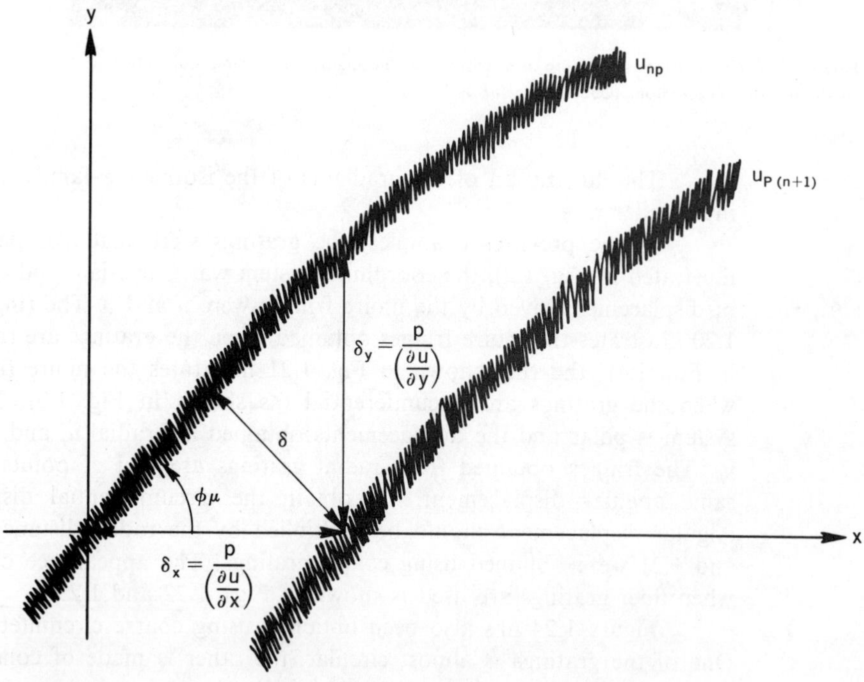

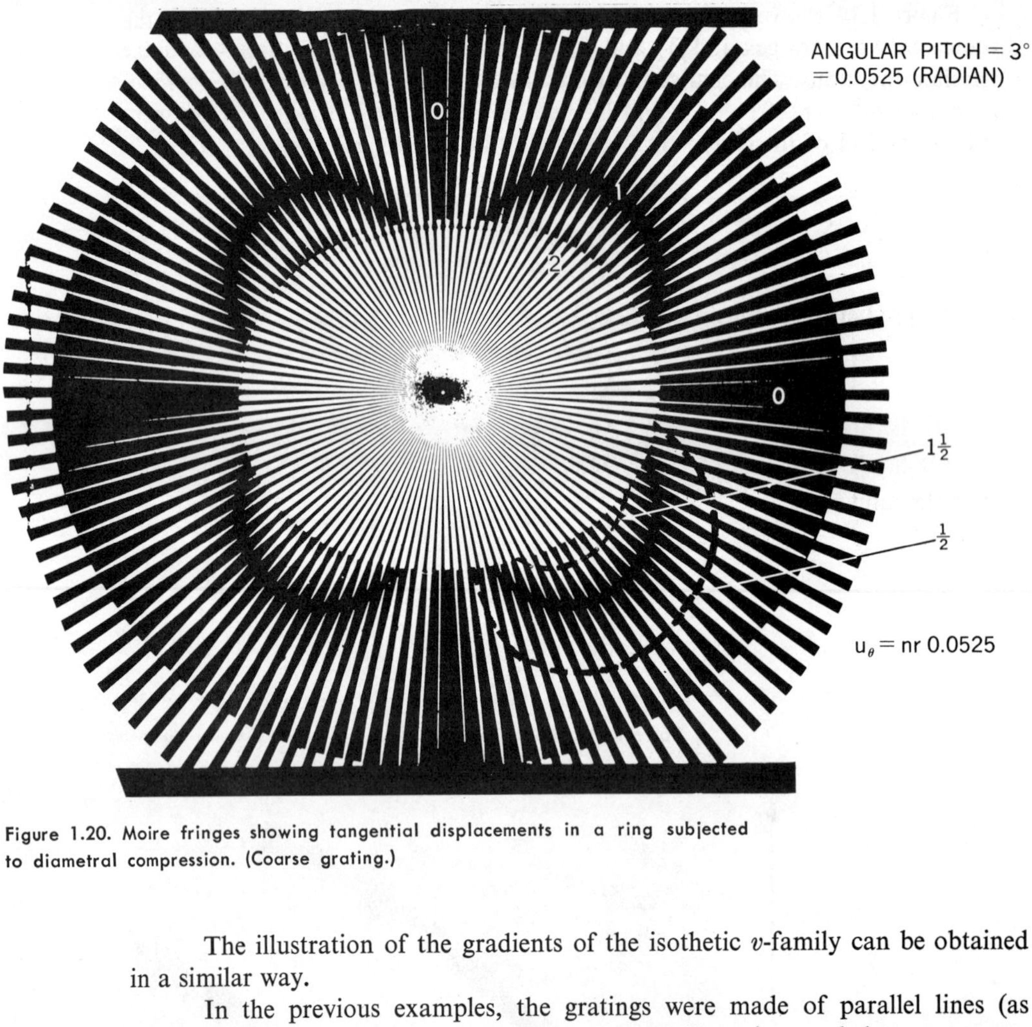

Figure 1.20. Moire fringes showing tangential displacements in a ring subjected to diametral compression. (Coarse grating.)

The illustration of the gradients of the isothetic *v*-family can be obtained in a similar way.

In the previous examples, the gratings were made of parallel lines (as illustrated in Fig. 1.2), the coordinate system was Cartesian, and the components of displacement given by the moire fringes were *u* and *v*. The ring shown in Fig. 1.20 illustrates the moire fringes obtained when the gratings are radial (as shown in Fig. 1.4); the ring shown in Fig. 1.21 illustrates the moire fringes obtained when the gratings are circumferential (as shown in Fig. 1.3). The coordinate system is polar and the displacements obtained are radial u_r and circumferential u_θ. The fringes obtained from radial gratings are loci of points exhibiting the same angular displacement. To obtain the circumferential displacement, the angular displacement should be multiplied by the radial distance. Figures 1.20 and 1.21 were obtained using coarse gratings. The appearance of the isothetics when finer gratings are used is shown in Figs. 1.22 and 1.23.

Figure 1.24 has also been obtained using coarse circumferential gratings. One of the gratings is almost circular, the other is made of concentric ellipses.

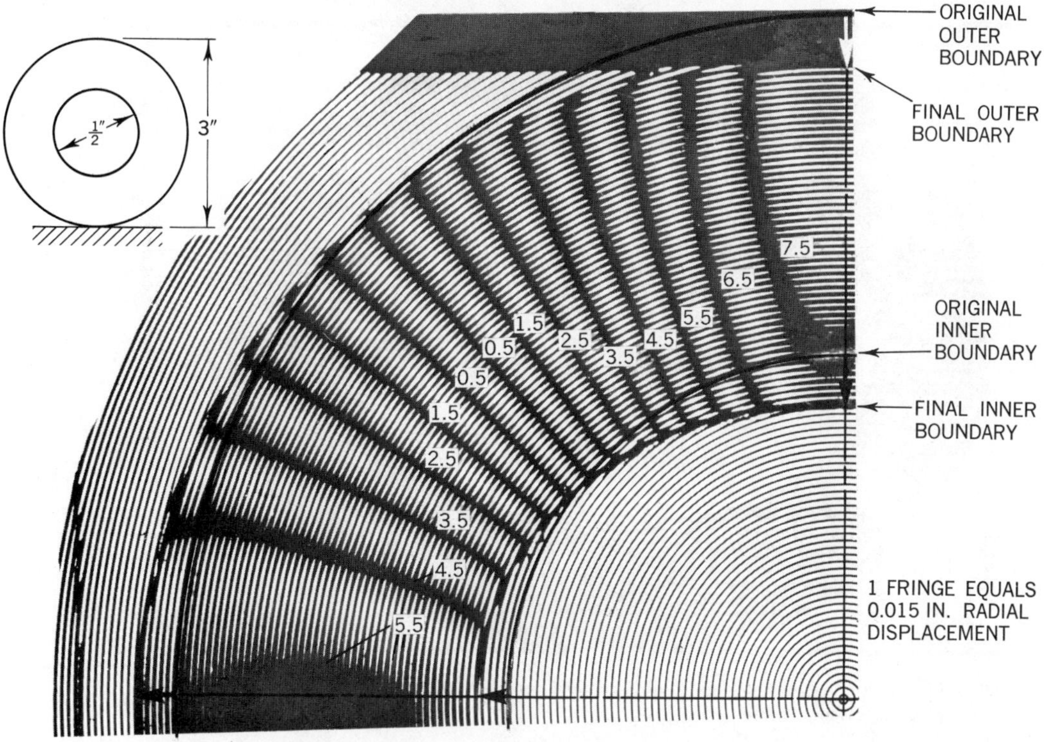

Figure 1.21. Moire fringes of radial displacement in a quadrant of a ring subjected to diametral compression. The intersection of the boundary of the undeformed specimen with the boundary of the deformed specimen locates the point of zero radial displacement.

Figure. 1.22. Moire fringes showing tangetial displacements in a ring subjected to diametral compression. (Dense grating.)

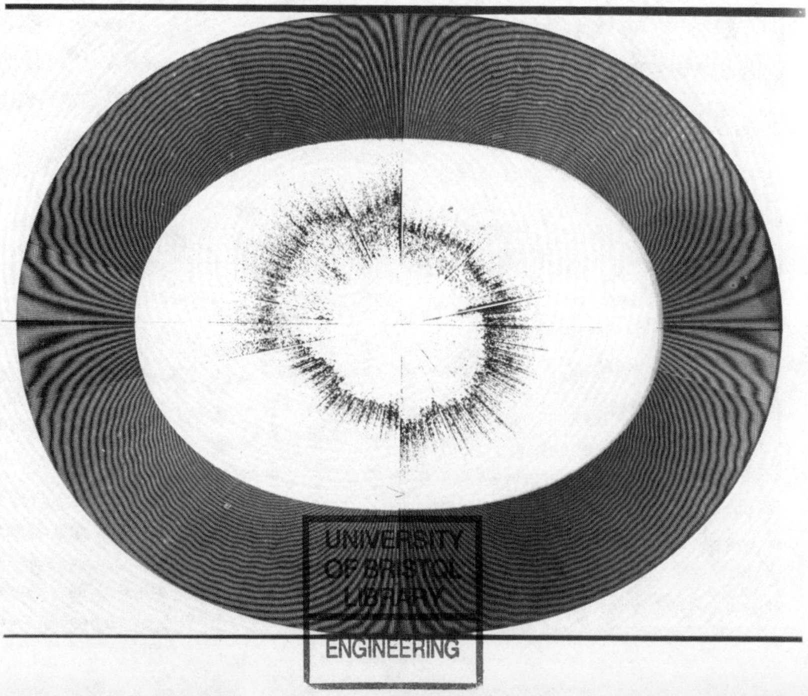

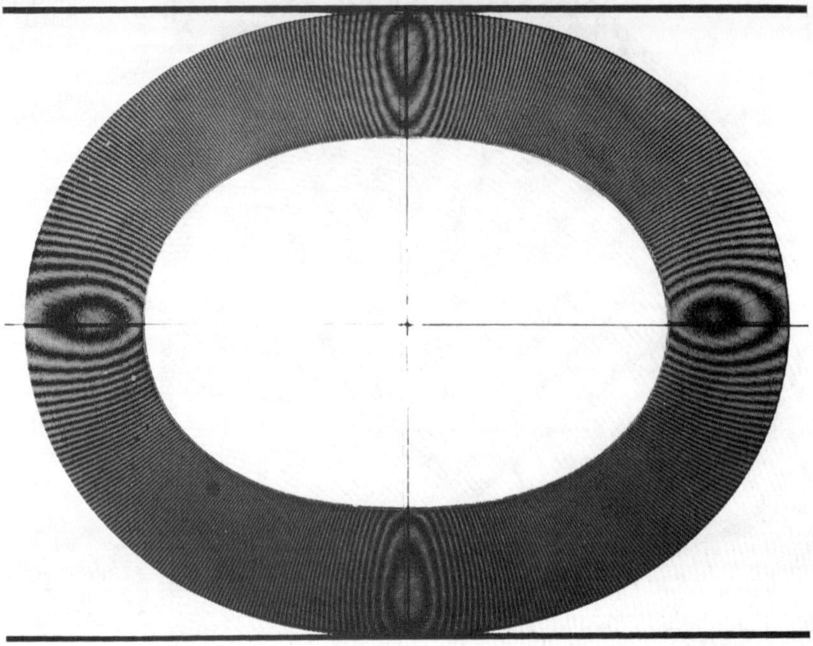

Figure 1.23 Moire fringes showing radial displacements in a ring subjected to diametral compression. (Dense grating.)

Figure 1.24. Moire fringes showing radial displacement between an almost circular grating and an elliptical grating concentrically superposed.

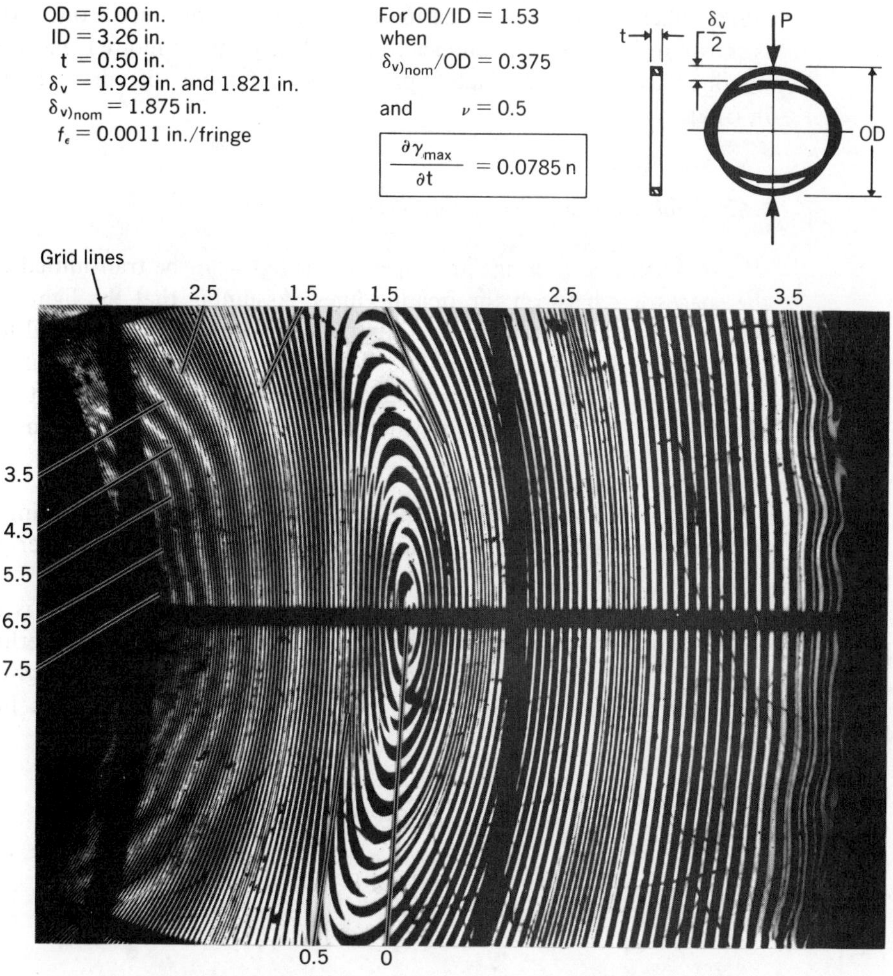

Figure 1.25. Isotachics of isochromatics in a circular ring subjected to diametral compression. (Large deformation.)

The fringes represent components of displacements perpendicular to the circular grating, or—if the elliptical grating is thought of as the reference grating—components of displacements perpendicular to the elliptical grating.

It is also possible to produce moire with gratings of no well-defined geometry. It will be seen in Sec. 7.8 how a moire pattern can be used as a grating to produce a moire-of-moire effect that permits an automatic differentiation of the displacements. In Sec. 10.2 it will also be seen how the moire pattern can be used as a grating to produce another second-order moire, the one corresponding to the velocities of the components of displacements. It should be mentioned here that not only moire patterns but also photoelasticity isochromatic patterns, when sufficiently dense, can be used as gratings to produce moire. An illustration is shown in Fig. 1.25. The interpretation of this type of moire is more conveniently obtained using the parametric properties of families of curves to be explained in Chapter 2.

1.6. Light Intensity Between Fringes

When two gratings are superposed, light can be transmitted only through the spaces left between the opaque lines. Assuming that the light source is uniform, the amount of transmitted light will be, locally, proportional to the surface of the space between the opaque lines. The phenomenon is illustrated in Fig. 1.26. In this figure it is assumed that the specimen is subjected to a linear displacement in the direction perpendicular to the grating lines. Thus,

$$u = 0.1x \tag{1.16}$$

It is also assumed that the first line of the specimen grating (on the left of the figure) has not moved and that it falls between two master-grating lines, producing a dark fringe.

The width of the bars in the central portion of Fig. 1.26 is proportional to the actual intensity of light being transmitted through the interline.

On the graph at the top of Fig. 1.26, the intensity of light transmitted through each interline has been averaged over the pitch and a line has been

Figure 1.26. Light intensity between fringes of a moire pattern, based on rectilinear optics. The average values show a linear rise and fall.

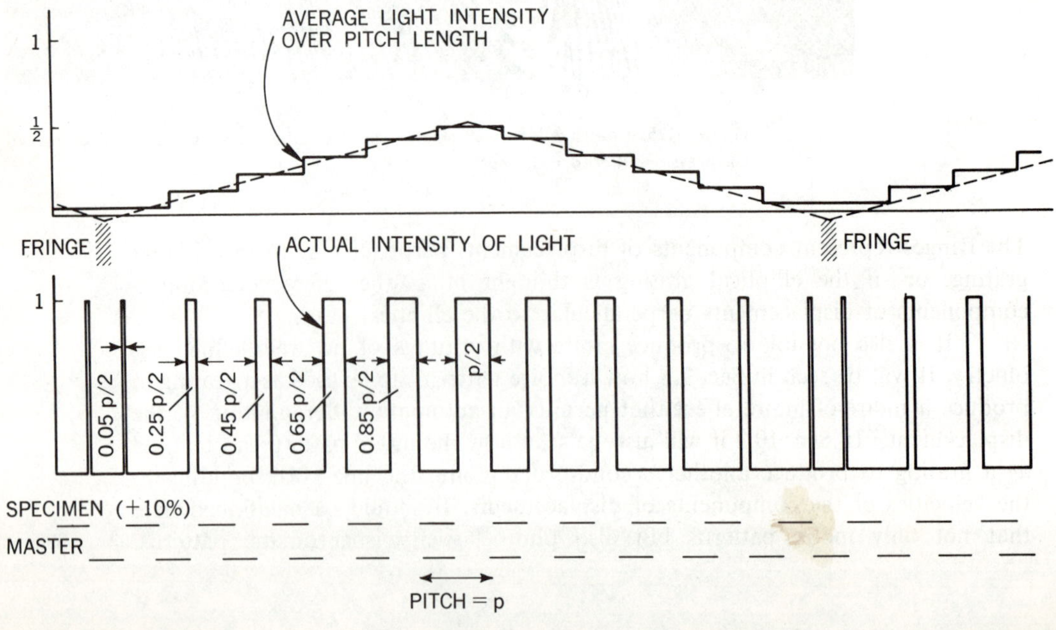

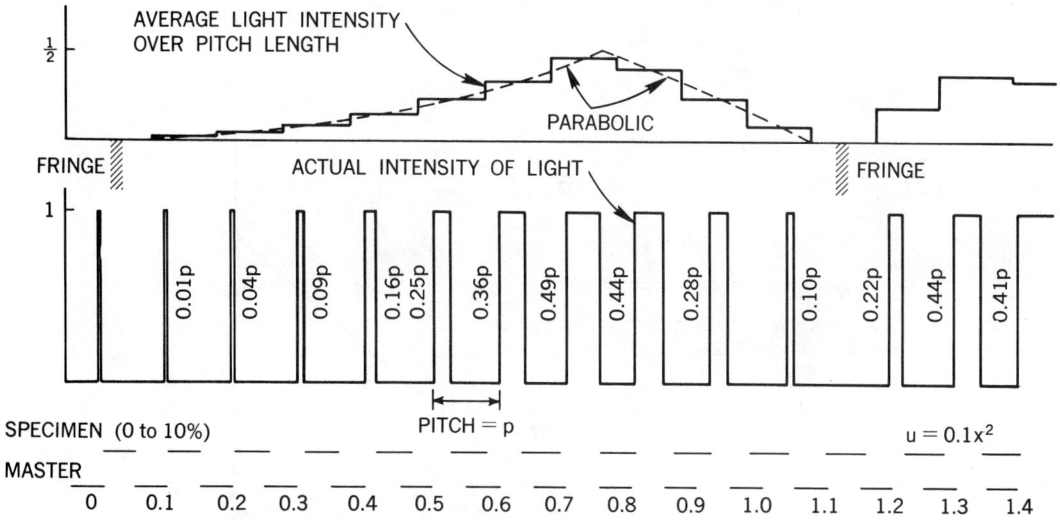

Figure 1.27. Light intensity between the fringes of a moire pattern. The average value shows a parabolic rise and fall.

drawn through the center of the step-wise representation. That line represents the average light intensity transmitted through the interline when the relative displacement of the gratings is expressed by Eq. (1.16).

Figure 1.27 illustrates the phenomenon in a case of nonlinear displacement —that is, where

$$u = 0.1x^2 \tag{1.17}$$

The same characteristics shown in Fig. 1.26 are evident in Fig. 1.27. A dark fringe occurs when one grating line falls between two lines of the other grating. Maximum light intensity occurs when one grating line has moved a distance of one-half the pitch and falls in front of a line of the other grating. In between those two extremes, the light intensity is proportional to the displacement:

$$u = p\left(n \pm \frac{1}{2}\frac{I}{I_{max}}\right)$$

where I_{max} is the maximum light intensity in between fringes and I is the intensity at the particular point under consideration. The plus-or-minus sign corresponds to the fact that the point may be between np and $(np + \frac{1}{2}p)$ or between $[(n - 1)p + \frac{1}{2}p]$ and np.

1.7. Displacements in the Direction Perpendicular to the Master-Grating Plane

If the two gratings (the master and the specimen) do not lie exactly one on top of the other, a pattern will be produced associated with the distance

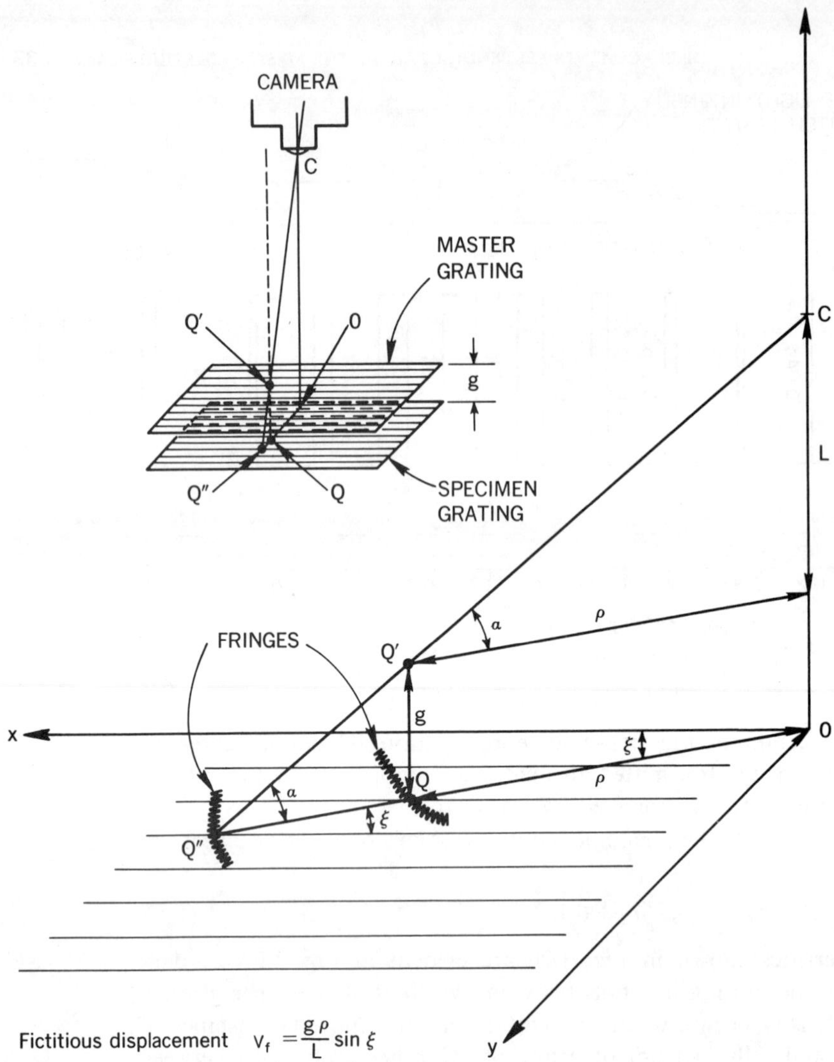

Figure 1.28. Effect of the lack of coincidence of the planes containing the specimen and master gratings.

between the gratings, in the direction perpendicular to their planes. It will be seen that the phenomenon is due to the perspective of the picture and that it will disappear if the light going through both gratings is parallel. A study of this aspect of the phenomenon is conducted using Fig. 1.28. A more detailed study including diffraction properties will be conducted in Sec. 1.8.6.

In this figure, $\overline{CO}$ is the optical axis and C the optical center of the camera. The points Q' and Q of the master and specimen gratings should be coincident but, due to the gap g existing between the two gratings, the point Q' is coincident in projection with the point Q''. Therefore, the camera "sees" Q as having a fictitious displacement v_f in the direction of the principal section y. If L is the distance between the optical center of the camera and the plane of the master grating, and ρ is the distance from the point Q' to the optical axis, then

$$\tan \alpha = \frac{L}{\rho} \tag{1.18}$$

From Fig. 1.28 the apparent displacement is

$$\overline{QQ''} = \frac{g}{\tan \alpha} \tag{1.19}$$

Replacing Eq. (1.18) in Eq. (1.19) gives

$$\overline{QQ''} = \frac{g}{L}\rho \tag{1.20}$$

The fictitious displacement in the y-direction is given by

$$v_f = \frac{g\rho}{L} \sin \xi \tag{1.21}$$

where ξ is the azimuth of the plane $CQ''O$. For all the points located in the circle of radius ρ, the minimum fictitious displacement (equal to zero) takes place at $\xi = 0$. The maximum fictitious displacement takes place at $\xi = \pi/2$ and is given by $v_f = g\rho/L$.

From Eq. (1.21) it is easy to compute the apparent strain introduced by the lack of coincidence of the specimen and grating planes. Assuming $\xi = \pi/2$ and $\rho = y$,

$$\epsilon_f = \frac{\partial v}{\partial y} = \frac{\partial}{\partial y}\left(\frac{g}{L}y\right) = \frac{g}{L} + \frac{1}{L}\left(\frac{\partial g}{\partial y}y\right) \tag{1.22}$$

The apparent strain is independent of the position of the point when the azimuth is $\xi = \pi/2$. To reduce the error when displacements in the plane of the master grating are measured, L must be as large as possible.

This analysis justifies the use of embedded gratings for the three-dimensional analysis of strain.[1]

1.8. Diffraction Properties[2]

1.8.1. Introduction

The discussion up to this point has dealt with light as rays moving in a straight line. This way of treating light properties is often termed *ray optics*, *rectilinear optics*, or *geometric optics*.

Certain light phenomena are better explained by considering the wave

---

[1] See C. Sciammarella, "The Moire Method Applied to Three-Dimensional Elastic Problems," *Exp. Mech.*, Vol. 4, No. 11 (Nov. 1964), pp. 313–319.

[2] Instructors using this book as a text may find it convenient to introduce the student to some basic aspect of light-diffraction properties following the elementary approach of this section. If the reader is familiar with wave properties, the reading of Sec. 1.8.1 and 1.8.2 may not be necessary.

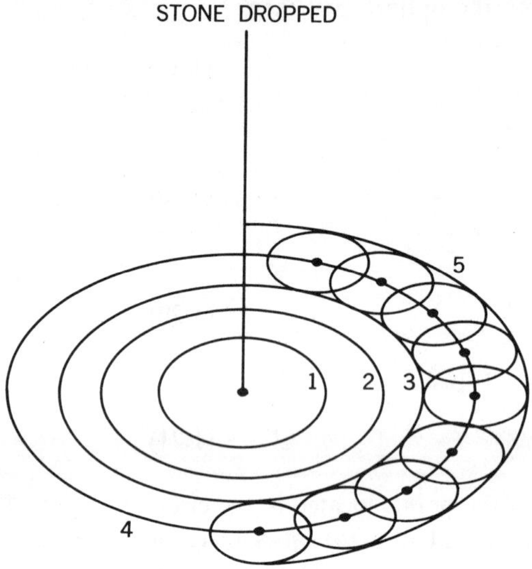

Figure 1.29. Analogy to illustrate a wave emanating from a point source.

character of light. This approach has the name of *physical optics*. In physical optics light is treated as continually emanating from every point in all directions, as the ripples of water from a dropped stone.

After a stone is dropped in water, it can be seen that at some distance from the stone every point in the water is disturbed. It is understood that from every one of these new points, disturbances can emanate in all directions on the surface of the water, just as the stone disturbed the water initially. The wave associated with the stone serves two purposes in this two-dimensional analogy: First, it represents how a disturbance may propagate, and second the stone is just the initial source for many point disturbances all of which, in turn, act like stones. However, despite the innumerable individual point sources, we see that the overall pattern of disturbance still takes the form of concentric circles. These waves actually represent the combined influence of all the local disturbances which act in all directions and form the series of concentric circles. Each wave is therefore essentially an envelope of the disturbances and is also called a *wavefront*.

This analogy is shown schematically in Fig. 1.29. Lines 1, 2, 3, and 4 represent the top or bottom of concentric disturbances. Ripple 4 occurs as a disturbance at some distance from the center and all the water at that distance is disturbed, say, upward. Simultaneously, each particle of the ripple imparts its disturbance to all the particles around it. The total effect is to create another ripple as an envelope for the myriad of local disturbances. And so the wavefront extends away from the source. The idea of developing the wavefront by construction of this series of small equal radii was conceived by Huygens. Suppose in what follows that the distance from the wavefront is long enough for the portions of circles to be coincident with planes; this is called a *plane wavefront*.

1.8.2. Diffraction Patterns

The diffraction pattern produced by a grating can be seen by looking at a small light source, such as that shown in Fig. 1.30a, through the grating. Instead of a single image, a series of images can be seen as in part (b) of the figure. If the light source is small enough, the images will appear as small dots completely separated from one another (Fig. 1.30c). A simple explanation of diffraction patterns which follows from the assumption that light travels in waves is given below.

Consider the optical system shown schematically in Fig. 1.31. A parallel beam of light enters the optical system from the left side traveling in the z-direction. Assume, for simplicity, that the light is monochromatic with wavelength λ and of uniform intensity. The light is represented by straight lines or rays in the z-direction.

Figure 1.30. Illustration of diffraction by a grating. When a small light source (a) is viewed through a grating, multiple images (b) are seen. If the light source is sufficiently small and the grating fine enough, the images will appear as a row of separated dots (c).

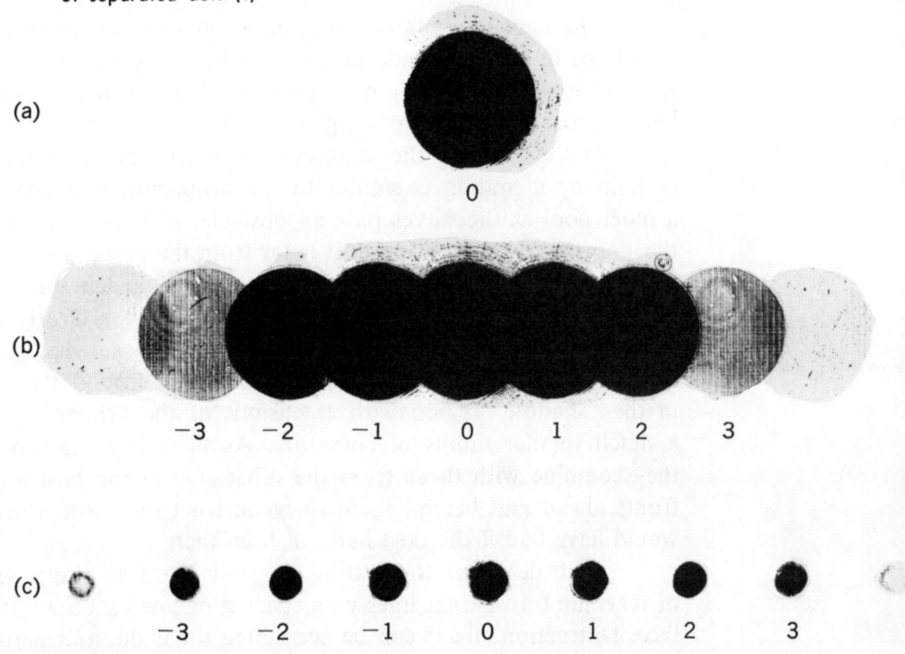

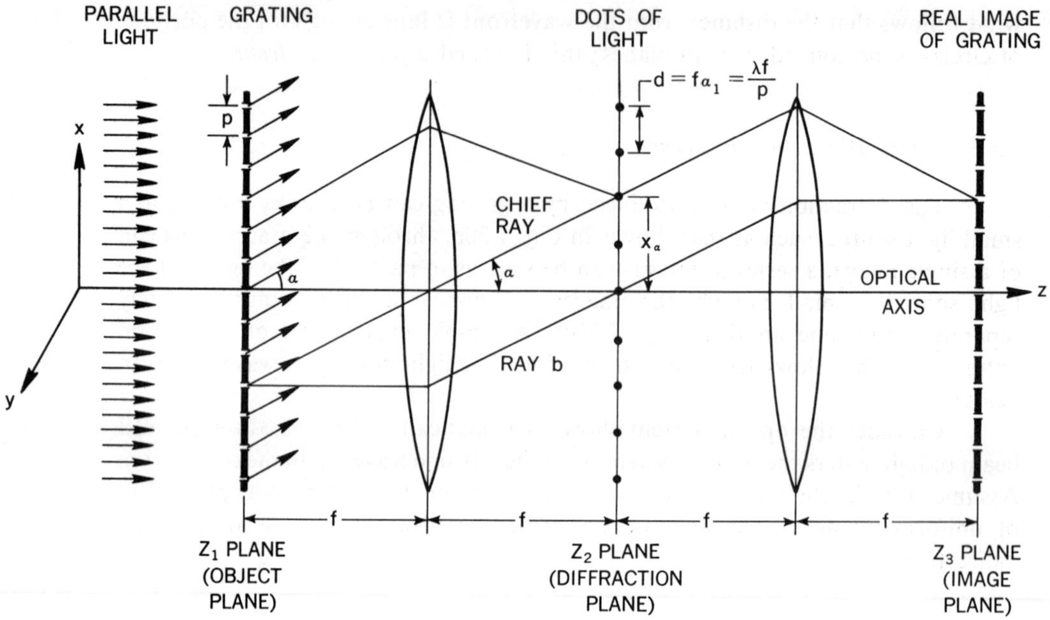

Figure 1.31. Passage of a beam of initially parallel light through an optical system which has a grating in the object plane.

The light illuminates a grating located in the z_1-plane. The grating causes variations in the amplitude of the light beam by reflecting or absorbing light in some regions and allowing it to pass in others. It also diffracts the light so that light leaving the grating propagates in different directions.

Going back to the analogy used in the previous section, the diffraction of light by a grating is similar to the propagation of waves on the surface of a quiet pool as the waves pass an obstacle, such as a post standing vertically in the pool (Fig. 1.32). If the post is far from the point where the waves began, the wavefront will be nearly straight just before it reaches the post. After passing the post, the wavefront does not appear to leave a clearly defined quiet region beyond the post like the shadow cast by the post when it obstructs the sun's rays. Instead, the wavefront appears to bend around the post and quickly fill in the "shadow" region with extensions of the primary wavefront which have a much smaller radius of curvature. As these new wavefront extensions grow, they combine with those from the other side of the post and those from wavefronts ahead and behind them to reconstruct the form which the incident wave would have had if the post had not been there.

Consider again the grating shown in Fig. 1.31. Suppose the grating consists of very thin transparent lines with spacing or pitch p on an otherwise opaque surface. Diffraction effects can be accounted for if the transparent lines are replaced by lines of small monochromatic lights of the same wavelength in phase with

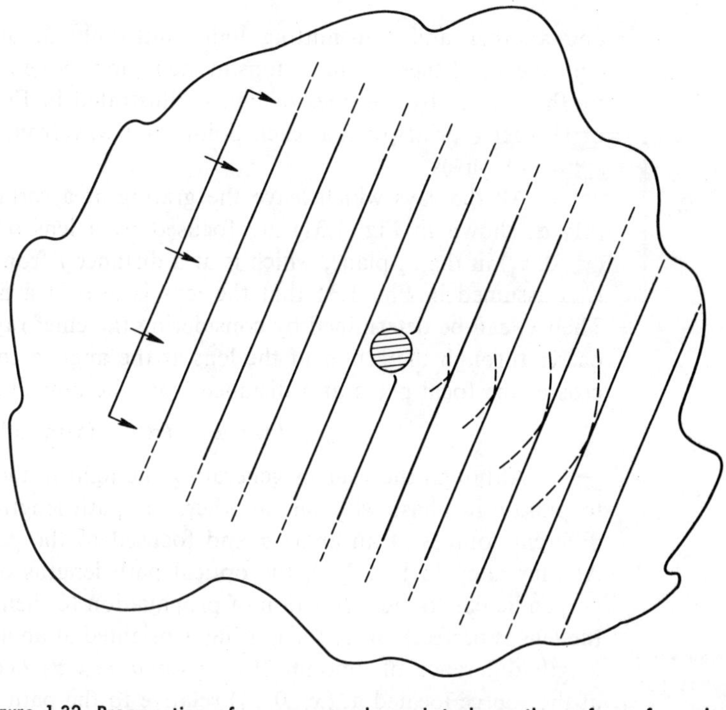

Figure 1.32. Propagation of waves around an obstacle on the surface of a calm pool.

Figure 1.33. Diffracted light propagates away from a grating with very narrow slits just as if it were generated by small lights in the grating.

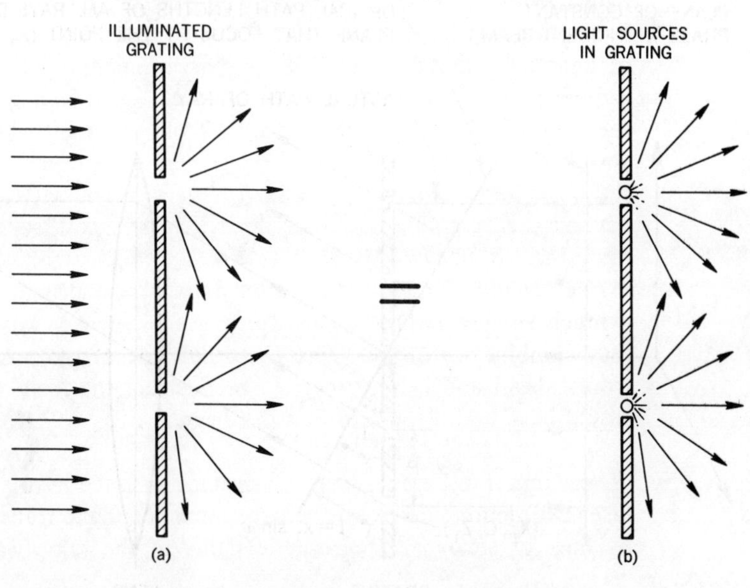

one another and transmitting light uniformly in all directions. The physical equivalence of these illuminations in the region to the right of the grating produced by the slits or by the lines of lights (illustrated in Fig. 1.33) is a particular case of Huygens' principle that each point on a wavefront may be regarded as a new source of waves.

All the rays which leave the grating at a certain angle to the z-direction (say α, shown in Fig. 1.31) are focused by a lens of focal length f on a point $(x_\alpha, 0, z_2)$ in the z_2-plane, which is at a distance f from the lens. For convenience, it is assumed in Fig. 1.31 that the lens is also at a distance f from the grating. Then x_α can be determined by considering the chief ray shown in Fig. 1.31, which passes through the center of the lens at the angle α to the optical axis and hence crosses the focal plane at a distance from the optical axis of:

$$x_\alpha = f \tan \alpha \approx f\alpha \qquad (\alpha \text{ in radians}) \tag{1.23}$$

Although the sources generating the light in the grating plane are assumed to pulsate in phase with one another, the path lengths of rays emitted from the different sources at an angle α and focused to the point $(x_\alpha, 0, z_2)$ are different. As illustrated in Fig. 1.34, the optical path lengths of rays drawn from a plane perpendicular to their direction of propagation to their point of focus are equal (if the lens is perfect). Since the grating is oriented at an angle α to this plane, a path-length difference of amount ($l = x_1 \sin \alpha \approx x_1\alpha$) occurs for the ray beginning at the source located at $(x_1, 0, z_1)$ relative to the path length of the ray beginning at $(0, 0, z_1)$, chosen as a reference ray.

Figure 1.34. Path-length difference for rays emitted at the diffraction grating.

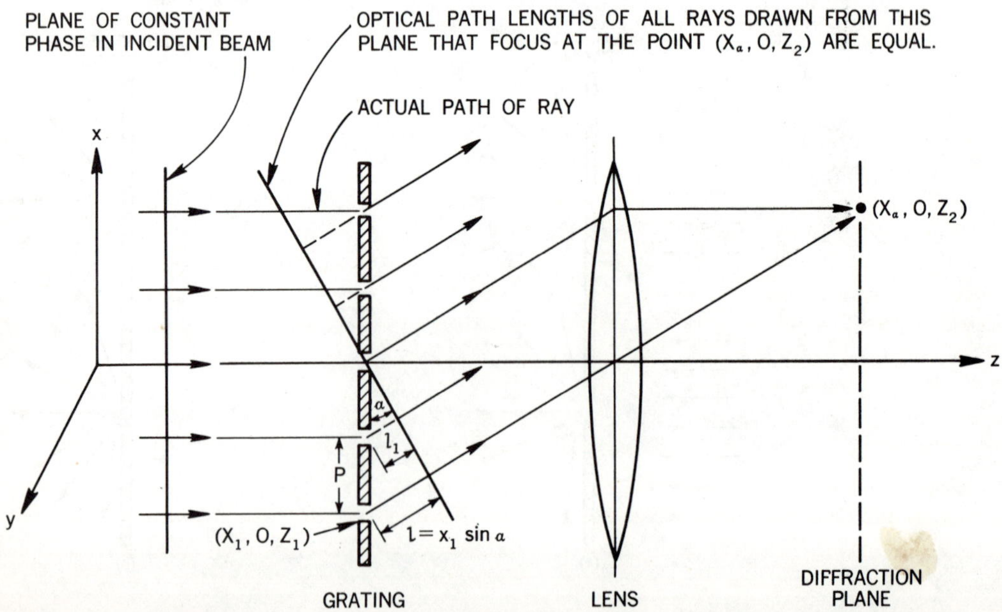

Looking at the light disturbance now as a wave of wavelength λ propagating in the direction of a ray, the path-length difference is associated with an angular phase shift of the light wave arriving at the focal point. This shift relative to the phase of the reference ray will be $\phi = (2\pi/\lambda)\alpha x_1$. The total amplitude of the light at the point $(x_\alpha, 0, z_2)$ is equal to the vector sum of the contributions from each ray at a given instant. If the slits specified above are small compared to the pitch and are of the order of the wavelength, it can be shown that the total amplitude is nearly zero except close to focal points, where all rays arrive in phase. The condition is seen to occur when the difference in path length between two adjacent slits is a multiple of the wavelength:

$$l_n = n\lambda = p \sin \alpha_n \qquad \text{or} \qquad \alpha_n \approx \sin \alpha_n = \frac{n\lambda}{p} \qquad (n = 1, 2, 3 \ldots) \tag{1.24}$$

as illustrated in Fig. 1.34 for the case $n = 1$. The lens also directs rays not in the x-z plane to the same focal point $(x_\alpha, 0, z_2)$. Hence, the light focused on the z_2-plane is a row of dots spaced a distance d apart where

$$d = f\alpha = \frac{\lambda f}{p} \tag{1.25}$$

The pattern of light observed in the z_2-plane is then a row of dots. It is at the focus of rays which left the grating parallel to one another rather than at the focus of rays which left the grating from the same point. The pattern is called the *diffraction pattern* of the grating and the z_2-plane is called the *diffraction plane.*

If the incident light came from a distant point source, the dots observed in the z_2-plane would each be an image of the point source, as shown in Fig. 1.30.

If no obstructions are placed in the diffraction plane, the rays may continue through a second lens (Fig. 1.31) with the same focal length located a distance f from the diffraction plane. The rays then form a real image of the grating at a distance f beyond the second lens, as is demonstrated by constructing the path of a second ray b from an arbitrary point of the object grating to its image.

As noted above, the occurrence of diffraction patterns does not depend critically upon location of the first lens relative to the grating. The diffraction pattern will be formed 1 focal length away from the lens if a converging lens is used in parallel light. So long as the second lens is also converging and is located at least 1 focal length away from the grating, a real image of the grating will be formed but, in general, it will be of different size.

Purely monochromatic parallel light beams are an idealization only approximately realized in practice. However, even a white, diffused light source will diffract light in a continuous series of overlapping spectrums, corresponding to successive n orders like those shown in Fig. 1.30. If the dimensions of the image formed by the lens are smaller than the average d-value in the diffraction pattern, nonoverlapping images may be viewed. The diffraction pattern which approaches the ideal is achieved, of course, when a highly monochromatic, parallel, and coherent light (as from a laser source) is used for the illumination.

1.8.3. Moire Fringes

Gratings commonly used in experimental strain analysis are of the type shown in Fig. 1.2 (bar and slit of equal width). The intensity of light going through this grating is a step function similar to the one shown in Fig. 1.26. The mathematical analysis of this function could be made using Fourier analysis which shows that it is the sum of sinusoidal functions. For simplicity in what follows the special sinusoidal type of grating will be considered which is important for the analysis of optical filtering systems to be described in the next section.

A transmission coefficient t is defined as the output intensity I_{out} divided by the input intensity I_{in}:

$$t = \frac{I_{out}}{I_{in}}$$

If the input intensity is assumed constant, then the transmission coefficient

Figure 1.35. Light transmission through two superposed, parallel gratings producing moire.

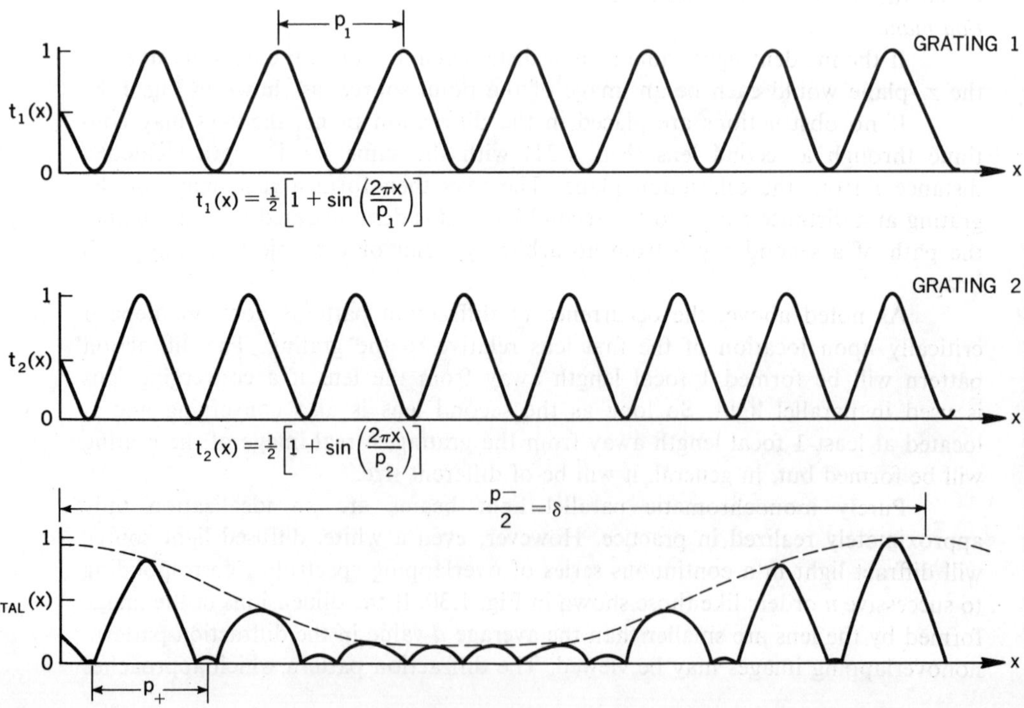

of a grating, the light intensity of which varies sinusoidally between zero and one, can be defined as a function of some coordinate of position such as:

$$t(x) = \frac{1}{2}\left(1 + \sin\frac{2\pi x}{p}\right) \tag{1.26}$$

Consider two such gratings as shown in Fig. 1.35. Moire fringes are observed if these two gratings having slightly different pitches are superposed and oriented parallel to one another. Later the case of two gratings of the same pitch and one rotated with respect to the other also will be studied. These are the two fundamental cases presented in Secs. 1.3 and 1.4.

The transmission coefficient t_{total} for light transmitted through the combination is the product of the two individual transmission coefficients:

$$\begin{aligned} t_{\text{total}}(x) &= \frac{I_{\text{out}}}{I_{\text{in}}} = [t_1(x)][t_2(x)] \\ &= \frac{1}{4}\left(1 + \sin\frac{2\pi x}{p_1} + \sin\frac{2\pi x}{p_2} + \sin\frac{2\pi x}{p_1}\sin\frac{2\pi x}{p_2}\right) \end{aligned} \tag{1.27}$$

Applying the trigonometric identities—

$$\sin A + \sin B = 2\sin\tfrac{1}{2}(A + B)\cos\tfrac{1}{2}(A - B)$$

$$\sin A \sin B = \tfrac{1}{2}(\cos 2(A - B)\cos 2(A + B)$$

—the transmission coefficient is cast in a more revealing form:

$$t_{\text{total}}(x) = \frac{1}{4}\left[1 + \left(\sin\frac{2\pi x}{p^+}\right)\left(2\cos\frac{2\pi x}{p^-}\right) + \frac{1}{2}\left(\cos\frac{4\pi x}{p^-} - \cos\frac{4\pi x}{p^+}\right)\right] \tag{1.28}$$

where p^+ and p^- are defined by

$$\frac{2}{p^+} = \frac{1}{p_1} + \frac{1}{p_2}$$

$$\frac{2}{p^-} = \frac{1}{p_1} - \frac{1}{p_2}$$

If the difference in pitch between the two gratings is small, it is clear that the first and second terms on the right-hand side of Eq. (1.28) describe a sine wave with a pitch which is the average of the original grating pitches and is modulated slowly by the cosine term. This conclusion is illustrated by the plot of the total transmission coefficient shown in Fig. 1.35. The $(2\cos 2\pi x/p^-)$ factor in Eq. (1.28) varies much slower than the $(\sin 2\pi x/p^+)$ factor. As shown in Fig. 1.35, its effect is to modify the sine wave by a slow modulation of the total transmission coefficient corresponding to the physical observation of moire fringes when the superimposed gratings are viewed. The distance between fringes δ is equal to one-half the pitch of the modulation term p^-. The equations give the same results for the fringe spacing as the one obtained geometrically [Eq. (1.6)]:

$$\frac{1}{\delta} = \frac{1}{p_1} - \frac{1}{p_2} \tag{1.29}$$

The other terms in Eq. (1.28) are harmonics of $(\sin 2\pi x/p^+)(2\cos 2\pi x/p^-)$. The

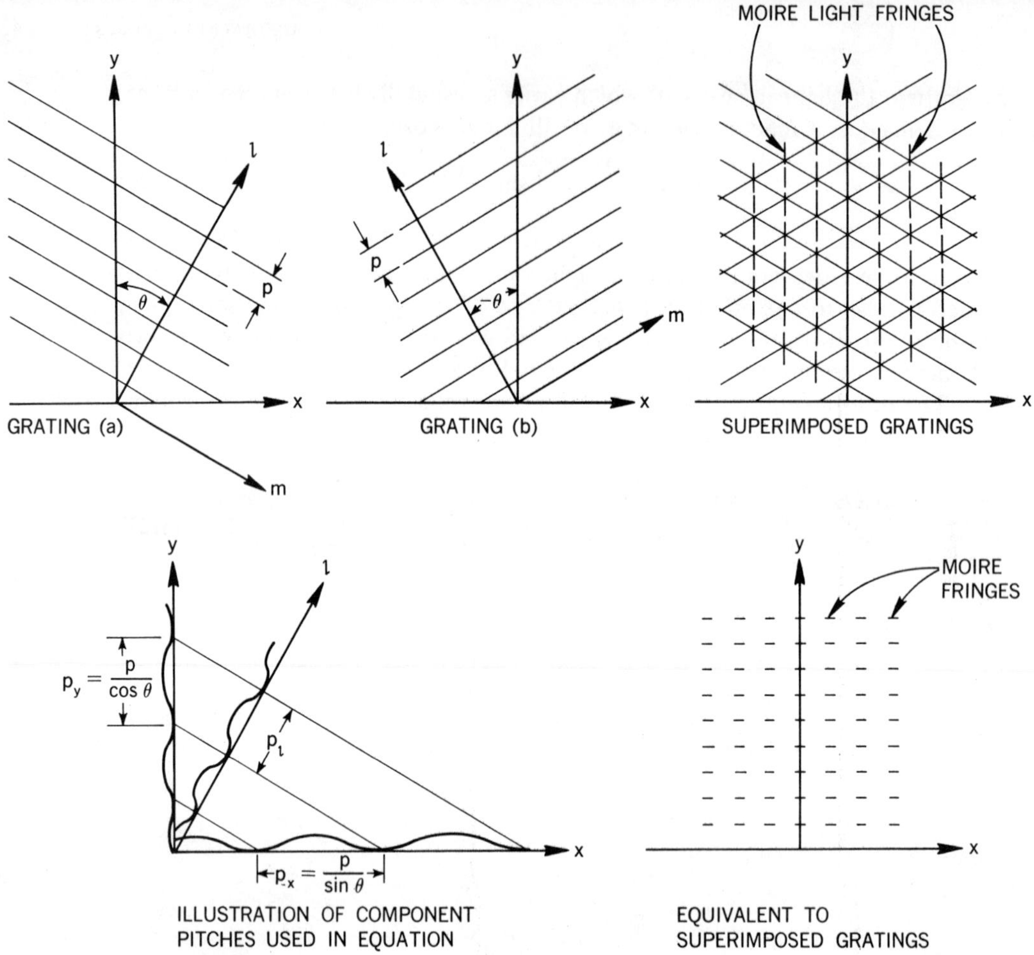

Figure 1.36. Light transmission through two superposed, inclined gratings of the same pitch, producing moire.

harmonics modify the lower frequency factors to produce the detailed structure plotted in Fig. 1.35.

The second fundamental case of moire-fringe formation occurs when two gratings, a and b of the form described by Eq. (1.26) both with the same pitch, are rotated in opposite directions in the x-y plane through a small angle θ, as shown in Fig. 1.36. (The angle is exaggerated in the figure.)

The transmission coefficient characterizing grating a referred to the principal coordinate system l-m, of the grating, is

$$t_a(l, m) = \frac{1}{2}\left(1 + \sin 2\frac{\pi l}{p}\right)$$

This coefficient, formulated with reference to the axes x and y as shown in Fig. 1.36, becomes—

$$t_a(x, y) = \frac{1}{2}\left[1 + \sin 2\pi\left(\frac{y\cos\theta + x\sin\theta}{p}\right)\right] \tag{1.30}$$

Similarly, the coefficient in the x-y coordinate system for grating b is:

$$t_b(x, y) = \frac{1}{2}\left[1 + \sin 2\pi\left(\frac{y \cos\theta - x \sin\theta}{p}\right)\right] \tag{1.31}$$

As before, the transmission coefficient of the combined gratings is

$$t_{\text{total}}(x, y) = \frac{I_{\text{out}}}{I_{\text{in}}} = [t_a(x, y)][t_b(x, y)] \tag{1.32}$$

Applying the same trigonometric identity as before and assuming θ is small,

$$t_{\text{total}}(x, y) = \frac{1}{4}\left[1 + \left(\sin\frac{2\pi y}{p}\right)\left(2\cos\frac{2\pi\theta x}{p}\right) + \frac{1}{2}\left(\cos\frac{4\pi\theta x}{p} - \cos\frac{4\pi y}{p}\right)\right] \tag{1.33}$$

we see that the ($\sin 2\pi y/p$) factor of the transmission coefficient describes a grating with the same pitch as that of the original gratings but oriented so that its lines are perpendicular to the y-axis bisecting the perpendiculars to the systems of original grating lines. In addition, the lines are modulated by the ($2\cos 2\pi\theta x/p$) factor—but in this case the modulation occurs in the x-direction as illustrated in Fig. 1.36. Moire fringes correspond to the modulation of the fine-pitch grating and thus are predicted to appear as lines parallel to the y-axis. The distance between fringes δ is equal to one half the pitch of the modulation term,

$$\delta = \frac{p}{2\theta} \tag{1.34}$$

If θ is redefined as the total angle between the rotated gratings, this agrees with the result shown geometrically in Sec. 1.4.

The analysis in both cases, parallel gratings of different pitch, and inclined gratings of the same pitch, suggests that moire fringes can be considered a modulation, or fading in and out, of an "equivalent" grating with the average pitch of the two original gratings.

1.8.4. Optical Filtering

The optical system illustrated schematically in Fig. 1.31 can be used to obtain modified images of fringes or gratings. These modifications include: (1) images where moire fringes are seen without the original grating lines being present, (2) images where the density of grating lines is multiplied, (3) images where the density of moire fringes is multiplied and (4) images of a different kind of moire fringes. These last fringes appear when two gratings are separated by a gap and are a measure of the gap distance. In this section, the first type of image modification mentioned above is described.

In many practical applications the ratio of moire fringe spacing to grating pitch will be of the order of at least 25. For cases where this ratio is large, the analysis is simplified if we consider a small region of the equivalent grating as shown in Fig. 1.37. Consider the diffraction pattern which results from the small region. Regardless of the location of the region, the diffraction pattern for the

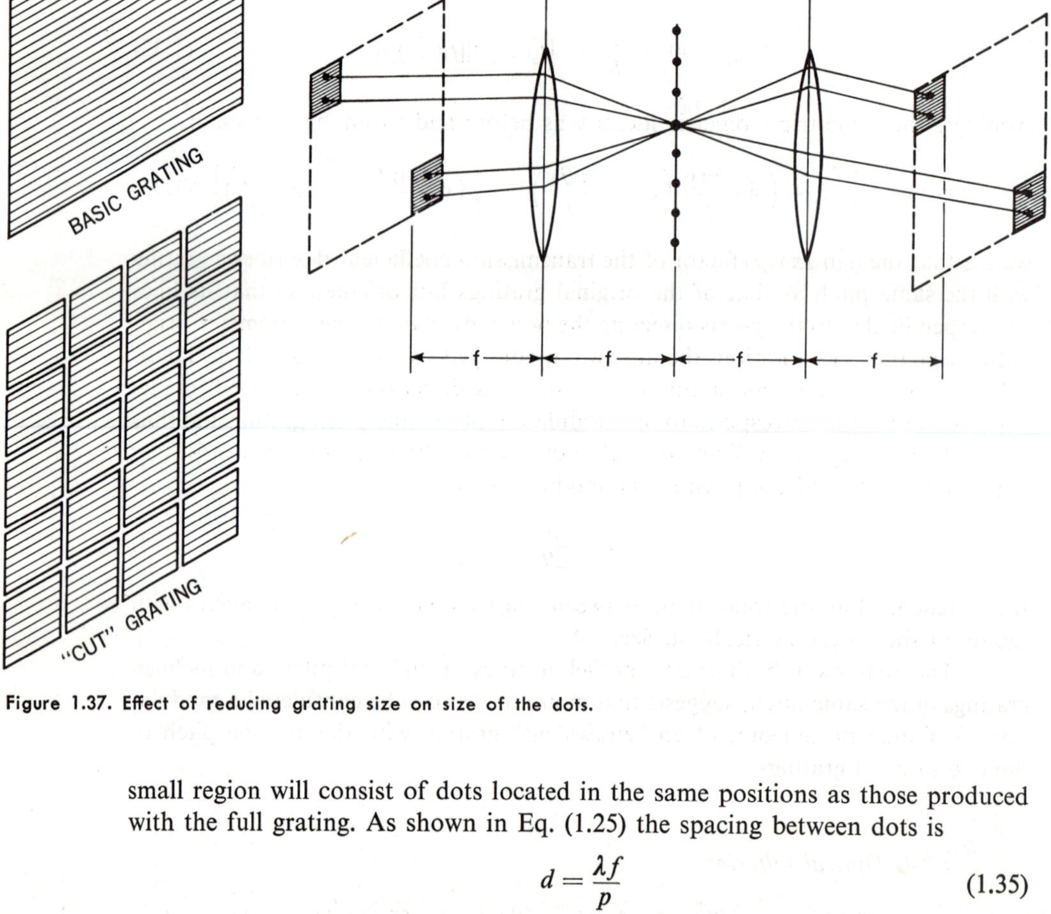

Figure 1.37. Effect of reducing grating size on size of the dots.

small region will consist of dots located in the same positions as those produced with the full grating. As shown in Eq. (1.25) the spacing between dots is

$$d = \frac{\lambda f}{p} \tag{1.35}$$

where λ is the wavelength of light and f is the focal length of the lens.

It can be shown that the effect of using smaller pieces of a grating is to slightly enlarge the size of each dot. The radius of a dot is related approximately to the grating size by a similar formula:

$$r \approx \frac{\lambda f}{s} \tag{1.36}$$

where s is the width of the region. Hence a square piece of a grating need be only a few times wider than the distance between grating lines to produce a diffraction pattern with separated dots such as those shown in Fig. 1.30.

Consider the optical system shown in Fig. 1.31 with a grating consisting of very thin transparent lines on an otherwise opaque surface. If the dots in Fig. 1.31 are not obstructed in the diffraction plane, the light rays will travel through the second lens as shown in Fig. 1.37 and form an image which corresponds in its location and form to that of the object grating.

The formation of this "equivalent" grating image has been described above as the result of focusing all rays which left a given point of the object grating into a corresponding point of the image grating. The following alternative explanation will prove useful in describing the operation of an optical filtering system.

Instead of constructing rays in one step from the object to the image, consider the system in two steps. First determine the location of dots in the diffraction pattern as described above. Then consider the dots as new coherent sources of light which in combination form an interference pattern in the grating image plane. If light from all dots is allowed to pass through an aperture in the diffraction plane, then the resultant interference pattern is identical to the grating image. The advantage of this two-step method of analyzing the optical system is that the images which are observed in the grating image plane when light from only selected dots is allowed to pass can be predicted by constructing the interference pattern produced by light from the selected dots. As an illustration in the optical system shown in Fig. 1.38 a light stop has been added in the diffraction plane which allows the light from only the three dots closest to the optical axis of the system (identified as the 0, +1, and −1 dots) to pass on to form the grating image. After passing through the second lens, the rays from each dot propagate in parallel beams.

Figure 1.38. Optical filtering of higher-order dots.

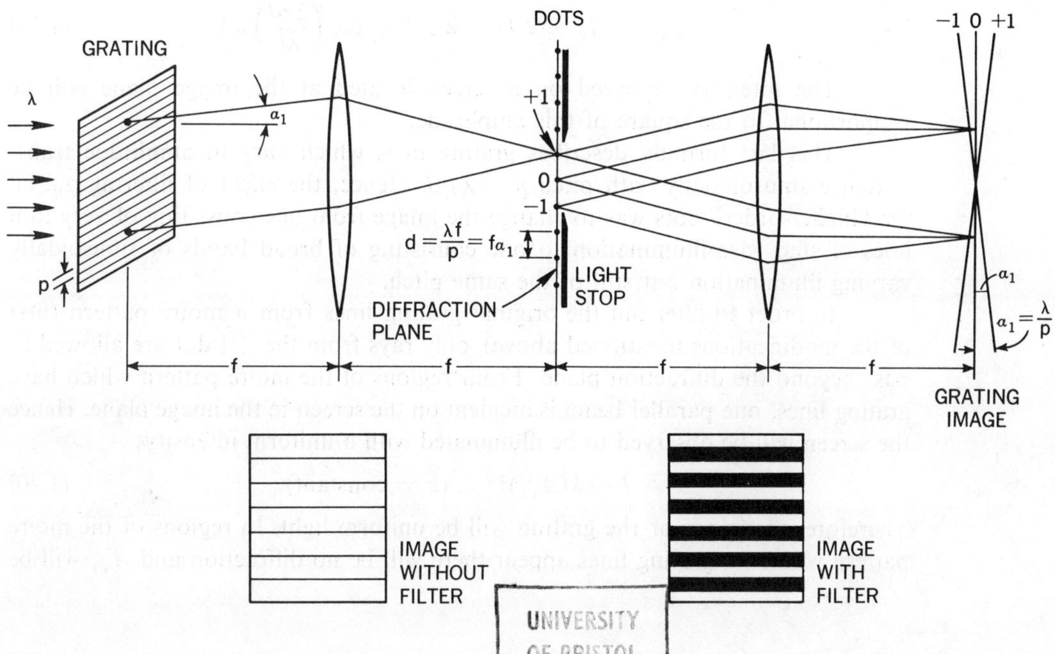

These beams propagate in slightly different directions until they become superposed in the image plane.

The interference pattern produced by the combination of beams in the grating image plane could be observed, for example, by scattering the light off a ground glass. The observed pattern can be predicted by computing the total amplitude of the three beams. By reversing the development in Sec. 1.8.2 it can be seen that if the spacing between two dots is $d = f\alpha$, then after passing through a collimating lens, the two beams will be out of phase (relative to each other) by an angle $\phi = (2\pi/\lambda)\alpha x$. To facilitate vector addition, this phase difference can also be described by a factor $e^{(i2\pi/\lambda)\alpha x}$. Then the total amplitude in the grating image plane of the three beams from the three dots is given by the following formula:

$$A_{\text{total}} = A_0 + A_{+1}e^{(i2\pi/\lambda)\alpha_{+1}x} + A_{-1}e^{(i2\pi/\lambda)\alpha_{-1}x} \tag{1.37}$$

where:

$$\alpha_{+1} = \frac{d}{f} \qquad \alpha_{-1} = \frac{-d}{f}$$

From symmetry the amplitudes of the $+1$ and -1 dots are the same. Call them A. Then the above formula can be simplified to

$$A_{\text{total}} = A_0 + A[(e^{+i(2\pi d/\lambda f)x} + e^{-i(2\pi d/\lambda f)x})] \tag{1.38}$$

Using the identity

$$\cos x = \frac{e^{ix} + e^{-ix}}{2}$$

we obtain

$$A_{\text{total}} = (A_0 - 2A) + 2A\left[1 + \cos\left(\frac{2\pi d}{\lambda f}\right)x\right] \tag{1.39}$$

The intensity observed on a screen located at the image plane will be proportional to the square of this amplitude.

This last formula describes grating lines which vary in amplitude transmittance sinusoidually with pitch $p = \lambda f/d$. Hence, the effect of filtering out all the "higher-order" dots was to change the image from one consisting of very thin lines of step wise illumination to one consisting of broad bands of sinusoidally varying illumination but still of the same pitch.

In order to filter out the original grating lines from a moire pattern (first of the modifications mentioned above), only rays from the $+1$ dot are allowed to pass beyond the diffraction plane. From regions of the moire pattern which have grating lines, one parallel beam is incident on the screen in the image plane. Hence the screen will be observed to be illuminated with a uniform intensity:

$$I = k(A_{+1})^2 \qquad (k = \text{constant}) \tag{1.40}$$

Therefore, the image of the grating will be uniform light. In regions of the moire pattern where no grating lines appear there will be no diffraction and A_{+1} will be

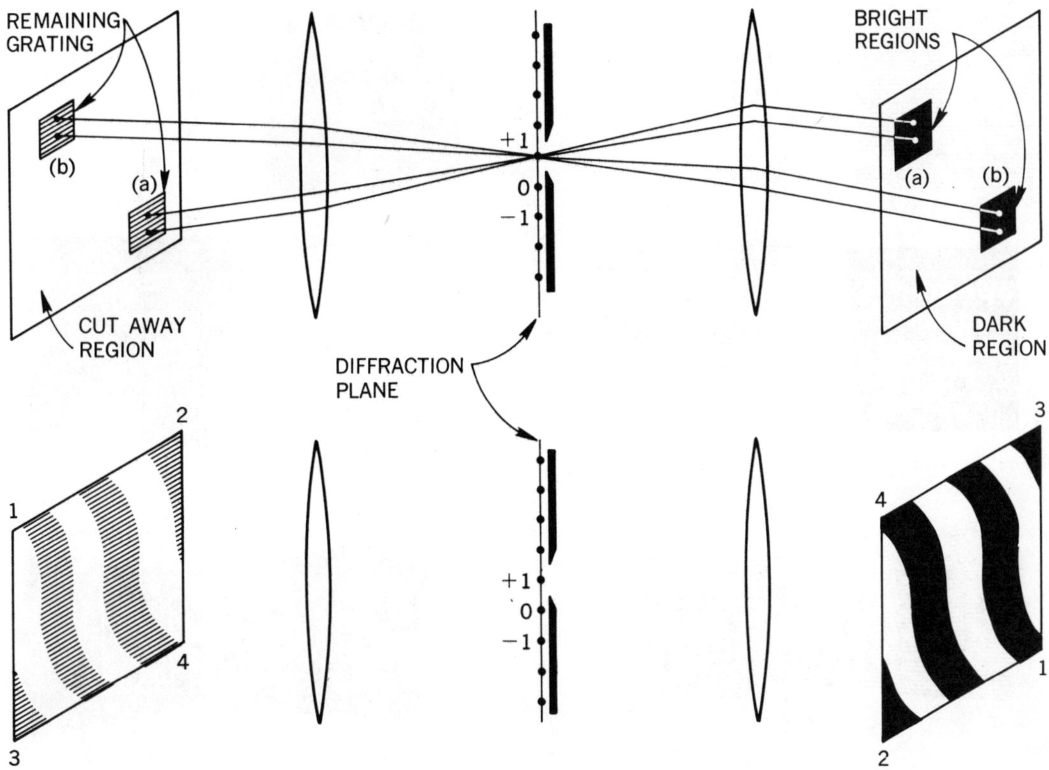

Figure 1.39. Filtering out of fine-pitch gratings. Schematic diagram of optical system for fringe multiplication.

zero. All light from these regions will come to the zero-order dot and be stopped by the aperture. There is darkness. No illumination will occur in corresponding regions of the image. This is illustrated in Fig. 1.39.

This effect is demonstrated in Fig. 1.40. On the left-hand side of the figure a photograph of two pieces of single gratings, an opaque region (solid white) and two superposed gratings are shown. (Transparent regions are black.) On the right-hand side a photograph of the image obtained after filtering out by an aperture all but the +1 diffraction order dot is shown. It can be concluded from the figure that corresponding to regions where the basic grating is sharply defined are regions of uniform illumination in the image plane. Corresponding to points where the grating fades away or becomes solidly opaque, are dark regions in the image plane. These conclusions apply both to the large, uniform regions and to the modulations occurring within the region of the superposed gratings. Note that in the latter case, the resultant effect is to keep only the moire fringes. The basic grating of finer pitch has been filtered out.

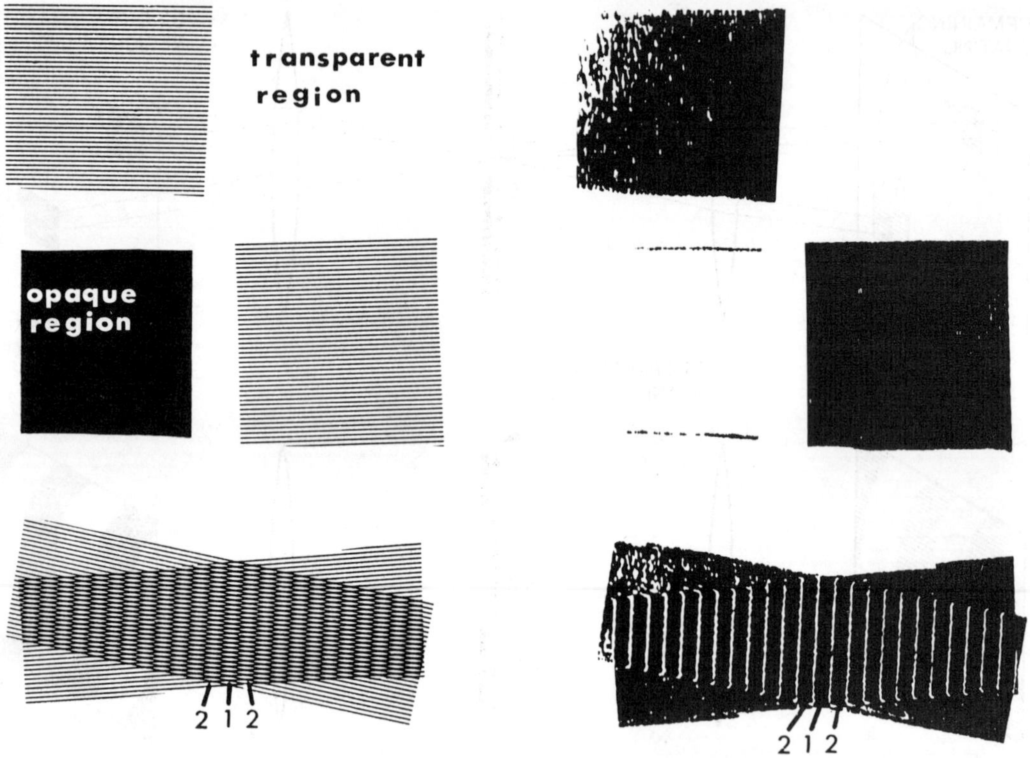

Figure 1.40. Illustration of optical filtering of gratings. Left side is a photograph before filtering. Right side is a photograph of the image obtained after filtering out all the diffraction orders but +1.

1.8.5. Multiplication of Grating Lines and Fringes

Two systems of fringe multiplication are presently being applied to problems in experimental stress analysis. One system is applicable to photoelastic and interferometric fringes as they appear in the materials. These fringes can be multiplied by placing the specimens between two highly reflective but slightly transparent mirrors mounted at a slight angle to one another so as to produce multiple reflections of an incident beam. Because the light beam passes through the model several times, the image formed by the beam exhibits several times as many fringes as are observed with conventional polariscopes or interferometers.[3]

Another system is applicable to gratings or photographic recordings of

[3] D. Post, "Isochromatic Fringe Sharpening and Fringe Multiplication in Photoelasticity," *Proc. SESA*, Vol. 12, No. 2 (1955), pp. 143–156.

fringe patterns. Its principles of operation can be illustrated with the same optical system analyzed earlier in this section. This is the only system that will be studied here.

Consider the grating of very thin transparent lines with pitch p introduced in Sec. 1.8.2. As shown in that section, a long row of dots will be observed if a screen is placed in the diffraction plane of the optical system. The dots will be separated by a distance $d = \lambda f/p$.

As shown in Sec. 1.8.4, if a filter is placed in the diffraction plane which allows only rays from the three "lowest-order" dots to pass through to the image plane, the image which they produce can be predicted by considering the interference pattern which results when the three beams associated with the lowest-order dots are superposed on a screen in the image plane.

In a similar manner, the images which result when other dots are allowed to pass can be predicted. Consider the case where only the +1 and −1 dots are not blocked. The total amplitude of the two beams in the image plane is given by

$$A_{\text{total}}^{+1,-1} = A_{+1}e^{(i2\pi/\lambda)\alpha_{+1}x} + A_{-1}e^{(i2\pi/\lambda)\alpha_{-1}x} \tag{1.41}$$

Assume that the amplitudes of the two beams are the same. This assumption is considered to be realistic based on the observation of many typical diffraction patterns and can be derived theoretically for any grating which has a symmetric form about an axis parallel to the grating lines. Applying the following identity—

$$\cos x = \frac{e^{ix} + e^{-ix}}{2}$$

—the following expression is obtained:

$$A_{\text{total}}^{+1,-1} = 2A \cos\left(\frac{2\pi d}{\lambda f}\right) x \tag{1.42}$$

The observed intensity is proportional to the square of the amplitude. As illustrated in Fig. 1.41, the intensity of the image grating fluctuates periodically with one-half the pitch of the object grating. The number of grating lines has been multiplied by a factor of 2.

An example of the multiplication of grating lines is given in Fig. 1.42. On the left-hand side of the figure the diffraction pattern produced by the coarser grating is shown. To obtain the image with finer grating lines only light from the +1 and −1 diffraction order dots was allowed to pass through an aperture.

If a filter is inserted in the diffraction plane of the optical system, which allows only the rays from the +2 and −2 dots to pass, the image amplitude is given by

$$\begin{aligned} A_{\text{total}}^{+2,-2} &= A_{+2}e^{(i2\pi/\lambda)\alpha_{+2}x} + A_{-2}e^{(i2\pi/\lambda)\alpha_{-2}x} \qquad \left(\alpha_{+2} = \frac{2d}{f}\right) \\ &= 2A_2 \cos\frac{4\pi dx}{\lambda f} \qquad \left(\alpha_{-2} = \frac{-2d}{f}\right) \end{aligned} \tag{1.43}$$

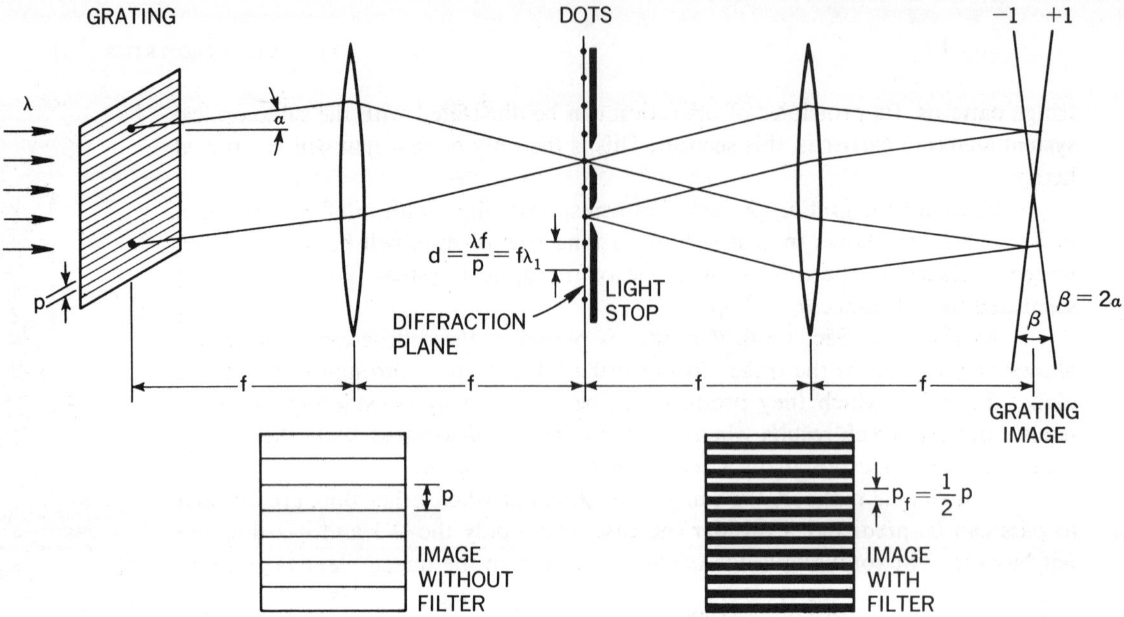

Figure 1.41. Schematic diagram of an optical system for grating multiplication.

Figure 1.42. Illustration of grating multiplication. By filtering out all light except that from the +1 and −1 diffraction orders (as shown in the diffraction pattern on the left), an image of grating lines with twice the density of the original grating is obtained.

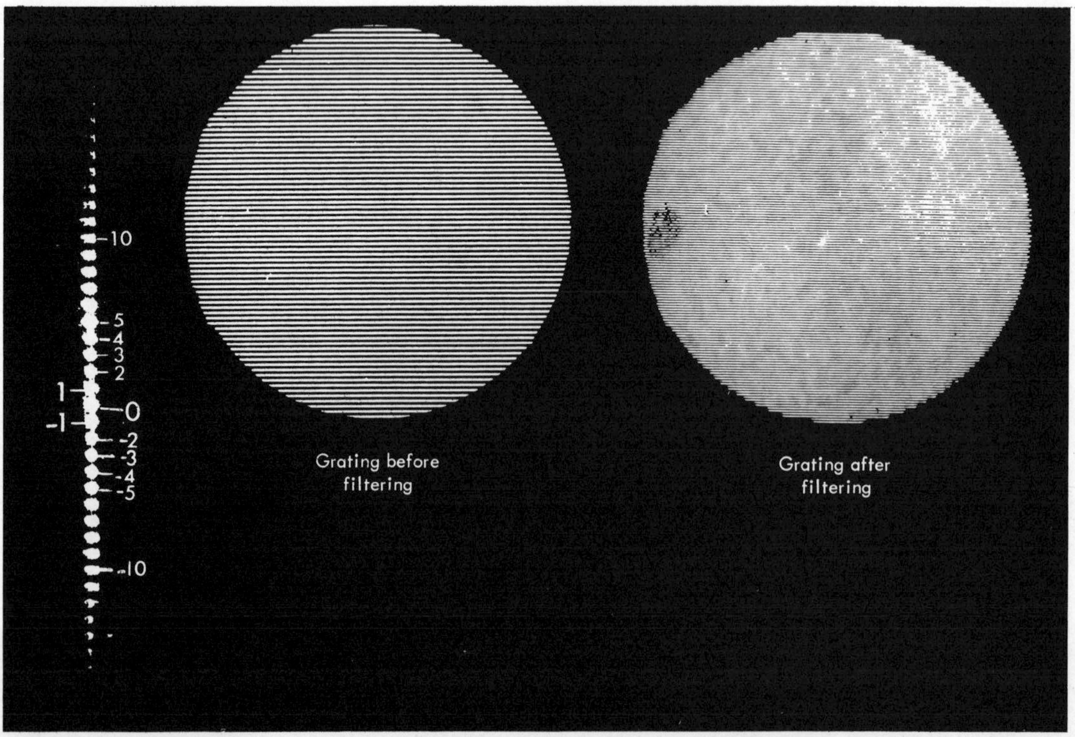

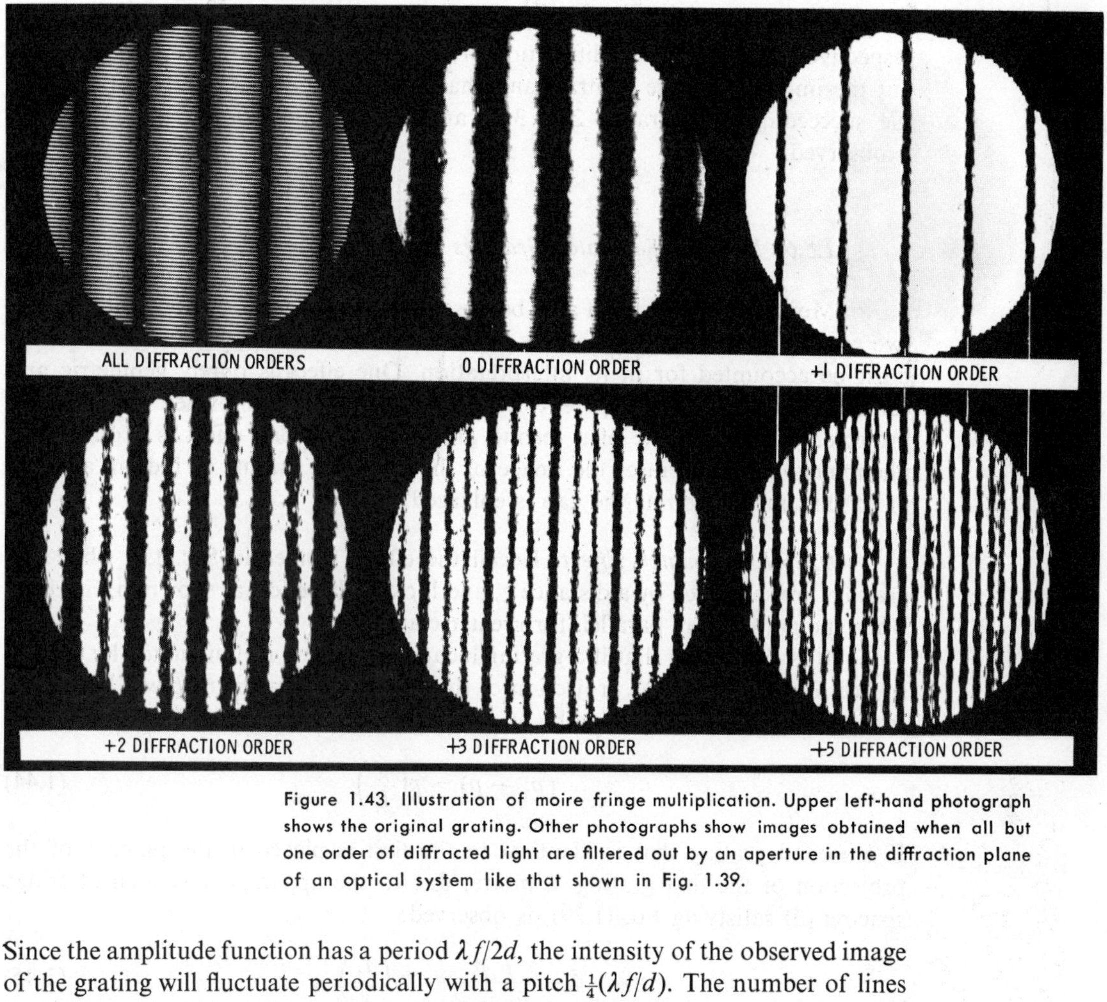

Figure 1.43. Illustration of moire fringe multiplication. Upper left-hand photograph shows the original grating. Other photographs show images obtained when all but one order of diffracted light are filtered out by an aperture in the diffraction plane of an optical system like that shown in Fig. 1.39.

Since the amplitude function has a period $\lambda f/2d$, the intensity of the observed image of the grating will fluctuate periodically with a pitch $\frac{1}{4}(\lambda f/d)$. The number of lines in the image is 4 times the number observed in the grating.

In a similar manner, images with greater multiples of grating line densities can be obtained if illumination from pairs of higher-order dots in the diffraction pattern are the only ones passed by the filter.

The multiplication of moire fringes produced by two superposed gratings besides being accomplished by interferometric methods, as mentioned at the beginning of this section, can also be accomplished by filtering out all diffraction orders but one.[4] The process is identical to the optical filtering described in Sec. 1.8.4 except that light from a higher diffraction order is allowed to pass through the aperture. An example of moire fringe multiplication is shown in Fig. 1.43. The original image of the crossed gratings and their moire is shown together with

[4] A detailed discussion is given by D. Post, "Analysis of Moire Fringe Multiplication Phenomena," *Applied Optics,* Vol. 6, No. 11 (November 1967), 1938–1942.

images obtained with light from the 0, +1, +2, +3, and +5 diffraction orders, respectively. The 0 and +1 diffraction order photographs show no multiplication, but the improved fringe contrast and sharpening is apparent in the +1 case. In the succeeding photographs 2×, 3×, and 5× multiplication of fringe density is observed.

1.8.6. Moire of Separated Gratings

Moire phenomena can also be observed when a gap is introduced between two gratings. Two additional effects not present in the moire of gratings in contact must be accounted for in its interpretation. One effect is purely geometric and can be considered as an apparent deformation of one grating when its shadow is projected onto the plane of a second grating. The second effect is a diffraction phenomenon which causes the image of the projected grating to fade in and out or to shift in position as the gap is enlarged.

1.8.6.1. Mismatch Effect. Recall the effect, studied in Sec. 1.7, when two gratings are separated by a distance g, and look at this effect as shown in Fig. 1.44, when the gratings are parallel, for greater clarity.

It is easily seen that if a grating located at plane a is illuminated by a point source S, the pitch of the projection of the grating at some plane b will be greater than that of the object grating by an amount:

$$(p_1 - p) = p\left(\frac{g}{L}\right) \tag{1.44}$$

If a second grating that is identical to the first is placed in the plane b of the projection of the first grating, a moire, due to the apparent mismatch of fringe spacing (δ) satisfying Eq. (1.29), is observed:

$$\delta = \frac{p_1 p}{p_1 - p} = p_1\left(\frac{L}{g}\right) \tag{1.45}$$

The component of a fictitious strain in the direction normal to the grating lines and in the plane of the grating would be, from Eq. (1.7),

$$\epsilon_f = \frac{p_1}{\delta} = \left(\frac{g}{L}\right) \tag{1.46}$$

in agreement with Eq. 1.22 derived previously, for the particular case where the gap distance g does not vary in the direction normal to the grating lines.

Although the moire described above could be observed with the optical system shown in Fig. 1.44b, the moire fringes can also be viewed or photographed with the simpler arrangement shown in Fig. 1.44c if a sufficiently small aperture is used. From the geometry it can be seen that the aperture stops all rays

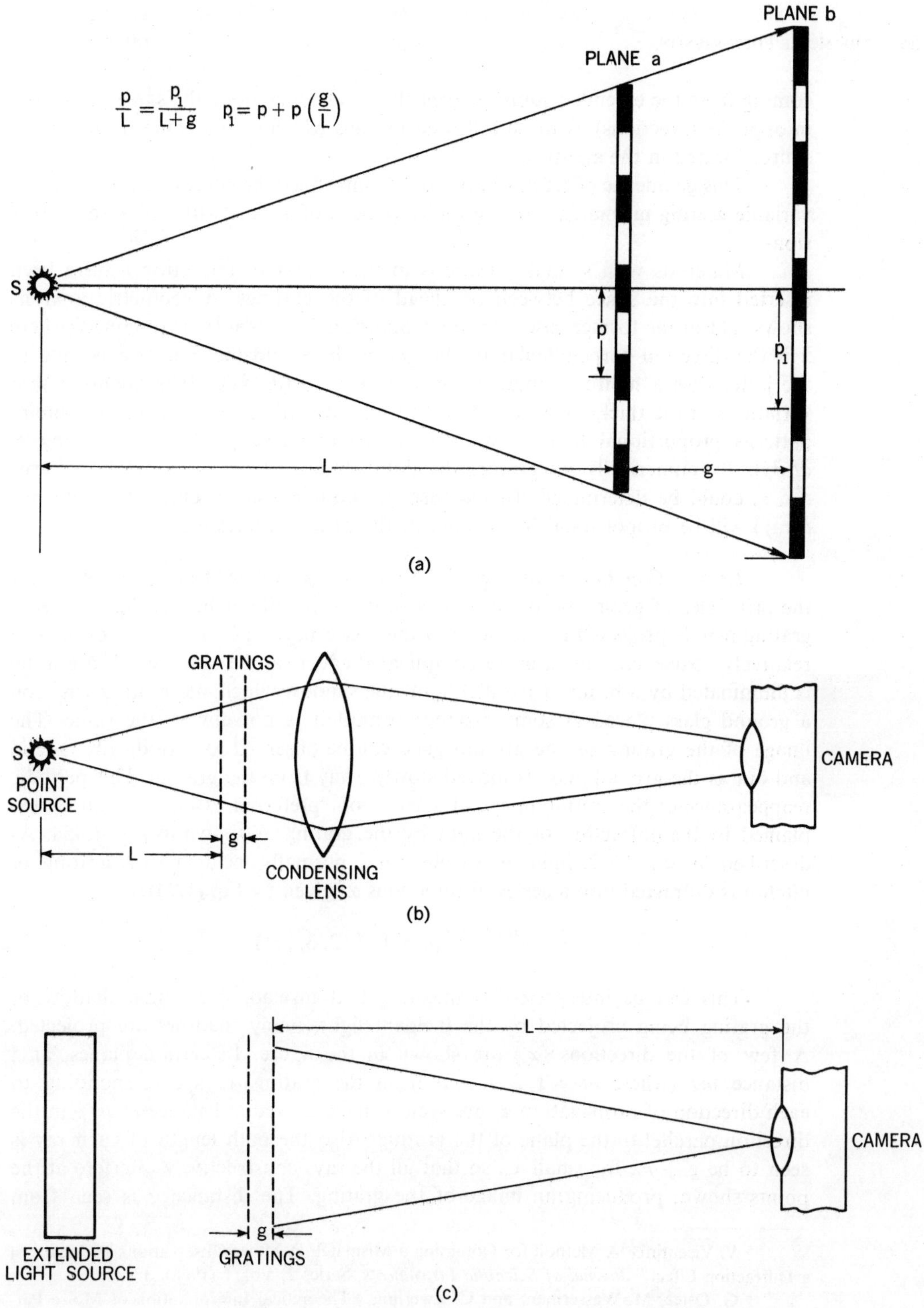

Figure 1.44. Mismatch effect associated with the gap between two gratings.

coming from the extended source except those traveling along the same paths (but in opposite directions) as those followed by rays which would come from a point source located in the aperture.

This geometric effect has been used to introduce the effect of a continuously variable grating mismatch into the moire pattern of a plate with holes under tensions.[5]

Materials with spatial variations in their index of refraction n have been inserted into the space between or ahead of the gratings. A geometric analysis shows that in the former case a moire proportional to $n(\partial n/\partial x)$ is obtained (where x is the direction perpendicular to the grating lines and the optical axis) and in the latter case a moire is obtained proportional to $(\partial n/\partial x)$.[6] It is apparent that variations in the thickness ($t(x, y)$) of transparent models will also produce moire patterns proportional to a component of the thickness gradient (for example, $\partial t/\partial x$). By numerically integrating the data obtained from such moire patterns $t(x, y)$ could be determined. In the case of two-dimensional elasticity problems, $t(x, y)$ will be proportional to the sum of the principal stresses.

1.8.6.2. Gap Effect. In Secs. 1.2 to 1.7 it was assumed in accordance with the principles of geometric optics that a beam of parallel light passing through a grating merely projected a shadow into the space beyond it. However, even with relatively coarse gratings a more complicated effect can be observed. If a grating is illuminated by a beam of parallel light, the shadow which the grating casts on a ground glass placed at some distance behind it is not always the same. The image of the grating on the ground glass can be observed to periodically fade in and out as the ground glass is moved slowly away from the grating. This periodic reappearance of the grating image at a series of "preferred distances" can be explained by the diffraction of the light by the grating as shown in Fig. 1.45a. As described in Sec. 1.8.2, light of wavelength λ normally incident to a grating of pitch p is diffracted into a series of directions α_n given by Eq. (1.24):

$$\alpha_n = \frac{n\lambda}{p}, \qquad (n = 0, 1, 2, 3, \ldots)$$

This can be interpreted as meaning that instead of a single shadow of the grating being projected by the incident light, many shadows are projected. A few of the directions (α_n) are shown in the figure. In certain planes, at a distance mg (where $m = 1, 2, 3, \ldots$) from the grating, rays corresponding to each direction of propagation α_n are seen to have moved a distance $m \cdot n \cdot p$ in the direction parallel to the plane of the grating. Also the path length of each ray is seen to be $g + n\lambda$ for small α_n, so that all the rays constructively interfere at the points shown, producing an image of the grating. The distance g is seen from

[5] V. Vicentini, "A Method for Obtaining a Mismatch in Moire Observations by Means of a Diffraction Effect," *Journal of Scientific Instruments*, Series 2, Vol. 1 (1968), 354–355.

[6] G. Oster, M. Wasserman, and C. Zwerling, "Theoretical Interpretation of Moire Patterns," *Journal of the Optical Society of America*, Vol. 54, No. 2 (1964), 169–175.

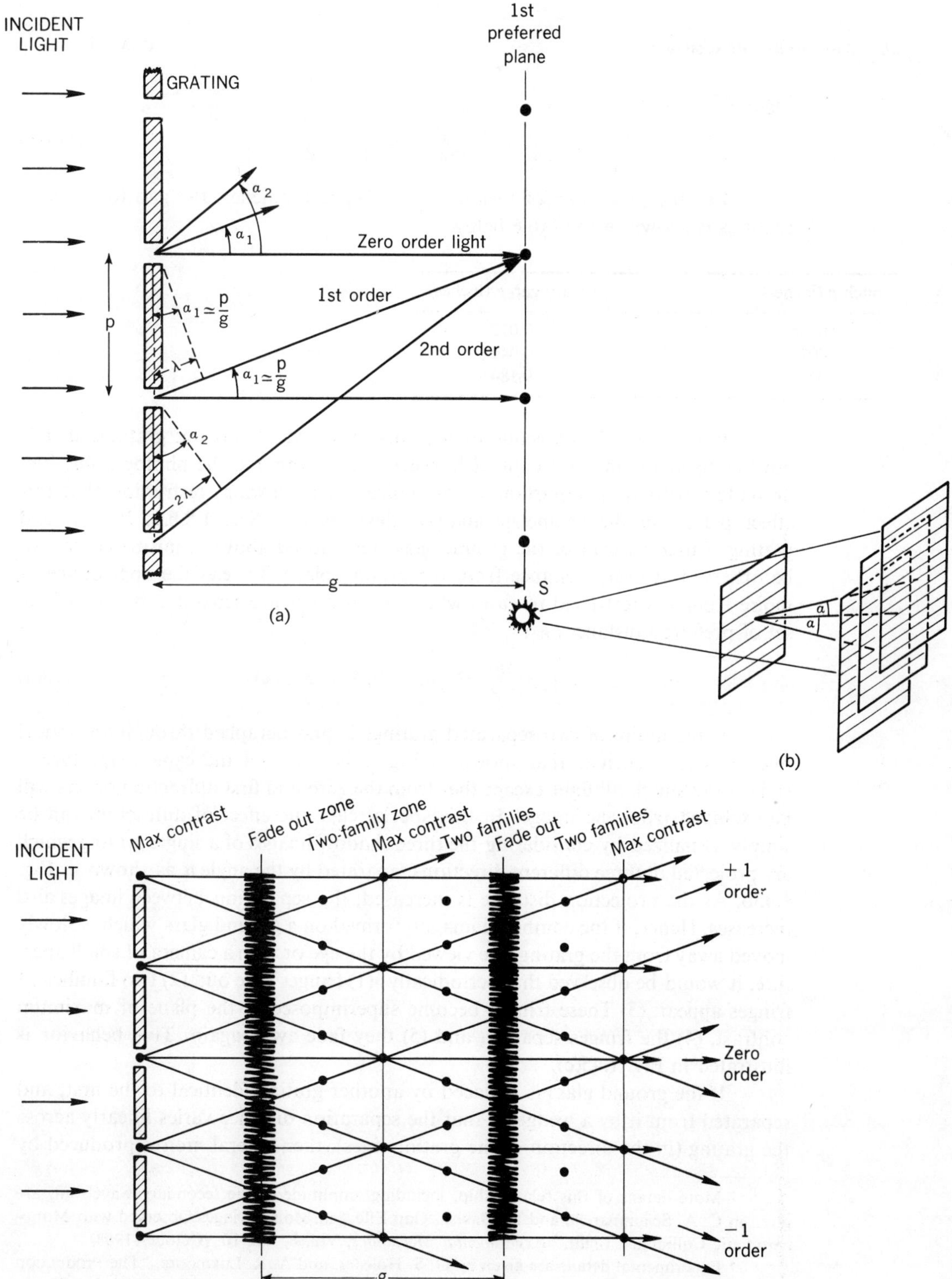

Figure 1.45. Periodic fading of a grating image as it moves away from the grating.

Fig. 1.45 to be:[7]

$$g = \frac{p}{\tan \alpha_1} = \frac{p}{\alpha_1} = \frac{p^2}{\lambda} \tag{1.47}$$

For the green line of mercury, $\lambda = 21\frac{1}{2}$ microinches, the gap for various gratings is shown in the table below.

pitch p (inches)	primary gap g (inches)
0.0005	0.012
0.001	0.046
0.002	0.186

It may be advantageous to use this effect when copying gratings if it is difficult to maintain good contact between the grating and the photographic film or plate.[8] Also it is important to recognize the (following) restriction that this effect places on the geometric analysis developed in Sec. 1.8.6.1. If a second grating is used instead of the ground glass introduced above, a mismatch cannot be observed at every distance from the grating plane. Instead the appearance of moire fringes is restricted to cases where the gratings are separated by one of the set of preferred distances g_n:

$$g_n = \frac{np^2}{\lambda} \qquad (n = 0, 1, 2, 3, \ldots) \tag{1.48}$$

If the moire of two separated gratings is photographed through an optical filtering system such as that shown in Fig. 1.44c, then, if the camera aperture is sufficiently small, all light except that from the zero and first diffraction orders will be excluded from the image. In this special case the effect of diffraction can be simply visualized by considering the three shadow images of a single grating which are projected in three different directions separated by the angle α as shown in Fig. 1.45b. As the projection distance is increased, the separation between images also increases. Hence, if the combined images, formed on a ground glass which is slowly moved away from the grating, are viewed by the eye or with a camera of small aperture, it would be observed that periodically: (1) fringes fade out, (2) two families of fringes appear. (3) These fringes become superimposed in the plane of maximum contrast, (4) the fringes separate, and (5) they fade away again. This behavior is illustrated in Fig. 1.45(c).

If the ground glass is replaced by another grating identical to the first, and separated from it by a wedge so that the separation distance varies linearly across the grating (in the direction of the grating lines), then several moires produced by

---

[7] More details of this relationship, including amplitude of the secondary wavefront, are given in C. A. Sciammarella and D. Davis, "Gap Effect in Moire Fringes Observed with Monochromatic Collimated Light," *Experimental Mechanics*, Vol. 8, No. 10, (October 1968).

[8] Experimental details are given by G. S. Holister, and A. R. Luxmoore, "The Production of High Density Moire Grids," *Experimental Mechanics*, Vol. 8, No. 5 (May 1968) 210–216.

Figure 1.46. Moire effect produced by two gratings of parallel horizontal lines, separated by a wedge-shaped gap.

the combination of images of the first grating with the second grating can be observed.

Moires of two identical horizontal gratings ($p = 0.001$ in.) which are separated by a wedge varying from zero separation on the left-hand side to $\frac{1}{4}$ in. on the right-hand side and illuminated with green light ($\lambda = 21.5 \times 10^{-6}$ in.) are shown in Figs. 1.46 and 1.47. Nearly horizontal fringes are observed which correspond to

Figure 1.47. Moire effect produced by two gratings of parallel horizontal lines, separated by a wedge-shaped gap (oblique view).

the combination of moires discussed above. The moire produced by the superposition of the $+1$ diffracted image of the first grating with the second grating can be distinguished from the moire produced by the superposition of the -1 diffracted image of the first grating with the second grating, as shown in Fig. 1.48. The observed patterns in Figs. 1.46 and 1.47 agree with those predicted by the

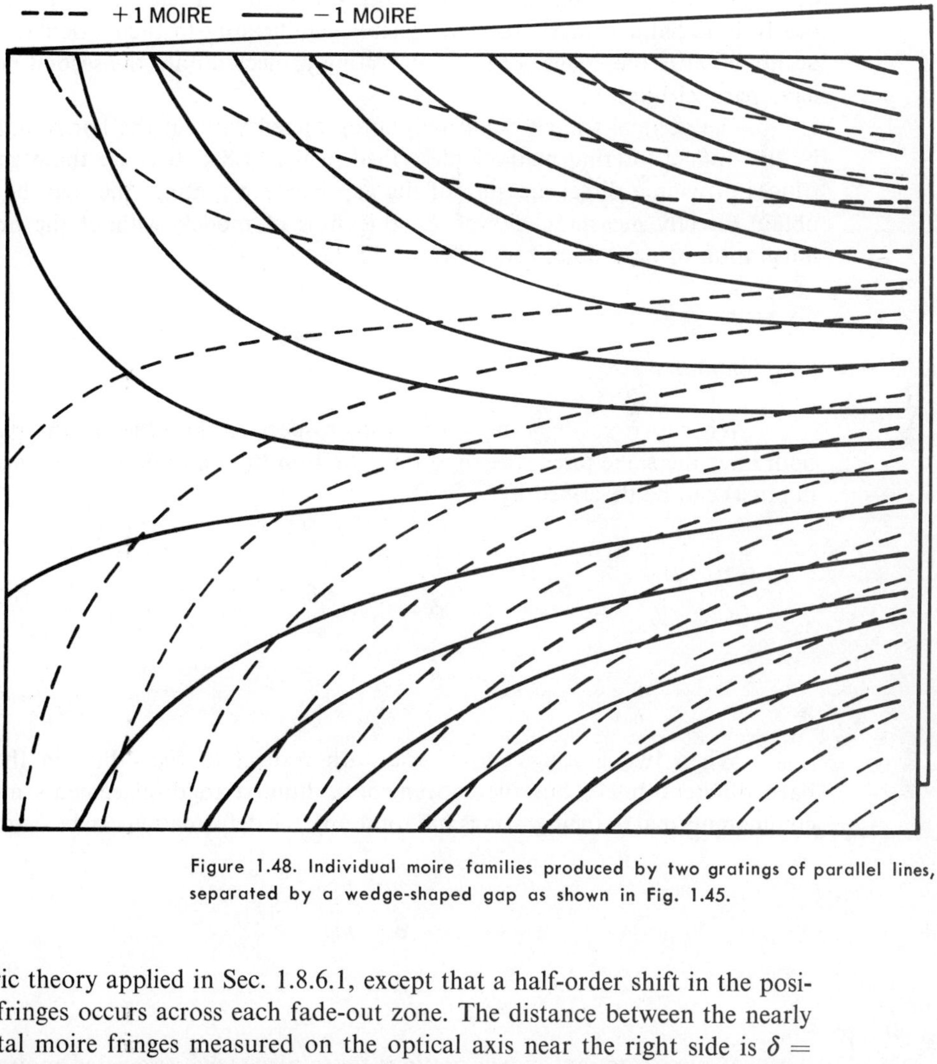

Figure 1.48. Individual moire families produced by two gratings of parallel lines, separated by a wedge-shaped gap as shown in Fig. 1.45.

geometric theory applied in Sec. 1.8.6.1, except that a half-order shift in the position of fringes occurs across each fade-out zone. The distance between the nearly horizontal moire fringes measured on the optical axis near the right side is $\delta =$ 0.24 in. The value predicted by Eq. (1.45) is:

$$\delta = p_1\left(\frac{L}{g}\right) = 0.001 \text{ in.} \times \frac{59.5 \text{ in.}}{0.25 \text{ in.}} = 0.238 \text{ in.}$$

The periodic shift in fringe location as the gap increases is also apparent in the figures. Vertical fringes appear at intervals of $1\frac{3}{16}$ in. Since the wedge angle is known to be $\alpha = \arctan \frac{1}{24}$, the gap increase between vertical fringes is $g =$ 0.0495 in. This can be compared with the value predicted by Eq. (1.47): $g = p^2/\lambda$ $= 0.0465$ in.

In Fig. 1.47 the same gratings and the same wedge-shaped gap as in Fig. 1.46 were used but the combination was viewed at an oblique angle. The vertical fringes are seen to remain in the same position independently of the direction of observation (confirming their dependence on pitch and gap only). The formerly nearly horizontal fringes are seen to shift significantly in their orientation. This occurs because the images of the first grating projected onto the second no longer have parallel lines.

The vertical fringes could be photographed without the horizontal fringes by the optical filtering methods described in Sec. 1.8.4. Because these gap effect fringes provide a direct measure of the gap between gratings they can be used to obtain directly measurements of the thickness of models without the numerical integration of data described in Sec. 1.8.6.1.

1.9. Summary

When two gratings move one with respect to the other in the plane and both have the same pitch, the rigid-body motion (no deformation, i.e., no change in pitch) can be described by

$$v = np \tag{1.1}$$

$$\phi = \frac{\pi}{2} \pm \frac{\theta}{2} \tag{1.10}$$

$$\theta = \frac{p}{\delta} \tag{1.9}$$

When two gratings move, one with respect to the other, in the plane, have different pitch, but the movement is limited to displacements along the common normal to their gratings (no rotation), the deformation can be described by

$$v = np \tag{1.1}$$

$$\delta = mp \tag{1.3}$$

$$\delta = \frac{p}{\epsilon^E} \tag{1.7}$$

If pure rotation of two gratings takes place, the associated moire fringes are practically perpendicular to the direction of the gratings. If displacement perpendicular to the common direction of the two gratings takes place (without rotation), the associated moire fringes are parallel to the gratings.

The diffraction pattern of a grating of parallel lines of pitch p can be observed, as shown in Figs. 1.30 and 1.31. It consists of a row of dots oriented perpendicular to the direction of the grating lines and parallel to the plane of the grating. The dots are separated by a distance:

$$d = \frac{\lambda f}{p} \tag{1.25}$$

where λ is the wavelength of the incident light and f is the focal length of the lens used to form the diffraction pattern.

When two identical parallel line gratings of pitch p, separated by a distance g, are illuminated by a point light source a distance L from the first grating, a geometric mismatch effect occurs which produces a moire spacing δ given by the relation:

$$\delta = p\left(\frac{g}{L}\right) \tag{1.45}$$

Also, a diffraction effect occurs which causes the grating lines to fade in and out periodically as the distance g is gradually increased. The distance between successive regions of fade out is given by the relation:

$$g = \frac{p^2}{\lambda} \tag{1.47}$$

2

MOIRE FRINGES AS PARAMETRIC CURVES

2.1. Introduction

The two gratings mentioned in the previous chapter can be considered in a more general way as two families of curves which can be expressed parametrically. The moire fringes produced by these two families of curves can also be considered as a family of parametric curves. The study of the properties of these systems of lines will help in the understanding of the moire phenomenon. It will also be useful in many strain-analysis applications.[1]

2.2. Moire-Fringe Equations

The moire pattern and the two grating motifs are interrelated, so that a knowledge of any two allows the determination of the third. And if an analytic expression is known for any two, the analytic expression for the third is defined. For example, consider the simple but very useful pattern (Fig. 2.1) produced by a set of parallel opaque lines alternated with transparent spaces of equal width superposed on a second similar set of opaque lines and spaces still of equal width but of slightly larger width than the first set. Assume that the direction of the lines in the first set differs somewhat from the direction of the lines in the second set. (This type of moire pattern is typical of that used to describe displacement

[1] The subject matter of this chapter has been presented in A. J. Durelli and V. J. Parks, "Moire Fringes as Parametric Curves," *Exp. Mech.*, Vol. 7, No. 3 (March 1967), pp. 97–104.

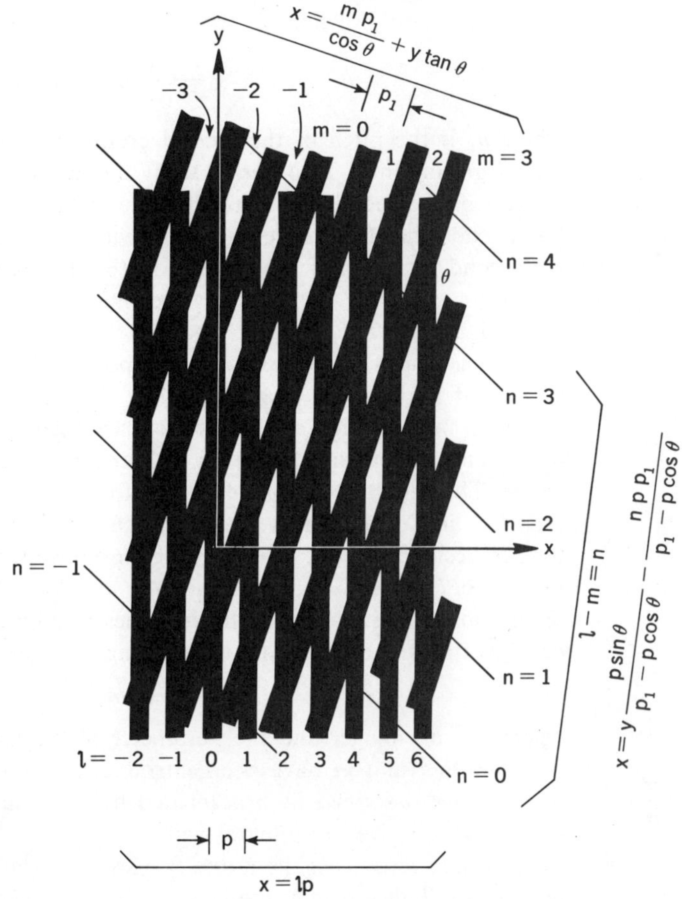

Figure 2.1. Equations of two rectilinear gratings of parallel lines and the associated moire fringes.

of the second grating with respect to the first, in strain or motion analysis. The considerations that follow are general, however, irrespective of what produces the difference between the two gratings.) If a Cartesian coordinate system is applied perpendicular and parallel to the lines in the first set as shown in the figure, the family of lines in the first set is described analytically by

$$x = lp \tag{2.1}$$

where p is the width between lines (the pitch of the grating) and l is the order number of each line, assuming that the line passing through the origin is given the number zero, and the lines away from the origin are ordered 1, 2, 3 in the positive x-direction and $-1, -2, -3$ in the negative x-direction. This is the same as Eq. (1.1) presented before, when applied to the moire fringes.

The family of lines in the second set is described analytically by

$$x = \frac{mp_1}{\cos\theta} + y\tan\theta \tag{2.2}$$

where p_1 is the pitch of the second grating and θ is the angle between the two sets of grating lines. In order to study the interference of the two gratings, moire fringes, or the locations where the dark areas intersect the light areas, the Cartesian coordinates have been chosen with the origin on a transparent space in the second grating and m will refer to the transparent spaces, with zero order at the origin and $m = 1, 2, 3, \ldots$ to the right and $m = -1, -2, -3, \ldots$ to the left in the figure.

The origin has been chosen at a point of interference and the orders chosen so that at the origin $l = m = 0$. Note that the interference at the origin is on the same moire fringe as the interference occurring at $l = m = 1$, and that the interferences occurring at $l = m = 2$ and at $l = m = 3$ are also on that same fringe. The general relation of l to m for the moire fringe through the origin is then $l = m$. Consider next the interference just above the origin. Here the interference occurs at $l = 0$, $m = -1$. Again, notice that the fringe continues through the points $l = 1$, $m = 0$; $l = 2$, $m = 1$; and $l = 3$, $m = 2$, at which points interference also takes place. The fringe is described in general by $l - m = 1$. Similar considerations show that the moire fringes in general are described by

$$l - m = n \tag{2.3}$$

where n is an integer and the parameter of the moire fringes.

Both l and m have been introduced as integers; however, note that the moire fringe described by n is located by any value of l and m (integer or noninteger) which gives an integral value of n. As such, Eq. (2.3) is continuous over the field and represents the moire fringes as a whole, not just the intersections of opaque and light areas in the field. In fact, the value of n can also be interpreted as a noninteger to give a continuous family of which the moire fringes are selected members.

This is essentially the same description of fringes considered in Sec. 1.5; however, here the description has been couched in analytic (mathematical) terms and this, in turn, allows the development of a mathematical expression for the fringes.

Substituting Eqs. (2.1) and (2.2) into Eq. (2.3) gives the equation of the moire fringe in terms of x, y, and n:

$$\frac{x}{p} - \frac{x\cos\theta - y\sin\theta}{p_1} = n \tag{2.4}$$

or in the form

$$Ax + By + C = 0$$

$$x(p_1 - p\cos\theta) - yp\sin\theta - np_1p = 0$$

The space between fringe lines can be computed using analytic geometry.

Thus, the distance from the origin to the fringe of order $n = 1$ is

$$\delta = \frac{|C|}{\sqrt{A^2 + B^2}} = \frac{pp_1}{\sqrt{p^2 \sin^2 \theta + (p_1 - p \cos \theta)^2}} \tag{2.5}$$

This is the spacing of fringes throughout the pattern. The slope of the fringes measured from the master grating is the negative of the coefficient of y divided by the coefficient of x:

$$\tan \phi = \frac{p \sin \theta}{p_1 - p \cos \theta} \tag{2.6}$$

The position of the second (or specimen) grating is described completely from the spacing δ, angle ϕ of the moire grating, and the pitch of the first (or master) grating p:

$$p_1 = \frac{\delta}{\sqrt{1 + \left(\frac{\delta}{p}\right)^2 + 2\left(\frac{\delta}{p}\right) \cos \phi}}$$

$$\theta = \arctan \frac{\sin \phi}{\frac{\delta}{p} + \cos \phi} \tag{2.7}$$

These equations are developed from a different point of view in Chapter 3 and nomographs given to assist in determining p_1 and θ. For $p_1 = p$ the slope reduces to $-\cot \frac{1}{2}\theta$, showing that for two gratings of equal pitch, the moire fringes will always be inclined at one-half the relative inclination of the two gratings. This property is also required from symmetry.

As mentioned above, this is an example useful in strain and motion analyses. It typifies a homogeneous two-dimensional field of strain in a body. If one grating (say, the one with pitch p) is considered as fixed, then the second grating can be considered as a grating which originally coincided with the fixed grating and had a pitch p, but which was strained until the pitch was p_1 and rotated from the fixed position, an angle θ. In strain analysis the fixed pitch is known, the moire pattern is obtained, and from these the pitch p_1 and angle θ can be determined. (The actual computation of strain may require some further analysis with other gratings to be explained later.) Although, strictly speaking, the above analysis is for a homogeneous field, it is applied to a small area in the nonhomogeneous field with the assumption that the small area is practically homogeneous.

2.3. *Superposition of Equidistant Straight Lines on Equiangular Straight Lines*

In a second example a moire pattern will be considered which is not homogeneous. Consider a set of equiangular radial lines superposed on a set of equidistant parallel lines (Fig. 2.2). In strain analysis radial lines are found when initially parallel vertical lines have been marked on a beam which is subjected

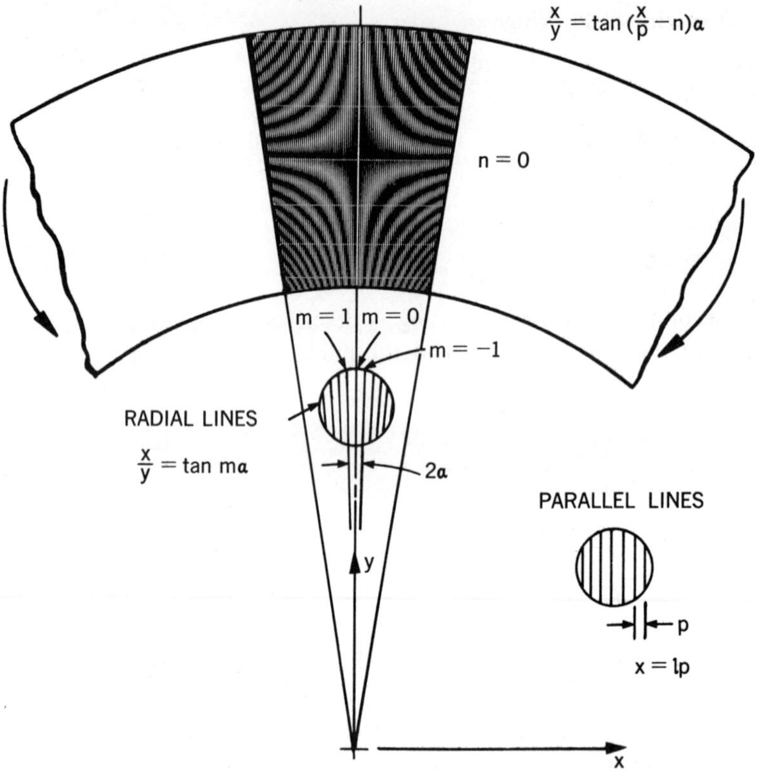

Figure 2.2. Equations of two rectilinear gratings (one of parallel lines and one of radial lines) and the associated moire fringes.

to bending. Then the center of the curvature of the beam is the point of intersection of the radial lines. The parallel lines on which the radial lines are superposed constitute a fixed-reference grating.

As in Eq. (2.1), the parallel lines are described by

$$x = lp \tag{2.8}$$

with all the terms having the same definitions as in Eq. (2.1).

The radial lines are expressed by

$$\frac{x}{y} = \tan m\alpha \tag{2.9}$$

where m is the order of the radial lines (zero along the positive y-axis, plus values going clockwise, and minus values going counterclockwise) and α is the angle between each of the radial lines.

Again substituting into Eq. (2.3) gives

$$\frac{x}{p} - \frac{\arctan \dfrac{x}{y}}{\alpha} = n \qquad \text{or} \qquad \frac{x}{y} = \tan\left(\frac{x}{p} - n\right)\alpha \tag{2.10}$$

Figure 2.2 shows an area on the y-axis where the distance between radial lines is about equal to the pitch p. The cross at the center is the fringe described by $n = 0$.

The above two examples follow the approach of Oster, Wasserman, and Zwerling,[2] who combine a number of analytic expressions and obtain the expression of the corresponding moire patterns. The approach is not used directly in strain analysis since, in general, only one grating has an analytic expression, and the moire pattern is measured to determine the shape of the second grating. However, the concept of a homogeneous field, as shown in the first example, is used to interpret the strain at a point.

2.4. Other Fringe Families

The parametric approach to the study of the moire phenomenon permits the understanding of a difficulty that in some cases may be important. This difficulty is the tendency of some moire-fringe patterns to fade out and, in more severe circumstances, the tendency for a new, different fringe pattern to form. Practically speaking, this can come from two causes (which, in some cases, may act together). The first cause is too large an angle between the grating lines; the second cause is too large a variation between the pitches of the two gratings.

To demonstrate these conditions, consider a more general form of Eq. (2.3):

$$l - rm = n \tag{2.11}$$

where r is any rational number.

In Sec. 2.4.1 it will be shown that different angles between the grating lines of two superposed gratings can give fringe patterns corresponding to both $r = +1$ and $r = -1$ or

$$l - m = n$$
$$l + m = n$$

In Sec. 2.4.2 it will be shown that if the two grating pitches are multiples, the value of r will be associated with that multiple.

2.4.1. Large Angles Between Gratings

Figure 2.3 shows a grating of equidistant parallel lines set at different angles on a fixed master grating of the same pitch. The fringe pattern visible in the range between $+30°$ and $-30°$ is the one described in Sec. 2.3. As the rotation

[2] G. Oster, M. Wasserman, and C. Zwerling, "Theoretical Interpretation of Moirè Patterns," *Jnl. of the Optical Society of America*, Vol. 54, No. 2 (Feb. 1964), pp. 169–175.

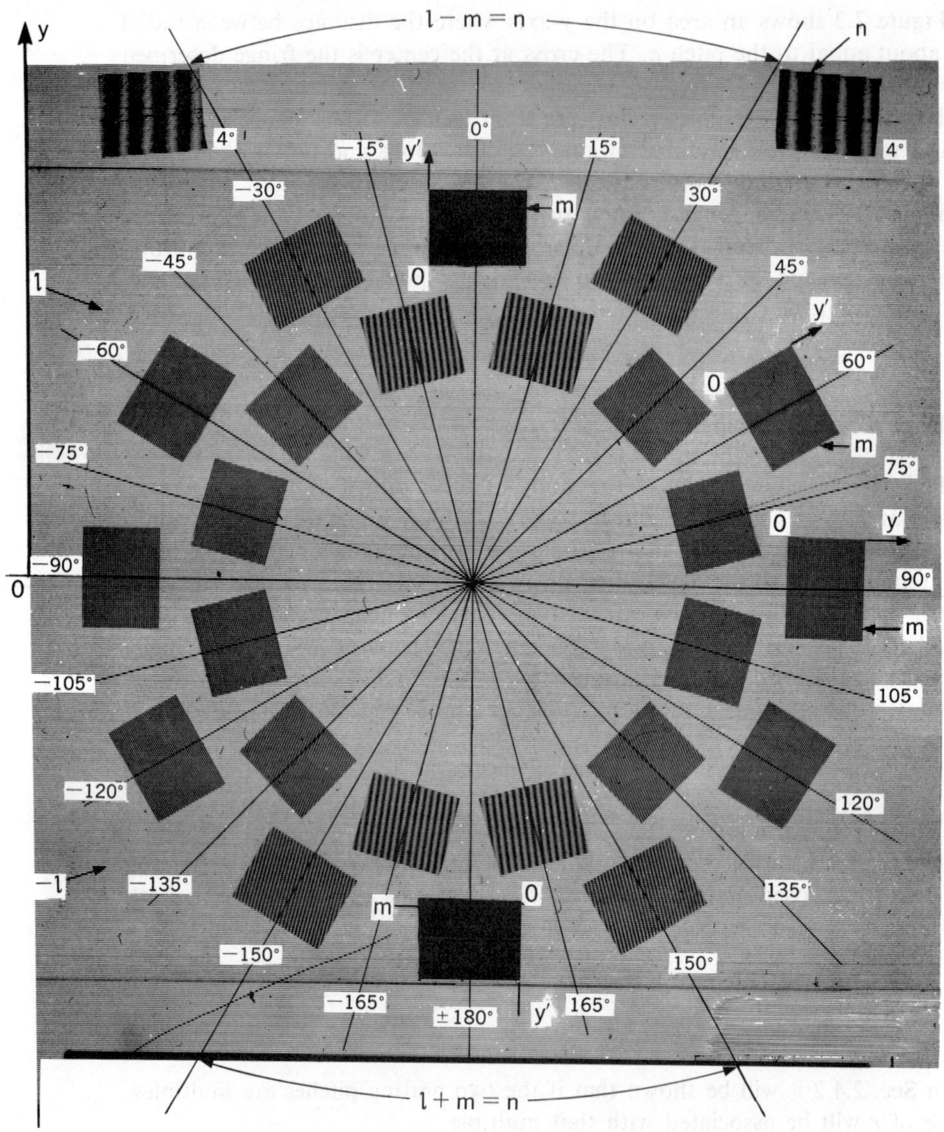

Figure 2.3. Each small rectangle shows a moire produced by a grating of parallel lines rotated by the indicated angle with respect to the master grating of horizontal parallel lines. Both gratings have the same pitch.

of the movable grating is increased beyond this angle, the fringes come closer and closer; and when the rotation is $\pm 45°$, it is difficult to distinguish the pattern. The fringes are very close together and jagged. The fading of this fringe pattern continues up to a rotation of $\pm 180°$. A new, similar fringe pattern is formed in the range (180° ± 30°). This fringe pattern is a mirror image of the first one and also fades out the further the angle between the gratings deviates from 180°.

Each fringe pattern is predominant to the eye on its own side of 90°. When the gratings are at 90° with respect to each other, both fringe patterns have equal strength.

The fringes noted in the range ±30° are defined by Eq. (2.3), $l - m = n$, provided the values of l and m increase in the same direction at zero rotation. If the grating is rotated 180° and the orders l and m are held, then the two gratings again coincide, but the grating orders increase in opposite directions, and the fringes produced in the range (180° ± 30°) will be described—by reasoning similar to that used to obtain Eq. (2.3)—by the expression

$$l + m = n \tag{2.12}$$

Figure 2.4. Two superposed gratings of equispaced concentric circles with large distance between their centers. (Fringes in the region between the centers are additive; fringes in the region to the left of the left center, and to right of the right center are subtractive.)

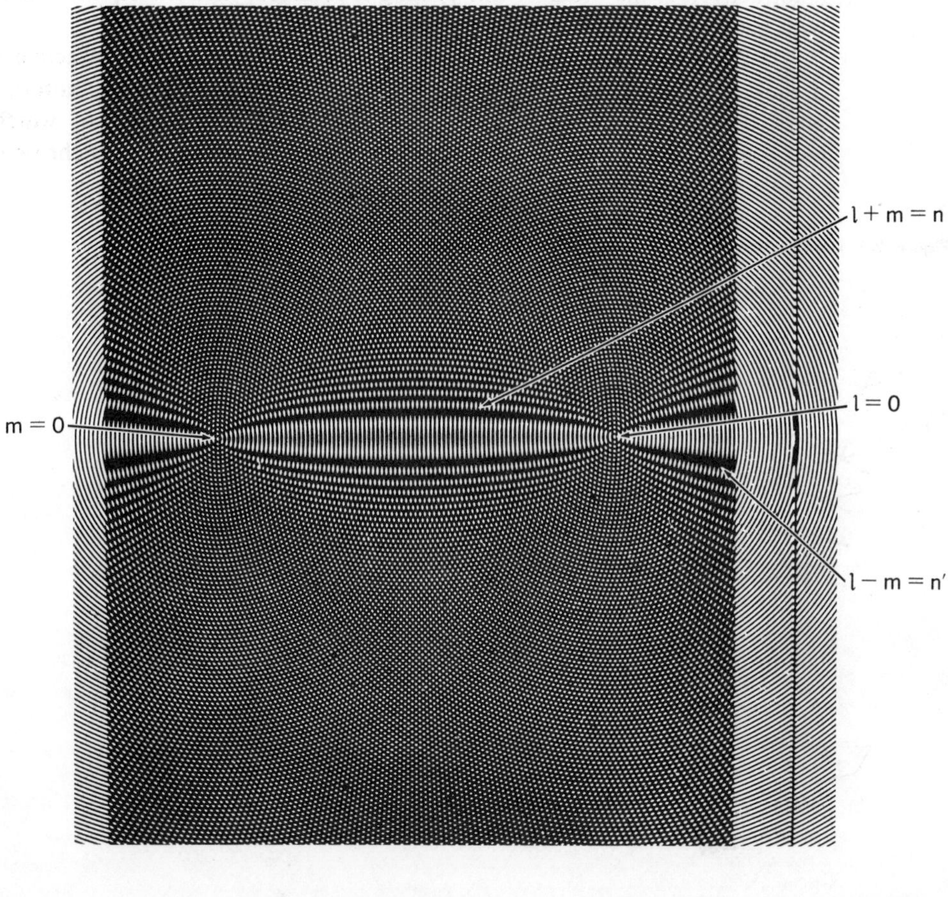

Although the fringe patterns associated with each expression can be interchanged simply by reversing the order of one of the gratings, it can be seen, by following the fringe pattern with changes in the angle, that the two expressions represent two distinct fringe patterns that are everywhere present in the field (though not everywhere visible), and that in the regions $\pm 30°$ and $(180° \pm 30°)$, one of the fringe patterns predominates over the other.

In Fig. 2.4 both of these types of patterns can be seen. One can be called subtractive (Eq. 2.3) and the other additive (Eq. 2.12). Both are visible in the same field. Figure 2.4 is the result of superposing two gratings of equispaced concentric circles eccentric to each other. Between the centers of the two gratings, one family of fringes is formed, and outside the two centers the other family is formed. To the eye, the two families do not overlap, and are separated by an area with no fringes. Mathematically, however, both the additive and the subtractive processes extend over the whole field, and in this mathematical sense, the fringe families do overlap each other. If the concentric circles in both gratings are ordered from the center out, the fringes between the two centers will be described by Eq. (2.12) (additive), and the fringes outside the centers by Eq. (2.3) (subtractive). If the order of one grating is reversed, then the equation associated with each of the families will be reversed.

Photoelasticians may be familiar with this additive-and-subtractive method if they have separated stresses (or strains) with an isochromatic pattern, which gives the loci of constant $(\sigma_1 - \sigma_2)$, and an isopachic pattern, which gives $(\sigma_1 + \sigma_2)$. As shown in Fig. 2.5, the loci of intersections of the isochromatic and

Figure 2.5. Construction of σ_1 and σ_2 isobars from isopachics and isochromatics.

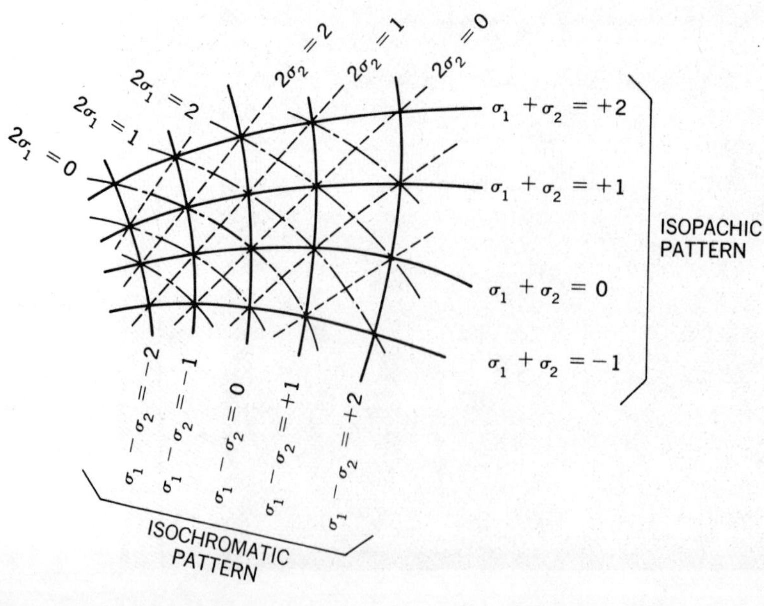

isopachic patterns in one direction are one family of isobars (lines of constant σ_2) and are obtained by the following subtractive process:

$$(\sigma_1 + \sigma_2) - (\sigma_1 - \sigma_2) = 2\sigma_2$$

The loci of the same intersections of the isochromatic and isopachic patterns in the other direction are the other isobars (lines of constant σ_1) and are obtained from the following additive process:

$$(\sigma_1 + \sigma_2) + (\sigma_1 - \sigma_2) = 2\sigma_1$$

2.4.2. Large Variations Between Pitches

Figures 2.6a, b, and c illustrate the moire patterns produced by gratings of pitches which are multiples or near-multiples of each other. The three patterns are all produced by straight-line gratings superposed one on the other at the same angle in the three cases. The pattern in Fig. 2.6b is produced by two gratings, both with the same pitch. In Fig. 2.6c an identical fringe pattern is produced by one grating having the same pitch as the gratings which produced the pattern in Fig. 2.6b and one grating with twice that pitch ($p_1 = 2p$). The pattern in Fig. 2.6a is produced by a grating with the same pitch as one of the gratings which produced the pattern in Fig. 2.6b and a grating with one-half that pitch ($p_1 = \frac{1}{2}p$). In the pattern in Fig. 2.6a twice as many fringes are produced as in the other two, but at the same angle.

Note that in the general case, for large variations between pitches, the moire pattern corresponds to the pattern that would be seen if the coarser grating's pitch were divided by the integer which would bring the coarse-grating pitch to about the same values as the fine-grating pitch.

The terms in Eq. (2.11) are then defined as follows:

l is the order of the denser grating

m is the order of the coarser grating

n is the moire-fringe order

r is the small whole number or a fraction greater than unity, composed of two small whole numbers, that is near the value of the ratio of the coarser pitch to the denser pitch

Again, as with large angular variations, the negative values of r denote the same fringe pattern as that denoted by the corresponding positive values of r with the sign of one of the grating orders reversed.

Another way of noting the fringe formation is to consider the transparent rhomboids produced in the patterns. These rhomboids define the light or transparent fringes. The light fringe is a series of rhomboids. Each rhomboid will tend to be associated with the two nearest rhomboids on either side of it. Thus, by noting the center-to-center distance between rhomboids, it can be decided where the moire fringe lies. If the pitch p_1 in the pattern in Fig. 2.6b were

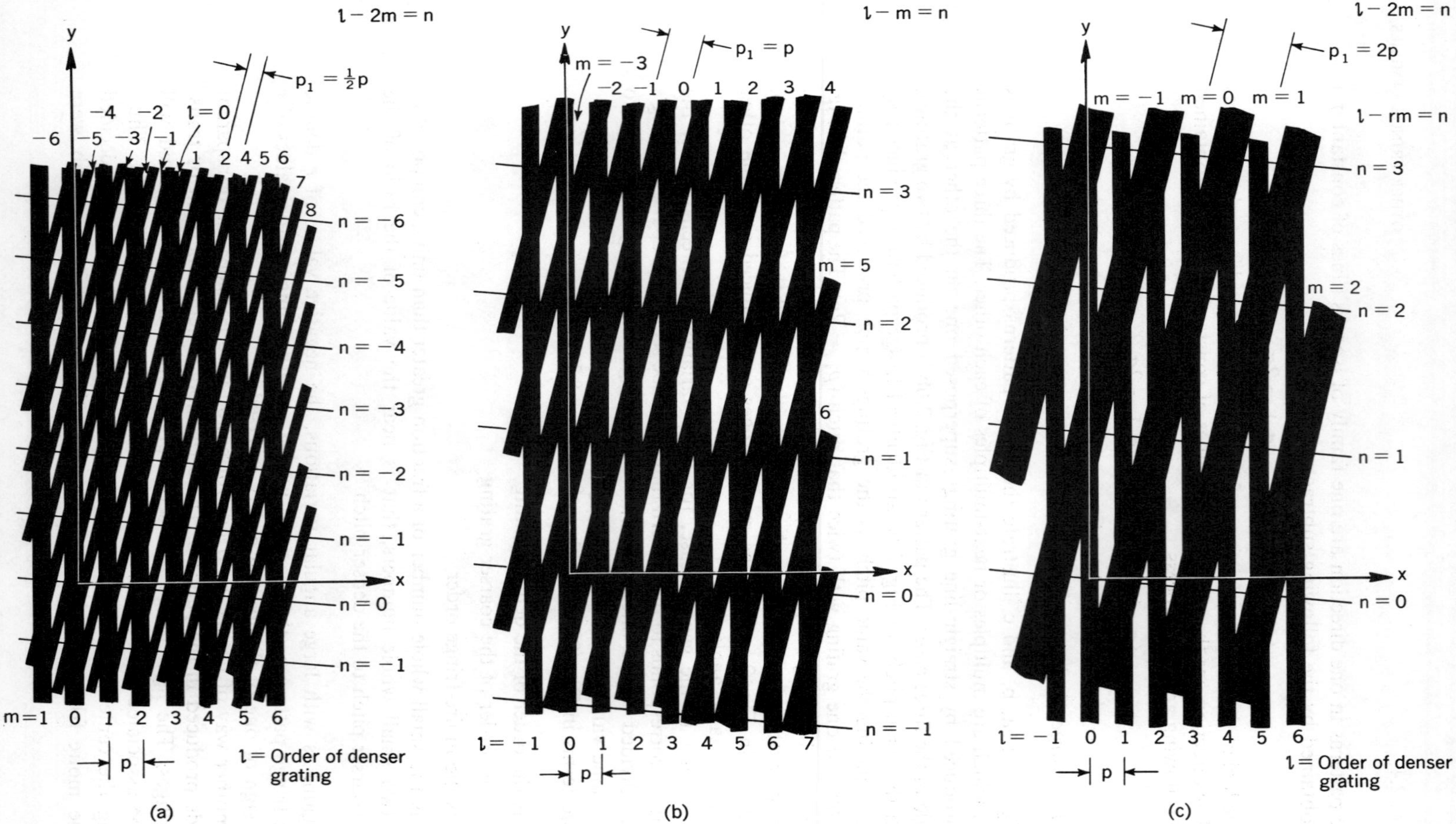

Figure 2.6. The same moire fringe pattern can be obtained from two parallel-line gratings of equal pitch and from two parallel-line gratings where one has twice the pitch of the other.

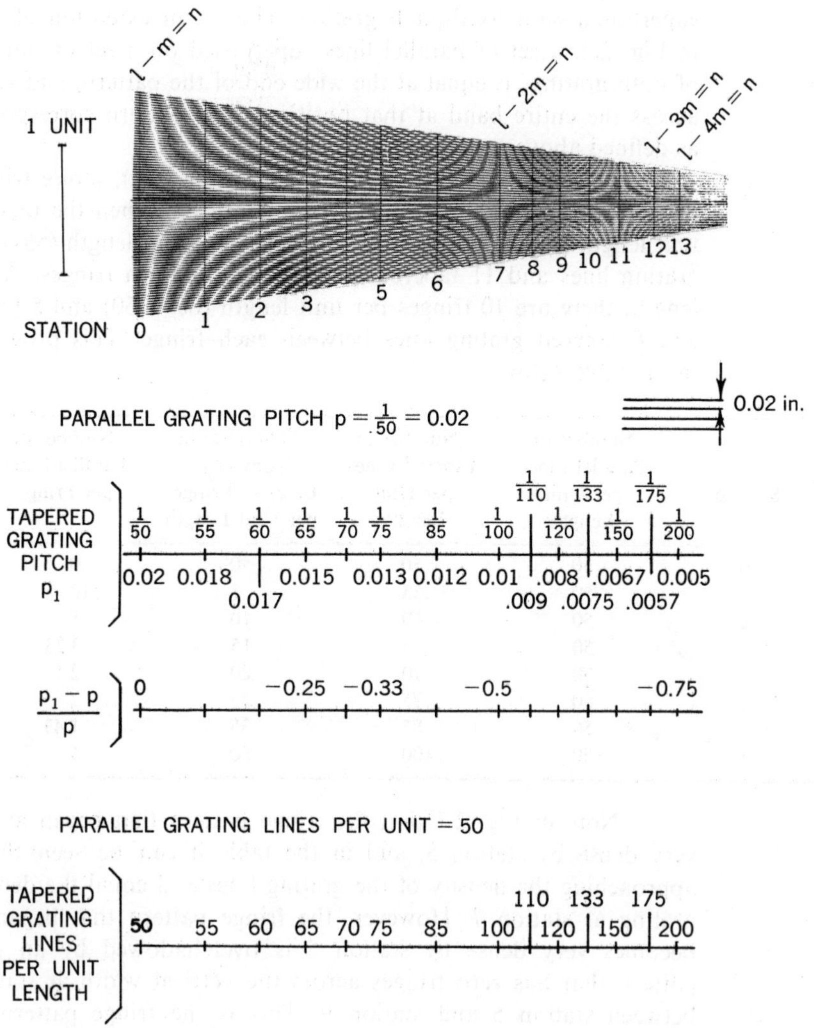

Figure 2.7. Illustration of a multiple moire fringe pattern obtained by superposing a set of coarse parallel lines and a set of denser tapered lines.

increased from its value of p, it would be seen that the associated rhomboids in each of the light fringes would be drawn farther and farther apart, until a critical point is reached at which each rhomboid would be equally spaced from four rhomboids to form a triangular motif. Beyond the critical point, as the pitch is opened further, each rhomboid would be closer to a new pair of rhomboids, and the pattern shown at the right (in Fig. 2.6c) would evolve.

To demonstrate that each of these moire patterns is distinct, Fig. 2.7 shows an example of the fringe patterns produced by a varying-pitch grating superposed on a fixed-pitch grating. This is an extension of the pattern shown in Fig. 2.2, a set of parallel lines superposed on a set of radial lines. The pitch of both gratings is equal at the wide end of the pattern, and so there is no fringe across the entire band at that position. This pattern corresponds to $l - m = n$, as defined above.

As the radial grating lines taper to the right, moire fringes are produced. As shown in the figure, the tapering is severe. When the taper has 55 lines per unit length (station 1), there are 5 fringes per unit length (55 − 50) or 10 parallel grating lines and 11 tapered grating lines between fringes. At 60 lines per unit length, there are 10 fringes per unit length (60 − 50) and 5 parallel grating lines and 6 tapered grating lines between each fringe. This progression is extended in the table below.

Station	Number of Parallel Lines per Unit Length	Number of Tapered Lines per Unit Length	Difference in Number of Lines = Fringes per Unit Length	Number of Parallel Lines per Fringe	Number of Tapered Lines per Fringe
0	50	50	0	0	0
1	50	55	5	10	11
2	50	60	10	5	6
3	50	65	15	3.33	4.33
4	50	70	20	2.5	3.5
5	50	75	25	2	3
6	50	85	35	1.43	2.43
7	50	100	50	1	2

Note in Fig. 2.7 that the fringe pattern that began at station 0 becomes very dense by station 5, and in the table it can be seen that the fringes are approaching the density of the grating lines and equal the density of the parallel grating at station 7. However, the fringe pattern that began at station 0 and becomes very dense by station 5 is overshadowed by an independent fringe pattern that has zero fringes across the vertical width at station 7 and extends between station 5 and station 9. This is the fringe pattern corresponding to $l - 2m = n$, where l corresponds to the tapered-grating orders, m to the parallel-grating orders, and n is any integer. A third independent pattern appears at station 11, corresponding to $l - 3m = n$, and a fourth at station 13 ($l - 4m = n$). The patterns tend to overlap, as can be seen in Fig. 2.4, and become less distinct as r varies from unity.

A careful study of the pattern also indicates a star about station 5. This corresponds to an r-value of $\frac{3}{2}$.

Note that, although the star patterns around stations 0, 7, 11, and 13 become less distinct, the density of the fringes in each of the star patterns increases with respect to the other star patterns in the proportion of 1: 2: 3: 4. This agrees with the observation in Fig. 2.6 that the denser grating determines the moire pattern.

Figure 2.8 shows the same basic combination of gratings as Fig. 2.7, a radial grating superposed on a straight-line grating. The radial grating is the same as the one shown in Fig. 2.7, but the straight-line grating has a density of 300 lines per inch.

The governing fringe equation is again

$$l - rm = n$$

But now l refers to the straight lines (which here form the denser grating) and m to the tapered lines (which here form the coarser grating).

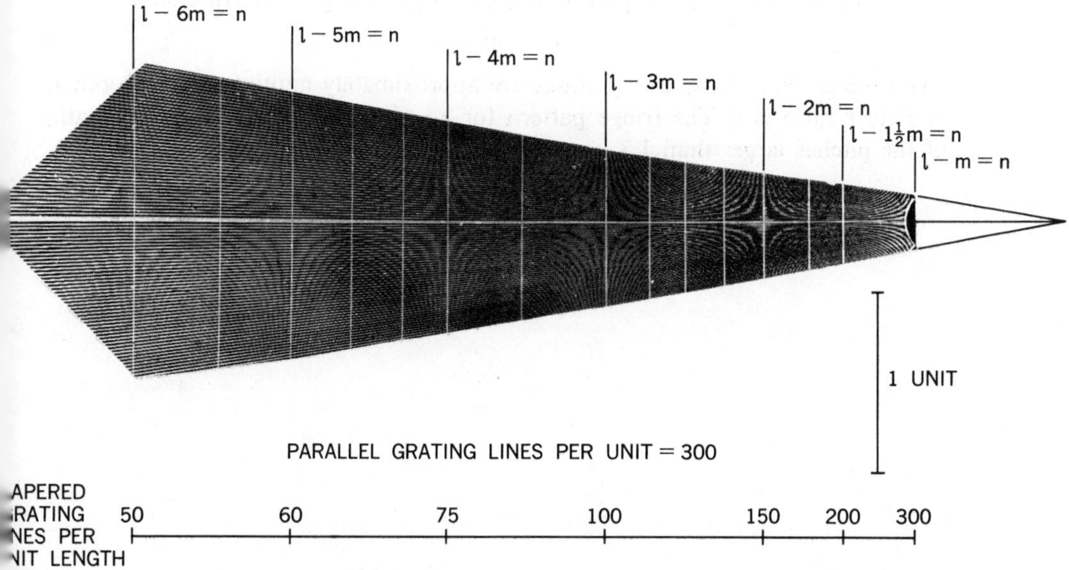

Figure 2.8. Illustration of a multiple moire fringe pattern obtained by superposing a set of dense parallel lines on a set of coarser tapered lines.

In the figure there are 7 star patterns corresponding to $r = 6, 5, 4, 3, 2, 1\frac{1}{2}$, and 1. The basic and most distinct pattern (corresponding to $r = 1$) is seen just beginning at the right end of the figure. In this wedge, all the star fringe patterns have the same density, since l now is the same for all the stars and fixes the fringe orders n.

In concluding, it is suggested that for angles greater than 30° between gratings and for pitches whose ratio is greater than 1.3, the fringe pattern will tend to fade; and that for rotations greater than 150° and pitches whose ratio is greater than 1.5, new fringe patterns will appear which must be studied carefully to avoid misinterpretation.

2.5. Summary

Call l the order of a line of a grating and m the order of a line of a second superposed grating, and take the origin at the intersection of a line and an interline; then the order n of the resulting interference fringe is given by

$$l - m = n \tag{2.3}$$

A fringe corresponding to the expression

$$l + m = n \tag{2.12}$$

also exists and confuses the first one if the angle of rotation is larger than approximately 30°.

A fringe may also be produced according to the general expression

$$l - rm = n \tag{2.11}$$

when the pitches of the two gratings are approximately multiples of each other, and their ratio is r. The fringe pattern for $r = 1$ will begin to fade for a ratio of the pitches larger than 1.3.

3

GEOMETRIC RELATIONSHIPS BETWEEN GRATINGS AND MOIRE FRINGES

In order better to understand the moire phenomenon, it is convenient to develop some geometric relationships between the moire fringes and the two gratings producing them. The considerations to be made in this chapter will be purely geometric and do not require any special technical background on the part of the reader. The use of these relationships to determine strains is more complicated and will be discussed in a subsequent chapter.

The relationships developed in this chapter are not necessary for the correct interpretation of moire patterns. The general approach based on displacements, on their derivatives, or on both of them is sufficient for all applications. Some readers may find the geometric approach more direct and, in some particular cases of strain analysis, easier to use. The reader who is more inclined toward the general mechanics of the continuum point of view of strains and displacements may wish to continue with the reading of the next chapter.[1]

---

[1] The subject matter of this chapter has been presented in S. Morse, A. J. Durelli, and C. A. Sciammarella, "Geometry of Moire Fringes in Strain Analysis," *J. Eng. Mech. Div.* Proc. A.S.C.E., EM4,No. 2736 (February 1961), pp. 55–74. See also *Trans. A.S.C.E.*, Vol. 127, Part II (1962), pp. 39–53.

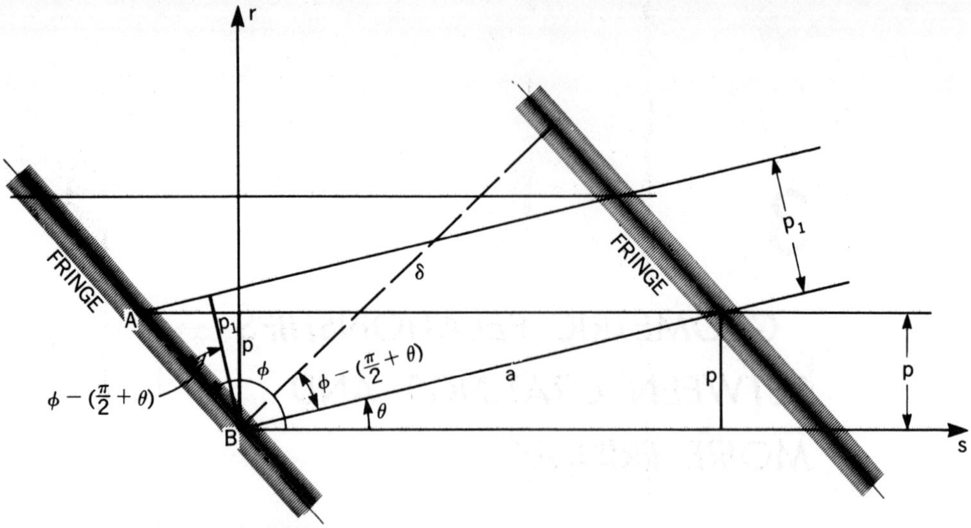

Figure 3.1. Sketch to obtain the inclination of the moire fringes and their distances as functions of the master grating pitch, the model grating pitch, and the rotation.

3.1. *Definitions*

The angle θ is defined at any point at which two grating lines intersect as the acute angle measured from the fixed master-grating lines to the specimen-grating lines. The angle from the fixed master-grating lines to a fringe at a point measured in the same direction as θ is designated as ϕ and may be either acute or obtuse. Figure 3.1 shows how to obtain the inclination ϕ of the moire fringes and their distances δ as functions of the master-grating pitch p, the specimen-grating pitch p_1, and the rotation θ of both gratings at a point.

Let us assume either the case of a homogeneous deformation and rotation (which may be accompanied by translation) or the case of a sufficiently small element in a nonhomogeneous field. Equations relating any desired sets of parameters for the moire phenomenon are easily derived from the simple geometry of fringe formation. Those for the normal fringe spacing δ and the fringe angle ϕ are perhaps the most fundamental. However, the information available for the analysis will usually be in the form of photographs on which δ and ϕ can be measured. Alternatively, the coordinate fringe spacings δ_r and δ_s can be measured, as will be shown in Fig. 3.4. The purpose of the analysis will be in general the determination of θ and p_1 at the desired points. Obviously the master-grating pitch p and the direction of its lines must be known.

3.2. *Inclination of the Fringes*

In Fig. 3.1,

$$\overline{AB} = \frac{p}{\cos\left(\phi - \frac{\pi}{2}\right)} = \frac{p}{\sin\phi} \tag{3.1}$$

and

$$\overline{AB} = \frac{p_1}{\cos\left(\phi - \frac{\pi}{2} - \theta\right)} = \frac{p_1}{\sin(\phi - \theta)} \tag{3.2}$$

Therefore

$$p_1 = p\frac{\sin(\phi - \theta)}{\sin\phi} \tag{3.3}$$

Because

$$p(\sin\phi\cos\theta - \sin\theta\cos\phi) = p_1\sin\phi \tag{3.4}$$

then

$$\tan\phi = \frac{p\sin\theta}{p\cos\theta - p_1} \tag{3.5}$$

3.3. *Distance Between Fringes*

Also from Fig. 3.1 and the foregoing,

$$a = \frac{p}{\sin\theta} \tag{3.6}$$

and

$$\delta = a\cos\left(\phi - \frac{\pi}{2} - \theta\right) = \frac{p\sin(\phi - \theta)}{\sin\theta} = \frac{p_1\sin\phi}{\sin\theta} \tag{3.7}$$

From Eq. (3.5),

$$\sin\phi = \frac{p\sin\theta}{\sqrt{p^2\sin^2\theta + (p\cos\theta - p_1)^2}} \tag{3.8}$$

and

$$\delta = \frac{pp_1}{\sqrt{p^2\sin^2\theta + (p\cos\theta - p_1)^2}} \tag{3.9}$$

3.4. *Rotation of the Specimen Grating in Terms of the Inclination of the Fringes and Their Spacing*

From Eq. (3.7),

$$\frac{\delta}{p}\sin\theta = \sin(\phi - \theta) = \sin\phi\cos\theta - \cos\phi\sin\theta \tag{3.10}$$

and

$$\tan\theta = \frac{\sin\phi}{\frac{\delta}{p} + \cos\phi} \tag{3.11}$$

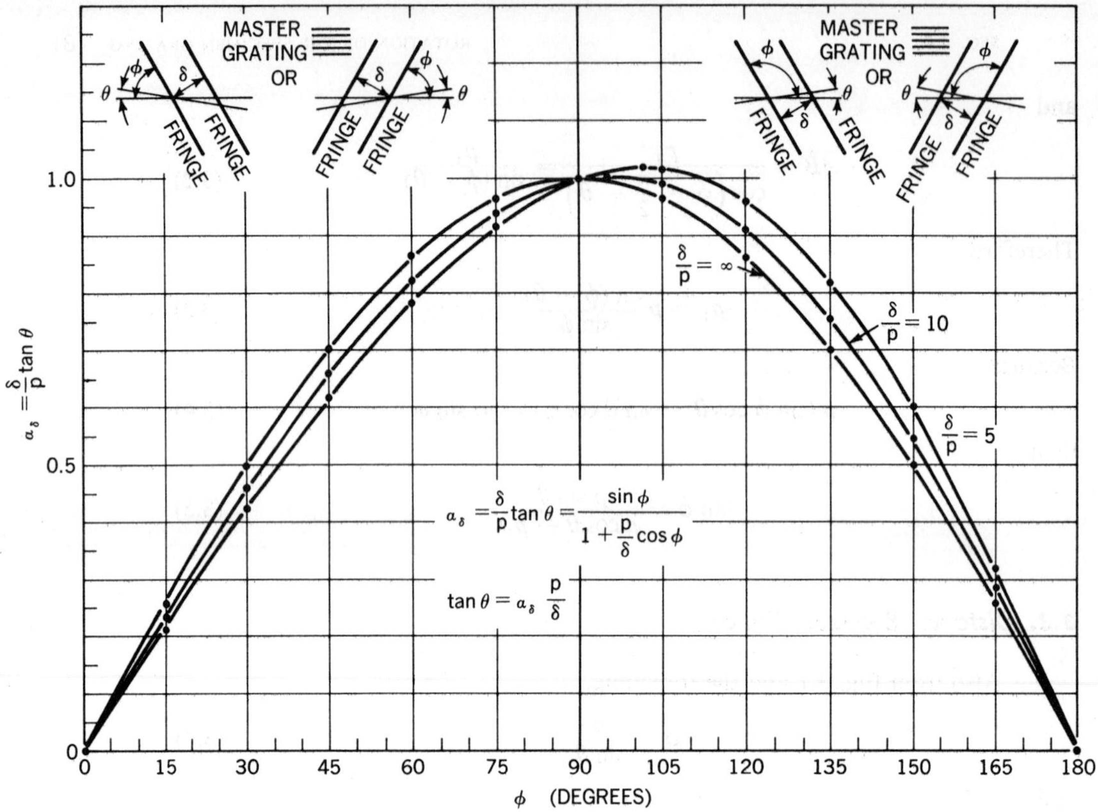

Figure 3.2. Rotation of specimen grating as a function of fringe angle and normal fringe spacing.

To represent graphically the relationship between θ and ϕ it is convenient to introduce the following definitions:

$$\alpha_\delta = \frac{\delta}{p} \tan\theta = \frac{\delta}{p}\left(\frac{\sin\phi}{\dfrac{\delta}{p} + \cos\phi}\right) = \frac{\sin\phi}{1 + \dfrac{p}{\delta}\cos\phi} \tag{3.12}$$

and

$$\tan\theta = \alpha_\delta \frac{p}{\delta} \tag{3.13}$$

Values of α_δ for the entire possible range of ϕ (0° to 180°) and for various values of δ/p are plotted in Fig. 3.2, permitting easy determination of the specimen rotation at a point without knowledge of p_1, the specimen pitch.

3.5. Specimen Pitch in Terms of ϕ and δ

Again, from Eq. (3.7),

$$p_1 = \frac{\delta \sin\theta}{\sin\phi} \tag{3.14}$$

From Eq. (3.11),

$$\sin\theta = \frac{\sin\phi}{\sqrt{\sin^2\phi + \left(\frac{\delta}{p} + \cos\phi\right)^2}} \tag{3.15}$$

$$p_1 = \frac{\delta}{\sqrt{1 + \left(\frac{\delta}{p}\right)^2 + 2\left(\frac{\delta}{p}\right)\cos\phi}} \tag{3.16}$$

and

$$\frac{p}{p_1} = \sqrt{\left(\frac{p}{\delta}\right)^2 + 2\left(\frac{p}{\delta}\right)\cos\phi + 1} \tag{3.17}$$

3.6. Determination of Change in Specimen Pitch

To compute the value of the change in specimen pitch using the measured quantities δ and ϕ, a graphical construction can be utilized. For this construction it is convenient to introduce the following definitions:

$$\beta_\delta = \frac{\delta}{p}\left(1 - \frac{p}{p_1}\right) = \frac{\delta}{p}\left[1 - \sqrt{\left(\frac{p}{\delta}\right)^2 + 2\left(\frac{p}{\delta}\right)\cos\phi + 1}\right] \tag{3.18}$$

and

$$\frac{p_1 - p}{p_1} = \beta_\delta\left(\frac{p}{\delta}\right) \tag{3.19}$$

Values of β_δ for all possible values of ϕ and for various values of δ/p have been plotted in Fig. 3.3, permitting easy determination of the specimen pitch, pitch ratio, or the quantity $1 - (p/p_1)$, free from the influence of rotation.

If Eq. (3.19) is converted to the form—

$$\frac{p_1 - p}{p} = \frac{\frac{\beta_\delta p}{\delta}}{1 - \frac{\beta_\delta p}{\delta}} \tag{3.20}$$

—the change in pitch with respect to the initial pitch is obtained.

3.7. Rotation at a Point of the Specimen Grating, in Terms of ϕ and the Distance Between Fringes Along the Coordinate Directions

In many cases it may be preferred to measure the fringe spacing along the coordinate directions or it may be desirable to have a second method for checking the specimen rotations. From Fig. 3.4 we have, for the two cases (a) ϕ acute, and (b) ϕ obtuse, respectively:

$$\delta = \delta_s \sin\phi \tag{3.21a}$$

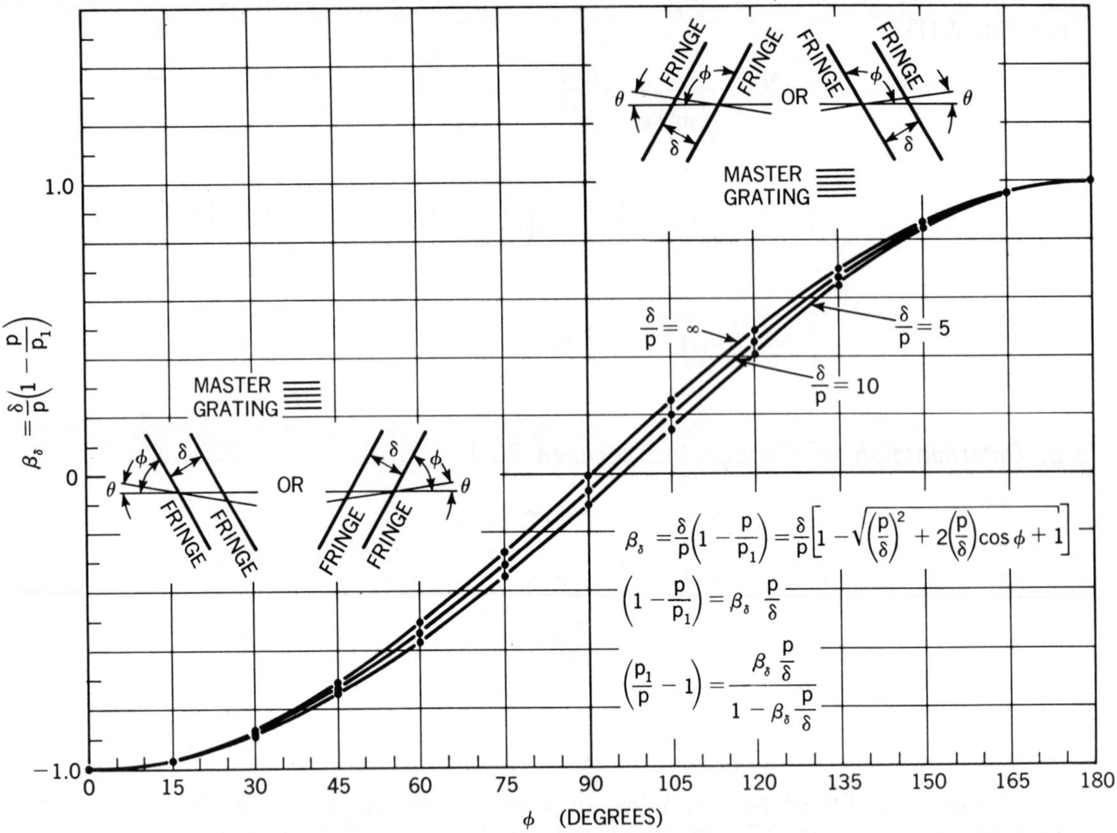

Figure 3.3. Specimen pitch as a function of fringe angle and normal fringe spacing.

and

$$\delta = \delta_s \sin(\pi - \phi) = \delta_s \sin\phi \tag{3.21b}$$

Substituting Eqs. (3.21) into Eq. (3.11) and extracting $\sin\theta$ by trigonometric transformation, we obtain

$$\sin\theta = \frac{1}{\sqrt{1 + \left(\frac{\delta_s}{p} + \cot\phi\right)^2}} \tag{3.22}$$

Similarly, from Fig. 3.4,

$$\delta = \delta_r \cos\phi \tag{3.23a}$$

$$\delta = \delta_r \cos(\pi - \phi) = \delta_r(-\cos\phi) \tag{3.23b}$$

To avoid having to give a negative sign to δ_r, a measured distance, these equations can be combined as

$$\delta = \delta_r |\cos\phi| \tag{3.24}$$

Again substituting in Eq. (3.11) and extracting $\sin\theta$,

$$\sin\theta = \frac{1}{\sqrt{1 + \left(1 + c\dfrac{\delta_r}{p}\right)^2 \cot^2\phi}} \tag{3.25}$$

where $c = +1$ for $0 < \phi < \pi/2$ and $c = -1$ for $\pi/2 < \phi < \pi$.

Following the same method used previously α_s and α_r can be defined as follows:

$$\alpha_s = \frac{\delta_s}{p}\sin\theta = \frac{\dfrac{\delta_s}{p}}{\sqrt{1 + \left(\dfrac{\delta_s}{p} + \cot\phi\right)^2}} \tag{3.26}$$

Figure 3.4. Sketches used to derive the value of fringe spacing δ_r and δ_s in the coordinate directions.

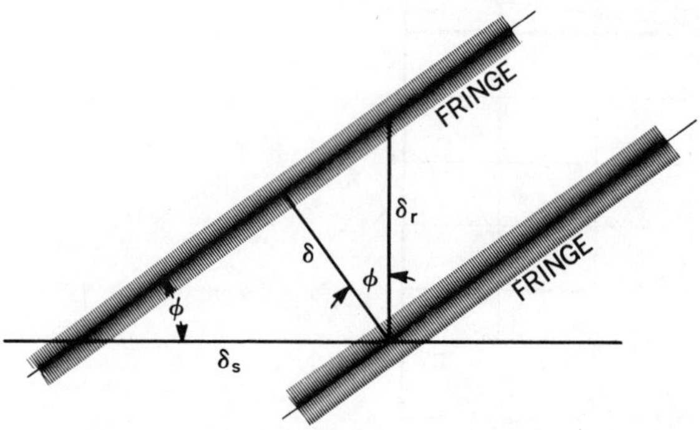

CASE (a), ϕ AN ACUTE ANGLE

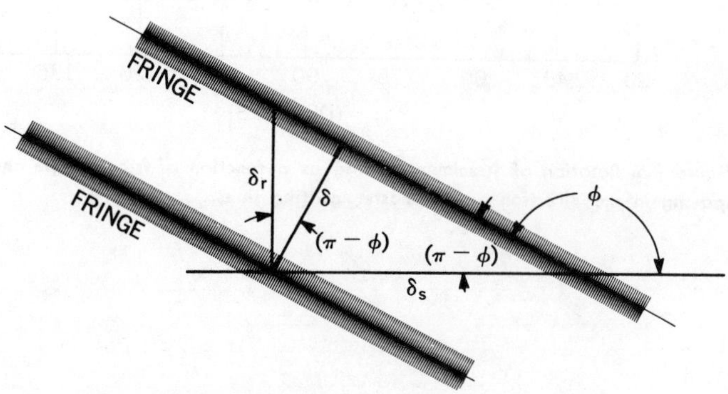

CASE (b), AN OBTUSE ANGLE

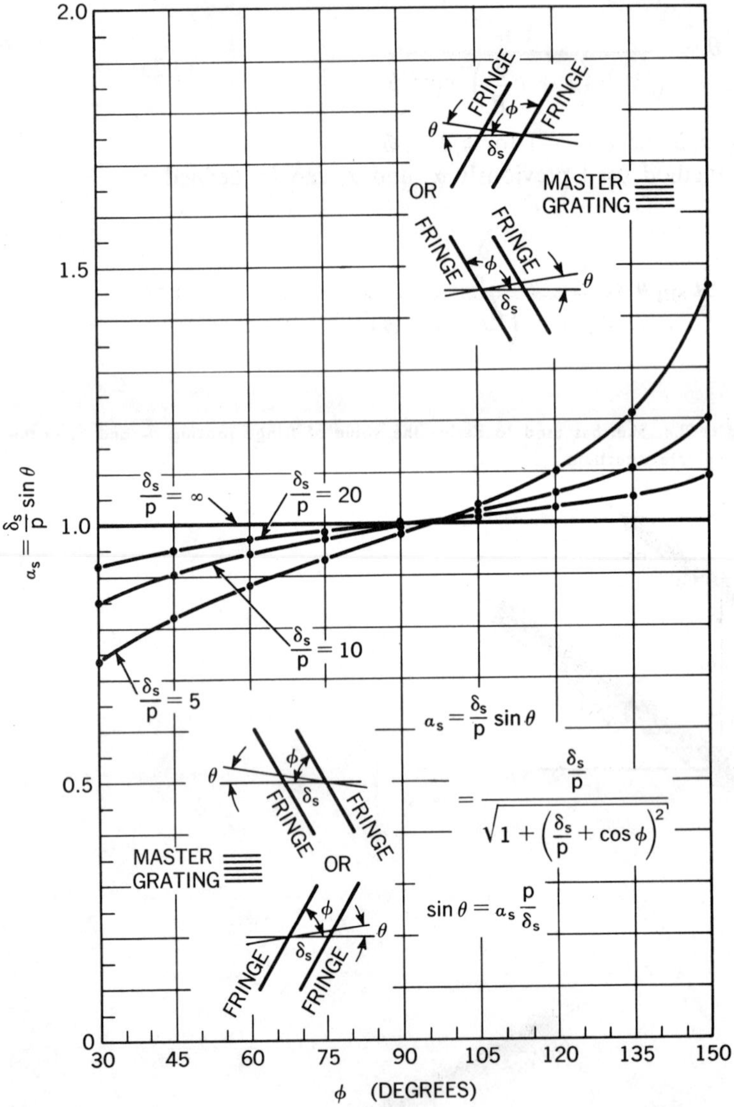

Figure 3.5. Rotation of specimen grating as a function of fringe angle and fringe spacing in the direction of the master grating lines.

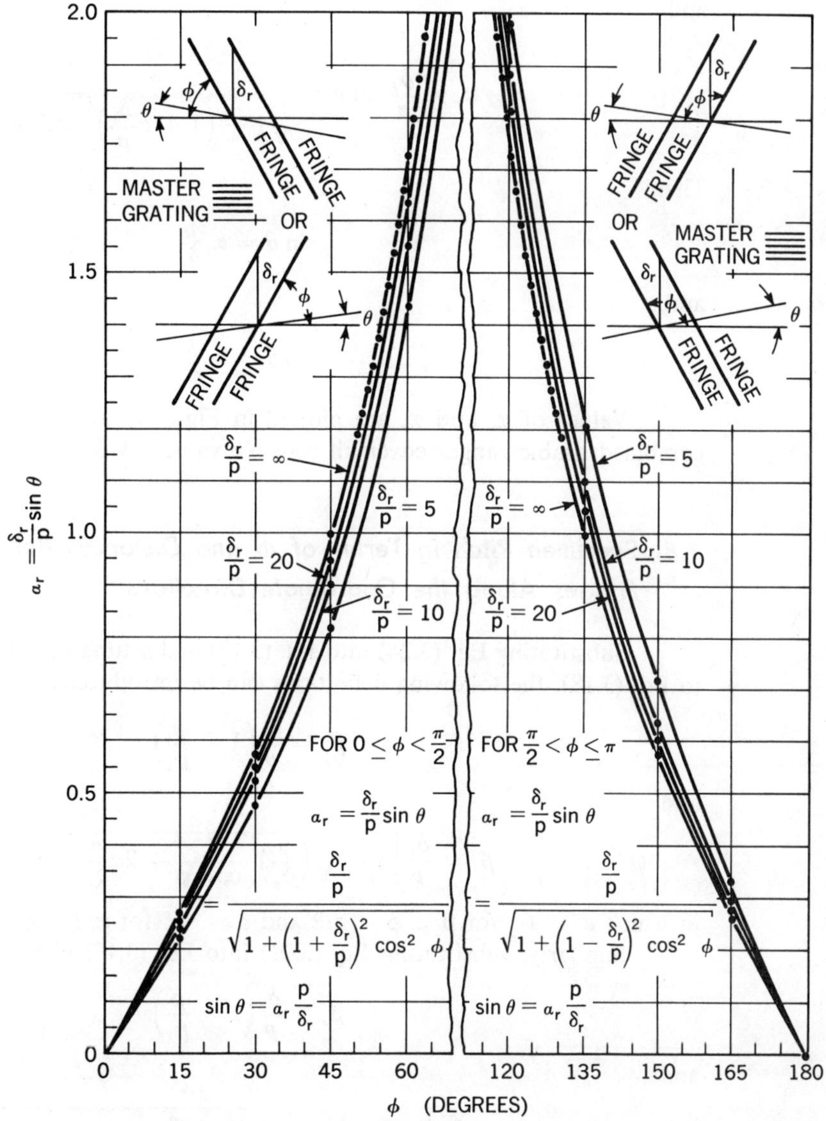

Figure 3.6. Rotation of specimen grating as a function of fringe angle and fringe spacing normal to the direction of the master grating lines.

and

$$\alpha_r = \frac{\delta_r}{p}\sin\theta = \frac{\dfrac{\delta_r}{p}}{\sqrt{1+\left(1+c\dfrac{\delta_r}{p}\right)^2\cot^2\phi}} \tag{3.27}$$

Then

$$\sin\theta = \alpha_s\frac{p}{\delta_s} \tag{3.28}$$

and

$$\sin\theta = \alpha_r\frac{p}{\delta_r} \tag{3.29}$$

Values of α_s and α_r are plotted in Figs. 3.5 and 3.6, respectively. Their combined usable ranges cover all possible values of ϕ.

3.8. Specimen Pitch in Terms of ϕ and Distances Between Fringes Along the Coordinate Directions

Substituting Eq. (3.24) into Eq. (3.17) and setting up a relationship similar to Eq. (3.18), the following definitions can be introduced:

$$\beta_r = \frac{\delta_r}{p}\left(1-\frac{p}{p_1}\right) \tag{3.30a}$$

and

$$\beta_r = \frac{\delta_r}{p}\left[1-\sqrt{\left(\frac{p}{\delta_r}\right)^2\frac{1}{\cos^2\phi}+2c\frac{p}{\delta_r}+1}\right] \tag{3.30b}$$

in which $c = +1$ for $0 < \phi < \pi/2$ and $c = -1$ for $\pi/2 < \phi < \pi$.

Similarly, substituting Eq. (3.21) into Eq. (3.17) gives

$$\beta_s = \frac{\delta_s}{p}\left(1-\frac{p}{p_1}\right) \tag{3.31a}$$

and

$$\beta_s = \frac{\delta_s}{p}\left[1-\frac{1}{\sin\phi}\sqrt{\left(\frac{p}{\delta_s}\right)^2+\frac{p}{\delta_s}\sin 2\phi+\sin^2\phi}\right] \tag{3.31b}$$

Then the change in pitch is given by

$$\frac{p_1-p}{p_1} = \beta_r\left(\frac{p}{\delta_r}\right) \tag{3.32}$$

and

$$\frac{p_1-p}{p_1} = \beta_s\left(\frac{p}{\delta_s}\right) \tag{3.33}$$

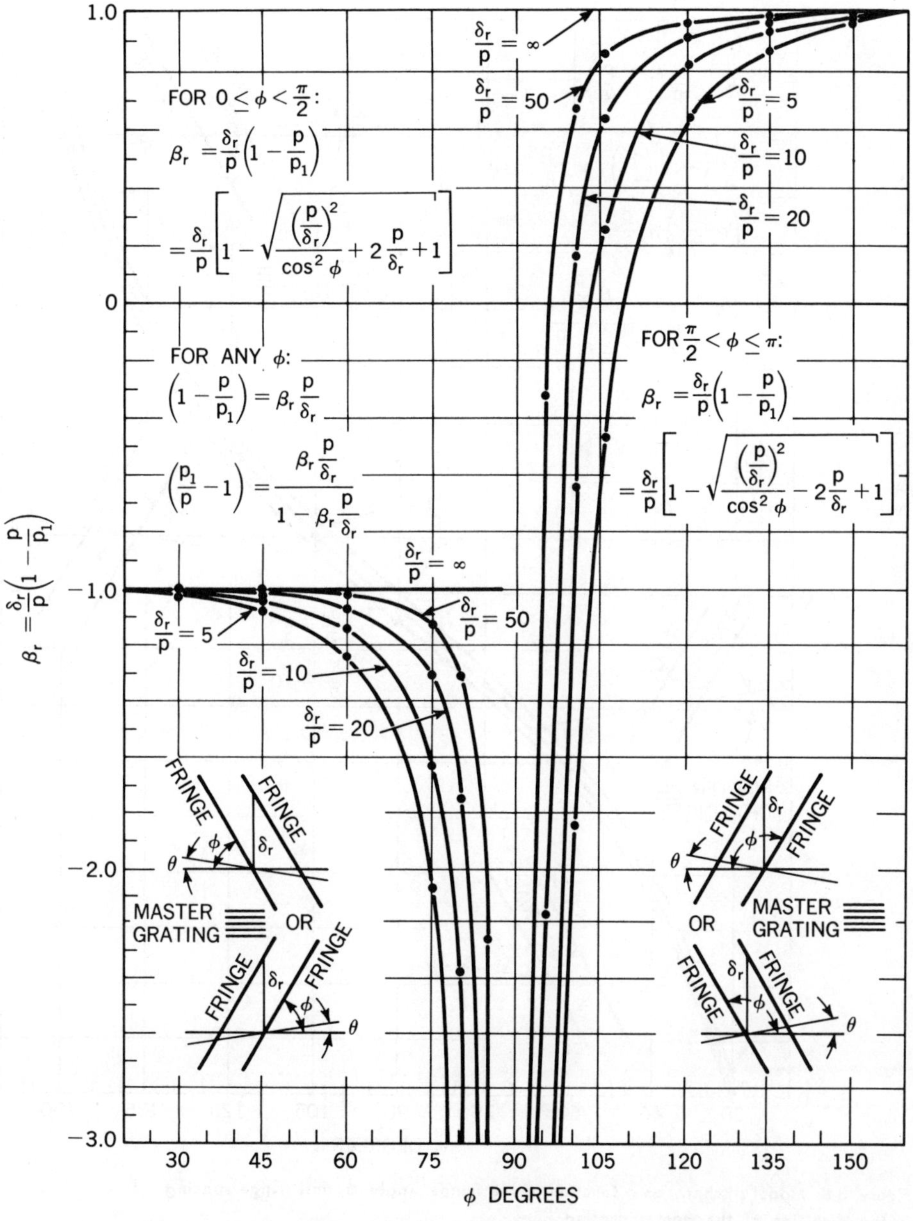

Figure 3.7. Specimen pitch as a function of fringe angle and fringe spacing perpendicular to the direction of the master grating lines.

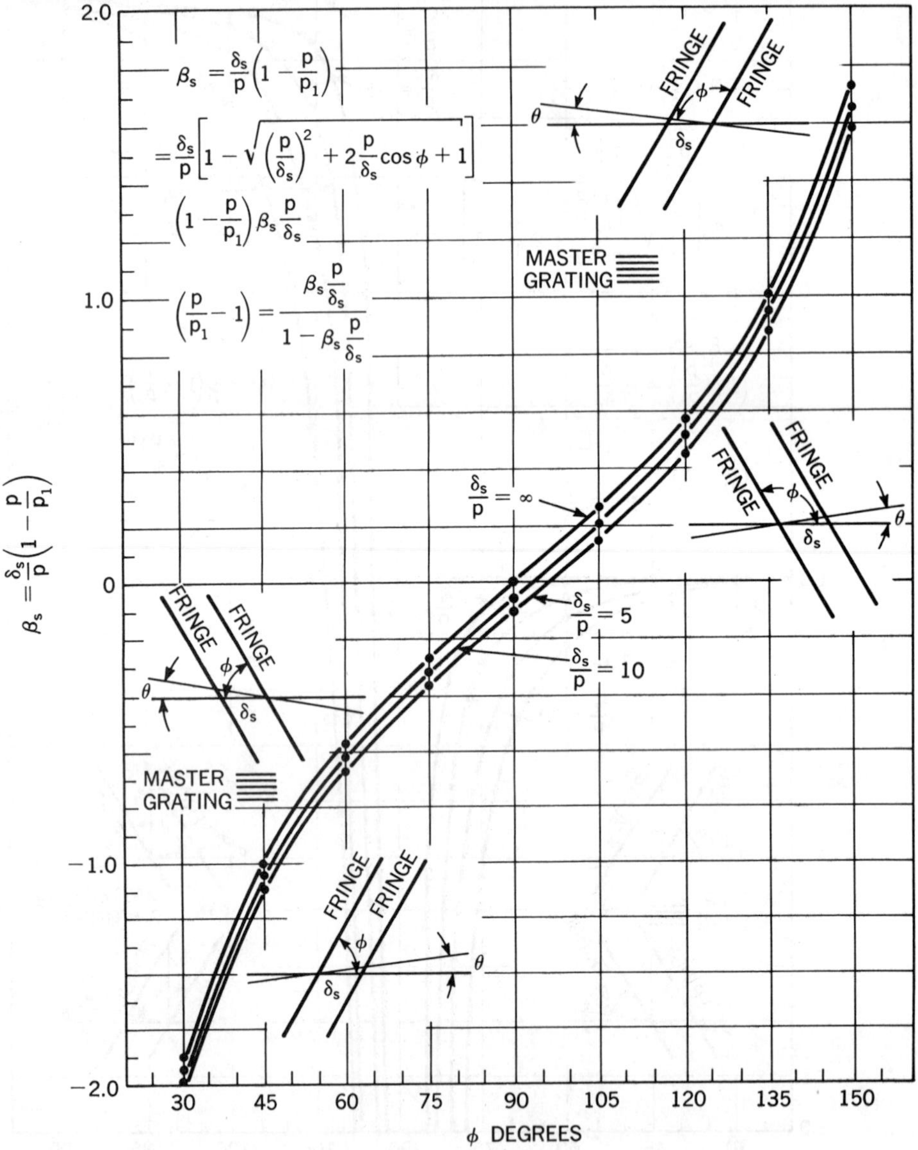

Figure 3.8. Model pitch, p_1, as a function of the fringe angle, ϕ, and fringe spacing in the direction of the master grating lines.

Values of β_r and β_s are plotted in Figs. 3.7 and 3.8, respectively. Their combined usable ranges cover all possible values of ϕ.

3.9. Simplified Equations for Small Rotations

In Eq. (3.5), if $p_1 \neq p$, as θ approaches zero, the angle ϕ approaches zero, or 180°. In this case, it is apparent from Fig. 3.7 that sufficient accuracy may result from taking $\beta_r = \pm 1$ and reducing Eq. (3.32) to

$$\frac{p_1 - p}{p_1} = \pm \frac{p}{\delta_r} \tag{3.34}$$

By simple transformation,

$$\frac{p_1 - p}{p} = \frac{\pm \frac{p}{\delta_r}}{1 \mp \frac{p}{\delta_r}} \tag{3.35}$$

In Fig. 3.5, if ϕ is sufficiently close to 90° (depending on the value of δ_s/p), the value of α_s is close to 1.0, and sufficient accuracy may result from a reduction of Eq. (3.28) to

$$\sin \theta = \frac{p}{\delta_s} \tag{3.36}$$

3.10. Determination of the Direction of the Fringe Angle

In order to determine the direction in which the fringe angle ϕ is to be measured in the foregoing equations and graphs, a small amount of information is needed in addition to a photograph of the fringe pattern. For example, the direction of rotation may often be deduced from the loading conditions or could be found by means of a coarse set of lines or dots on the model, in addition to the closely spaced array which produces the moire. Once the direction of θ is known, the direction of ϕ is determined by its definition and all ambiguity disappears.

Alternatively, all ambiguity is removed if the sign of the quantity $(1 - p/p_1)$ is known—that is, if it is known whether p/p_1 is greater or less than unity. This can usually be determined from the moire pattern if the actual loaded model and master gratings are available. Differentiating Eq. (3.5) leads to

$$\frac{d(\tan \phi)}{d\theta} = \frac{1 - \frac{p_1}{p}\cos \theta}{\left(\cos \theta - \frac{p_1}{p}\right)^2} \tag{3.37}$$

Because the denominator of Eq. (3.37) cannot be negative, the sign of the derivatives is determined by the sign of $[1 - (p_1/p) \cos \theta]$. If this expression is negative, then

$$\frac{p_1}{p} > \sec \theta \geq 1 \tag{3.38}$$

and ϕ decreases for an increase in θ. If the master grating is rotated in a given direction with respect to the model and the fringes rotate in the same direction, this condition is satisfied and $p_1/p > 1$. If the gratings were originally identical, this positively indicates tensile strain.

For $[1 - (p_1/p) \cos \theta]$ positive the situation is not quite so definite because

$$\frac{p_1}{p} < \sec \theta \geq 1 \tag{3.39}$$

can be satisfied in a number of ways when θ is a large angle. If, however, ϕ is not near 90° and the fringes rotate opposite to the rotation of the master grating, p_1/p may ordinarily be considered to be less than 1, indicating compression for originally identical gratings.

Other guides to the sign of $1 - (p/p_1)$ are sometimes available directly from a moire photograph. If ϕ is 90° at any point, $p/p_1 > 1$ from Eq. (3.17). Once a fringe is known to represent $p/p_1 > 1$ at any point, the same fringe cannot represent $p/p_1 < 1$ without going beyond 90°, actually through $\phi = \pi/2 + \theta/2$.

Fringes parallel to the master-grating lines represent the condition of no rotation (Eq. 3.11). Unless they are continuations from areas where the p/p_1 condition can be evaluated, it is impossible to distinguish from a photograph alone whether p/p_1 is greater than or less than unity.

4

STRAIN–DISPLACEMENT RELATIONSHIPS

4.1. Introduction

Experimental strain-and-stress analysts seldom have to compute rotations. This is a consequence of the fact that most gages follow the deformation of the specimen, and are not influenced by its rotation. That is the case of photoelasticity; mechanical, electrical, and optical strain gages; brittle coatings, etc. With grids and moires, the situation is different. Rotations do influence the measurement. If the measurement is analyzed without taking the rotations into account, an erroneous answer to the strain analysis may result.

It is also true that analysts using strain gages do not need a very elaborate understanding of the strain phenomenon. Their computations are usually limited to the application of the formula

$$\epsilon = \frac{l_f - l_i}{l_i}$$

where l_f is the length of a segment (the gage base length) after deformation, l_i is the length of the same segment before deformation, and ϵ is the direct or "normal" strain.

In a few cases the understanding of displacement–strain relations is required. If rotations are neglected, then the needed strain–displacement relations are as follows:

$$\epsilon_x = \frac{\partial u}{\partial x}$$

$$\epsilon_y = \frac{\partial v}{\partial y}$$

$$\epsilon_z = \frac{\partial w}{\partial z}$$

The evaluation of moire requires a more refined understanding of the relationship between strains and displacements. The following sections are directed to the reader who is not familiar with that part of the theory of strain.[1]

Beyond the material presented here, there is an extensive treatment of the topics of strains, displacements, and stresses which can be used in the analysis of moire and which is easily available in standard texts.

4.2. Usual Assumptions

The strain–displacement relations are obtained from a definition of strain put in terms of the positions of points in a body, as defined by some coordinate system, both before and after deformation. These relations are frequently obtained by considering a region, about a generic point in the body, sufficiently small so that the strain is sensibly homogeneous.[2,3]

In the discussion throughout the remainder of this chapter the whole body will be considered as subjected to homogeneous strain. This allows a direct geometric presentation and the use of simple examples taken from experimental analysis. It may be noted, however, that the development is the same as the one applied to the small sensibly homogeneous region about a point in the nonhomogeneous body.

A further assumption that is often made in developing the relations is that the partial derivatives of the displacements be small, so that the products and powers of the derivatives can be neglected with respect to the derivatives and so that the derivatives can be neglected with respect to unity.

This last assumption is often made in the theory of elasticity and restricts the analysis to small strains and small rotations. The restriction is a valuable simplification in many phases of the theory of elasticity. In what follows, however, the restrictions will not be used initially, and the more general form of the strain–displacement relations will be developed. Later, some of the restrictions will be applied in turn. First, it will be required that only the rotations be small. Then, it will be required that only the strains be small. Finally, it will be required that both rotations and strains be small. Each of these cases will, of course, find application in various strain problems. The cases are treated individually in the next four sections, followed by two sections of applications.

The development follows the general approach of the classic theory of strain. The presentation, the analysis of errors, and the application to specific

---

[1] Part of the material presented in this chapter has been published in V. J. Parks and A. J. Durelli, "Various Forms of the Strain–Displacement Relations Applied to Experimental Stress Analysis," *Exp. Mech.*, Vol. 4, No. 2 (Feb. 1964), pp. 37–47.

[2] A. E. H. Love, *A Treatise on the Mathematical Theory of Elasticity*, 4th ed. (New York: Dover Publications, Inc., 1944), p. 37.

[3] I. S. Sokolnikoff, *Mathematical Theory of Elasticity*, 2nd ed. (New York: McGraw-Hill Book Company, 1956), p. 6.

stress-analysis methods may be particularly useful to people in the experimental-stress-analysis field.

4.3. The Definitions of Strain

Table 4.1 shows ten different definitions of direct or "normal" strain. All are specified in terms of an initial fiber length and a final fiber length, and attempt to represent in one way or another the change in length that occurs in the fiber in nondimensional terms. For small strains and small rotations, the six definitions in the top half of the table give approximately the same value (essentially a percentage, or change per unit length of the fiber).

The four definitions in the bottom half give the same value as the one given by the six definitions in the top half, either added or subtracted from unity and squared in two cases. Thus the strain definition and strain–displacement relation given in Sec. 4.1 are sufficient in the case of small-strain–small-rotation.

Five of the strain definitions at the top of the table are shown by Seth[4] to be represented by the single equation

$$\epsilon_n = \frac{1}{n}\left[1 - \left(\frac{l_i}{l_f}\right)^n\right]$$

Table 4.1 Definitions of Direct Strain Along a Line

Lagrangian Strain (Engineering) $\epsilon^L = \frac{l_f - l_i}{l_i}$	Natural Strain $ln\frac{l_f}{l_i}$	Eulerian Strain (Engineering) $\epsilon^E = \frac{l_f - l_i}{l_f}$
Lagrangian Strain (Tensorial) $e^L = \frac{l_f^2 - l_i^2}{2l_i^2}$	A Hybrid Definition $\frac{l_f^2 - l_i^2}{2l_f l_i}$	Eulerian Strain (Tensorial) $e^E = \frac{l_f^2 - l_i^2}{2l_f^2}$
Green–St. Venant Deformation (Tensorial) $\frac{l_f^2}{l_i^2}$		Almansi–Hamel (Cauchy) Deformation (Tensorial) $\frac{l_i^2}{l_f^2}$
Stretch $\frac{l_f}{l_i}$		Reciprocal of Stretch $\frac{l_i}{l_f}$

l_i = initial fiber length
l_f = final fiber length

[4] B. R. Seth, "Generalized Strain and Transition Concepts for Elastic-Plastic Deformation-Creep and Relaxation," *Proceedings of the 11th Int. Congress of Appl. Mech.*, Munich, 1964 (*Springer-Verlag*, Berlin, 1966).

where n has values of -2, -1, 0, 1, and 2. This expression is convenient and unifies the presentation.

In the case of large strain or large rotation or both it becomes necessary to specify which of the definitions will be used. Throughout this chapter and most of the book the Lagrangian and Eulerian descriptions of engineering strain (at the upper left and upper right of the table) will be specified as exact and will be used. The four definitions specified as tensorial will be used only in Sec. 7.7. The other four are included only for reference.

Besides defining the direct strain, it will be also necessary to define the shear strain. *Shear strain* is defined here as the change in a right angle due to deformation. In orthogonal coordinates the right angle will be understood to be the angle between the coordinate axes. For other directions a right angle must be specified just as a fiber direction is specified when direct strain is determined.

The strains that will commonly be used are then formally defined in this text as

$$\begin{aligned} \epsilon^L &= \frac{l_f - l_i}{l_i}\bigg|_{l_i \to 0} \\ \epsilon^E &= \frac{l_f - l_i}{l_f}\bigg|_{l_f \to 0} \\ \gamma^L &= \frac{\pi}{2} - \xi_f \\ \gamma^E &= \xi_i - \frac{\pi}{2} \end{aligned} \tag{4.1}$$

where ϵ and γ are the direct and shear strains, respectively, in the Lagrangian or Eulerian description as specified by the superscripts L and E, respectively; l is the fiber length, either initial or final as specified by the subscripts i and f, respectively; ξ_f is the final size of an angle which was initially a right angle; and ξ_i is the initial size of an angle which after deformation is a right angle.

The requirement $l \longrightarrow 0$ is added to the definition of direct strain to refer the strain to a small neighborhood around the generic point, as mentioned in the previous section. (This does not mean the strain is small.) As indicated in the previous section, setting $l \longrightarrow 0$ is the same as studying a homogeneous field. In the subsequent sections, the base length l_f or l_i will be specified as a unit (1) to simplify the explanation. This is meant to be in full accord with the specification that $l \longrightarrow 0$.

The equations indicate that the difference between the Lagrangian and Eulerian definitions depends on the choice of a base length or base angle. When the strain is associated with a particular coordinate direction, this leads to a second distinction. A strain component in the Lagrangian description will refer to a line or right angle which initially lay in the coordinate directions.

Thus, the Lagrangian strain component, besides being the increase in length of a line in terms of the initial length, is also understood to be the strain along a line initially parallel to, say, the x-direction. The Eulerian strain component is the increase in length of a line finally parallel to the x-axis, referred to its final length.

In evaluating derivatives of a given coordinate direction both of these two distinctions must be considered. If the Largrangian description is used, displacement must be differentiated along a line which was initially parallel to a coordinate direction (on the undeformed body) and which has the initial dimensions of the body in that direction. If the Eulerian description is used, the dis-

Figure 4.1. Direct strain computed at the point P on a cantilever beam subjected to an applied displacement at the free end. Both Lagrangian and Eulerian descriptions are used to determine "horizontal" direct strain.

Illustrations of variation in meaning between the Lagrangian and Eulerian description of strain

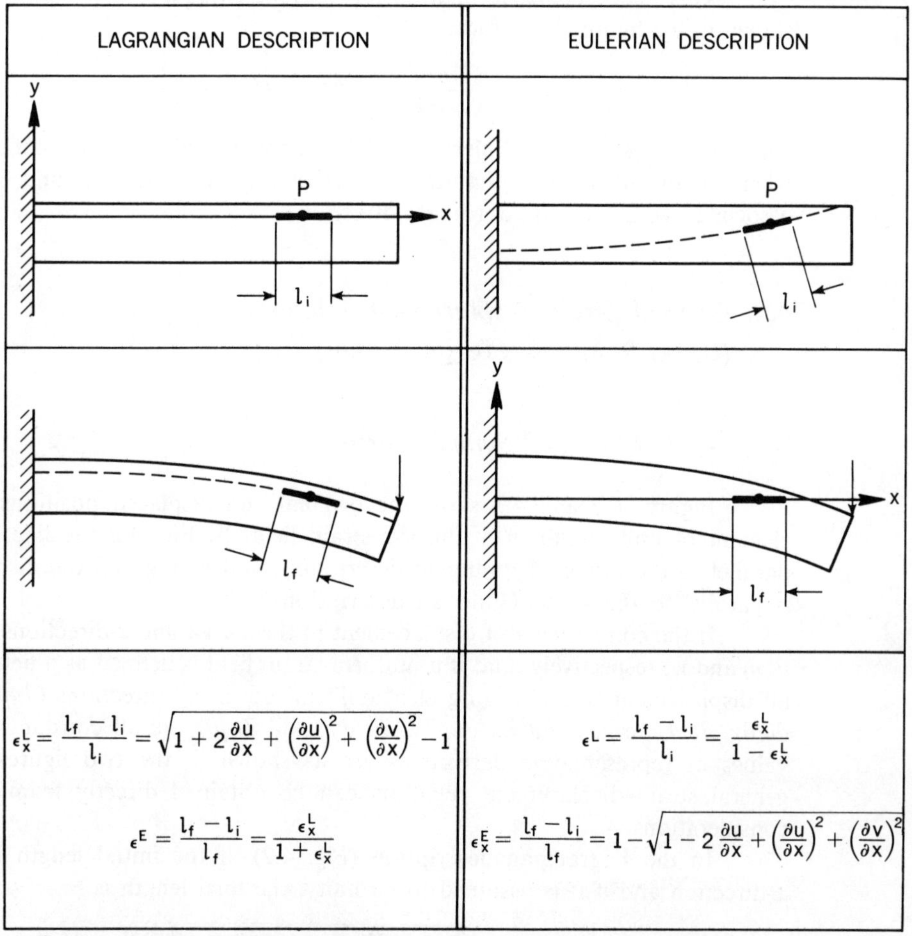

The line elements through the point P in the two descriptions are not the same. Thus the values of l_f and l_i in each description can be different, and in this case $\epsilon_x^L \neq \epsilon^L \quad \epsilon^E \neq \epsilon_x^E$

placement values must be differentiated along a line which is finally parallel to a coordinate direction (on the deformed body) and with the final dimensions of the line.

Figure 4.1 attempts to illustrate the distinction between the Lagrangian and Eulerian descriptions in a two-dimensional longitudinal strain analysis on a cantilever beam at the point *P*. The Lagrangian and Eulerian strain components are given as ϵ_x^L and ϵ_x^E; these components refer to two different lines, as well as having different base lengths. For the case of no rotation (not shown in the figure), the two terms ϵ_x^L and ϵ_x^E would refer to the same line and as such have the following fixed geometric relationship:

$$\epsilon_x^E = \frac{\epsilon_x^L}{1 + \epsilon_x^L}$$

This fixed geometric relationship may be extended to the general case as shown in the figure in the form

$$\epsilon^E = \frac{\epsilon^L}{1 + \epsilon^L} \qquad \text{or} \qquad \epsilon^L = \frac{\epsilon^E}{1 - \epsilon^E} \tag{4.2}$$

where ϵ^E and ϵ^L are the Eulerian and Lagrangian strains for the same line anywhere in the field. Note that the strains in Eq. (4.2) are not, and cannot be, associated with a given coordinate direction.

4.4. General Strain–Displacement Relations (Large Rotations—Large Strains)

4.4.1. *Direct (or "Normal") Strain*

Figures 4.2 and 4.3 show the original and displaced positions of a line element of unit length in a uniform-strain field. In Fig. 4.2 the line is initially parallel to the *x*-axis (Lagrangian description) and in Fig. 4.3 the displaced line is parallel to the *x*-axis (Eulerian description).

If the components of displacement in the *x*-, *y*-, and *z*-directions are called *u*, *v*, and *w*, respectively, and if a uniform-strain field is defined as a field in which all displacements are changing at a uniform rate in all directions ($\partial u/\partial x$, $\partial u/\partial y$, $\partial u/\partial z$, $\partial v/\partial x$, $\partial v/\partial y$, $\partial v/\partial z$, $\partial w/\partial x$, $\partial w/\partial y$, and $\partial w/\partial z$ are all constant), then the values of representative derivatives are as shown in the two figures, and the general strain–displacement relations can be obtained directly from geometric considerations.

In the Lagrangian description (Fig. 4.2), if the initial length lies in the *x*-direction and if it is assumed to be unity, the final length is

$$l_f = \sqrt{\left(1 + \frac{\partial u}{\partial x}\right)^2 + \left(\frac{\partial v}{\partial x}\right)^2 + \left(\frac{\partial w}{\partial x}\right)^2} \tag{4.3}$$

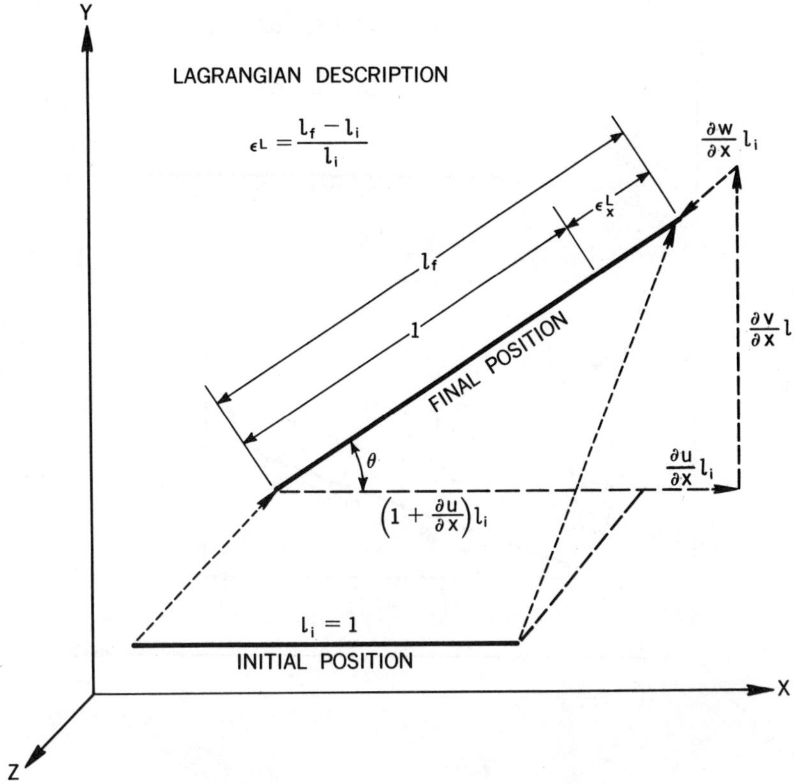

Figure 4.2. A line initially of unit length and parallel to the x-axis is shown subjected to a uniform strain field. The Cartesian components of the final length are given in terms of the partial derivatives of the displacements.

and from above and since $l_i = 1$, the strain component in the x-direction from Eq. (4.1) is

$$\epsilon_x^L = \sqrt{1 + 2\frac{\partial u}{\partial x} + \left(\frac{\partial u}{\partial x}\right)^2 + \left(\frac{\partial v}{\partial x}\right)^2 + \left(\frac{\partial w}{\partial x}\right)^2} - 1 \tag{4.4}$$

In the Eulerian description, if the final length lies in the x-direction and is unity, the initial length is computed as

$$l_i = \sqrt{\left(1 - \frac{\partial u}{\partial x}\right)^2 + \left(\frac{\partial v}{\partial x}\right)^2 + \left(\frac{\partial w}{\partial x}\right)^2} \tag{4.5}$$

and the strain component along a line lying finally in the x-direction, from Eq. (4.1), is

$$\epsilon_x^E = 1 - \sqrt{1 - 2\frac{\partial u}{\partial x} + \left(\frac{\partial u}{\partial x}\right)^2 + \left(\frac{\partial v}{\partial x}\right)^2 + \left(\frac{\partial w}{\partial x}\right)^2} \tag{4.6}$$

Similar expressions for the strain components along the other two orthogonal

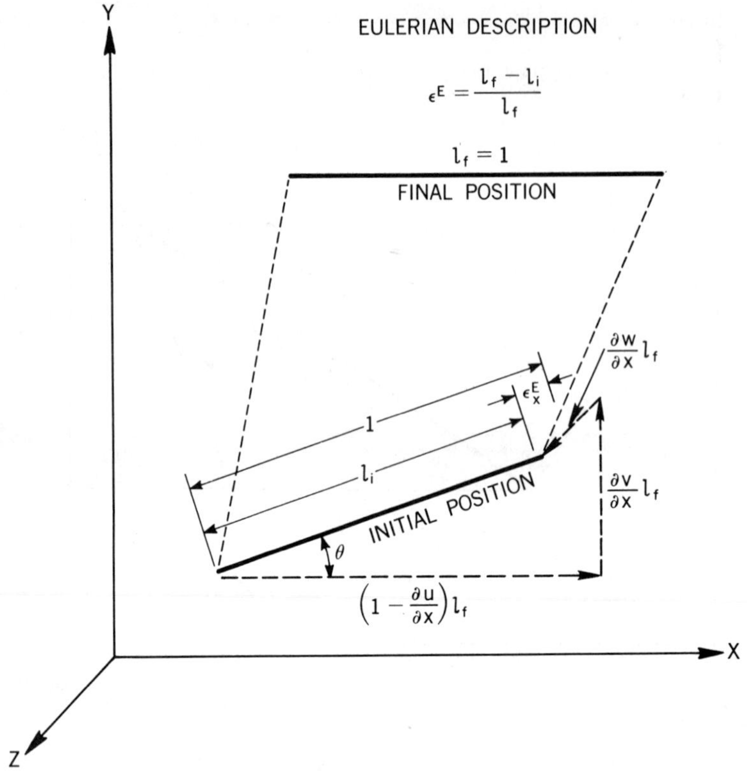

Figure 4.3. A line is shown subjected to a uniform strain such that it has a final length of unity and its direction is parallel to the x-axis. The Cartesian components of the initial length are given in terms of the partial derivatives of the displacements.

directions can be obtained in each of the systems. The following are the two complete sets of equations:

$$\begin{aligned}
\epsilon_x^L &= \sqrt{1 + 2\frac{\partial u}{\partial x} + \left(\frac{\partial u}{\partial x}\right)^2 + \left(\frac{\partial v}{\partial x}\right)^2 + \left(\frac{\partial w}{\partial x}\right)^2} - 1 \\
\epsilon_y^L &= \sqrt{1 + 2\frac{\partial v}{\partial y} + \left(\frac{\partial u}{\partial y}\right)^2 + \left(\frac{\partial v}{\partial y}\right)^2 + \left(\frac{\partial w}{\partial y}\right)^2} - 1 \\
\epsilon_z^L &= \sqrt{1 + 2\frac{\partial w}{\partial z} + \left(\frac{\partial u}{\partial z}\right)^2 + \left(\frac{\partial v}{\partial z}\right)^2 + \left(\frac{\partial w}{\partial z}\right)^2} - 1
\end{aligned} \tag{4.7}$$

$$\begin{aligned}
\epsilon_x^E &= 1 - \sqrt{1 - 2\frac{\partial u}{\partial x} + \left(\frac{\partial u}{\partial x}\right)^2 + \left(\frac{\partial v}{\partial x}\right)^2 + \left(\frac{\partial w}{\partial x}\right)^2} \\
\epsilon_y^E &= 1 - \sqrt{1 - 2\frac{\partial v}{\partial y} + \left(\frac{\partial u}{\partial y}\right)^2 + \left(\frac{\partial v}{\partial y}\right)^2 + \left(\frac{\partial w}{\partial y}\right)^2} \\
\epsilon_z^E &= 1 - \sqrt{1 - 2\frac{\partial w}{\partial z} + \left(\frac{\partial u}{\partial z}\right)^2 + \left(\frac{\partial v}{\partial z}\right)^2 + \left(\frac{\partial w}{\partial z}\right)^2}
\end{aligned} \tag{4.8}$$

Note that the derivatives have a somewhat different meaning in each of the sets.

These strain components along a specified direction are sometimes called *direct strains* and more commonly, although less properly, "normal" strains.

4.4.2. Shear Strain

The relations for shear strains can be obtained considering two line elements of unit length, each of them parallel to a coordinate direction and forming a right angle (either before or after loading, similarly to what was done above for the normal strain). Figures 4.4 and 4.5 show the line elements for the Lagrangian and Eulerian cases, respectively. From Eq. (4.1),

$$\gamma^L = 90° - \angle ABC \tag{4.9}$$

$$\gamma^E = \angle DEF - 90° \tag{4.10}$$

or

$$\sin \gamma^L = \sin (90° - \angle ABC) = \cos \angle ABC$$

$$\sin \gamma^E = \sin (\angle DEF - 90°) = -\cos \angle DEF$$

From geometry, the cosine of an angle is the sum of the pair-wise products of the cosines of each of the lines of the angle with respect to any coordinate system. From Fig. 4.4, the cosines of the two sides of the angle $\angle ABC$ with respect to the x-y-z coordinate system are

$$\cos (l_f, x) = \frac{1 + \dfrac{\partial u}{\partial x}}{1 + \epsilon_x^L}$$

$$\cos (l_f, y) = \frac{\dfrac{\partial v}{\partial x}}{1 + \epsilon_x^L}$$

$$\cos (l_f, z) = \frac{\dfrac{\partial w}{\partial x}}{1 + \epsilon_x^L}$$

$$\cos (m_f, x) = \frac{\dfrac{\partial u}{\partial y}}{1 + \epsilon_y^L}$$

$$\cos (m_f, y) = \frac{1 + \dfrac{\partial v}{\partial y}}{1 + \epsilon_y^L}$$

$$\cos (m_f, z) = \frac{\dfrac{\partial w}{\partial y}}{1 + \epsilon_y^L} \quad \text{etc.}$$

(Note that the Lagrangian definition for direct strain has been used to save space in writing the denominators.)

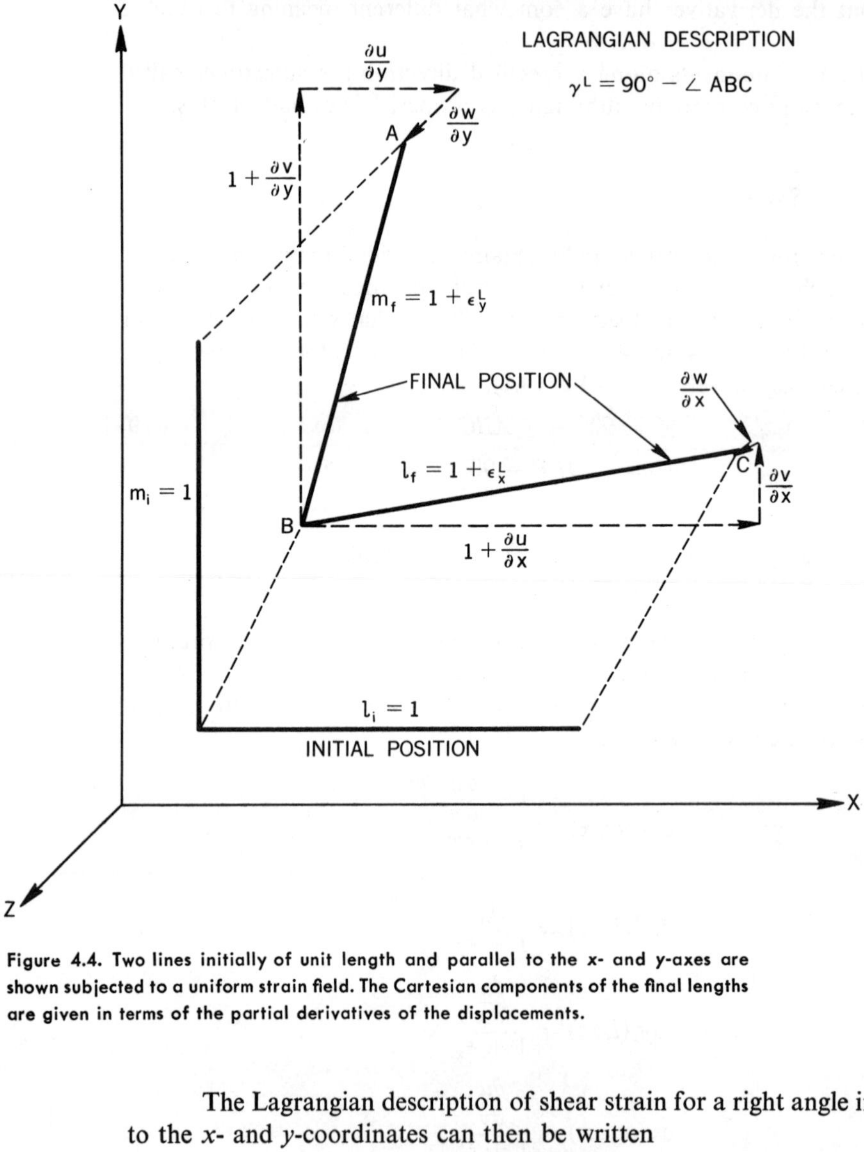

Figure 4.4. Two lines initially of unit length and parallel to the x- and y-axes are shown subjected to a uniform strain field. The Cartesian components of the final lengths are given in terms of the partial derivatives of the displacements.

The Lagrangian description of shear strain for a right angle initially parallel to the x- and y-coordinates can then be written

$$\gamma_{xy}^L = \arcsin\left\{\left[\left(1 + \frac{\partial u}{\partial x}\right)\left(\frac{\partial u}{\partial y}\right)\Big/(1 + \epsilon_x^L)(1 + \epsilon_y^L)\right]\right.$$
$$+ \left[\left(\frac{\partial v}{\partial x}\right)\left(1 + \frac{\partial v}{\partial y}\right)\Big/(1 + \epsilon_x^L)(1 + \epsilon_y^L)\right]$$
$$\left. + \left[\left(\frac{\partial w}{\partial x}\right)\left(\frac{\partial w}{\partial y}\right)\Big/(1 + \epsilon_x^L)(1 + \epsilon_y^L)\right]\right\}$$

or more concisely,

$$\gamma_{xy}^L = \arcsin \frac{\dfrac{\partial u}{\partial y} + \dfrac{\partial v}{\partial x} + \left(\dfrac{\partial u}{\partial x}\right)\left(\dfrac{\partial u}{\partial y}\right) + \left(\dfrac{\partial v}{\partial x}\right)\left(\dfrac{\partial v}{\partial y}\right) + \left(\dfrac{\partial w}{\partial x}\right)\left(\dfrac{\partial w}{\partial y}\right)}{(1+\epsilon_x^L)(1+\epsilon_y^L)}$$

$$\gamma_{yz}^L = \arcsin \frac{\dfrac{\partial v}{\partial z} + \dfrac{\partial w}{\partial y} + \left(\dfrac{\partial u}{\partial y}\right)\left(\dfrac{\partial u}{\partial z}\right) + \left(\dfrac{\partial v}{\partial y}\right)\left(\dfrac{\partial v}{\partial z}\right) + \left(\dfrac{\partial w}{\partial y}\right)\left(\dfrac{\partial w}{\partial z}\right)}{(1+\epsilon_y^L)(1+\epsilon_z^L)} \qquad (4.11)$$

$$\gamma_{zx}^L = \arcsin \frac{\dfrac{\partial w}{\partial x} + \dfrac{\partial u}{\partial z} + \left(\dfrac{\partial u}{\partial x}\right)\left(\dfrac{\partial u}{\partial z}\right) + \left(\dfrac{\partial v}{\partial x}\right)\left(\dfrac{\partial v}{\partial z}\right) + \left(\dfrac{\partial w}{\partial x}\right)\left(\dfrac{\partial w}{\partial z}\right)}{(1+\epsilon_z^L)(1+\epsilon_x^L)}$$

Figure 4.5. Two lines are shown subjected to a uniform strain field such that their final lengths are equal to unity and their final directions are parallel to the axes. The Cartesian components of the initial lengths are given in terms of the partial derivatives of the displacements.

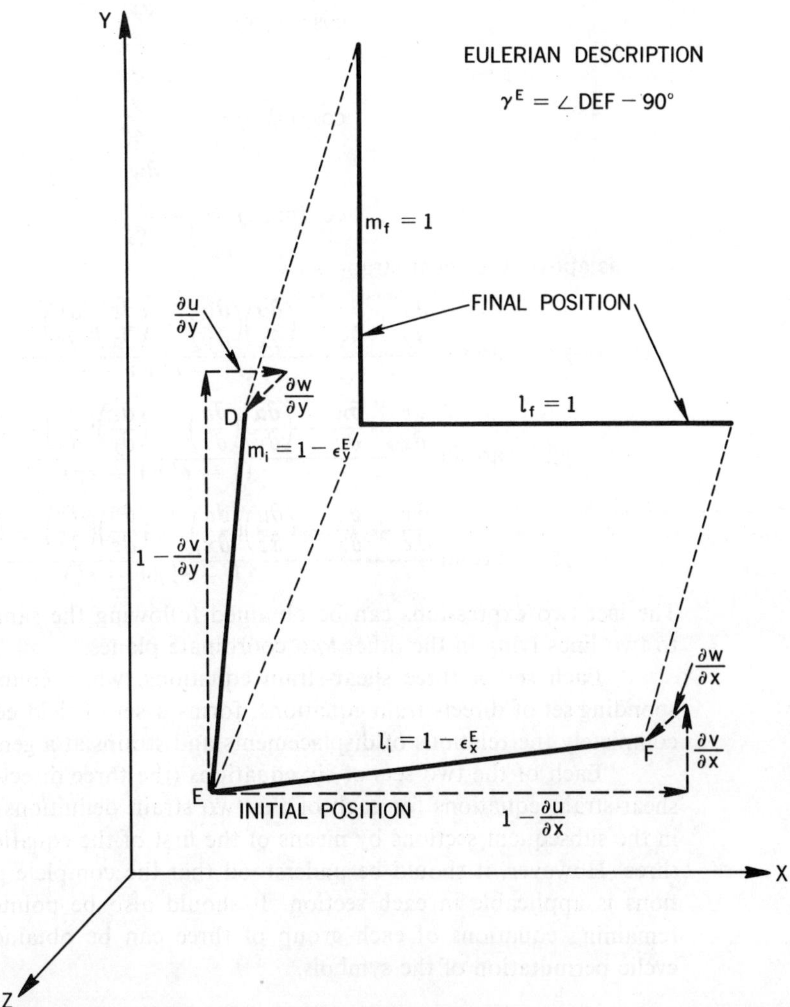

The last two expressions can be obtained following the same approach, applied to two lines lying in the other two coordinate planes.

From Fig. 4.5 (Eulerian description), the cosines of the two sides of $\angle DEF$ with respect to the x-y-z coordinate system are

$$\cos(l_i, x) = \frac{1 - \dfrac{\partial u}{\partial x}}{1 - \epsilon_x^E}$$

$$\cos(l_i, y) = \frac{-\dfrac{\partial v}{\partial x}}{1 - \epsilon_x^E}$$

$$\cos(l_i, z) = \frac{-\dfrac{\partial w}{\partial x}}{1 - \epsilon_x^E}$$

$$\cos(m_i, x) = \frac{-\dfrac{\partial u}{\partial y}}{1 - \epsilon_y^E}$$

$$\cos(m_i, y) = \frac{1 - \dfrac{\partial v}{\partial y}}{1 - \epsilon_y^E}$$

$$\cos(m_i, z) = \frac{-\dfrac{\partial w}{\partial y}}{1 - \epsilon_y^E}$$

and, as above, the shear strain is

$$\gamma_{xy}^E = \arcsin \frac{\dfrac{\partial u}{\partial y} + \dfrac{\partial v}{\partial x} - \left(\dfrac{\partial u}{\partial x}\right)\left(\dfrac{\partial u}{\partial y}\right) - \left(\dfrac{\partial v}{\partial x}\right)\left(\dfrac{\partial v}{\partial y}\right) - \left(\dfrac{\partial w}{\partial x}\right)\left(\dfrac{\partial w}{\partial y}\right)}{(1 - \epsilon_x^E)(1 - \epsilon_y^E)}$$

$$\gamma_{yz}^E = \arcsin \frac{\dfrac{\partial v}{\partial z} + \dfrac{\partial w}{\partial y} - \left(\dfrac{\partial u}{\partial y}\right)\left(\dfrac{\partial u}{\partial z}\right) - \left(\dfrac{\partial v}{\partial y}\right)\left(\dfrac{\partial v}{\partial z}\right) - \left(\dfrac{\partial w}{\partial y}\right)\left(\dfrac{\partial w}{\partial z}\right)}{(1 - \epsilon_y^E)(1 - \epsilon_z^E)} \tag{4.12}$$

$$\gamma_{xz}^E = \arcsin \frac{\dfrac{\partial w}{\partial x} + \dfrac{\partial u}{\partial z} - \left(\dfrac{\partial u}{\partial z}\right)\left(\dfrac{\partial u}{\partial x}\right) - \left(\dfrac{\partial v}{\partial z}\right)\left(\dfrac{\partial v}{\partial x}\right) - \left(\dfrac{\partial w}{\partial z}\right)\left(\dfrac{\partial w}{\partial x}\right)}{(1 - \epsilon_z^E)(1 - \epsilon_x^E)}$$

The last two expressions can be obtained following the same approach, applied to two lines lying in the other two coordinate planes.

Each set of three shear-strain equations, when coupled with the corresponding set of direct-strain equations, forms a set of field equations that defines completely the relations of displacements and strains at a generic point in a body.

Each of the two sets of six equations (the three direct-strain and the three shear-strain equations for each of the two strain definitions) will be represented in the subsequent sections by means of the first of the equations of each group of three. However, it should be understood that the complete group of three equations is applicable in each section. It should also be pointed out that the two remaining equations of each group of three can be obtained from the first by cyclic permutation of the symbols.

4.5. The Strain–Displacement Relations for Small Rotations

In what follows, the general expressions developed above (Eqs. 4.7, 4.8, 4.11, and 4.12) will be simplified so that they will be in a form more readily applicable to particular cases. The first simplification will deal with the case of small rotations.

4.5.1. Direct Strain

With reference to Eqs. (4.7) and (4.8), it is obvious that an important simplification could be obtained if the cross-derivatives were sufficiently small so that their square will be negligible with respect to the first direct derivative. With this assumption, the direct-strain equations take the following form:

$$\epsilon_x^L = \frac{\partial u}{\partial x} \tag{4.15}$$

$$\epsilon_x^E = \frac{\partial u}{\partial x} \tag{4.16}$$

with the derivatives having the variance in meaning noted in the previous section.

It will be shown below that these equations give practically the same results as the general equation when rotations are small. It will also be shown for which combinations of strain and rotation these equations give sufficiently accurate results.

4.5.2. Shear Strain

In the equations for shear strains (Eqs. 4.11 and 4.12) each term contains a cross-derivative, so that the expression does not simplify as readily. If the reduced expressions for normal strains (Eqs. 4.15 and 4.16) are substituted in Eqs. (4.11) and (4.12), they become

$$\gamma_{xy}^L = \arcsin \frac{\dfrac{\partial u}{\partial y}\left(1 + \dfrac{\partial u}{\partial x}\right) + \dfrac{\partial v}{\partial x}\left(1 + \dfrac{\partial v}{\partial y}\right) + \left(\dfrac{\partial w}{\partial x}\right)\left(\dfrac{\partial w}{\partial y}\right)}{\left(1 + \dfrac{\partial u}{\partial x}\right)\left(1 + \dfrac{\partial v}{\partial y}\right)}$$

and

$$\gamma_{xy}^E = \arcsin \frac{\dfrac{\partial u}{\partial y}\left(1 - \dfrac{\partial u}{\partial x}\right) + \dfrac{\partial v}{\partial x}\left(1 - \dfrac{\partial v}{\partial y}\right) - \left(\dfrac{\partial w}{\partial x}\right)\left(\dfrac{\partial w}{\partial y}\right)}{\left(1 - \dfrac{\partial u}{\partial x}\right)\left(1 - \dfrac{\partial v}{\partial y}\right)}$$

and some simplification is indicated.

The term $[(\partial w/\partial x)(\partial w/\partial y)]$ and similar terms in the other equations can be dropped, and the expressions for shear strain become

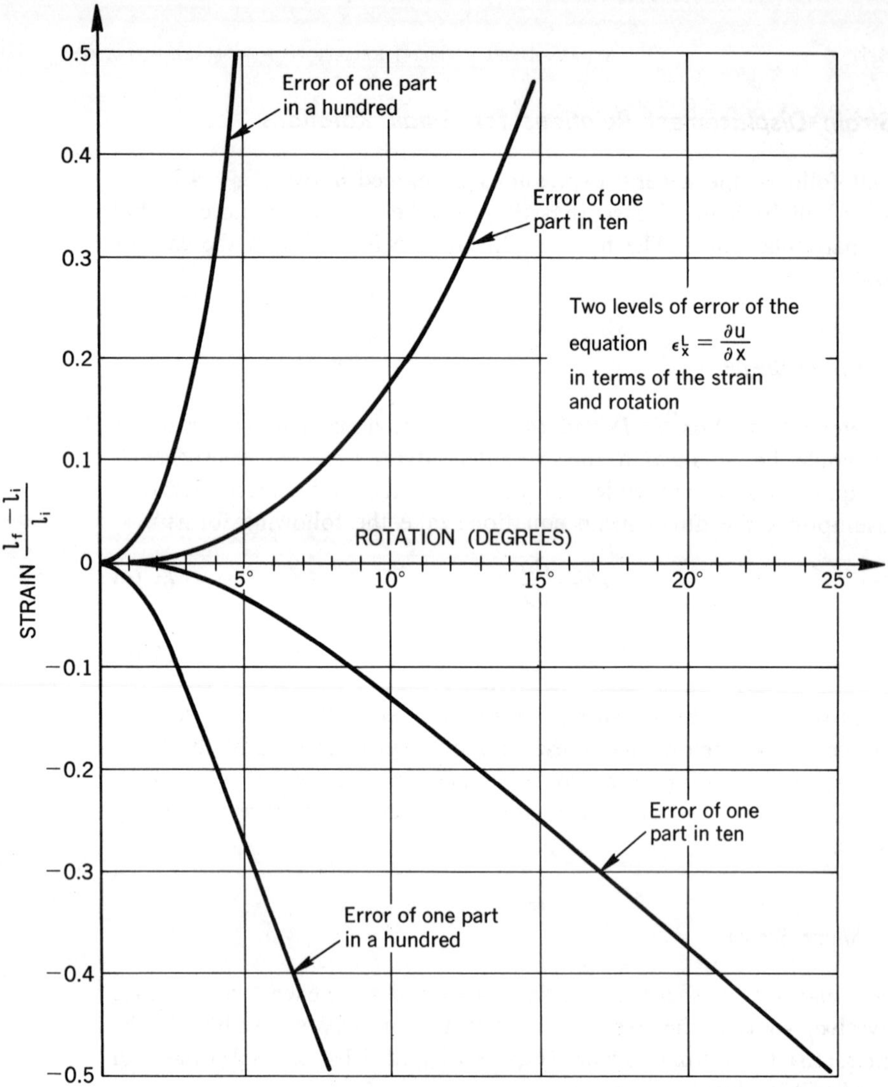

Figure 4.6. Error made when equation $\epsilon_x^L = \partial u/\partial x$ is used.

$$\gamma_{xy}^L = \arcsin\left(\frac{\dfrac{\partial u}{\partial y}}{1 + \dfrac{\partial v}{\partial y}} + \frac{\dfrac{\partial v}{\partial x}}{1 + \dfrac{\partial u}{\partial x}}\right) \tag{4.17}$$

and

$$\gamma_{xy}^E = \arcsin\left(\frac{\dfrac{\partial u}{\partial y}}{1 - \dfrac{\partial v}{\partial y}} + \frac{\dfrac{\partial v}{\partial x}}{1 - \dfrac{\partial u}{\partial x}}\right) \tag{4.18}$$

4.5.3. *Errors Introduced When the Simplified Expressions Are Used*

The errors introduced when the simplified equations are used can be determined by computing the differences between the results obtained using the general equations and the simplified equations. For better understanding of the strain phenomenon and the field of applications of the simplified equations, it is convenient to represent graphically the error as a function of the rotation and strain. This is done in what follows for the direct strains in the Lagrangian description.

From Fig. 4.2, it is obvious that if a line originally parallel to the x-direction is strained an amount ϵ_x and rotated through an angle θ from the x-direction, then

$$\cos\theta = \frac{1 + \dfrac{\partial u}{\partial x}}{1 + \epsilon_x^L}$$

$$\frac{\partial u}{\partial x} = (1 + \epsilon_x^L)\cos\theta - 1$$

Thus, for various strains ϵ_x^L and angles of rotation θ, we can compute $\partial u/\partial x$, and note the difference between $\partial u/\partial x$ and ϵ_x^L. This difference is the error made when the first derivative is taken for the strain.

$$\text{Error} = \frac{\partial u}{\partial x} - \epsilon_x^L = (1 + \epsilon_x^L)(\cos\theta - 1)$$

The true strain and angle of rotation which will give errors of 10 and 1 per cent with the simplified Lagrangian definition of strain are shown in Fig. 4.6. The errors are small for large strains and small rotations.

4.6. The Strain–Displacement Relations for Small Strains

The general equations (Eqs. 4.7, 4.8, 4.11, and 4.12) can also be simplified when the strains are small. It will be seen that these equations for small strains are still quite general and apply irrespective of the amount of rotation.

4.6.1. *Direct Strain*

The general strain–displacement relations (Eqs. 4.7 and 4.8) can be written

$$(\epsilon_x^L + 1)^2 = 1 + 2\frac{\partial u}{\partial x} + \left(\frac{\partial u}{\partial x}\right)^2 + \left(\frac{\partial v}{\partial x}\right)^2 + \left(\frac{\partial w}{\partial x}\right)^2$$

and

$$(\epsilon_x^E - 1)^2 = 1 - 2\frac{\partial u}{\partial x} + \left(\frac{\partial u}{\partial x}\right)^2 + \left(\frac{\partial v}{\partial x}\right)^2 + \left(\frac{\partial w}{\partial x}\right)^2$$

If the strain is small enough so that the second power of the strain can be neglected with respect to the strain itself, then the equations can be written as follows:

$$\epsilon_x^L = \frac{\partial u}{\partial x} + \frac{1}{2}\left[\left(\frac{\partial u}{\partial x}\right)^2 + \left(\frac{\partial v}{\partial x}\right)^2 + \left(\frac{\partial w}{\partial x}\right)^2\right] \tag{4.19}$$

$$\epsilon_x^E = \frac{\partial u}{\partial x} - \frac{1}{2}\left[\left(\frac{\partial u}{\partial x}\right)^2 + \left(\frac{\partial v}{\partial x}\right)^2 + \left(\frac{\partial w}{\partial x}\right)^2\right] \tag{4.20}$$

4.6.2. Shear Strain

This definition of small strain can be written as

$$\epsilon_x^2 \ll \epsilon_x$$

which requires that $\epsilon_x \ll 1$. This indicates that ϵ_x can be neglected with respect to unity and allows the approximation

$$1 + \epsilon_x \simeq 1$$

The shear-strain expressions (Eqs. 4.11 and 4.12) can then be written

$$\gamma_{xy}^L = \arcsin\left[\frac{\partial u}{\partial y} + \frac{\partial v}{\partial x} + \left(\frac{\partial u}{\partial x}\right)\left(\frac{\partial u}{\partial y}\right) + \left(\frac{\partial v}{\partial x}\right)\left(\frac{\partial v}{\partial y}\right) + \left(\frac{\partial w}{\partial x}\right)\left(\frac{\partial w}{\partial y}\right)\right]$$

and

$$\gamma_{xy}^E = \arcsin\left[\frac{\partial u}{\partial y} + \frac{\partial v}{\partial x} - \left(\frac{\partial u}{\partial x}\right)\left(\frac{\partial u}{\partial y}\right) - \left(\frac{\partial v}{\partial x}\right)\left(\frac{\partial v}{\partial y}\right) - \left(\frac{\partial w}{\partial x}\right)\left(\frac{\partial w}{\partial y}\right)\right]$$

If from the definition of small strains it is further assumed, as for the direct strains, that γ_{xy}^2 can be neglected with respect to γ_{xy}, then the power-series expansion

$$\sin \gamma = \gamma - \frac{\gamma^3}{3!} + \frac{\gamma^5}{5!} - \frac{\gamma^7}{7!} + \cdots$$

can be reduced to $\sin \gamma \simeq \gamma$ and the equation giving the shear strains can be written

$$\gamma_{xy}^L = \frac{\partial u}{\partial y} + \frac{\partial v}{\partial x} + \left(\frac{\partial u}{\partial x}\right)\left(\frac{\partial u}{\partial y}\right) + \left(\frac{\partial v}{\partial x}\right)\left(\frac{\partial v}{\partial y}\right) + \left(\frac{\partial w}{\partial x}\right)\left(\frac{\partial w}{\partial y}\right) \tag{4.21}$$

and

$$\gamma_{xy}^E = \frac{\partial u}{\partial y} + \frac{\partial v}{\partial x} - \left(\frac{\partial u}{\partial x}\right)\left(\frac{\partial u}{\partial y}\right) - \left(\frac{\partial v}{\partial x}\right)\left(\frac{\partial v}{\partial y}\right) - \left(\frac{\partial w}{\partial x}\right)\left(\frac{\partial w}{\partial y}\right) \tag{4.22}$$

4.6.3. Error Introduced When the Simplified Expressions Are Used

As for the case of the formulas for small rotations, the error here can be determined by making differences between the results obtained using the simplified

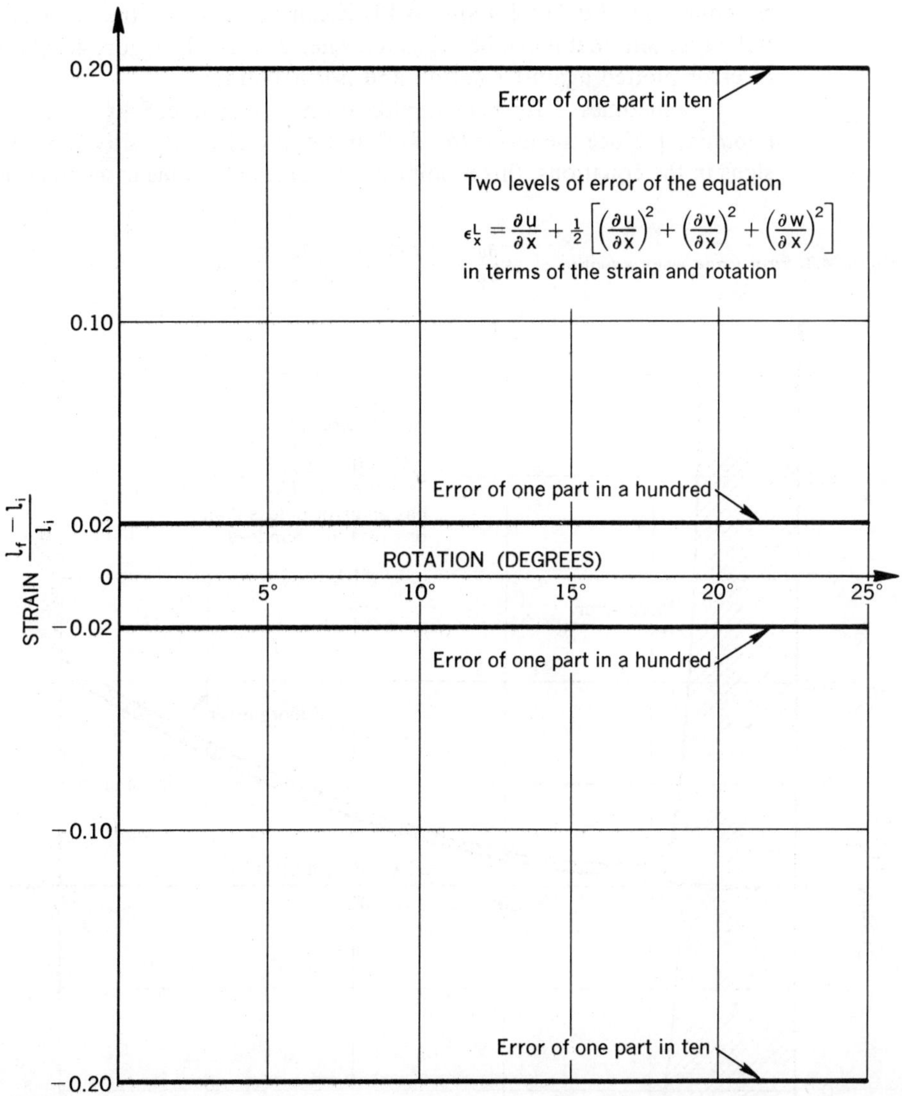

Figure 4.7. Error made when equation $\epsilon_x^L = \frac{\partial u}{\partial x} + \frac{1}{2}\left[\left(\frac{\partial u}{\partial x}\right)^2 + \left(\frac{\partial v}{\partial x}\right)^2 + \left(\frac{\partial w}{\partial x}\right)^2\right]$ is used.

equations for small strains and the general equations. The procedure of representing graphically the error will be illustrated for the case of the normal strain in the Lagrangian description. For the purpose of error analysis, Eq. (4.7) can be written

$$\epsilon_x^L = \frac{\partial u}{\partial x} + \frac{1}{2}\left[\left(\frac{\partial u}{\partial x}\right)^2 + \left(\frac{\partial v}{\partial x}\right)^2 + \left(\frac{\partial w}{\partial x}\right)^2\right] - \frac{1}{2}(\epsilon_x^L)^2 \qquad (4.23)$$

By comparing Eqs. (4.23) and (4.19) it can be seen that the error is of the parabolic type and that it can be neglected when ϵ_x is small. Figure 4.7 shows two levels of error plotted against rotation and normal strain.

Sometimes a further simplification is introduced by neglecting the term $[\frac{1}{2}(\partial u/\partial x)^2]$. Since the terms $(\partial u/\partial x)^2$ and ϵ_x^2 are always positive and have opposite signs in the equations, this modification can lead in some cases to a more accurate

Figure 4.8. Error made when equation $\epsilon_x^L = \frac{\partial u}{\partial x} + \frac{1}{2}\left[\left(\frac{\partial v}{\partial x}\right)^2 + \left(\frac{\partial w}{\partial x}\right)^2\right]$ is used.

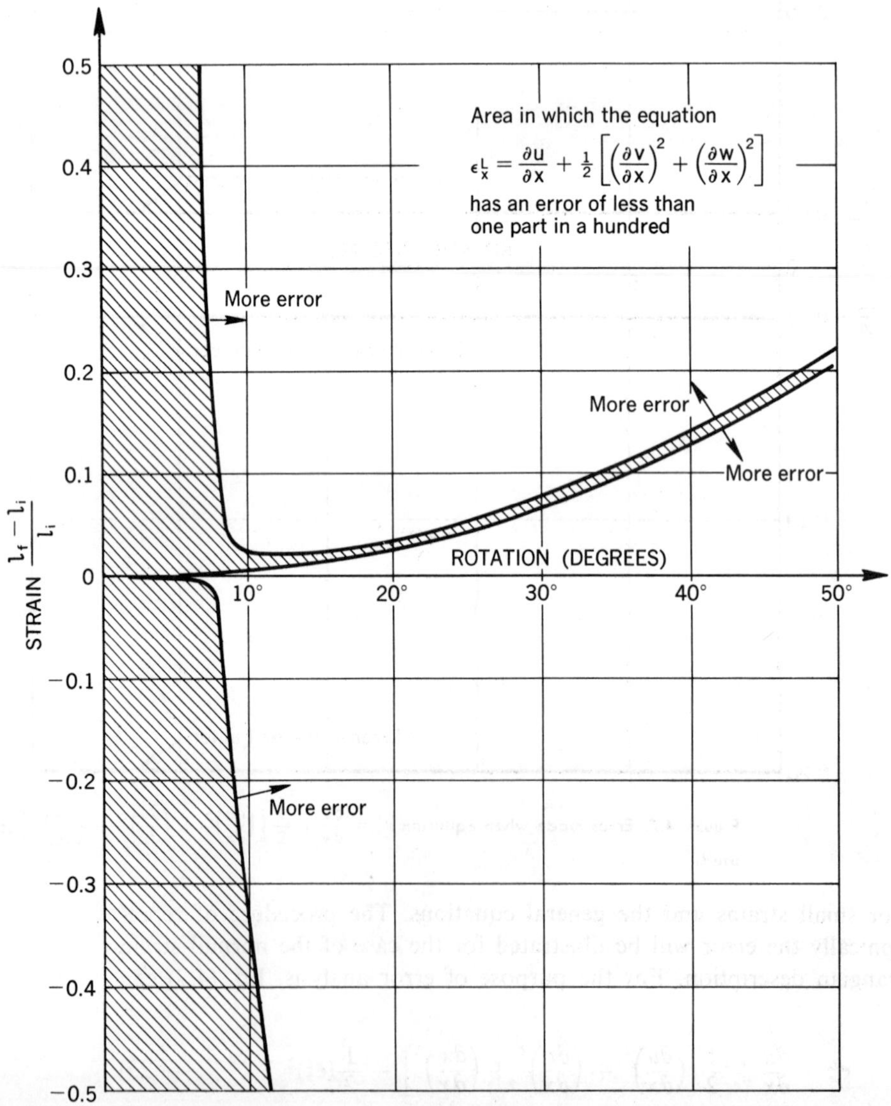

value of strain than the one obtained from Eq. (4.19), above. The accuracy of the approximation is dependent on both the angle of rotation and the strain. The modification has the form

$$\epsilon_x^L = \frac{\partial u}{\partial x} + \frac{1}{2}\left[\left(\frac{\partial v}{\partial x}\right)^2 + \left(\frac{\partial w}{\partial x}\right)^2\right] \tag{4.24}$$

The values of strain and rotation which will give an error smaller than 1 part in 100 are shown in Fig. 4.8.

4.7. The Strain–Displacement Relations for Small Strains and Small Rotations

If both the restrictions introduced in the previous two sections are made simultaneously, then the classic form of the strain–displacement relations used in theory of elasticity is obtained.

4.7.1. Direct Strain

Equations (4.15) and (4.16) give the strains in their simplest form. However, they were developed using the restriction that cross-derivatives be sufficiently small for their squares to be negligible with respect to the direct derivatives. In the case of ϵ_x, this requires that

$$\left(\frac{\partial v}{\partial x}\right)^2 \ll \frac{\partial u}{\partial x}$$

$$\left(\frac{\partial w}{\partial x}\right)^2 \ll \frac{\partial u}{\partial x}$$

If this is the case, $\partial u/\partial x$ was shown to be equal to ϵ_x. The restriction then can be written

$$\left(\frac{\partial v}{\partial x}\right)^2 \ll \epsilon_x$$

$$\left(\frac{\partial w}{\partial x}\right)^2 \ll \epsilon_x$$

In Section 4.6.2 it was required that

$$\epsilon_x^2 \ll \epsilon_x$$

Thus, it can be stated that the strain must be small enough so that its square can be neglected with respect to the strain itself, and that the rotations must be such that the powers of the cross-derivatives can be neglected with respect to *that* strain. The direct strain is then defined as

$$\epsilon_x = \frac{\partial u}{\partial x} \tag{4.25}$$

4.7.2. Shear Strain

Here, again, the small-rotation equations (Eqs. 4.17 and 4.18) are considered:

$$\gamma_{xy}^{L} = \arcsin\left(\frac{\dfrac{\partial u}{\partial y}}{1 + \dfrac{\partial v}{\partial y}} + \frac{\dfrac{\partial v}{\partial x}}{1 + \dfrac{\partial u}{\partial x}}\right)$$

Figure 4.9. Error made when equation $\epsilon_x = \partial u/\partial x$ is used (range of small strains and rotations).

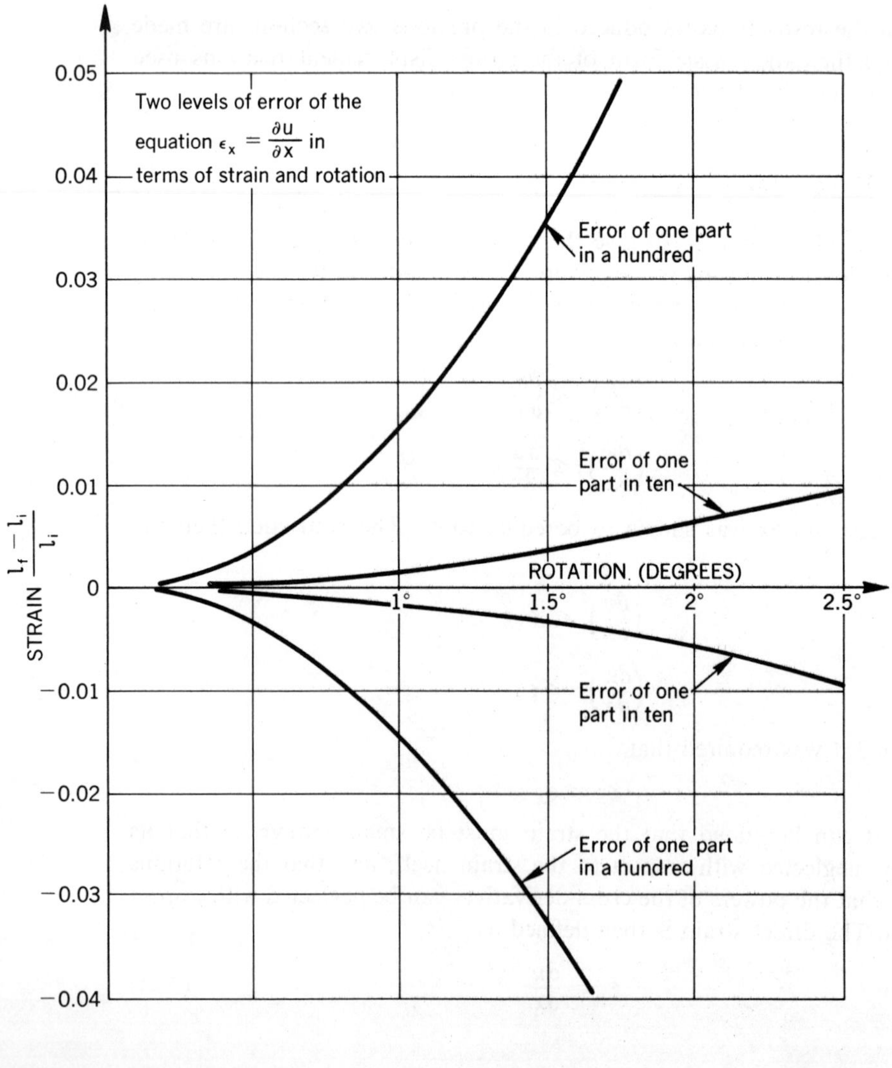

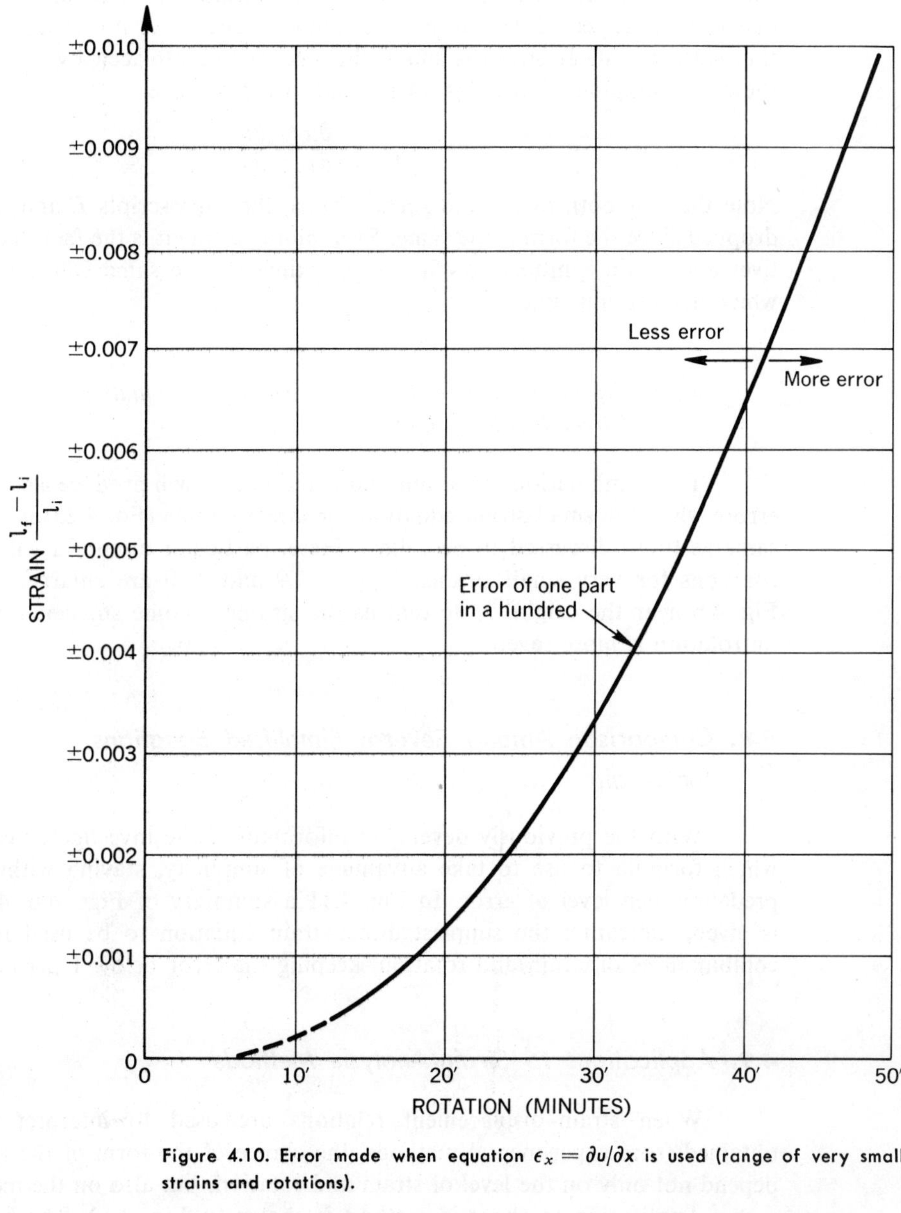

Figure 4.10. Error made when equation $\epsilon_x = \partial u/\partial x$ is used (range of very small strains and rotations).

$$\gamma_{xy}^E = \arcsin\left(\frac{\dfrac{\partial u}{\partial y}}{1 - \dfrac{\partial v}{\partial y}} + \frac{\dfrac{\partial v}{\partial x}}{1 - \dfrac{\partial u}{\partial x}}\right)$$

The additional requirement that the direct strain be small allows $\partial u/\partial x$ and $\partial v/\partial y$ to be neglected with respect to unity. In Sec. 4.6.2 it was also pointed out that since the shear strain is small, the sine can be replaced by the angle. With these two simplifications, Eqs. (4.17) and (4.18) become

$$\gamma_{xy} = \frac{\partial u}{\partial y} + \frac{\partial v}{\partial x} \tag{4.26}$$

Note that for both direct and shear strains, the superscripts E and L have been dropped since the form is the same. Even more pertinent is the fact that the derivatives and strain symbols now have approximately the same values regardless of which definition is used.

4.7.3. Errors Introduced When the Simplified Equation for Direct Strain Is Used

The combinations of strain and rotation that will produce a certain level of errors when the small-strain equation for direct strains (Eq. 4.25) is used are the same as those discussed in Sec. 4.5.3. Here, particular attention will be paid to rotations for very small strains. Figures 4.9 and 4.10 are enlarged portions of Fig. 4.6 near the origin. Note that as the strains become smaller, the limitation on rotation is more severe.

4.8. Comparison Among Several Simplified Equations for Strain

With the previously developed information, the investigator could decide which formula to use to take advantage of simplicity, staying within a certain predetermined level of error. In Fig. 4.11 a summary of Figs. 4.6, 4.7, and 4.8 is given, indicating the simplest direct-strain equation to be used for different combinations of strain and rotation, keeping the error below 1 per cent.

4.9. Applications to Strain-Analysis Methods

When strain–displacement relations are used to interpret the results obtained from experimental methods, the choice of the form of the relation will depend not only on the level of strain and rotation, but also on the method used.

Application to the grid method is shown in Fig. 4.12. The figure shows a straight beam with an embedded grid subjected to a large curvature. For analysis, it is convenient to place the photographs of the unloaded and loaded beams on the stage of a microscope and to measure displacements with a calibrated screw that moves the stage. The direction of the undeformed grid is used as a reference, and measurements are made in both grid directions following the distorted shape

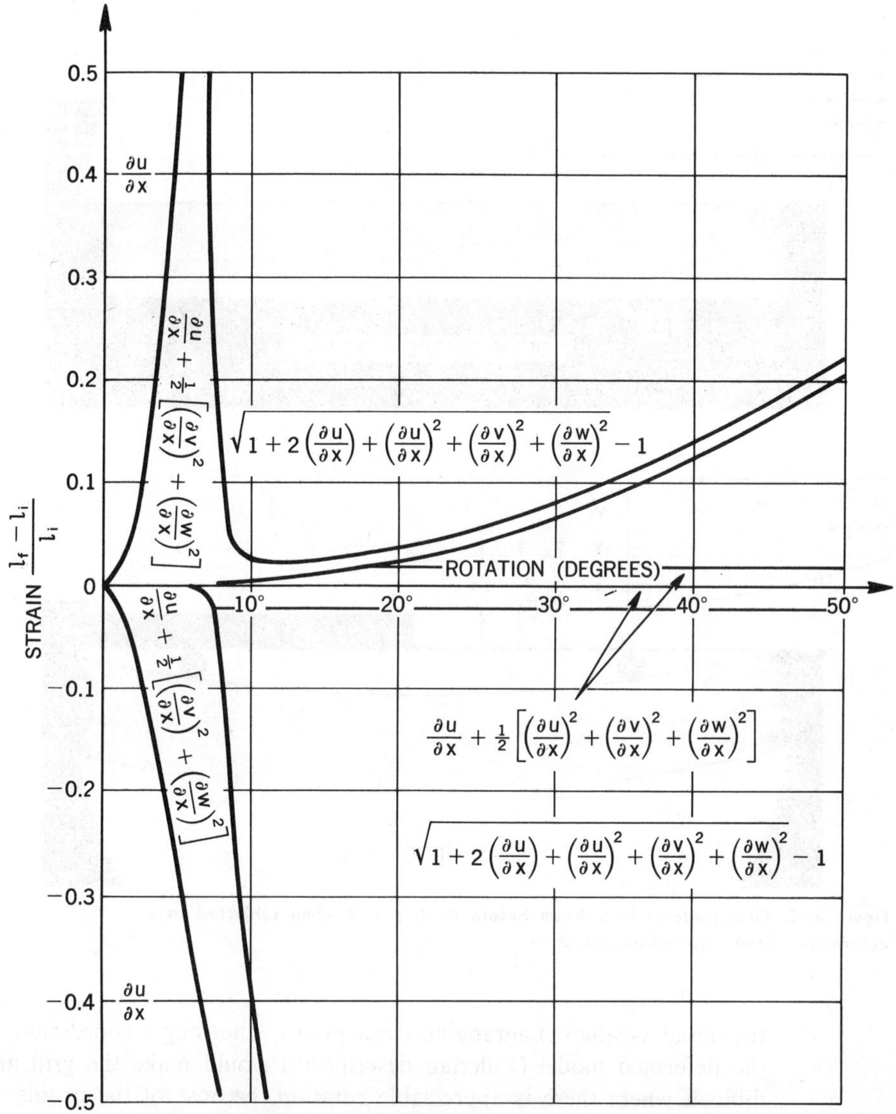

Figure 4.11. Simplest Lagrangian direct strain equation to obtain an accuracy of 1% or better for various strains and rotations.

of the grid lines which initially were straight. A sample of the measurements made along one line of the undeformed model is shown in the figure as x_1, x_2, x_3, and x_4. Since the initial y-values were zero, the y'-values directly give v-values.

From these data the derivatives could be obtained in either the Lagrangian or the Eulerian description; however, as shown in Fig. 4.1, the choice of a line initially in the x-direction suggests that the derivatives be taken with respect to

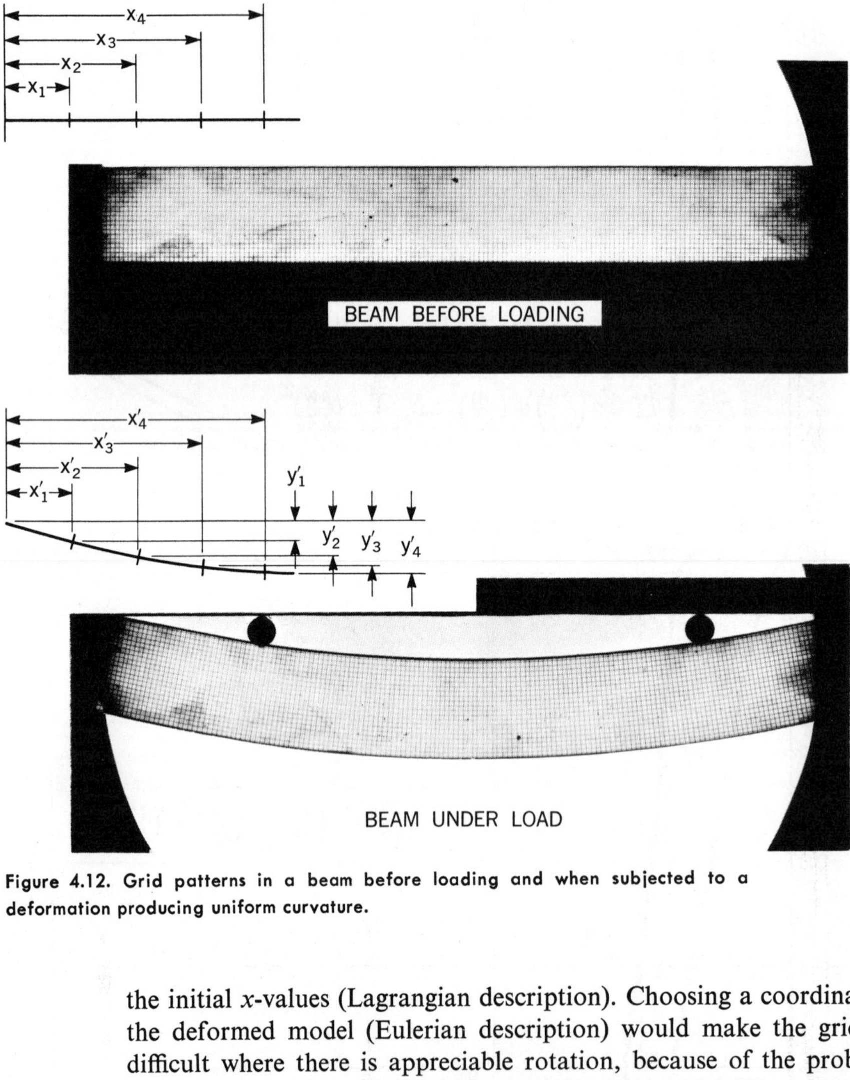

Figure 4.12. Grid patterns in a beam before loading and when subjected to a deformation producing uniform curvature.

the initial x-values (Lagrangian description). Choosing a coordinate system along the deformed model (Eulerian description) would make the grid analysis more difficult where there is appreciable rotation, because of the problem in locating the corresponding points on the undeformed model. Thus the grid method in areas of large rotation requires the use of Eqs. (4.7) and (4.11), or of Eqs. (4.19) and (4.21).

The use of mechanical and optical strain gages is much like the grid method in that they give the change in length along a base line. However, unlike the grids, they do not represent a field and, in general, do not allow the practical determination of derivatives. Equations (4.1) and (4.2) are used to determine the strains at individual points. The Lagrangian description is the most common.

The same reasoning applies to electrical-resistance strain gages which, in principle, give Lagrangian strain directly. A potentiometer measures the change in resistance from some base resistance, and the ratio of change in resistance to base resistance is proportional to Lagrangian strain. The question of how well this proportionality holds for large strains has not yet been completely clarified.

The choice of strain–displacement relations in a typical moire problem will be shown in what follows. Figure 4.13 shows moire patterns obtained from a 1000-lines/in. grating on a cantilever beam subjected to a fixed downward end displacement. The two patterns give the complete u- and v-fields along x- and y-coordinates which are parallel and perpendicular to the straight boundaries of the underformed beam. The convenient way to analyze these fields is the opposite of that used in the grid method. The u- and v-fields are fixed on the deformed model. Values of u and v and their derivatives are taken along lines in the deformed model which are parallel to the x- and y-directions and which, in general, were not straight in the undeformed model. Since this is a small-strain problem, Eqs. (4.20) and (4.22) should be used in the areas of large rotation. In areas of small rotation, Eqs. (4.25) and (4.26) can be used. The Lagrangian equations could be used efficiently along the boundaries which in the undeformed model were lines parallel to the x- or y-direction, since they automatically suit the requirements of the

Figure 4.13. Moire patterns of the u-field and v-field of a cantilever beam subjected to a displacement applied at the free end. The master grating has 1000 lines/in.

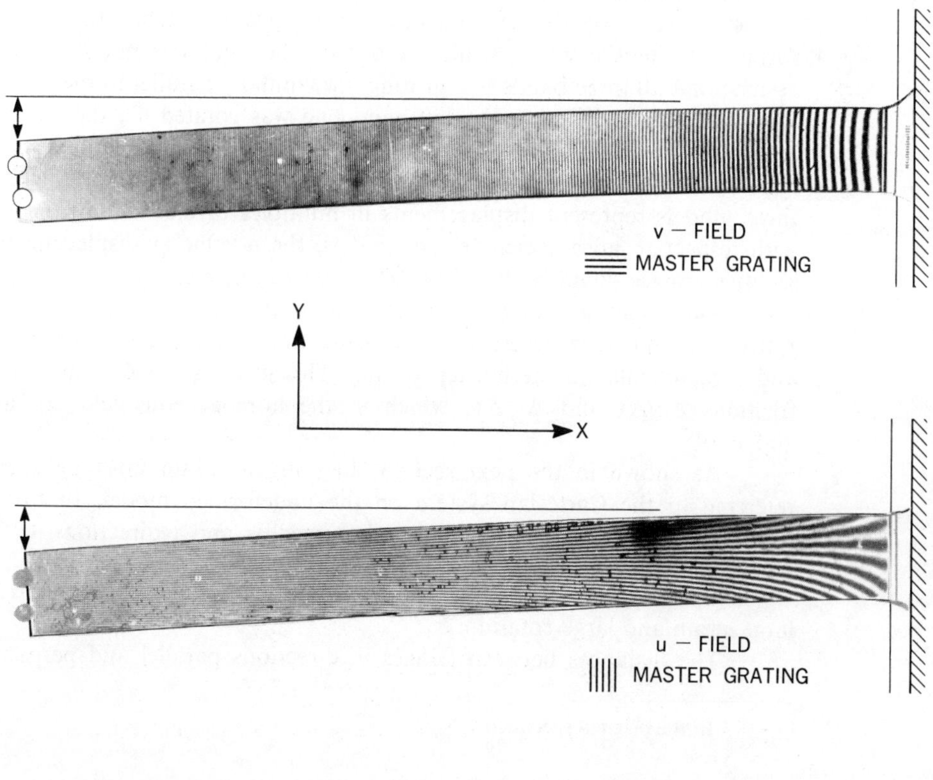

Lagrangian analysis. Besides taking the u- and v-values along the boundary, it would be necessary to measure the length along these curved lines and, to be completely accurate, adjust the length values with the corresponding displacement (u- and v-) values. Equations (4.19) and (4.21) could then be applied.

4.10. Numerical Examples of the Equations Applied to Moire Patterns

If it is assumed that the out-of-plane displacements can be neglected, the two components of the in-plane displacements can be treated as the u- and v-fields discussed in Sec. 4.4, with $w = 0$. Thus, using the various forms of the strain–displacement equations shown above, the complete strain field in a plane surface can be computed from two properly oriented moire patterns.

To illustrate the application of the various forms of the equation, once again recourse is made to the homogeneous-strain field. Figure 4.14 shows a long area of equispaced lines crossed by three smaller areas. Consider the long band as the master. Consider further that the x-direction is perpendicular to the master-grating lines. The long band and the band on the right have approximately $71\frac{1}{2}$ lines/in. and the short bands on the left and in the middle have approximately 62 lines/in.[1] Assume that all three short bands initially had $71\frac{1}{2}$ lines/in. and that the two short bands on the left and in the center were stretched to 62 lines/in. uniformly by a tensile strain, while the one on the right was not strained. Further, assume that all three bands had grating lines initially parallel to the master grating and that, as can be seen, the left-hand one was rotated 4°, the center one not at all, and the right-hand one about $6\frac{1}{2}$°.

The pitch of the master grating is $1/71.5 = 0.014$ in. All the fringes on the three models represent displacements in multiples of 0.014 in. in the x-direction with respect to an adjacent fringe; that is, the u-value (x-displacement) of consecutive fringes could be 0, 0.014, 0.028, 0.042 in., etc.

This gives a ready value $\Delta u = 0.014$ in. between fringes. To compute Eulerian strain, it is necessary to measure the distance between fringes parallel and perpendicular to the master grating. This gives Δx or Δy, and so forms the fractions $\Delta u/\Delta x$ and $\Delta u/\Delta y$, which in the homogeneous field are also $\partial u/\partial x$ and $\partial v/\partial y$.

As shown in the next section, Lagrangian strain must be computed by referring to the Cartesian system on the undeformed model. In this case, the boundaries of the model give the undeformed x- and y-directions.

Case A. The left-hand pattern in Fig. 4.14 can be considered a case of large strain and large rotation.

The distances between fringes in directions parallel and perpendicular to

---

[1] In the original specimen.

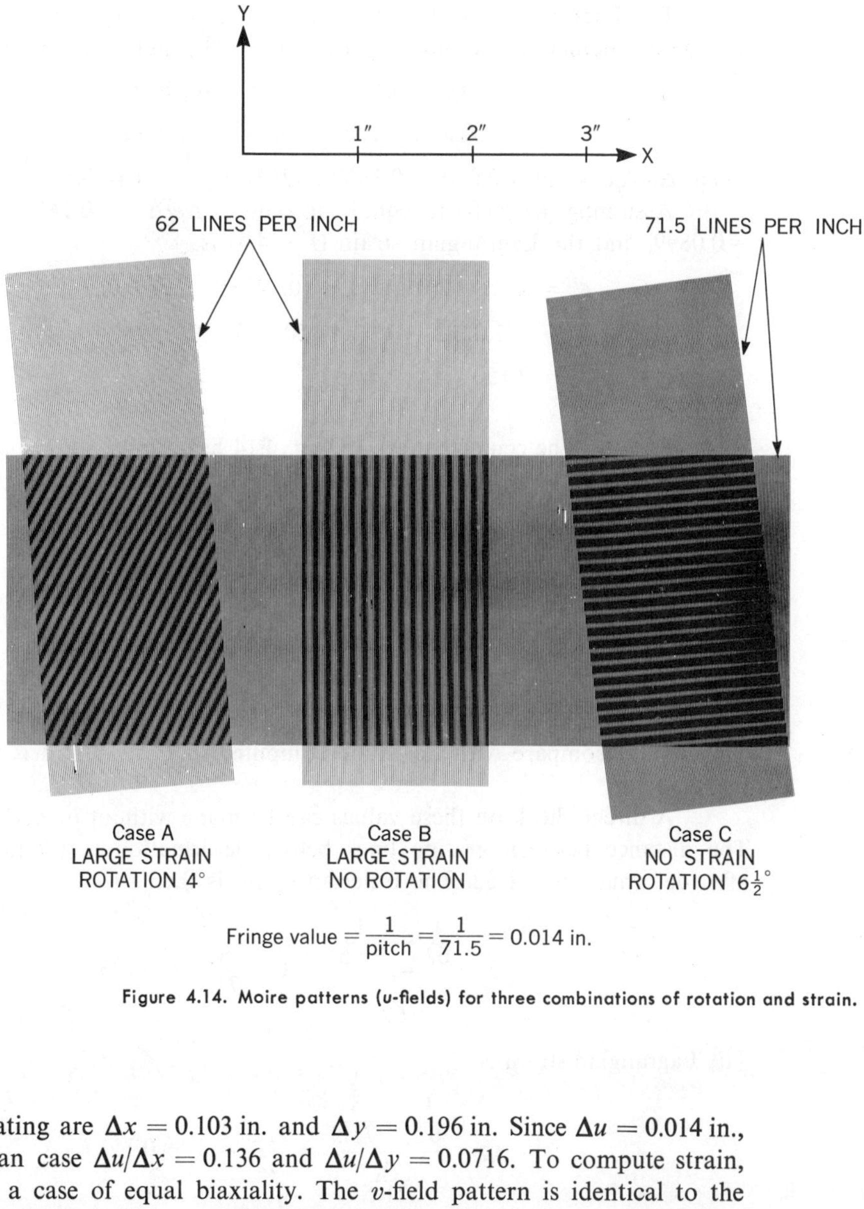

Figure 4.14. Moire patterns (u-fields) for three combinations of rotation and strain.

the master grating are $\Delta x = 0.103$ in. and $\Delta y = 0.196$ in. Since $\Delta u = 0.014$ in., for the Eulerian case $\Delta u/\Delta x = 0.136$ and $\Delta u/\Delta y = 0.0716$. To compute strain, let us assume a case of equal biaxiality. The v-field pattern is identical to the u-field pattern. Then, $\Delta v/\Delta x = -0.0716$ and $\Delta v/\Delta y = 0.136$. The Eulerian strain (Eq. 4.6) is

$$\begin{aligned}\epsilon_x^E &= 1 - \sqrt{1 - 2(0.136) + (0.136)^2 + (-0.0716)^2} \\ &= 1 - \sqrt{0.7516} = 1 - 0.867 \\ &= 0.133\end{aligned}$$

For Lagrangian strain, the distances between fringes in directions parallel and perpendicular to the model grating are 0.110 and 0.170 in.; Δx and Δy are

$$\Delta x = 0.110 - 0.014 = 0.096 \text{ in.}$$

$$\Delta y = 0.170 - 0.014 = 0.156 \text{ in.}$$

Then, $\Delta u/\Delta x = 0.014/0.096 = 0.1459$ and $\Delta u/\Delta y = 0.014/0.156 = 0.0899$.

Assuming as before, equal biaxiality, $\Delta v/\Delta y = 0.1459$ and $\Delta v/\Delta x = -0.0899$, and the Lagrangian strain (Eq. 4.4) is

$$\begin{aligned}\epsilon_x^L &= \sqrt{1 + 2(0.1459) + (0.1459)^2 + (-0.0899)^2} - 1 \\ &= \sqrt{1.3212} - 1 = 1.150 - 1 \\ &= 0.150\end{aligned}$$

Case B. The center pattern in Fig. 4.14 has a large strain but no rotation. The distance between fringes is $\Delta x = 0.106$ in. Computing the Eulerian strain (Eq. 4.16),

$$\epsilon_x^E = \frac{\partial u}{\partial x} = \frac{\Delta u}{\Delta x} = \frac{0.014}{0.106} = 0.132$$

The Lagrangian strain (Eq. 4.15) is computed as

$$\epsilon_x^L = \frac{\partial u}{\partial x} = \frac{\Delta u}{\Delta x} = \frac{0.014}{0.106 - 0.014} = \frac{0.014}{0.092} = 0.152$$

This should compare with the value computed in Case A and is within experimental error.

A direct check on these values can be made without involving the fringes. The distance between grating lines before deformation is 1/71.5, the distance after deformation is 1/62. The Eulerian strain is then

$$\epsilon_x^E = \frac{\dfrac{1}{62} - \dfrac{1}{71.5}}{\dfrac{1}{62}} = 1 - \frac{62}{71.5} = 0.133$$

The Lagrangian strain is

$$\epsilon_x^L = \frac{\dfrac{1}{62} - \dfrac{1}{71.5}}{\dfrac{1}{71.5}} = \frac{71.5}{62} - 1 = 0.153$$

Case C. The pattern on the right side in Fig. 4.14 is a case of rotation with no strain. The large-rotation–small-strain equation applies here and, since the strain is zero, the equation is exact. The v-field pattern is exactly the same as the u-field.

For the Eulerian strain (Eq. 4.20), $\Delta x = 1.65$ in., $\Delta y = 0.1087$ in., and the strain equation is

$$\epsilon_x^E = \frac{0.014}{1.65} - \frac{1}{2}\left[\left(\frac{0.014}{1.65}\right)^2 + \left(-\frac{0.014}{0.1087}\right)^2\right]$$
$$= \left\{\frac{1}{1.65} - \frac{1}{2}\left[\left(\frac{1}{1.165}\right)^2 + \left(-\frac{1}{0.1087}\right)^2\right] 0.014\right\} 0.014$$
$$= [0.606 - \tfrac{1}{2}(0.367 + 85.5)0.014]0.014$$
$$= (0.606 - 0.601)0.014$$
$$= 0.00007$$

The variation of ϵ_x^E from zero is of the order of the experimental error.

The Lagrangian equation (Eq. 4.19) is the same as the Eulerian for this case, but with the opposite sign.

4.11. Principal Strains

To obtain principal strains from Cartesian strains, the tensor form can be used. The principal strains are obtained from the three roots a, b, c of the determinant—

$$\begin{vmatrix} \left(d - \begin{matrix} a \\ b \\ c \end{matrix}\right) & g & h \\ g & \left(e - \begin{matrix} a \\ b \\ c \end{matrix}\right) & i \\ h & i & \left(f - \begin{matrix} a \\ b \\ c \end{matrix}\right) \end{vmatrix} = 0$$

—or in its two-dimensional form (Mohr's circle) as

$$a, b = \tfrac{1}{2}[(d + e) \pm \sqrt{(d - e)^2 + g^2}]$$

Each of the symbols has the relation to strain indicated in the table on the following page, depending on whether small strain, Lagrangian strain, or Eulerian strain is used.

The double angle associated with Mohr's circle in the Eulerian description is the angle between the principal directions and the Cartesian directions in the deformed state. In the Lagrangian description the Cartesian directions are associated with the undeformed state, so that the angle is the angle between the Cartesian directions and the fiber which will be subjected to the principal strain. However, this fiber may also be rotated far from its undeformed position in the process of deforming.

	Small Strain	Lagrangian Strain	Eulerian Strain
a	ϵ_1	$(1 + \epsilon_1^L)^2$	$(1 - \epsilon_1^E)^2$
b	ϵ_2	$(1 + \epsilon_2^L)^2$	$(1 - \epsilon_2^E)^2$
c	ϵ_3	$(1 + \epsilon_3^L)^2$	$(1 - \epsilon_3^E)^2$
d	ϵ_x	$(1 + \epsilon_x^L)^2$	$(1 - \epsilon_x^E)^2$
e	ϵ_y	$(1 + \epsilon_y^L)^2$	$(1 - \epsilon_y^E)^2$
f	ϵ_z	$(1 + \epsilon_z^L)^2$	$(1 - \epsilon_z^E)^2$
g	$\frac{\gamma_{xy}}{2}$	$(1 + \epsilon_x^L)(1 + \epsilon_y^L) \sin \gamma_{xy}^L$	$(1 - \epsilon_x^E)(1 - \epsilon_y^E) \sin \gamma_{xy}^E$
h	$\frac{\gamma_{yz}}{2}$	$(1 + \epsilon_y^L)(1 + \epsilon_z^L) \sin \gamma_{yz}^L$	$(1 - \epsilon_y^E)(1 - \epsilon_z^E) \sin \gamma_{yz}^E$
i	$\frac{\gamma_{zx}}{2}$	$(1 + \epsilon_z^L)(1 + \epsilon_x^L) \sin \gamma_{zx}^L$	$(1 - \epsilon_z^E)(1 - \epsilon_x^E) \sin \gamma_{zx}^E$

5

PROPERTIES OF ISOTHETIC LINES

5.1. Introduction

Use will be made of the concept of the *isothetic line*, defined previously as the locus of points having the same value of displacement component in a given direction. For instance, if u and v are orthogonal components of the displacement at a point, there will be associated with the displacement field a family of u-isothetic lines and v-isothetic lines. Each isothetic of the u-family will correspond to a particular value of displacement in the u-direction, and similarly for the v-family. The difference in the value of displacement from one isothetic line to the next in the family, as determined by the moire pattern, will be a constant and will be equal to the pitch of the master grating.

Some properties of isothetic lines will be reviewed in this chapter.[1] Their main usefulness will become apparent when moire is used as a means of determining strains.

5.2. Properties of Isothetic Lines Resulting from Uniqueness and Continuity of Displacements

The displacements and components of displacements within the boundaries of a body are continuous and single-valued. They are continuous and, therefore,

[1] The subject matter of this chapter was dealt with in: A. J. Durelli, C. A. Sciammarella, and V. J. Parks, "Interpretation of Moire Patterns," *J. Eng. Mech. Div.*, Proc. A.S.C.E. EM2, No. 3487 (1963) pp. 71–88.

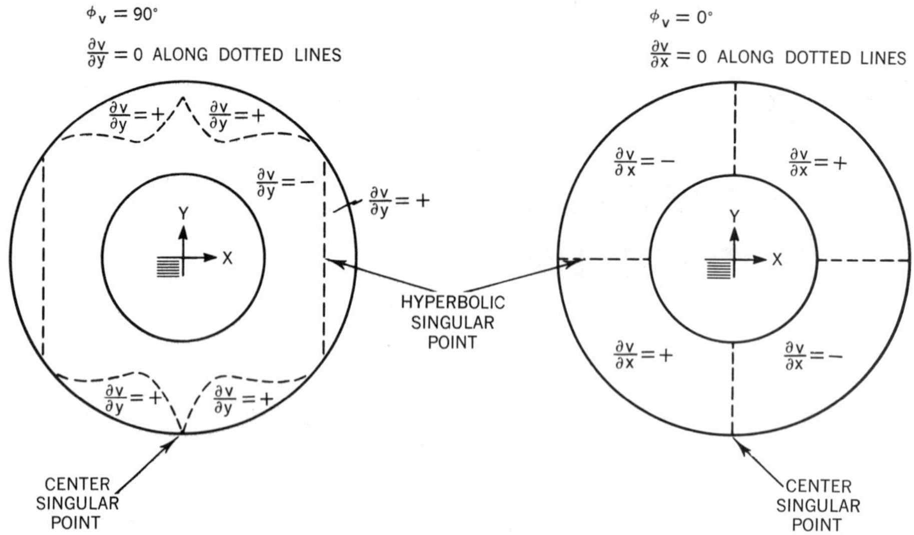

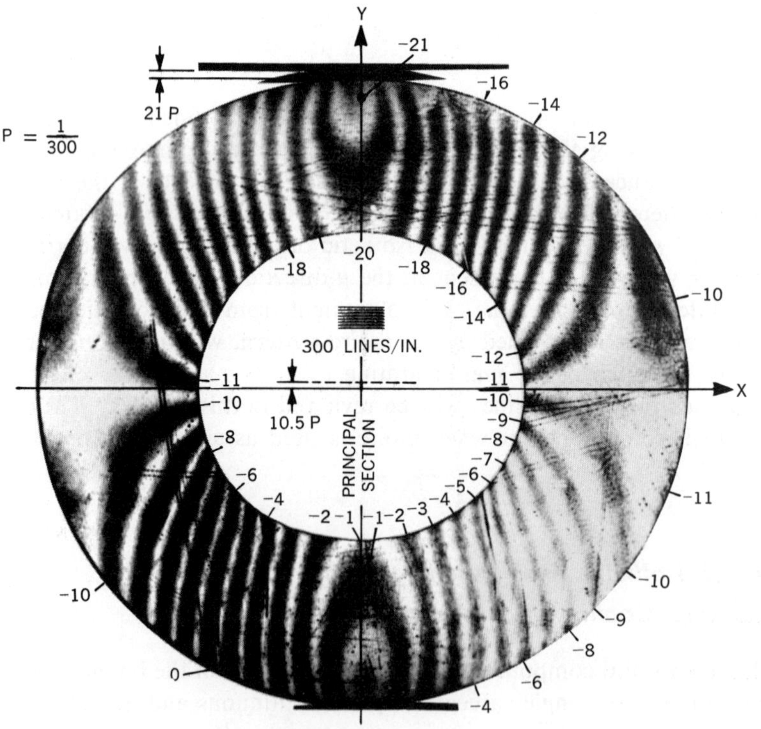

Figure 5.1. Moire pattern of the vertical displacements of a ring under diametral loading.

on a line between any two points with different displacements, the displacements with intermediate values must occur. They are single-valued and, therefore, no two points can occupy the same location on the deformed body. These considerations follow directly from the assumption of continuity. If the displacements were not continuous, gaps would exist in the body.

Continuity is illustrated by an analogous property that is found in the pattern of contours representing topological elevations. From physical considerations, it is obvious that, between two points with different elevations, the intermediate elevations must occur. The requirement that the isothetic pattern be single-valued is analogous to requiring that the topological surface have no vertical dropoffs (cliffs) within the field. The analogy may be carried one step further by noting that, because both patterns represent physical phenomena, there are no infinitely high or infinitely low values in either case. (Figure 1.17 may help in the understanding of these ideas.)

From the property of uniqueness, it follows that isothetic lines of different displacement values do not intersect. Because of continuity, the isothetic lines are similar to topological contours that end at the boundary of the pattern or form closed curves.

The property of continuity helps in ordering the moire fringes. Assigning a displacement value to the first fringe can be completely arbitrary, since it only defines the reference plane (which may be associated with a rigid-body motion).

Consecutive fringes will have a difference in displacement equal to the space between gratings. Thus, for a 1000-lines/in. grating, if a fringe is assigned the displacement 1.000 in., the consecutive fringes will be 1.001 in., 1.002 in., 1.003 in., and so forth.

From the first fringe value, it is sometimes possible to assign values to the entire pattern by a simple count. This is true if each fringe appears only once on the pattern. Sometimes considerations of the symmetry and the load aid in the determination of the displacement values. Thus, it is obvious that symmetrical fringes on opposite sides of the vertical axis in Fig. 5.1 have the same vertical-displacement values. It is also obvious that the diametral compressive loading will produce negative vertical displacements over much of the body.

It should be noted that consideration of some external evidence, such as the load, is necessary in order to determine the sign of the displacement. Even if the fringes are ordered with no difficulty, the sign cannot be known except by outside means.

5.3. Properties of Loci of Points with Zero Derivatives in the Direction of Principal and Secondary Sections, Respectively

A *principal section* is defined as a line perpendicular to the master-grating line. A line parallel to the grating lines is called a *secondary section.*

From observation of Fig. 1.19, the slope of the isothetic lines in the u-pattern with respect to the x-axis is governed by the equation

$$\tan \phi_u = -\frac{\frac{\partial u}{\partial x}}{\frac{\partial u}{\partial y}} \tag{5.1a}$$

Similarly, for the v-pattern

$$\tan \phi_v = -\frac{\frac{\partial v}{\partial x}}{\frac{\partial v}{\partial y}} \tag{5.1b}$$

in which, as shown in Fig. 3.4, ϕ represents the angle between the tangent to the isothetic line and the x-axis measured in the direction that places the positive y-axis 90° from the positive x-axis (the subscript refers to the pattern from which the isothetic line was obtained). From physical considerations, it has been noted that the derivatives in Eq. (5.1) must be finite.[2] However, the slope ($\tan \phi$) may become infinite. Thus, where the slope passes through infinity ($\phi = 90°$), the denominator must go through zero because the numerator cannot go through infinity, and the slope will change sign when the derivatives $\partial u/\partial y$ or $\partial v/\partial y$ change sign. Where the slope changes sign by passing through zero ($\phi = 0$), as the denominator cannot be infinite, it will do so when the derivatives $\partial u/\partial x$ or $\partial v/\partial x$ change sign. The case in which both derivatives from one pattern change sign simultaneously so that the slope does not change sign is a special case referred to as a "singular point" and is examined in greater detail subsequently. Despite the presence of singular points, it is possible to divide the isothetic pattern along lines of zero or infinite slope, so that areas of the same sign are established for $\partial u/\partial x$, $\partial u/\partial y$, $\partial v/\partial x$, and $\partial v/\partial y$.

There is also the special situation in which the slope goes to zero, but does not change sign. The case is not pertinent to the present analysis.

If the value of the displacement component is known for any two points in an area (by some external means), then the absolute sign of the derivative in the direction of the principal section ($\partial u/\partial x$ or $\partial v/\partial y$) can be determined, and by noting the sign of the slope, the sign of the derivative in the direction of the secondary section can be determined ($\partial u/\partial y$ or $\partial v/\partial x$). Once a sign is determined in one area, it can be determined for all areas on that pattern.

It will be seen that where rotations are zero or can be neglected, the direct derivatives are equal to strain, and if the strain at the point is known, this allows another means of determining signs.

Figures 5.1 and 5.2 show each of the isothetic patterns produced by grating

[2] A. E. H. Love, *A Treatise on the Mathematical Theory of Elasticity* (New York: Dover Publications, Inc., 1944), p. 65.

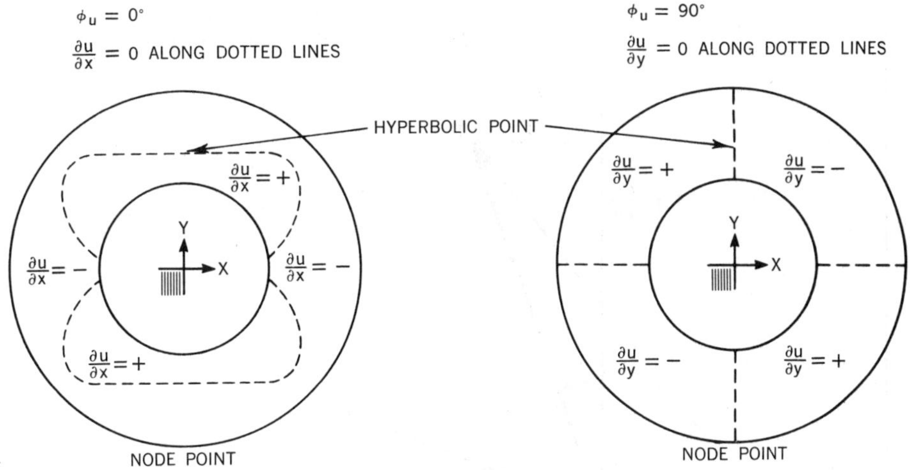

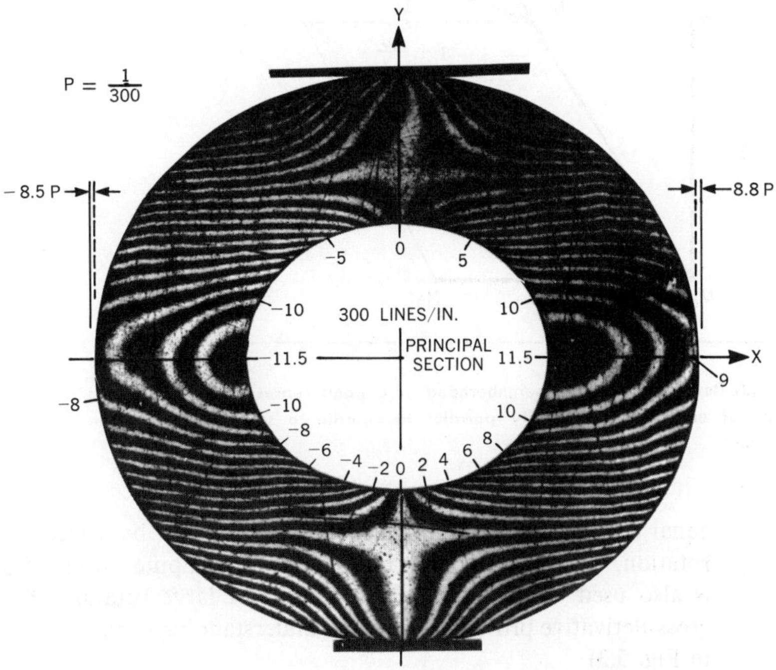

Figure 5.2. Moire pattern of the horizontal displacements of a ring under diametral loading.

lines parallel to each of the major axes of a ring under diametral compression. The lines of zero derivatives and the signs are sketched in Figs. 5.1 and 5.2. The direct derivatives are important because, as mentioned above, they are

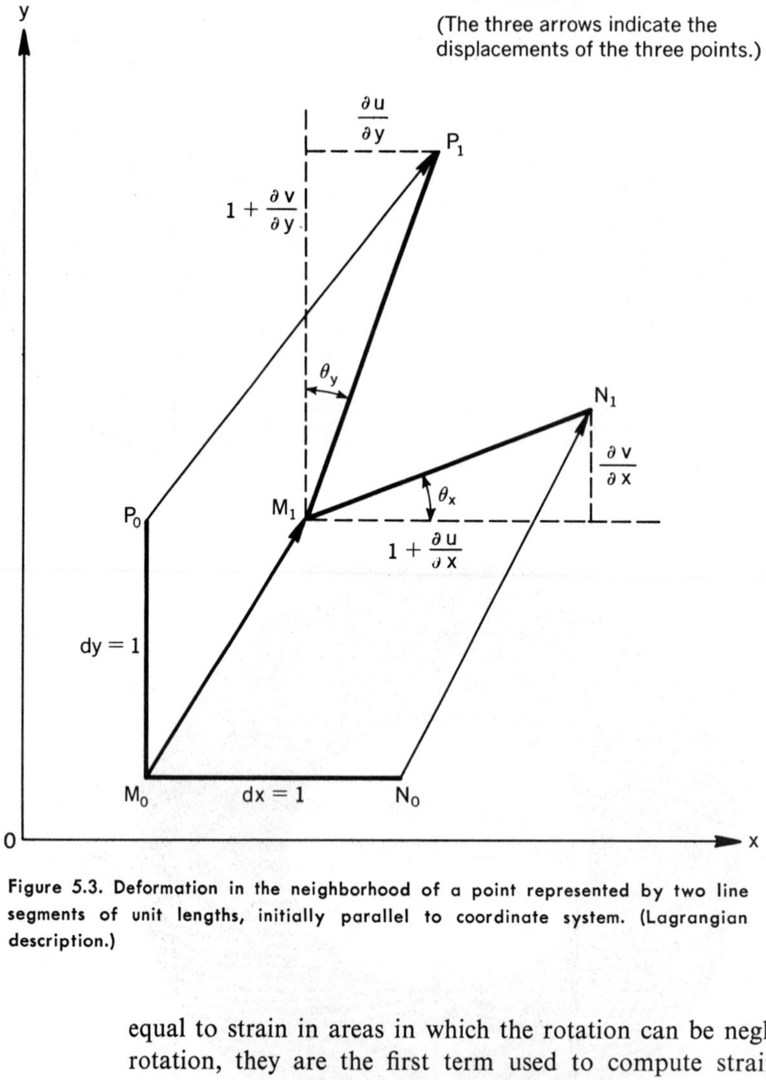

Figure 5.3. Deformation in the neighborhood of a point represented by two line segments of unit lengths, initially parallel to coordinate system. (Lagrangian description.)

equal to strain in areas in which the rotation can be neglected; in areas of large rotation, they are the first term used to compute strain. The cross-derivative is also used to compute strain in areas of large rotation. The zero line of the cross-derivative provides additional understanding of the pattern (this is illustrated in Fig. 5.3).

Figure 5.3 shows the deformation in the neighborhood of a point as it affects two lines initially parallel to the x- and y-axes. Consider the element $dx = 1$, which in the undeformed state is parallel to the x-axis. After deformation, it makes an angle θ_x with the x-axis such that

$$\tan \theta_x = \frac{\dfrac{\partial v}{\partial x}}{1 + \dfrac{\partial u}{\partial x}} \tag{5.2a}$$

Similarly the element $dy = 1$ makes an angle θ_y such that:

$$\tan \theta_y = \frac{\dfrac{\partial u}{\partial y}}{1 + \dfrac{\partial v}{\partial y}} \tag{5.2b}$$

Except for the extraordinary case of $\partial u/\partial x < -1$, the denominator in Eq. (5.2a) will always be positive and the sign of $\tan \theta_x$ and the sign of $\partial v/\partial x$ will be the same. From Fig. 5.3, it is concluded that the sign of $\tan \theta_x$ is the same as the sign of the

Figure 5.4. Surface representing displacement component and showing singular points.

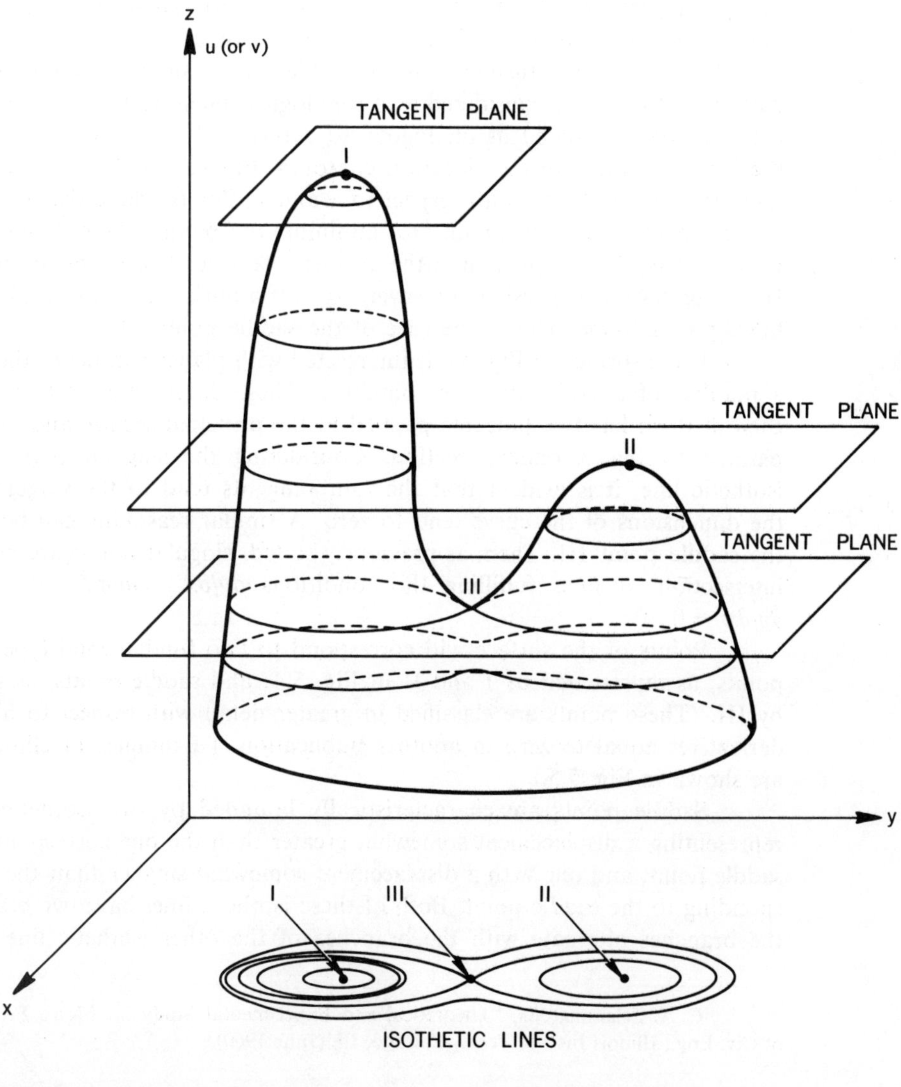

angle of rotation of the line parallel to the x-axis, and the sign of tan θ_y is opposite to the sign of the angle of rotation of the line parallel to the y-axis.

It can be concluded from Eq. (5.2) that, along the lines $\partial u/\partial y = 0$ and $\partial v/\partial x = 0$, the associated grating lines do not rotate.

The sign of the cross-derivative can be obtained by observing the sign of a change in displacement as a point moves in the positive coordinate direction. In the first quadrant of Fig. 5.1, as a point moves toward increasing x the fringe orders show an increase in the displacement component v.

The loci of the points of zero derivatives in the direction of the principal and secondary sections, respectively, intersect at points at which simultaneously $\partial u/\partial x = \partial u/\partial y = 0$ (for the u-pattern) or $\partial v/\partial y = \partial v/\partial x = 0$ (for the v-pattern). These points, as shown, are points at which the grating line does not rotate and the strain perpendicular to the grating line is zero. These points are called the *singular points* of the isothetic pattern, either u or v.

To understand further some properties of the singular points, the isothetic pattern will again be considered as a topological phenomenon. The displacement field can be considered as analogous to a series of hills and valleys for which the isothetic lines are the elevation contours. Figure 5.4 shows a surface representing u (or v) plotted with respect to x and y. Points where the components of displacement have a maximum or a minimum are points where the tangent plane is parallel to the x-y plane and the gradient vanishes (for example, points I and II on Fig. 5.4). It is possible, however, to have a horizontal tangent plane without having an extreme. This is the case of the saddle point III.

If the surface of Fig. 5.4 is intersected with planes parallel to the x-y plane, a number of closed curves are obtained. These curves are isothetic lines, and each of them has two tangents parallel to the principal section and two tangents parallel to the secondary section. Considering the singular point as a limit isothetic line, it is evident that the four tangents tend to the singular point as the dimensions of the curve tend to zero. A similar reasoning can be applied to the saddle point III. Then, as previously stated, singular points are points at the intersection of lines fulfilling the conditions $\partial u/\partial x = \partial u/\partial y = 0$ or $\partial v/\partial y = \partial v/\partial x = 0$.

Points of the surface will correspond to two fundamental types: elliptical points, as represented by I and II in Fig. 5.4; and saddle points, as represented by III. (These points are classified in greater detail with respect to higher-order derivatives equal to zero in another publication.[3]) Examples of elliptical points are shown in Fig. 5.5.

Saddle points are characteristically bounded by two isothetic lines, one representing a displacement somewhat greater than the one corresponding to the saddle point, and one with a displacement somewhat smaller than the one corresponding to the saddle point. Both of these isothetic lines have two branches, and the branches alternate with the branches of the other isothetic line to form a

---

[3] C. A. Sciammarella, "Theoretical and Experimental Study on Moire Fringes," Dept. of Civ. Eng., Illinois Inst. of Tech., Chicago, Ill. (June 1960).

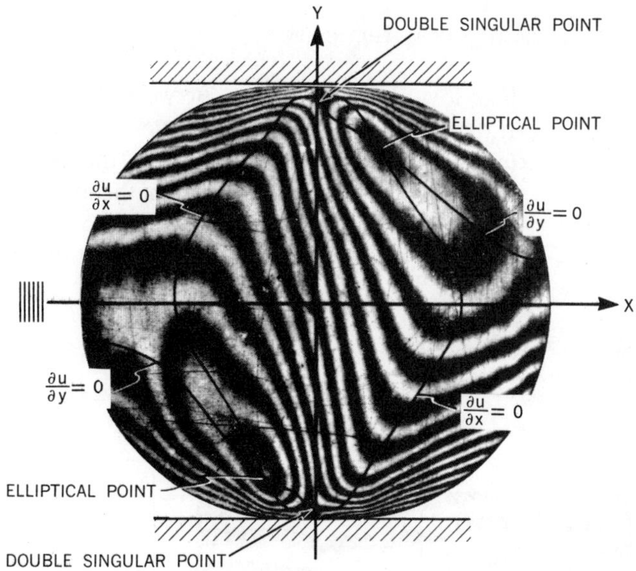

Figure 5.5. Disk under diametral load. The model grid forms a small angle with the direction of the load.

Figure 5.6. Moire pattern of the vertical displacements in the neighborhood of the loaded points in a large plate with two concentrated loads.

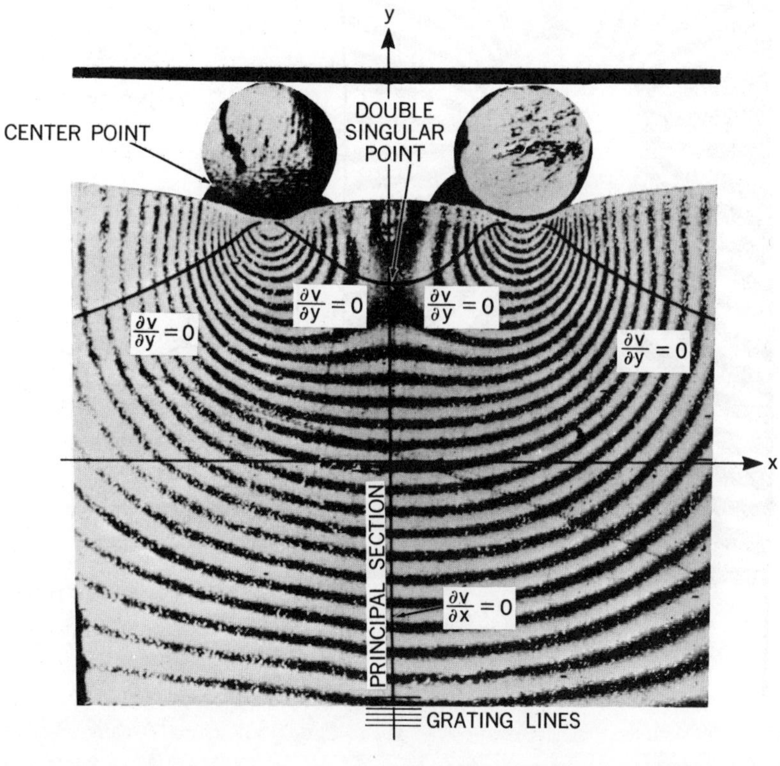

high-low-high-low pattern. Between the four branches, there are asymptotic lines. If the asymptotes fall along the coordinate axes, then the point is called *hyperbolic* (examples in Figs. 5.1 and 5.2). If the asymptotes do not fall on the coordinate axes, then the isothetic through the point has a double tangent and is called a *double singular point* (Figs. 5.5 and 5.6).

In counting fringes, in general, it is a great help to note these saddle points

Figure 5.7. Moire pattern of the vertical displacements of a bar with a hole under pure tension.

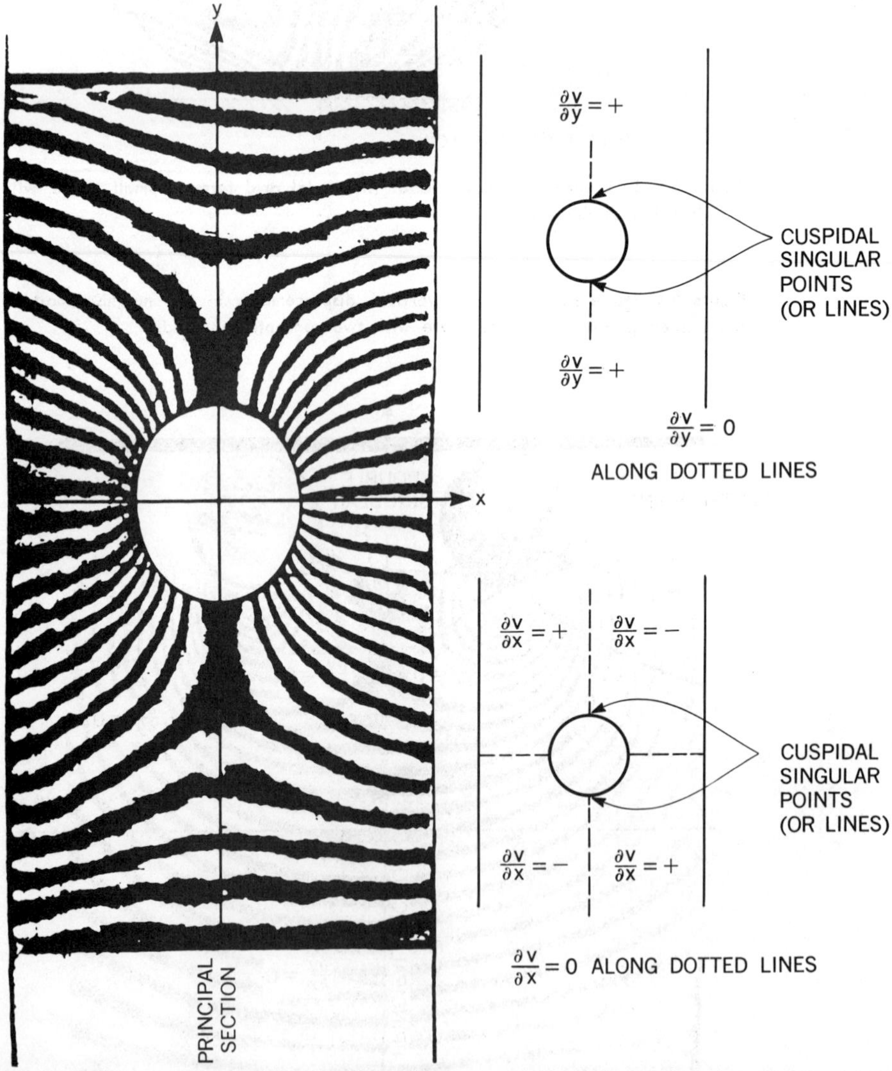

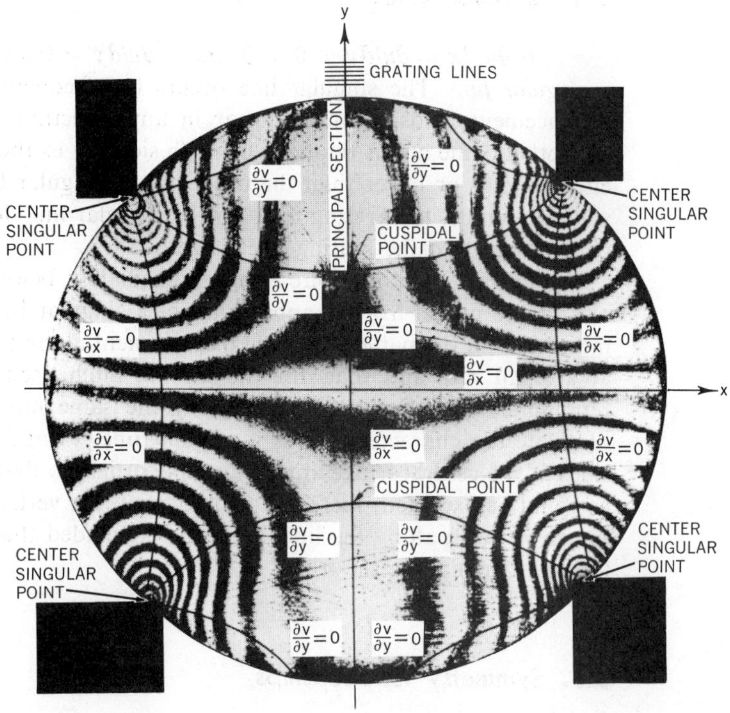

Figure 5.8. Moire pattern of the vertical displacements of a disk loaded with four concentrated forces.

and realize that the diagonally opposite branches must be assigned the same displacement value and that the other two branches have a value one unit larger or smaller.

A special case of a saddle point is shown in Figs. 5.7 and 5.8, where both derivatives are zero, but where only one changes sign. This requires branches on only one side of the point. In this case, the isothetic line presents a cusp at the singular point (this point is called *cuspidal*). In these two examples, the ordering of fringes must rely on symmetry to indicate that, about the cuspidal point, symmetrical fringes on either side of the axis have the same value.

There are two basic types of special points that are closely associated with singular points. These originate by the action of concentrated forces acting on free edges. The types are distinguished on the basis of the relative position of the principal section with respect to that line of action of the force. If the principal section is parallel to the force, the point called the *center point* is observed (Figs. 5.1, 5.6, and 5.8). This point, or sometimes a narrow region, is surrounded by concentric moire fringes. If the principal section is perpendicular to the force, the point is called a *node* (see Fig. 5.2).

5.4. Singular Lines

If $\partial u/\partial x = \partial u/\partial y = 0$ or $\partial v/\partial x = \partial v/\partial y = 0$ along a line, the line is called a *singular line*. The singular line occurs less frequently than singular points in displacement fields. It may appear in any direction on the pattern. It follows an isothetic line and is bounded on both sides by isothetic lines of the same order, either higher or lower than the order of the singular line. As is the case for the singular points, the grating lines on the singular line do not rotate and the strain perpendicular to the grating lines is zero.

In Fig. 5.7 the moire fringe immediately below the hole on the vertical axis of the model has the appearance of a singular line. However, there is some indication on the pattern that the displacement along the vertical axis may change at a small rate and that, for the grating pitch used, the displacement can be represented over a certain distance by the same moire fringe. The fringe does illustrate the characteristics of a singular line. Fringes on both the left and right sides of the central fringe are of the same order, so that $\partial v/\partial x = 0$ and, therefore, as could be expected, there is no rotation on the vertical axis. Because $\partial v/\partial y = 0$ (or $\partial v/\partial y \approx 0$) along the line, it can be concluded that the vertical strain is zero (or near zero) along the fringe.

5.5. Symmetry Relationships

Assume a model with an axis of symmetry and loads symmetrically distributed about this axis. The axis of symmetry cannot rotate. Therefore, along the axis of symmetry, when the grating lines are parallel to the axis of symmetry, the resultant moire fringes will be parallel to the axis of symmetry. This follows from Eq. (5.1), as the derivatives in the direction of the secondary section along the symmetry axis must be equal to zero. Examples are shown in Figs. 5.1, 5.2, 5.7, and 5.8. In all these cases, the patterns are symmetric.

Consider the case in which the axis of symmetry and the grating lines are neither parallel nor perpendicular. Figure 5.9a shows two points located symmetrically about an axis of symmetry. The principal directions are indicated by the symbols ① and ② and indicate that the strain patterns at the two points are mirror images. The mirror image of the left point is shown on the right point for comparison. The grating is in the same direction at both points, and not aligned with the axis of symmetry. Thus, the principal directions have a different orientation with respect to the grating.

In Sec. 7.4 a method of obtaining strains from the isothetic gradients by a Mohr's-circle construction is given. Here in Fig. 5.9b Mohr's circle is used just to indicate that the gradient vectors from the symmetrically located points, although different, give the same strains.

Figure 5.10 shows an example of a symmetrical body that is symmetrically loaded (a disk under diametral compression). The pattern corresponds to a

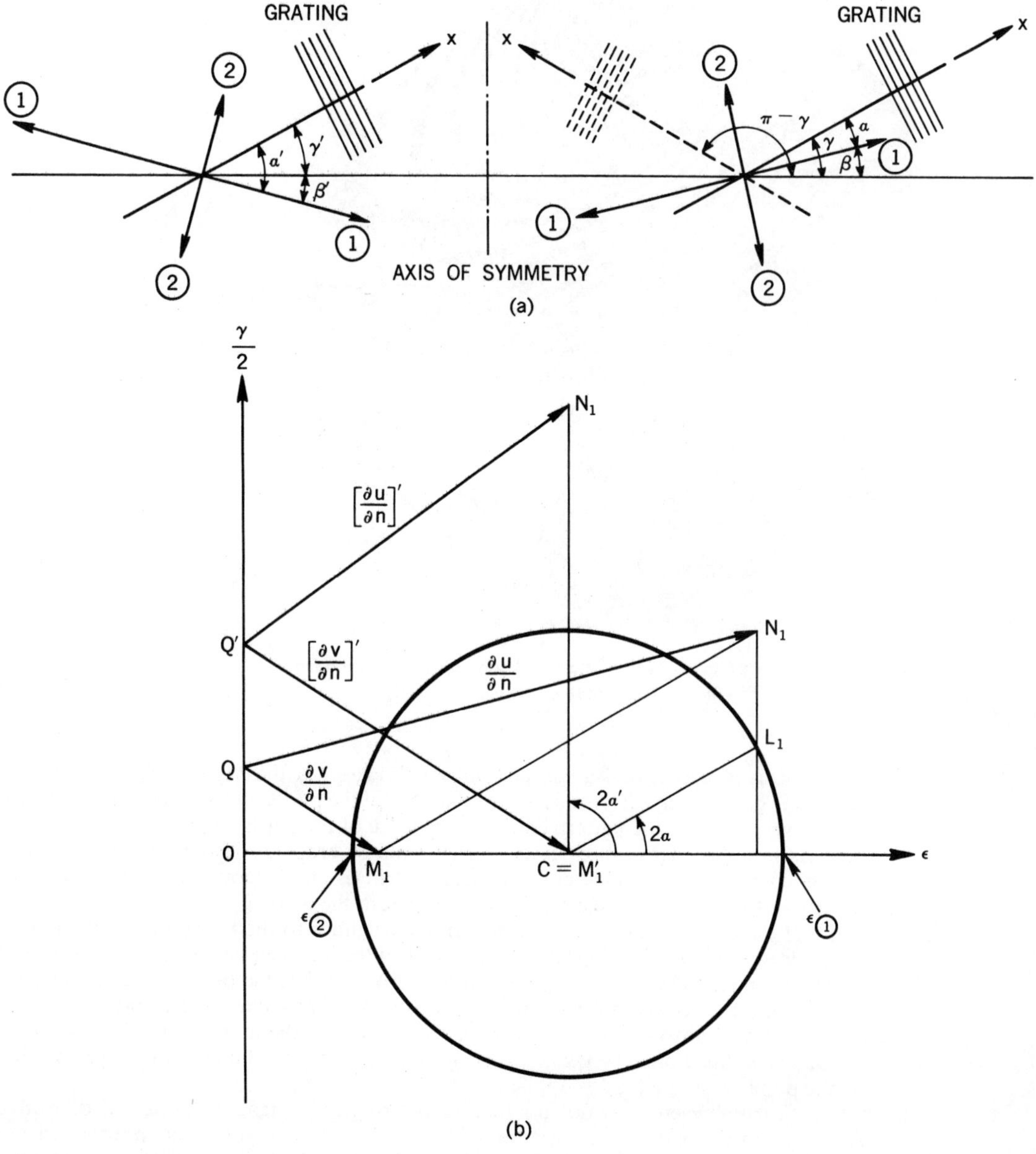

Figure 5.9. Symmetry relationships in moire pattern.

grating angle of $\gamma = 45°$. Because of the symmetry, the pattern to the left of the symmetry axis corresponds to the pattern that would have been obtained for the right side of the axis with a grating angle of 135°. Because the patterns on the left and right sides of the axis correspond to patterns of the same half of the disk taken with gratings 90° apart, this picture contains all the necessary information for the strain analysis.

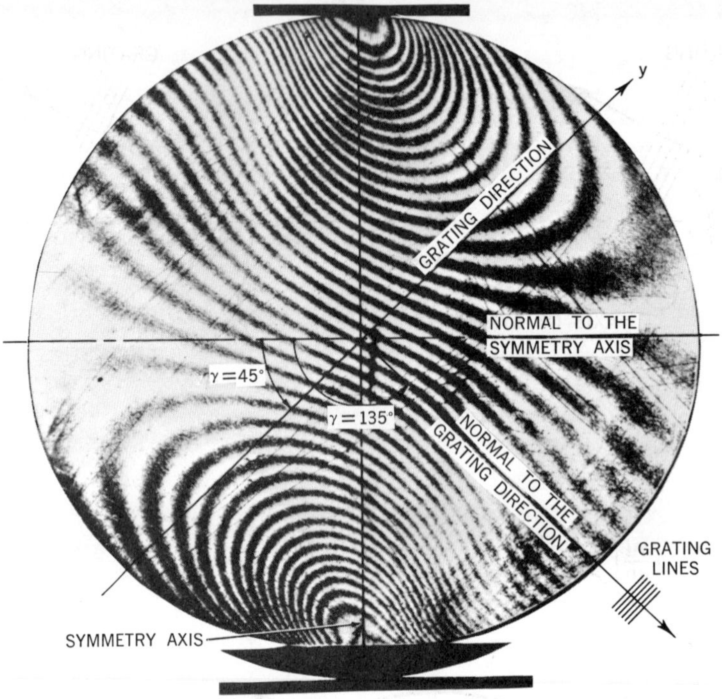

Figure 5.10. Moire pattern of a disk under diametral load obtained with a model grating at an angle of 45° with the horizontal direction.

Signs of partial derivatives and order to fringes. At the beginning of this chapter rules were given for the determination of the signs of the partial derivatives of the displacements everywhere in the pattern from a knowledge of the partial derivative at one point. It was also noted that in most cases fringes could be ordered without ambiguity. In this section comments are made about ordering of fringes and determination of the signs of the derivatives when these cannot be obtained from the loading conditions.

If there is doubt about whether a fringe neighbor to the fringe n should be order $n + 1$ or $n - 1$, the following rule is helpful when the load can be increased or decreased in the specimen. When the load is increased, assuming that the boundary conditions do not change, the fringe moves toward zones of lower absolute order, and away from zones of higher absolute order. If the load is decreased, the motion of the fringes is in the opposite direction. The fringe order must be equal to the number of fringes passing by the point when the load is applied.

In cases where it is not possible to determine the sign of the partial derivative anywhere on the pattern from the boundary conditions, several external approaches are available. Rotations of the master as described in Sec. 3.10 will in most cases give the relation of p/p' to unity, and thus the sign of the partial derivative. Another approach is to shift the master grating in the positive coordinate direction normal to the master grating lines. This will produce a shift of all fringes in the pattern toward a higher algebraic fringe order. This is illustrated in Chap. 16. The progression of fringe orders in a given coordinate direction specifies the sign of the partial derivatives of the displacement components. Fu-Pen Chiang [4] presents several other fringe ordering approaches.

---

[4] Fu-Pen Chiang, "Determination of Signs in Moire Method," *Journal of the Engineering Mechanics Division, Proc. ASCE,* Vol. 95, No. EM6, December 1969.

PART 2

MEASUREMENT OF DISPLACEMENTS AND DETERMINATION OF STRAINS

6

MOIRE USED TO MEASURE DISPLACEMENTS

6.1. Introduction

Two important uses of moire techniques, both of which involve measurement of displacements, will be described in this chapter. The first application is particularly useful in structural analysis; and the measured displacements, which are in general very large, are a means of determining redundant quantities in indeterminate structures. The second application deals with the measurement of much smaller displacements in bodies, usually of complicated shape. Sometimes all that is required is the knowledge of the displacement of points on the surface of these bodies. More frequently, however, the knowledge of these displacements is a means to the determination of strains, which will be described in the next chapter. The two types of problems have particular characteristics and will be described separately.

6.1.1. Determination of Deflection Curves

In structural analysis one is interested mostly in measuring displacements of the centerline of structural members. Since these displacements are relatively large, in general, no great sensitivity is required in the measuring instrument, and coarse gratings can be used for this purpose. The moire fringes are counted starting from a point of known displacement, preferably zero displacement. The final objective of these tests is either the determination of deflection curves (or "elastica") of frames and arches,* or the determination of influence lines.

Influence lines for certain stress functions are obtained by applying rotations and displacements characteristic of the stress functions studied. The deflection curve of the structure for a unit displacement or rotation represents the influence line of the redundant characterized by this displacement or rotation. In connection with this approach, the Beggs deformeter has been widely used as a sensitive apparatus for the application of small displacements. The induced, equally small displacements in the structure are measured by micrometric microscopes. This apparatus was used because of the requirement of the theory of elasticity that displacements and deformations be small. It was recognized for a long time that large displacements in structural models produce inadmissible errors often in excess of 10 per cent. The first successful demonstration of the applicability of the large-displacement method was made by Eney.[1] In this reference it is shown experimentally that if equal and opposite displacements or rotations are applied at one point of the model and if the algebraic differences of deflections in the structure are measured, the errors due to changes in geometry are eliminated. This was proved and stated in the form of a general theorem by Massonnet.[2] Thus, the use of finite displacements is perfectly valid and yields sufficiently accurate results, provided the displacements are accurately measured.

Structural models of beams and plane frames can be machined out of plastics such as Plexiglas. These models are mounted on plates, which can also be made of Plexiglas, by means of pins or screws through the points of support of the structures. Two ways of applying the gratings can be used. In one case, transparent sheets bearing prints of equidistant parallel lines are cemented to the models and Plexiglas plates. These sheets are commercially available as Artype or Zip-A-Tone sheets and are used by artists and draftsmen. Gratings of 50 and 60 lpi (lines per inch) are readily available in this form. Sometimes, in order to increase the accuracy, a denser grating is needed. This can be achieved by printing gratings of usually up to 300 lpi on Plexiglas. Structural models are machined from such printed plates. Applications to structural analysis will be described in Secs. 6.2 through 6.6.

---

* An early contribution to this subject was made by R. Weller and B. M. Shepard, "Displacement Measurement by Mechanical Interferometry," *Proc. SESA*, Vol. VI, No. 1, 1948, p. 35. More details about the method and the techniques used can be found in: A. J. Durelli and I. M. Daniel, "Structural Model Analysis by Means of Moire Fringes," *Trans. ASCE*, Vol. 127, 1962, Part II, pp. 39–48. An independent contribution was made by D. H. Pletta and J. H. Sword, "The Construction of Influence Lines with a Mechanical Interferometer," Bull. 124, Eng. Exp. St. Virginia Polytech. Inst., April 1958.

[1] W. J. Eney, "A Large Displacement Deformeter Apparatus for Stress Analysis with Elastic Models," *Proc. SESA*, Vol. VI, No. 2 (1949), pp. 84–93.

[2] Ch. Massonnet, "Détermination Experimentale des Lignes d'Influence des Constructions Hyperstatiques sans Emploi de Microscopes," Assn. Belge pour l'Étude, l'Essai et l'Emploi des Matériaux, A.B.E.M., No. 4, 1953.

6.1.2. Determination of Displacement Fields

When the displacements to be measured are smaller than the ones present in frames and arches when tested as described in Sec. 6.1.1, denser gratings are necessary. An example of determination of displacements in an epoxy ring, using gratings with 1000 lpi, will also be shown. The scope of the investigation in this case will be much broader. Instead of determining the deformed shape of a thin member, the whole displacement field will be determined and several methods of representing this field will be described (Sec. 6.7).

6.2. Simply-Supported Beam Under Pure Bending

A simple application of the moire method was made in the determination of the deflection line of a simply-supported beam under pure bending. A beam 1-in. deep and 12-in. long was machined from a $\frac{1}{4}$-in. thick Plexiglas plate with a grating of 300 lpi printed on it. The grating lines were parallel to the axis of the beam. A similarly printed plate was held in contact with the beam such that no fringes appeared at zero load. When the load was applied, the specimen grating moved with respect to the fixed-in-space master grating, and the fringe pattern shown in Fig. 6.1 appeared. The fringe orders start from zero at the

Figure 6.1. Deflection line of simply supported beam under pure bending obtained by means of moire fringes and theory.

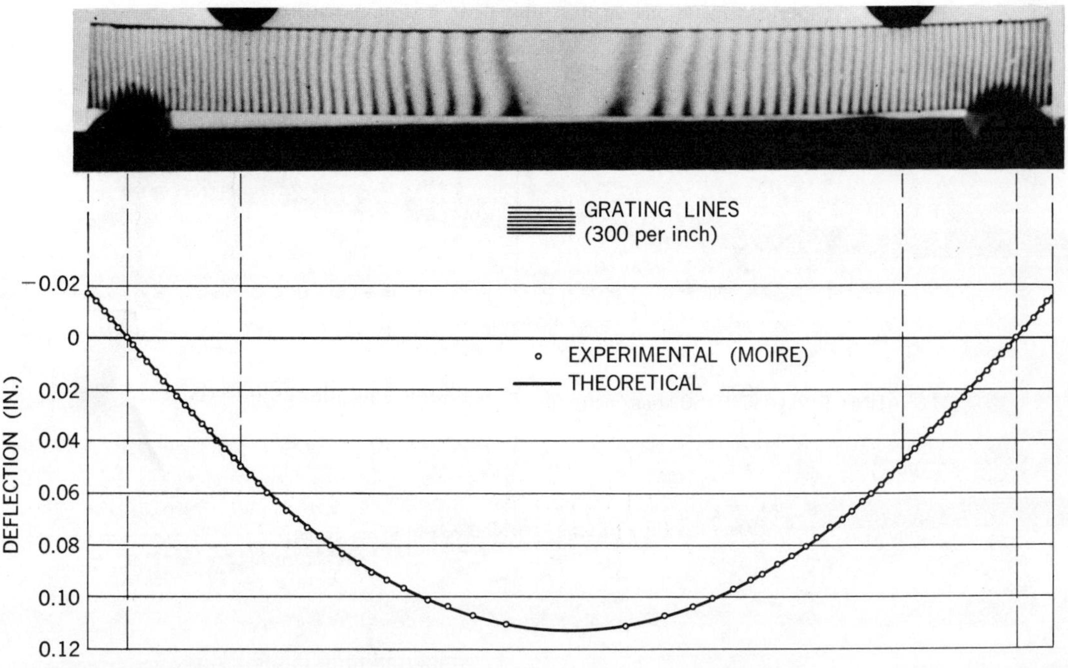

supports and increase toward the center. One fringe order corresponds to $\frac{1}{300}$ in. of displacement. A modulus of elasticity of 360,000 psi was determined from the maximum deflection at the center. Then the theoretical deflection curve was computed and plotted in Fig. 6.1. Nearly all the experimental points fall on the curve. These points were determined from the moire-fringe orders along the centerline of the beam. For small deflections and coarse gratings, the moire fringes would be straight and nearly vertical. In the present case, the shape and inclination of the fringes, especially near the center of the beam, may be due to one or several of the following: (1) the curvature of the specimen-grating lines, (2) vertical strains in the beam, or (3) the fact that the model is not viewed exactly in a direction perpendicular to the plane of the gratings. The curvature of the beam tends to tilt the fringes by making them perpendicular to the bisector of the angle of rotation. The vertical strains tend to produce fringes in the shape of elliptical arcs. The perspective effect can tilt fringes in either direction, depending on the angle of view. All these factors, however, do not seem to influence the fringe order along the neutral axis of the beam, as evidenced by the good agreement between theory and experiment.[3]

6.3. Cantilever Beam

Using the same technique described in the previous section, a Plexiglas beam was clamped at one end and subjected to a fixed displacement at the other. In this case the grating was not printed on the plastic, but was cemented to it

[3] Other considerations on this subject can be found in Chapter 2.

Figure 6.2. Isothetics in a cantilever beam subjected to a fixed displacement at the free end.

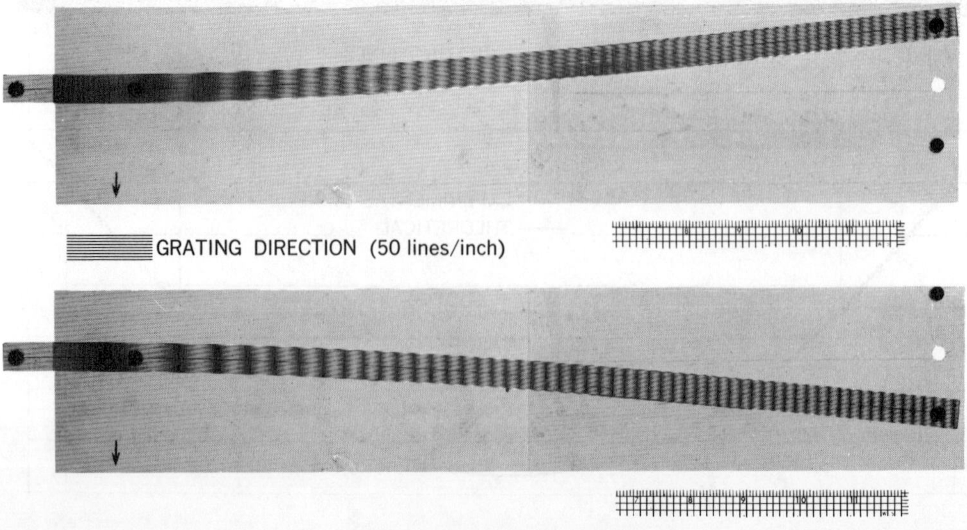

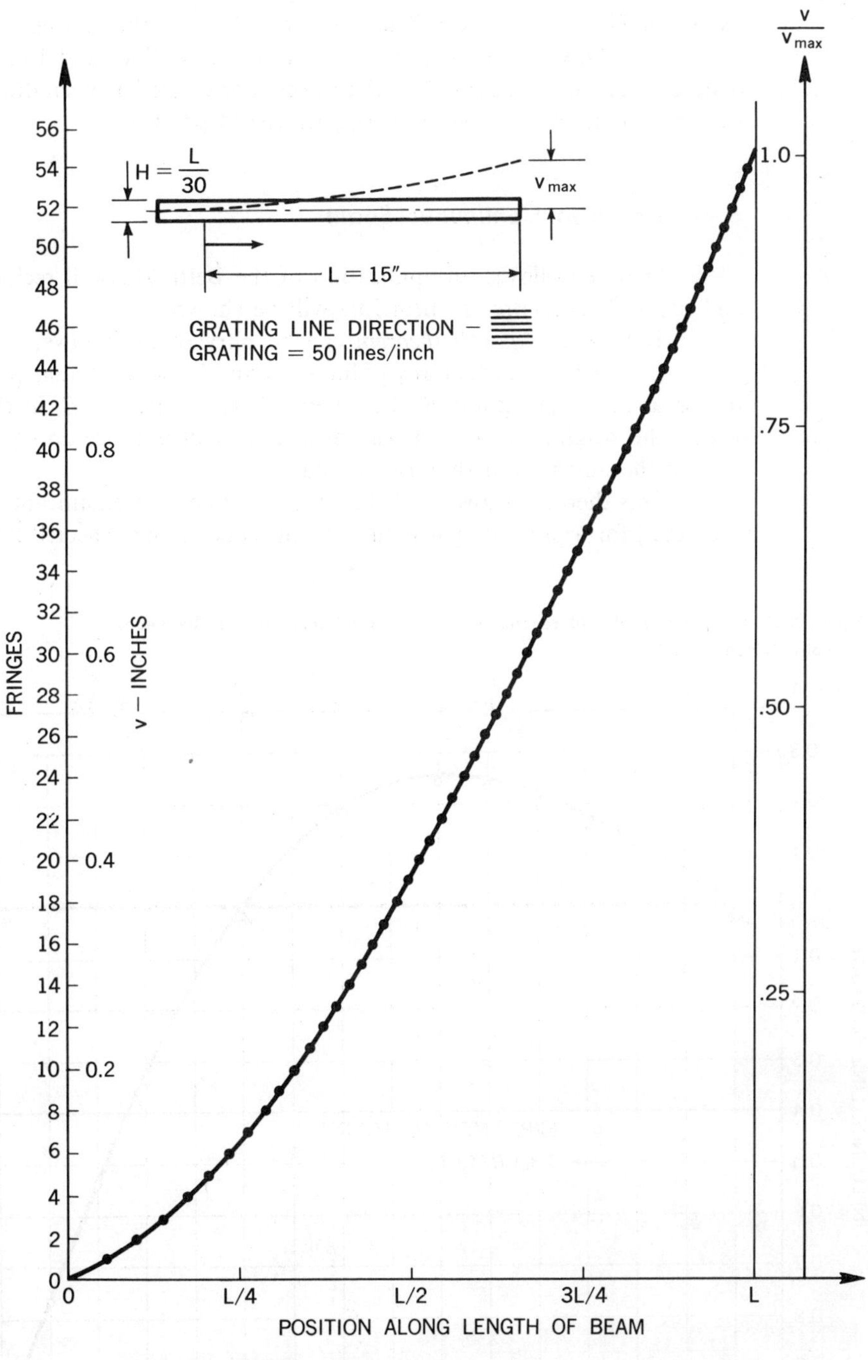

Figure 6.3. Vertical displacement of a cantilever beam subjected to a fixed displacement at the free end.

using the Zip-A-Tone sheet. The number of lines of the grating was 50 per inch. Figure 6.2 shows the photographs of the isothetic lines and Fig. 6.3 shows the deflection curve or "elastica" of the beam. The curve has been drawn using more than 50 points, which gives it a high degree of precision.

6.4. Two-Span Continuous Beam

In what follows, an application of the Betti–Maxwell reciprocity theorem and of the Muller–Breslau principle will be shown.

The *Betti–Maxwell theorem* can be phrased as follows:

If a unit load acts at any point A on an elastic body and produces a reaction δ at any other point B, then a unit load applied at B, in the direction in which the original deflection was measured, will produce an equal deflection δ at A in the direction of the original load.

This theorem shows that the influence line of a significant quantity (reaction, etc.) for a point in a structure, is the same as the deflected line of the struc-

Figure 6.4. Influence line of end reaction of a two-span beam obtained by means of moire fringes and theory.

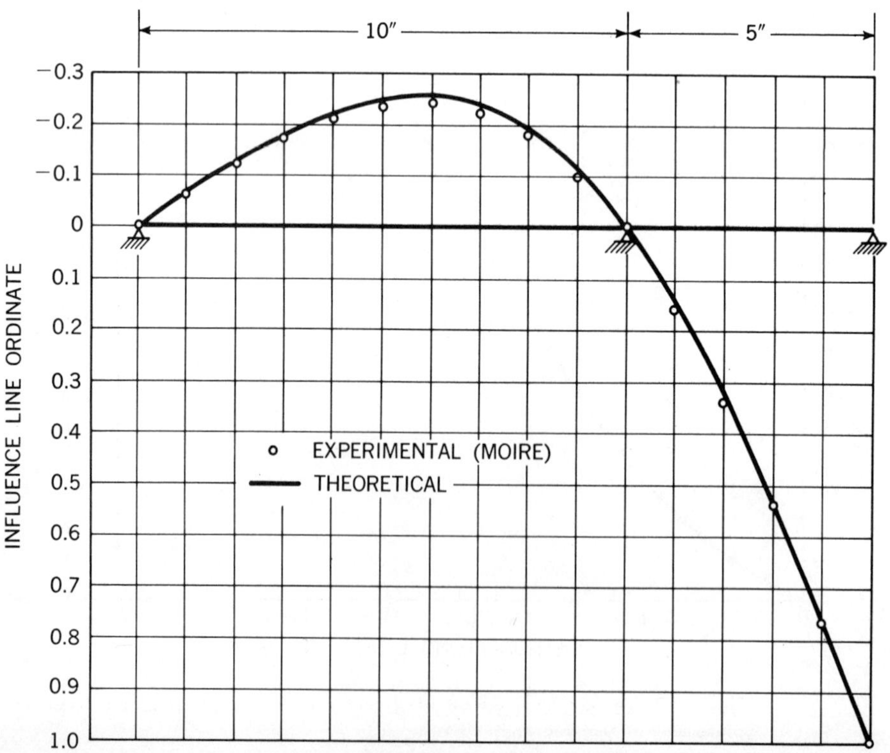

ture when a unit load is applied at the point for which the influence line is desired. This is called the *Muller–Breslau principle*.

A two-span continuous beam with spans 10-in. and 5-in. long was made out of Plexiglas and screwed to a Plexiglas plate at the points of supports (hinges). Horizontal gratings of 52 lpi were used and nearly equal vertical displacements in two directions were applied at the end of the short span of the beam. The fringe orders along the beam were plotted and converted to displacements, each fringe corresponding to $\frac{1}{52}$ in. of vertical displacement. These displacements divided by the total traverse of the end of the short span give the ordinates of the influence line for the reaction at that end. The experimental and theoretical influence lines for this reaction are shown in Fig. 6.4. The small discrepancy at the center of the large span may be due to some friction at the supports.

6.5. Frame with Haunches

Experimental methods in general gain advantage over the theoretical ones when the geometry of the structure becomes complicated. A two-bay frame with parabolic and tapered haunches and a tapered column was analyzed here for the case of a concentrated vertical load (Fig. 6.5). Gratings of 60 horizontal lines per inch were used as master and model gratings. Horizontal and vertical displacements in two opposite directions were applied at the ends of the outer

Figure 6.5. Geometry of the frame with haunches. Analyzed by moire fringes and theory.

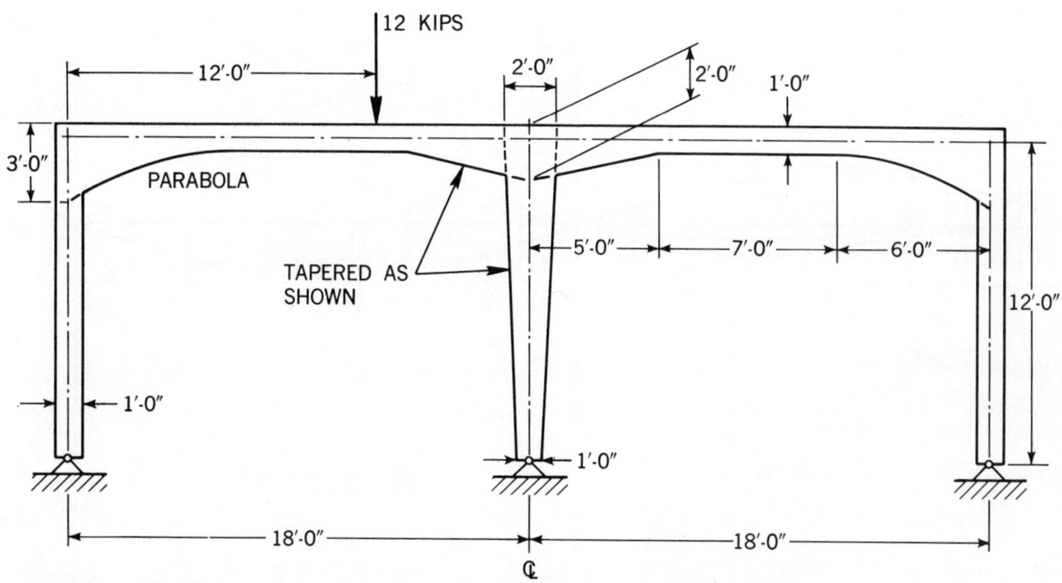

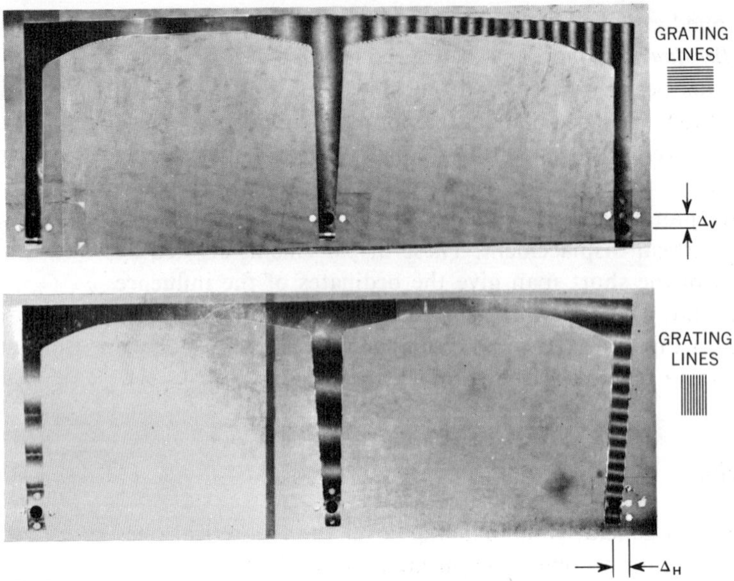

Figure 6.6. Moire patterns for vertical and horizontal displacements at the right-hand support of a two-bay frame with haunches.

Figure 6.7. Influence lines for vertical and horizontal reactions at support C and horizontal reaction at support *B*.

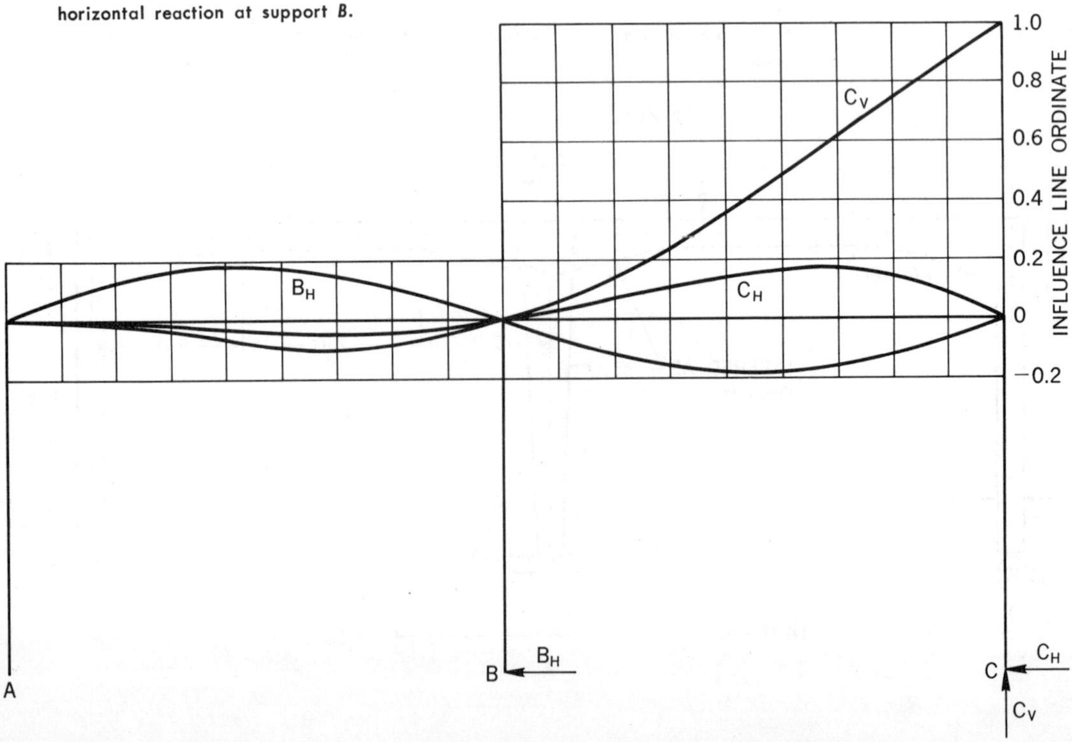

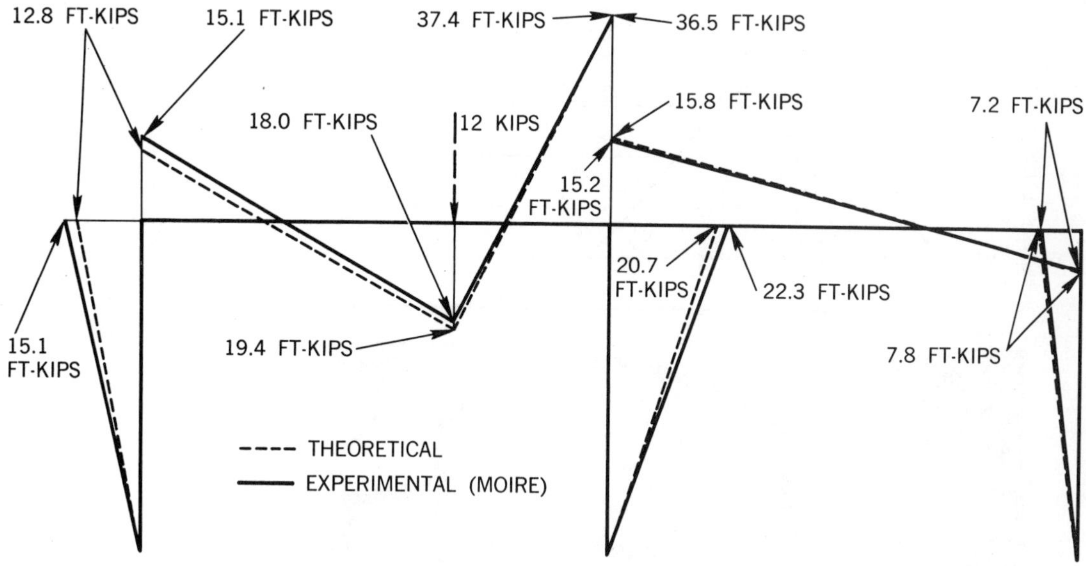

Figure 6.8. Moment diagram for a two-bay frame with haunches. Obtained by means of moire fringes and theory.

columns. Horizontal displacements were also applied at the end of the middle column. Fringe patterns for vertical and horizontal displacements at the end of an outer column are shown in Fig. 6.6. From such fringe patterns, by averaging effects of opposite displacements and taking advantage of symmetry when applicable, the influence lines of Fig. 6.7 were obtained. From these influence lines, the moment diagram corresponding to the given vertical load of 12 kips was determined and compared with the theoretical diagram in Fig. 6.8. Except for one point, the agreement seems to be satisfactory. The discrepancies may be due partly to experimental errors and partly to the fact that the theory does not take into account the true rigidity of the joints.

6.6. Three-Bay Frame

A three-bay frame of complicated geometry like the one shown in Fig. 6.9 would be exceedingly difficult to analyze theoretically. However, the moire method lends itself easily to the analysis of such a frame. A model was machined out of a Plexiglas plate with a horizontal grating of 133 lpi printed on it. The model was mounted on a similarly printed plate. Provision for the application of displacements and rotations at the points of support can be seen in Fig. 6.9. Figure 6.9 also shows the fringe pattern corresponding to a vertical displacement of support *B*. From this patttern, the influence line of the vertical reaction of support *B* is determined and plotted on the same figure. Fringe orders were counted starting

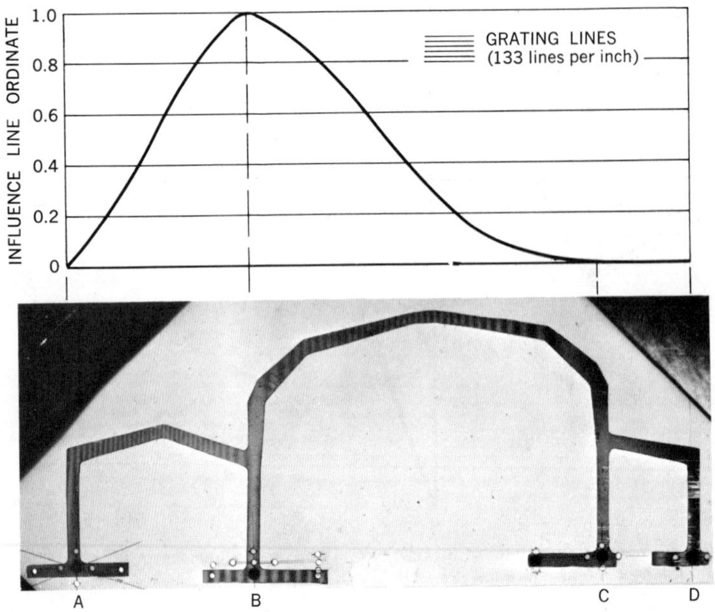

Figure 6.9. Moire fringe pattern in a frame subjected to a vertical displacement at one of the central supports. Influence line of vertical reaction at that support.

from the fixed supports *A* and *C*, where the displacement is zero. The light bands represent integral fringe orders and they progressively increase toward the column where the displacement was applied. An observation of the fringe patterns will show 34 fringes between *A* and *B* and 36 fringes between *C* and *B*. This is due to the fact that initially, before the application of a displacement at *B* there was some slight mismatch of the gratings, causing two fringes to appear between *B* and *C*.

If an equal and opposite displacement is applied at *B*, by using the upper hole as shown in the figure, 32 fringes will appear between *B* and *C*. This indicates that equal and opposite displacements eliminate not only the effects of large deformations (as mentioned previously), but also the effects of initial mismatch of the gratings.

6.7. Circular Ring

In this case the objective of the analysis is not the determination of the deflection curve of the centerline of a thin structural member, but the determination of the displacement field in a thick ring. In the previous examples it could be assumed that the stress distribution across the height of the member is linear. In a thick ring this assumption is not acceptable. It can be expected that the deflection curve of the centerline of the ring and the deflection curves of the boundaries will be appreciably different.

The dimensions of the ring are shown in Fig. 6.10. The material used was epoxy and a 1000-lpi grating was printed on the surface of the ring. A 1000-lpi grating had also been photographed on a film and used as a master.

The ring was loaded under diametral compression, with its grating vertical. An isothetic photograph of the *u*-family was obtained (Fig. 6.10). The ring and the master were rotated 90° and reloaded with its grating horizontal. An isothetic photograph of the *v*-family was obtained (Fig. 6.11).

Two isothetics are necessary to identify the displacement: the *u*-isothetic and the *v*-isothetic (when a Cartesian system is used). The difference in the value of displacement between points located in adjacent fringes is equal to the pitch

Figure 6.10. Isothetics (*u*-field) in a circular ring subjected to diametral compression.

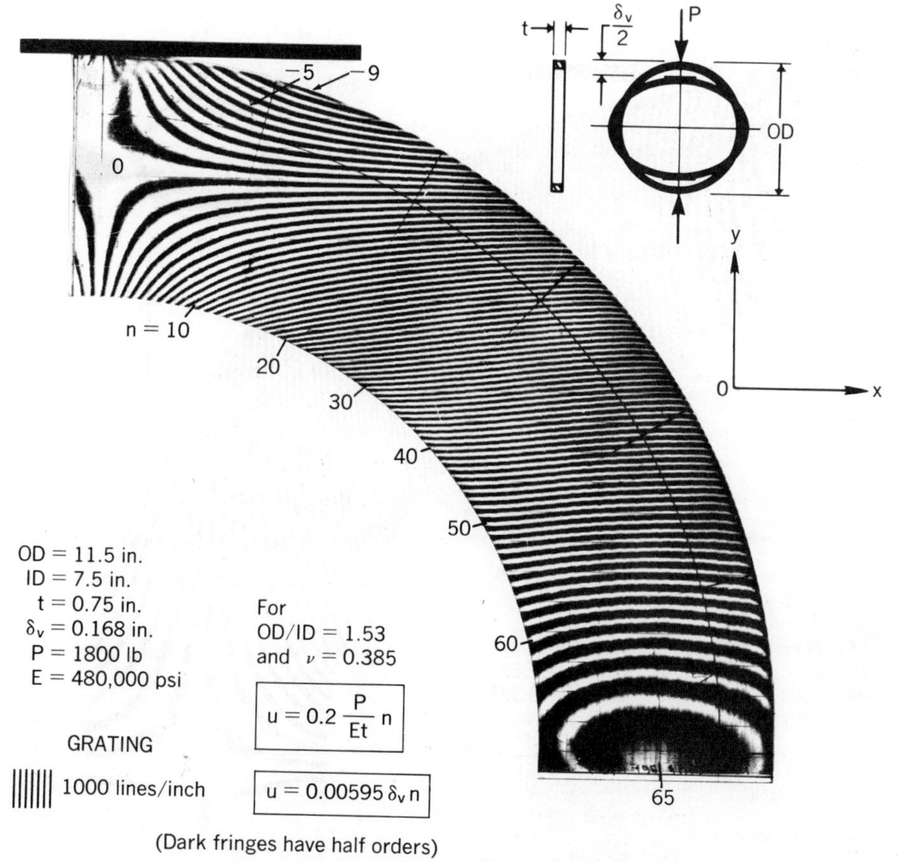

of the grating, 0.001 in. in this case. Therefore, the value of the u-displacement, or v-displacement, anywhere in the ring is given by the order of the corresponding moire fringe multiplied by the pitch of the grating. In the u-isothetic field, the fringe of zero order is the one the points of which have not moved horizontally. Because of the boundary conditions of the test, this fringe for the u-isothetics coincides with the vertical axis of the ring. It will be noticed in Fig. 6.10 that this is a light fringe. In the v-isothetic pattern the fringe of zero order coincides with the horizontal axis of the ring (the top platen of the testing machine moved down as much as the bottom platen moved up).

The specimen used for the moire effects shown in Figs. 6.10 and 6.11 is an epoxy ring ($E = 480{,}000$ psi) which was subjected to a load of 1800 lb. It is important that results be presented in generalized form so that they may be easily applied to all possible cases. Provided that the ratio of outside and inside diameters is equal to 1.53, that Poisson's ratio is equal to $\nu = 0.385$, and that the phenomenon

Figure 6.11. Isothetics (v-field) in a circular ring subjected to diametral compression.

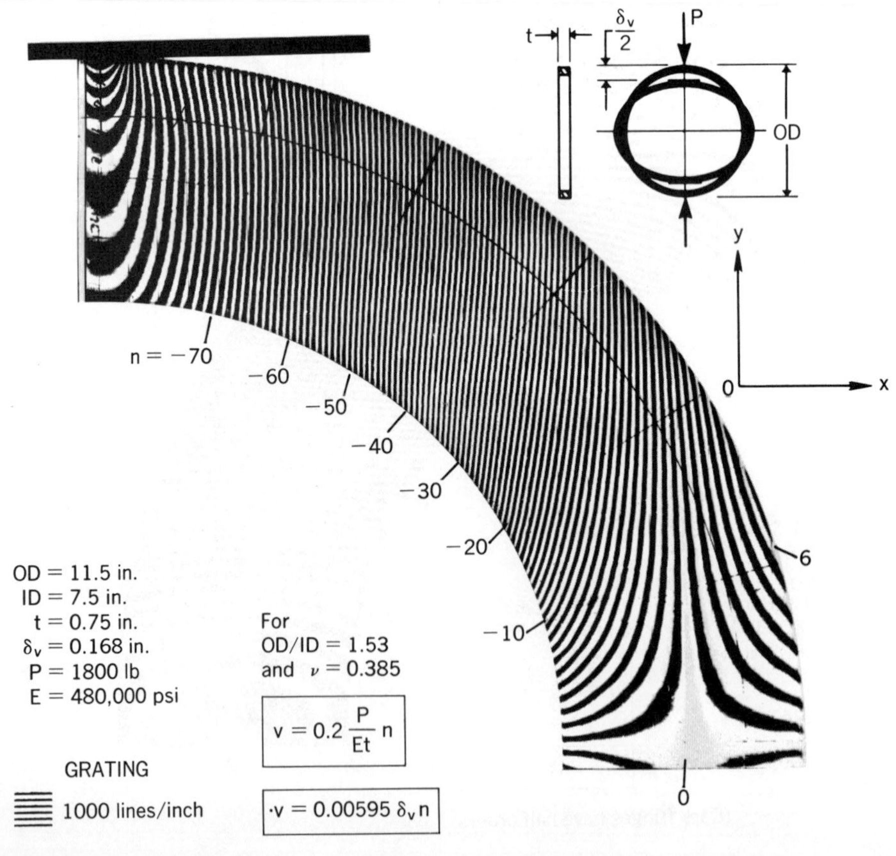

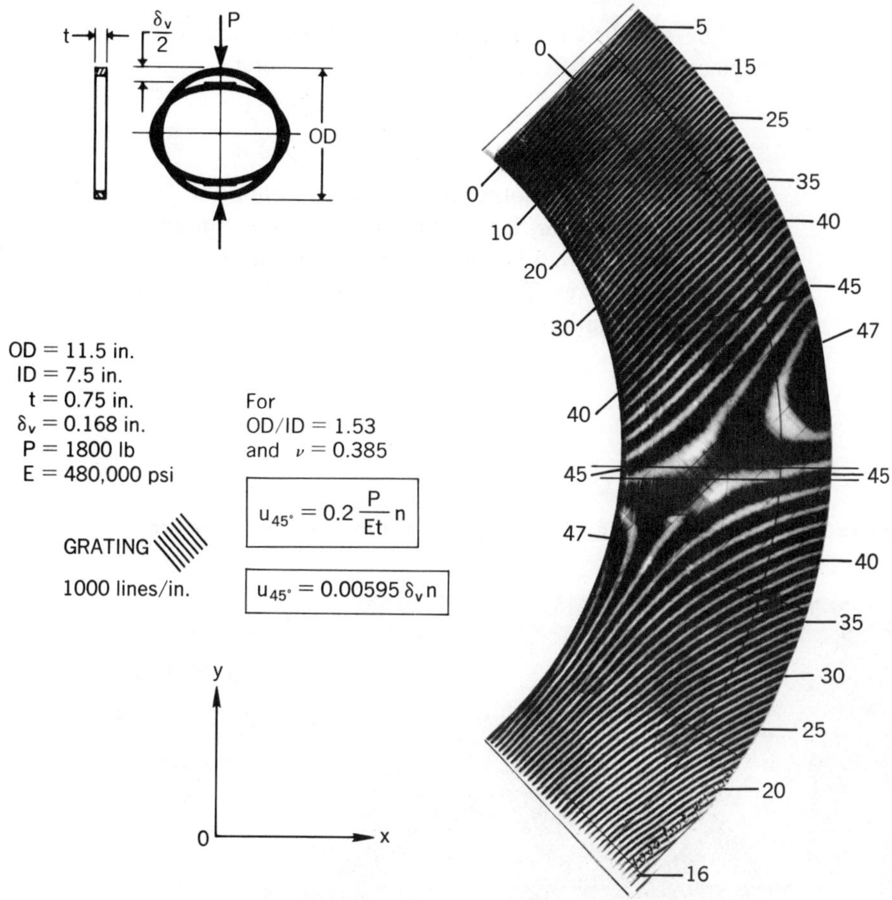

Figure 6.12. Isothetics (u_{45}-field) in a circular ring subjected to diametral compression.

is linear, all rings of any size will have u- and v-displacement given by

$$u = 0.2\frac{P}{Et}n \tag{6.1}$$

$$v = 0.2\frac{P}{Et}n \tag{6.2}$$

where n is the moire-fringe order of the corresponding pattern. This can be verified by replacing the values of P and E given above, and $t = 0.75$ in. in Eq. (6.1):

$$u = 0.2\frac{1800\text{ lb}}{480.000\text{ psi } 0.75\text{ in.}}n = 0.001n\text{ in.}$$

Expressions of the type of Eq. (6.1) are found useful by engineers who frequently determine displacements for known forces. The problem could, however, be specified in terms of displacements only. Calling δ_v the total applied displacement at the ends of the vertical diameter,

$$u = 0.005595\delta_v n \tag{6.3}$$

$$v = 0.005595\delta_v n \tag{6.4}$$

The u- and v-isothetics completely define the field of displacement in the plane body. However, it is sometimes convenient in analysis to have a third field of isothetics. Figure 6.12 shows the component of displacement in a direction 45° to the x- and y-axes. This will be called the 45° isothetic or $u_{45°}$ field. The general form of the equations is, then,

$$u_{45°} = 0.2\frac{P}{Et}n \tag{6.5}$$

$$u_{45°} = 0.005595\delta_v n \tag{6.6}$$

Figure 6.13. Isogonics (u_θ/r) in a circular ring subjected to diametral compression (integral orders).

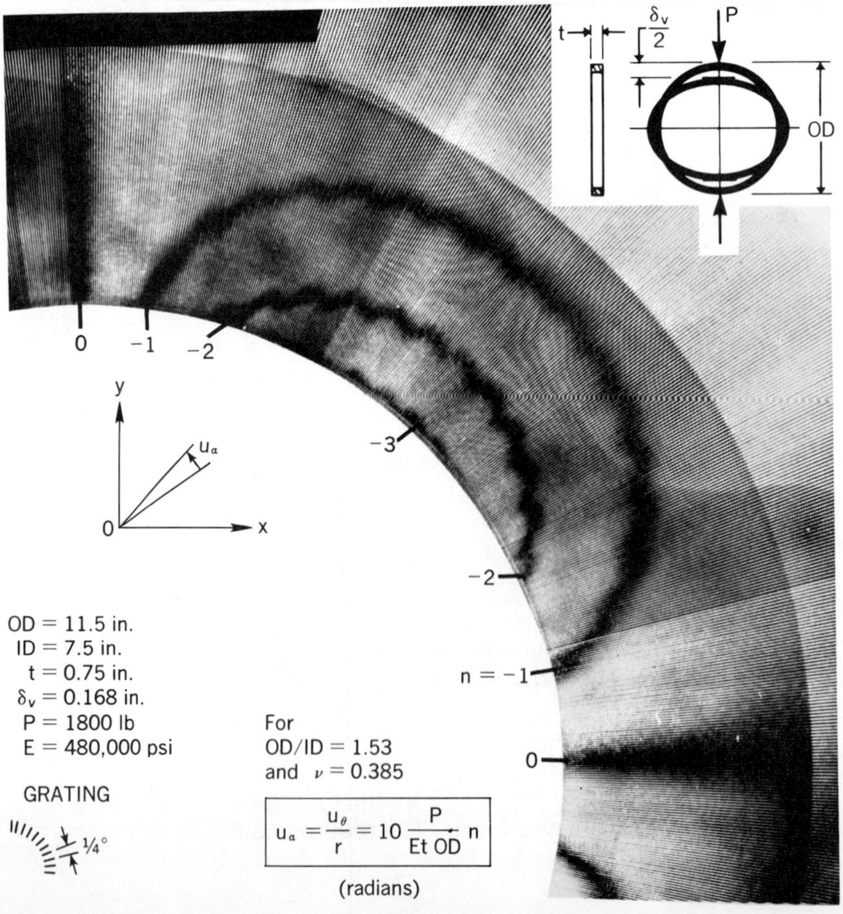

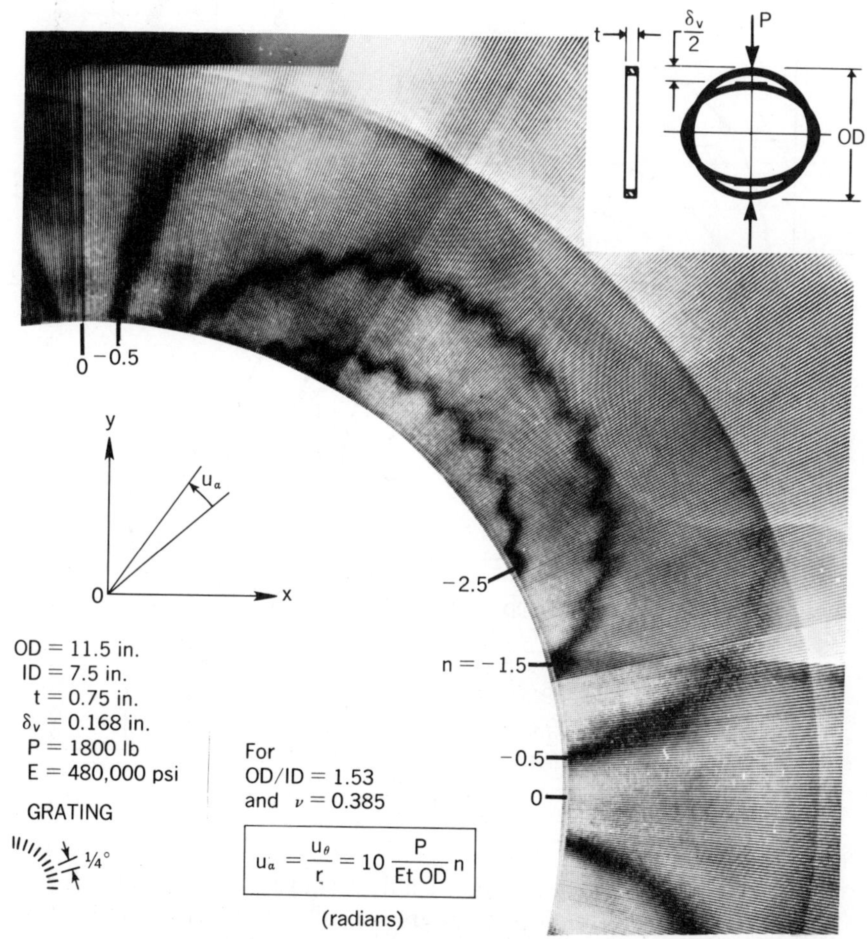

Figure 6.14. Isogonics (u_θ/r) in a circular ring subjected to diametral compression (half orders).

Sometimes it is more convenient to use radial and concentric gratings instead of Cartesian gratings. The fringes of a moire obtained from a radial grating are loci of points exhibiting the same angular displacement. (The angular pitch of the grating is constant and equal in this case to $\frac{1}{4}°$, but the circumferential pitch varies linearly with the distance from the center.) These fringes are called *isogonics* (Figs. 6.13 and 6.14). Since the radial grating used for this test is much coarser than the Cartesian grating, the number of fringes is appreciably smaller. To increase the response two photographs can be used, one giving the integral orders (Fig. 6.13) and the other giving the half orders (Fig. 6.14). The first pattern is obtained by matching the master grating and the deformed grating to produce a dark fringe at zero angular displacement; the second pattern is obtained by

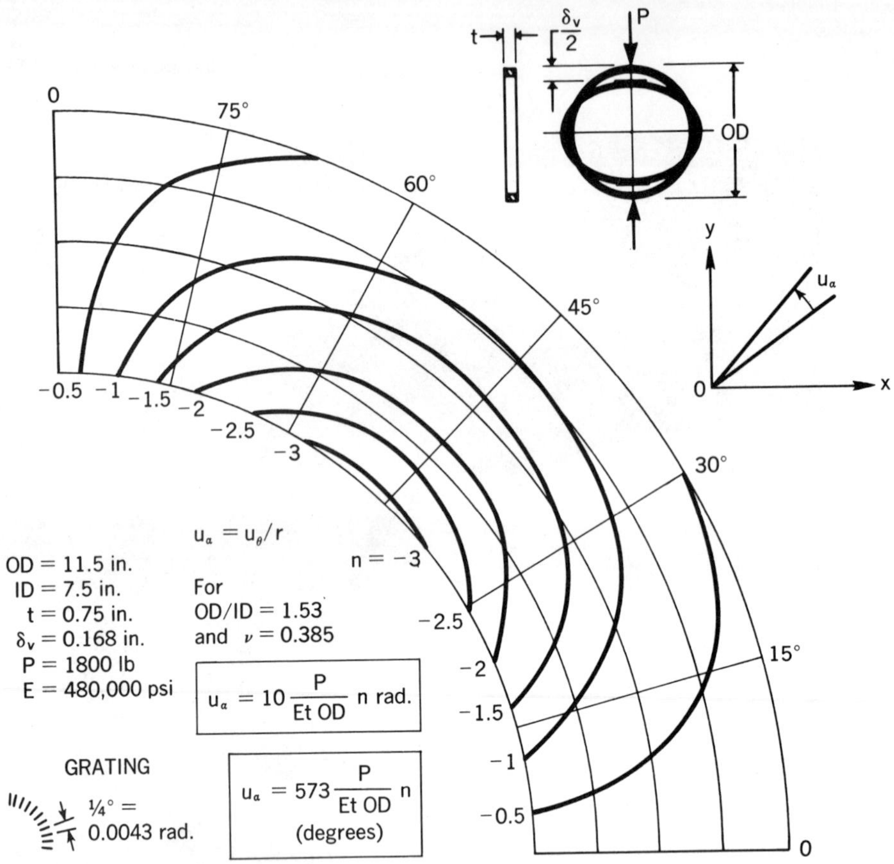

Figure 6.15. Isogonics (u_θ/r) in a circular ring under diametral compression.

matching them to produce a light fringe at zero angular displacement. The superposition of both patterns gives a sufficiently dense family of isogonics (Fig. 6.15). Since, as already pointed out, the axes of symmetry do not rotate, zero-order isogonics have to coincide with the vertical and horizontal axes of the ring.

The tangential displacement is obtained by multiplying the angular displacement by the radial distance. The result is the family of u_θ isothetics (Fig. 6.16).

Calling u_α the angular displacement, it is obvious that

$$u_\alpha = \frac{u_\theta}{r} \tag{6.7}$$

and the generalized scale for the moire patterns is

$$u_\alpha = 10\frac{P}{Et\,\mathrm{OD}}n \quad \text{(radians)} \tag{6.8}$$

$$u_\alpha = 573\frac{P}{Et\,\mathrm{OD}}n \quad \text{(degrees)} \tag{6.9}$$

$$u_\theta = 0.87\frac{P}{Et}n \quad \text{(units of length)} \tag{6.10}$$

In each case n is the fringe order in the corresponding pattern. In the case of u_θ (Fig. 6.16) the order has already been multiplied by the radius.

The fringes of a moire obtained from a circumferential grating are shown in Fig. 6.17. They are isothetics u_r and, since the radial pitch of the grating is 0.012 in., the radial displacement of any point in the ring is given by the fringe order multiplied by 0.012 in. In generalized form, the expression for the radial displacement is

$$u_r = 2.4\frac{P}{Et}n \tag{6.11}$$

or, in terms of the applied displacement,

$$u_r = 0.0715\delta_v n \tag{6.12}$$

The fringe with zero order is the one going through the intersection of the original and the deformed shapes of the boundaries.

Figure 6.16. Isothetics (u_θ) in circular ring subjected to diametral compression.

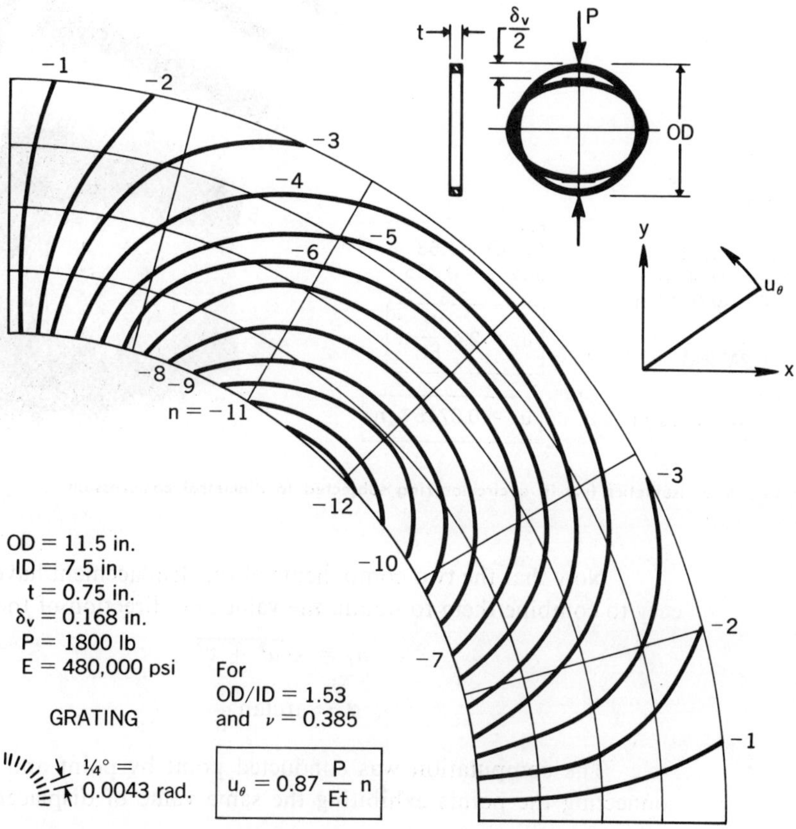

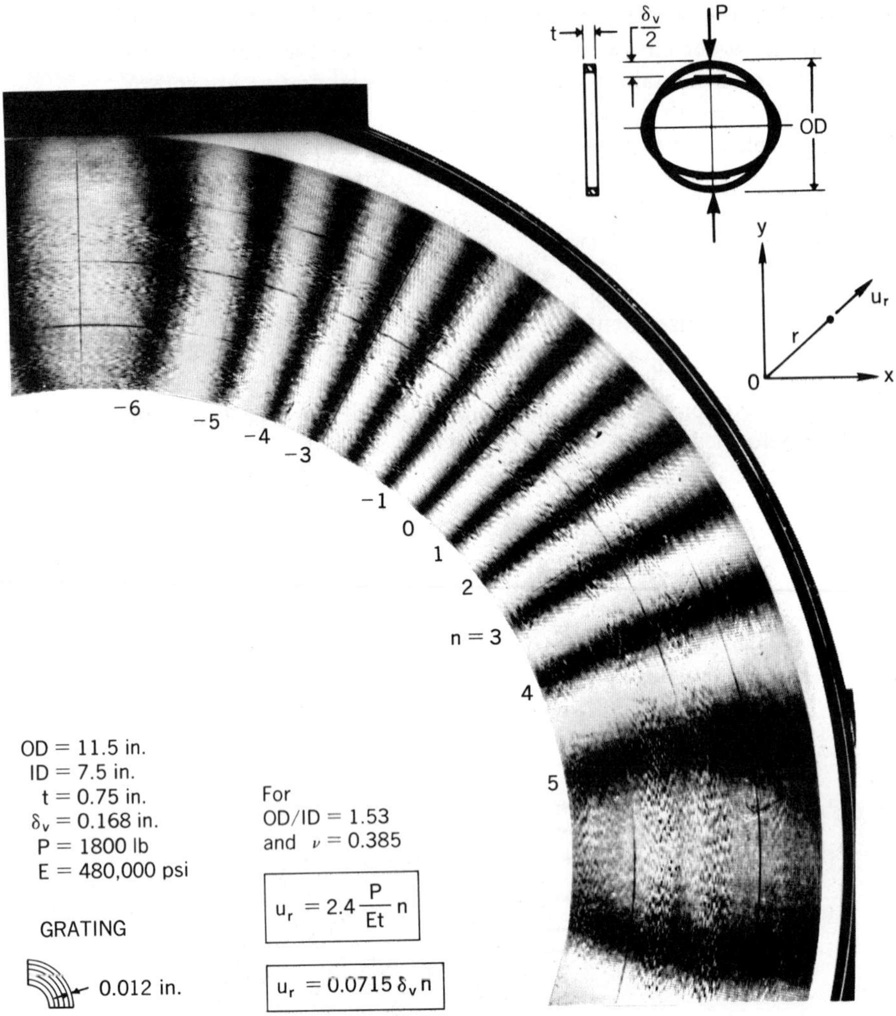

Figure 6.17. Isothetics (u_r) in a circular ring subjected to diametral compression.

Now that the two components of the displacement have been obtained, it is easy to combine them to obtain the value and direction of the total displacement:

$$u_T = \sqrt{u^2 + v^2} \tag{6.13}$$

$$\alpha = \arctan \frac{v}{u} \tag{6.14}$$

The computation was conducted point by point and the curves obtained connecting the points exhibiting the same value of displacement, or of its inclination, are shown in Figs. 6.18 and 6.19. It is suggested that the corresponding families of curves be called "isokinetics" and "isoclinics" of displacement, respectively. *Isokinetics* are therefore loci of points exhibiting the same value of displacement. *Isoclinics* of displacement are loci of points at which the displacement takes place in the same direction.

The direction of the displacement is independent of the size of the ring, but is a function of its modulus of elasticity and Poisson's ratio. The amounts

Figure 6.18. Isokinetics in a ring subjected to diametral compression.

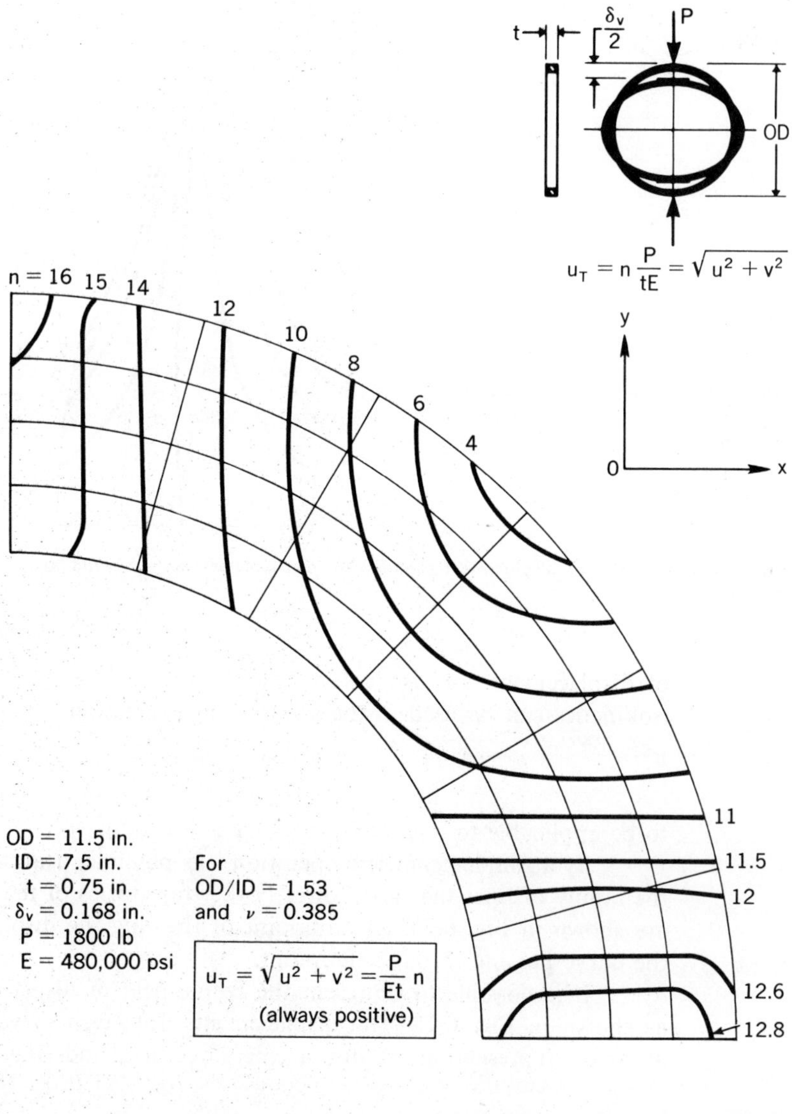

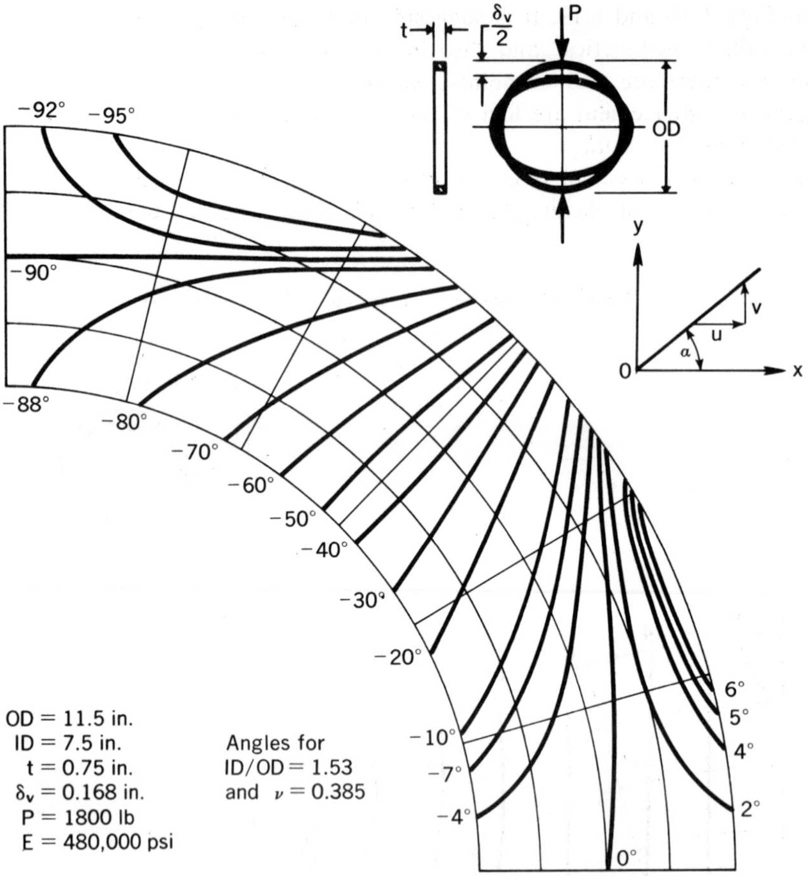

Figure 6.19. Isoclinics of displacement (arctan v/u) in a circular ring subjected to diametral compression.

of displacement are also a function of both constants. The parameters of the isokinetics can be used in the generalized expression

$$u_T = \frac{P}{Et} n \tag{6.15}$$

to be applicable to any ring (provided $\nu = 0.385$).

By a simple geometric operation it is possible, using the isoclinics, to draw the family of lines that gives directly the trajectories of the displacements. These are shown in Fig. 6.20. At any point of the ring the displacement is tangent to the curve passing through the point.

It is possible to represent the whole field of displacements in one graph, as shown in Fig. 6.21. The displacements are given in value and direction by an arrow representing the displacement vector at individual points.

Displacements taking place at points of selected lines, such as the boundaries, can be represented by two continuous curves as shown in Fig. 6.22.

As is obvious from the previous photographs, the moire effect in the present state of the art is a precise and sufficiently sensitive method to determine displacement in materials which have an allowable elastic strain of about 0.01. In more rigid materials, such as aluminum and steel, the moire response is not, in general, sufficiently large for determinations inside the elastic range, and artifices such as mismatching or multiplication should be used. On the other hand, the method is very well suited for determinations in the field of elastic finite strains (Figs. 6.23, 6.24, and 6.25) or plastic strains (Fig. 6.26).

Figure 6.23 is the *u*-field corresponding to a large deformation of a rubber ring. Figures 6.24 and 6.25 are isogonics for two levels of large deformation of

Figure 6.20. Trajectories of displacement in a circular ring subjected to diametral compression.

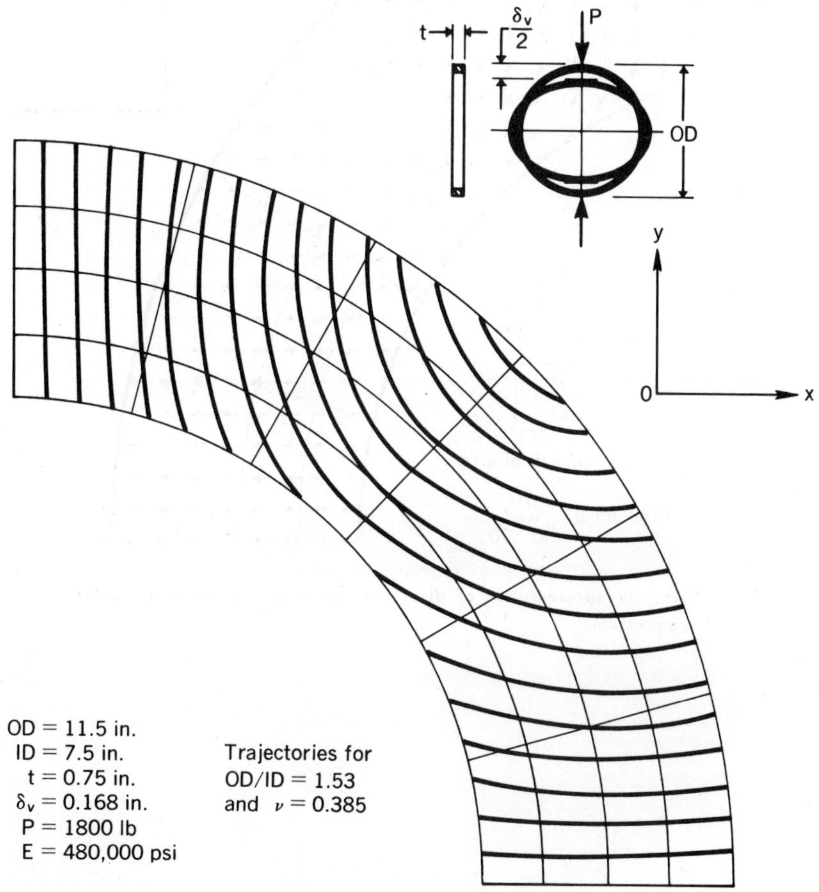

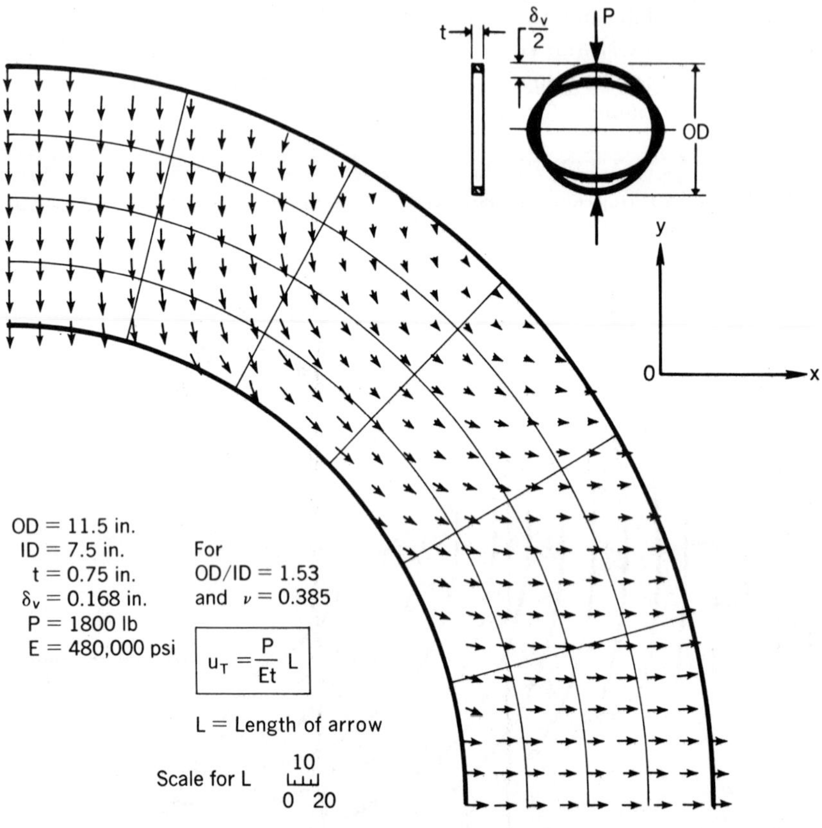

Figure 6.21. Vectorial representation of displacements in a circular ring subjected to diametral compression.

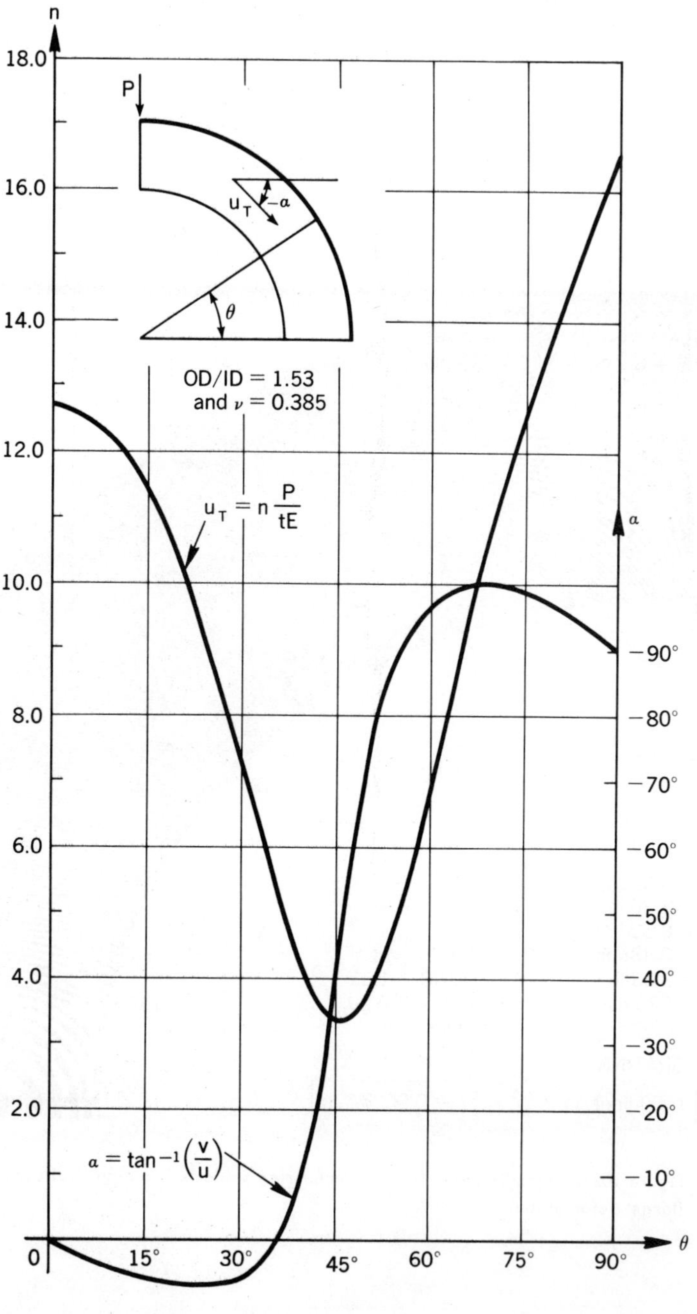

Figure 6.22. Magnitude and direction of displacement of the outer boundary of a circular ring subjected to diametral compression.

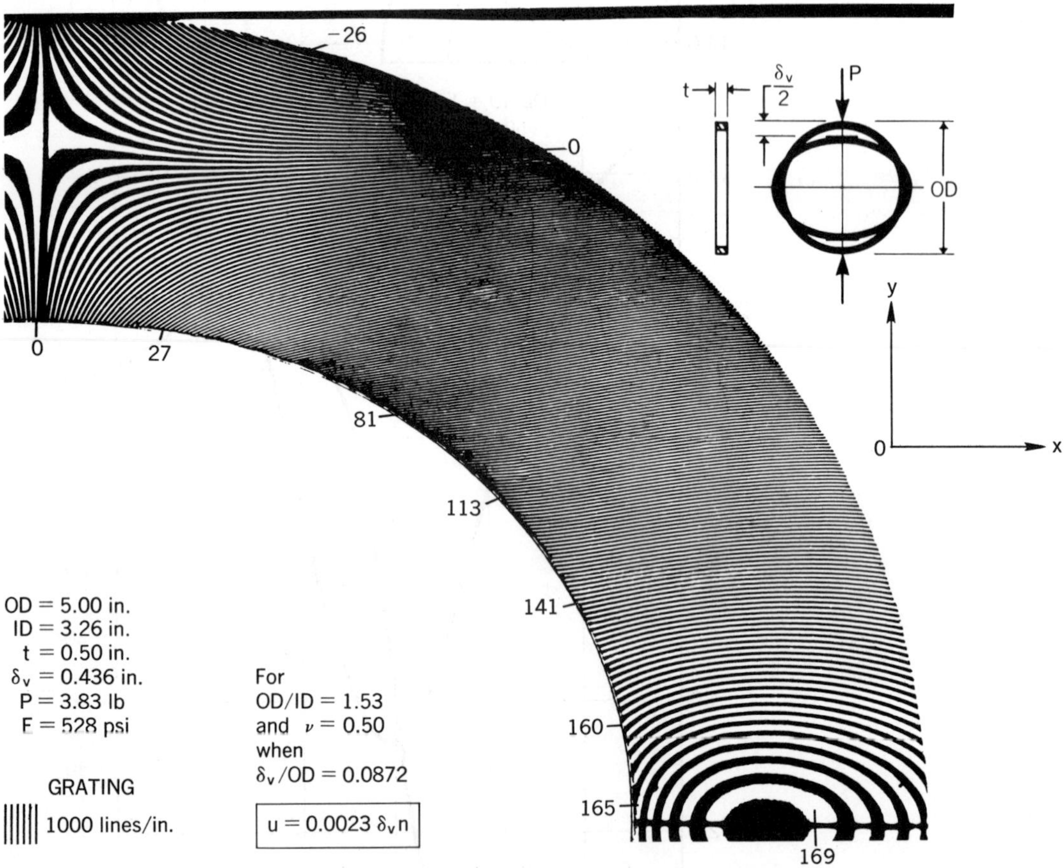

Figure 6.23. Isothetics (*u*-field) in a circular ring subjected to diametral compression (large deformation).

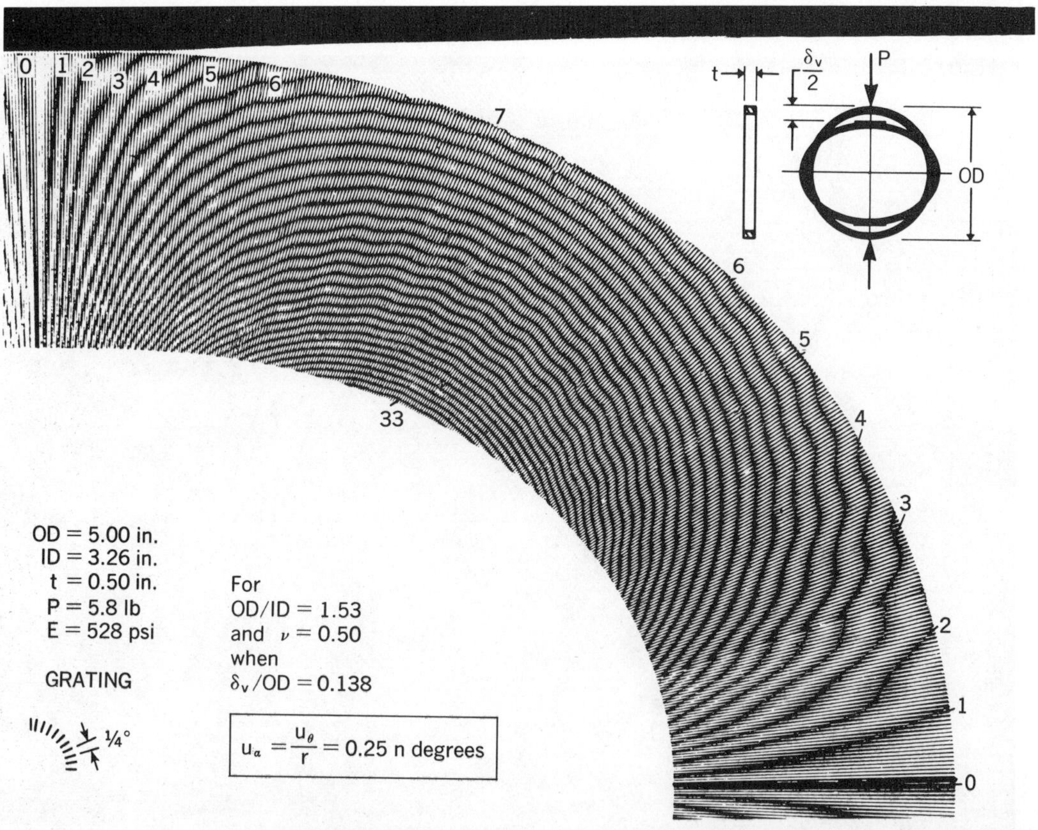

Figure 6.24. Isogonics in a circular ring subjected to diametral compression (large deformation).

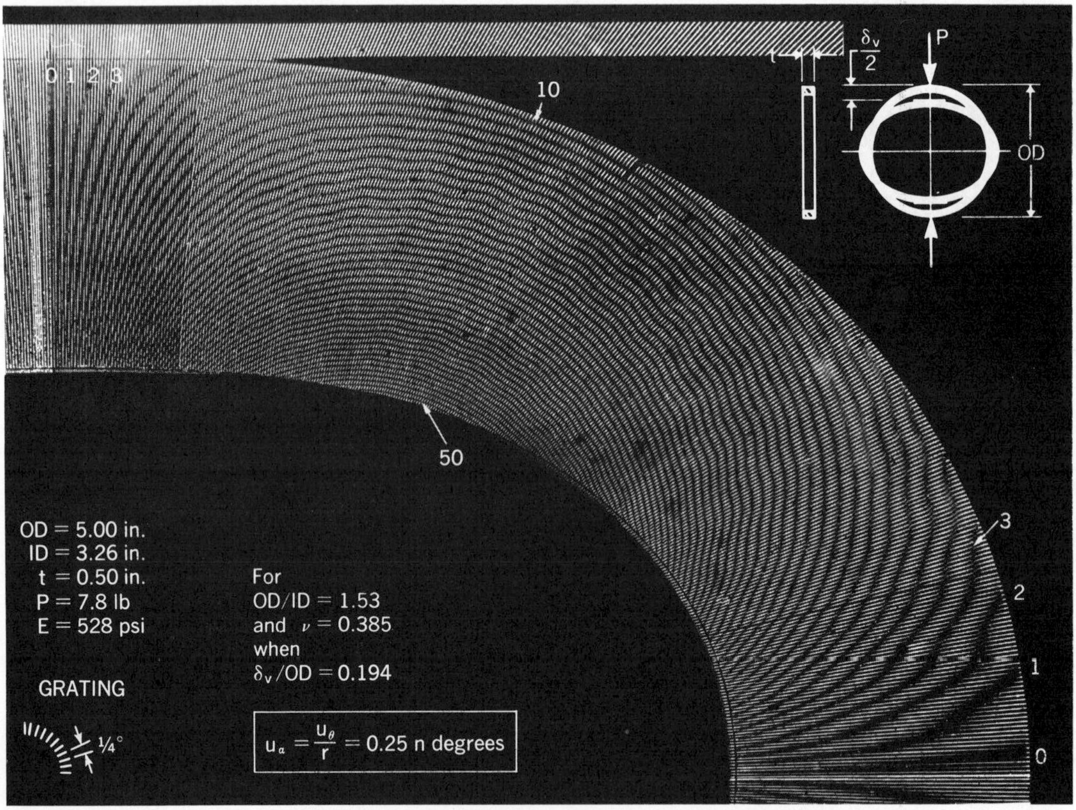

Figure 6.25. Isogonics in a circular ring subjected to diametral compression (large deformation). The pattern is obtained by superposing the master on photographic film of deformed grating.

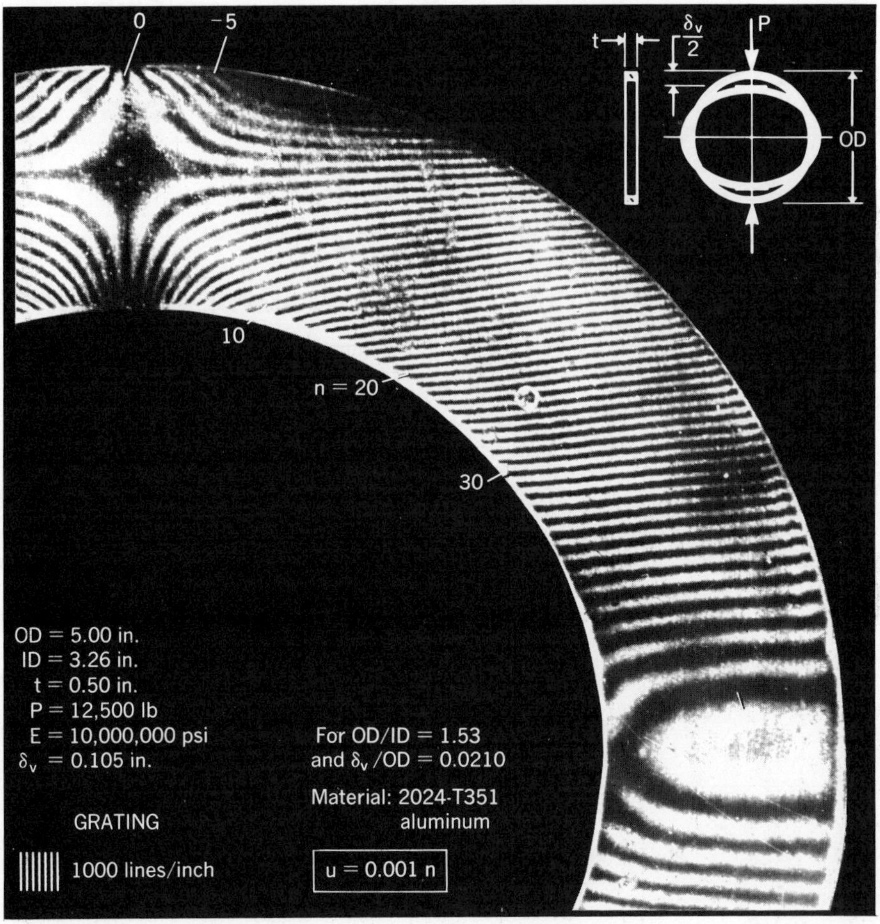

Figure 6.26. Isothetics (u-field) in an aluminum circular ring subjected to diametral compression beyond yielding.

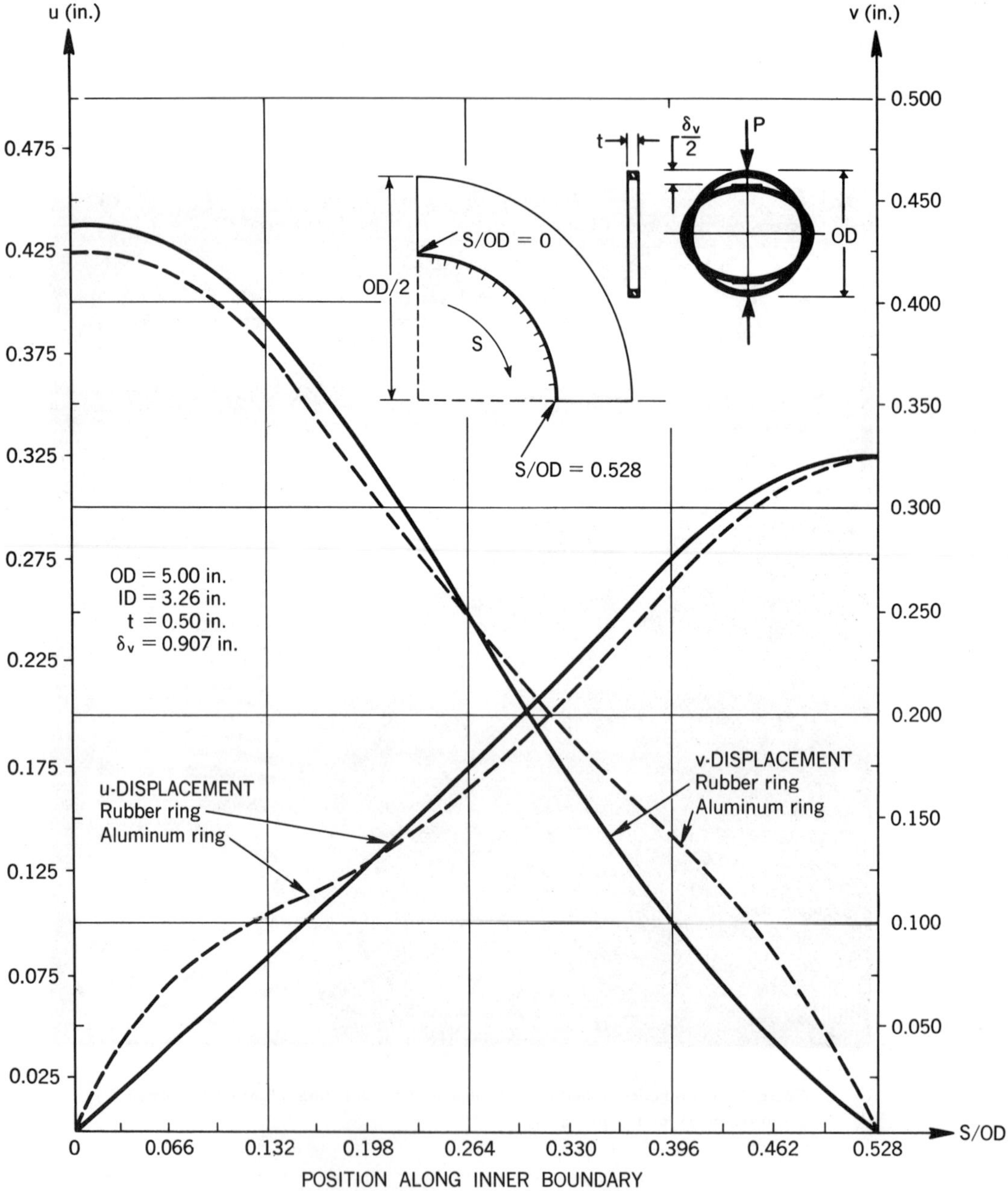

Figure 6.27. Vertical and horizontal displacements at points on the inner boundary of two circular rings made of rubber and aluminum, respectively. Subjected to the same vertical displacement (large deformation).

a rubber ring. At the loading shown in Fig. 6.25, the warpage of the surface was so large that it was impossible to place a master grating flat against it. The deformed model grating was photographed full size without a master, and the moire produced by placing the master on the film negative. This accounts for the different backgrounds of Figs. 6.24 and 6.25.

Figure 6.26 is the *u*-field of an aluminum model loaded beyond the yield point.

Figure 6.27 is a comparison of *u*- and *v*-displacements in rubber and aluminum rings of the same size, both subjected to the same vertical-displacement load. The *u*- and *v*-displacements along the inner diameter have been plotted for both models. The influence of yielding is apparent.

In some of the large deformation tests, the load could not be applied rigorously perpendicular or parallel to both the specimen and master gratings, and the results obtained from one quadrant are not always representative of the results obtained from the whole ring.

7

MOIRE USED TO DETERMINE STRAINS

Displacement-derivatives approach

7.1. Introduction

As shown in Chapter 4, strain can be defined in a general way as a function of the derivatives of the displacement of a point. It has also been shown, in Chapter 1, that the moire phenomenon is related to the displacement of a point. Therefore, the determination of the derivatives of the displacements obtained from the moire pattern can be used to compute strains. Considerations on several alternative procedures that can be followed for this purpose will be made in this chapter. In particular the following methods will be described: (1) graphical differentiation, (2) use of fringe gradient and fringe angle (3) photographic differentiation by shifting deformed gratings, and (4) photographic differentiation by shifting isothetic patterns (moire of moire). The technique of increasing precision by using an original mismatch will also be described.

In the next chapter, some of the same results will be obtained by using only the geometric considerations presented in Chapter 3.

7.2. Graphical Determination of the Derivatives of Displacement

The isothetic fields u and v are represented by the surface $U_i(x, y)$ as shown in Fig. 1.16. To compute the partial derivatives of $U_i(x, y)$, the curves of intersection of the surface with planes of equations $x = C_1$ and $y = C_2$ can be drawn, in which C_1 and C_2 are constants. The procedure used to obtain a partial derivative is illustrated in Fig. 7.1. The horizontal position of each point of intersection of the moire fringes with the line AB (trace of the intersecting plane) is first projected onto the base line CD. The orders are read from the moire fringes, and distances equal to np are scaled up from the base line. A line drawn through the points, thus plotted, defines the cross-section. The slope of this curve at a point gives the derivative $\partial v/\partial y$. The same procedure can be followed to obtain the other three partial derivatives. The values of these derivatives in Eqs. (4.8) and (4.12) give the strains.

7.3. Transformation to Lagrangian Description

If the preceding data are plotted directly from the moire pattern, the results obtained correspond to the Eulerian description. To plot in the Lagrangian description assume that the patterns corresponding to two orthogonal directions have been determined. A point P of initial coordinates x and y moves, after deformation, to point P' with coordinates x' and y'. If the components of the displacements experienced by the point P are u and v, then

$$x = x' - u \tag{7.1}$$

and

$$y = y' - v \tag{7.2}$$

By using the preceding relationships, it is possible to obtain the necessary data to replot the moire patterns referred to the initial configuration.

7.4. Mohr's-Circle Construction Using Moire-Fringe Gradients

In the preceding sections, the method of computing the derivatives of the displacements by using sections parallel to the coordinate axes was developed. Another possibility is to use the fringe gradient. This method can be used in connection with the construction of Mohr's circle at a point, when strains and rotations are small.

In order to find the gradient at a point, a section of the displacement surface in the direction of the gradient can be used. The direction of the gradient can be assumed to be that of the common normal to two neighboring fringes. If δ is the distance between fringes measured along the common normal, and p is the grating pitch, then

$$\text{modulus grad } U_i = \frac{p}{\delta} \tag{7.3}$$

The angle ψ can be measured directly with the protractor.

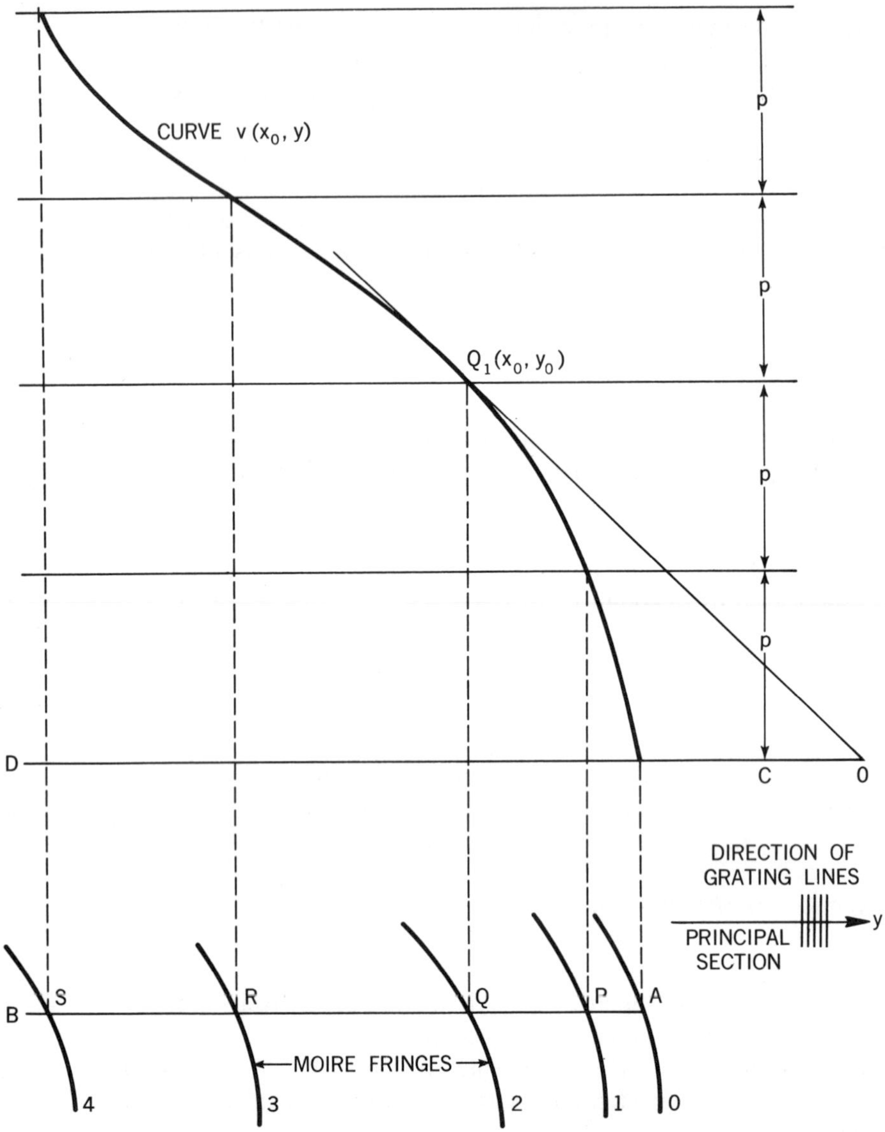

Figure 7.1. Construction of the intersection curve of the surface $v_2(x, y)$ with the plane $x = x_0$.

In the following, the gradient at a point will be indicated by the symbols $\partial u/\partial n$ or $\partial v/\partial n$, according to the case. The derivatives in the direction of the coordinate axes will be indicated by

$$\frac{\partial u}{\partial x} = \epsilon_x \tag{7.4}$$

$$\frac{\partial v}{\partial y} = \epsilon_y \qquad (7.5)$$

and

$$\frac{\partial u}{\partial y} + \frac{\partial v}{\partial x} = \gamma_{xy} \qquad (7.6)$$

These symbols represent, in general, the derivatives of the displacements in the direction of the principal sections and the sum of the derivatives in the direction of the secondary sections. They represent the components of the strain tensor only when the strains and the rotations are small, as shown in Sec. 4.5.

The construction of Mohr's circle is shown in Fig. 7.2. The ϵ-axis is made parallel to the x-axis.

Going from the physical plane to the ϵ-γ plane, the y-axis must be rotated clockwise $\pi/2$. The $\partial v/\partial n$ vector rotates rigidly attached to the y-axis.

In the ϵ-γ plane (Fig. 7.2b) a pole Q is taken, and from this pole the vector QN parallel to $\partial u/\partial n$ is drawn. From Q the vector QM, which is perpendicular to $\partial v/\partial n$, is also drawn. The direction of QM is obtained from $\partial v/\partial n$ by a clockwise rotation of $\pi/2$.

As shown in Fig. 1.16 the partial derivatives of $U_i(x, y)$ are given by

$$\frac{\partial U_i}{\partial x} = \text{grad } U_i(x, y) \cos \psi \qquad (7.7)$$

and

$$\frac{\partial U_i}{\partial y} = \text{grad } U_i(x, y) \sin \psi \qquad (7.8)$$

in which ψ is the angle between the x-axis and grad U_i. The tangent of ψ (Fig. 1.16) is given by

$$\tan \psi = \frac{\dfrac{\partial U_i}{\partial y}}{\dfrac{\partial U_i}{\partial x}} \qquad (7.9)$$

If the points M and N are joined, the vector MN is obtained, whose projections MP and NP are equal to

$$MP = \frac{\partial u}{\partial n} \cos \psi_u - \frac{\partial v}{\partial n} \sin \psi_v \qquad (7.10)$$

But

$$\frac{\partial u}{\partial n} \cos \psi_u = \epsilon_x \qquad (7.11)$$

and

$$\frac{\partial v}{\partial n} \sin \psi_v = \epsilon_y \qquad (7.12)$$

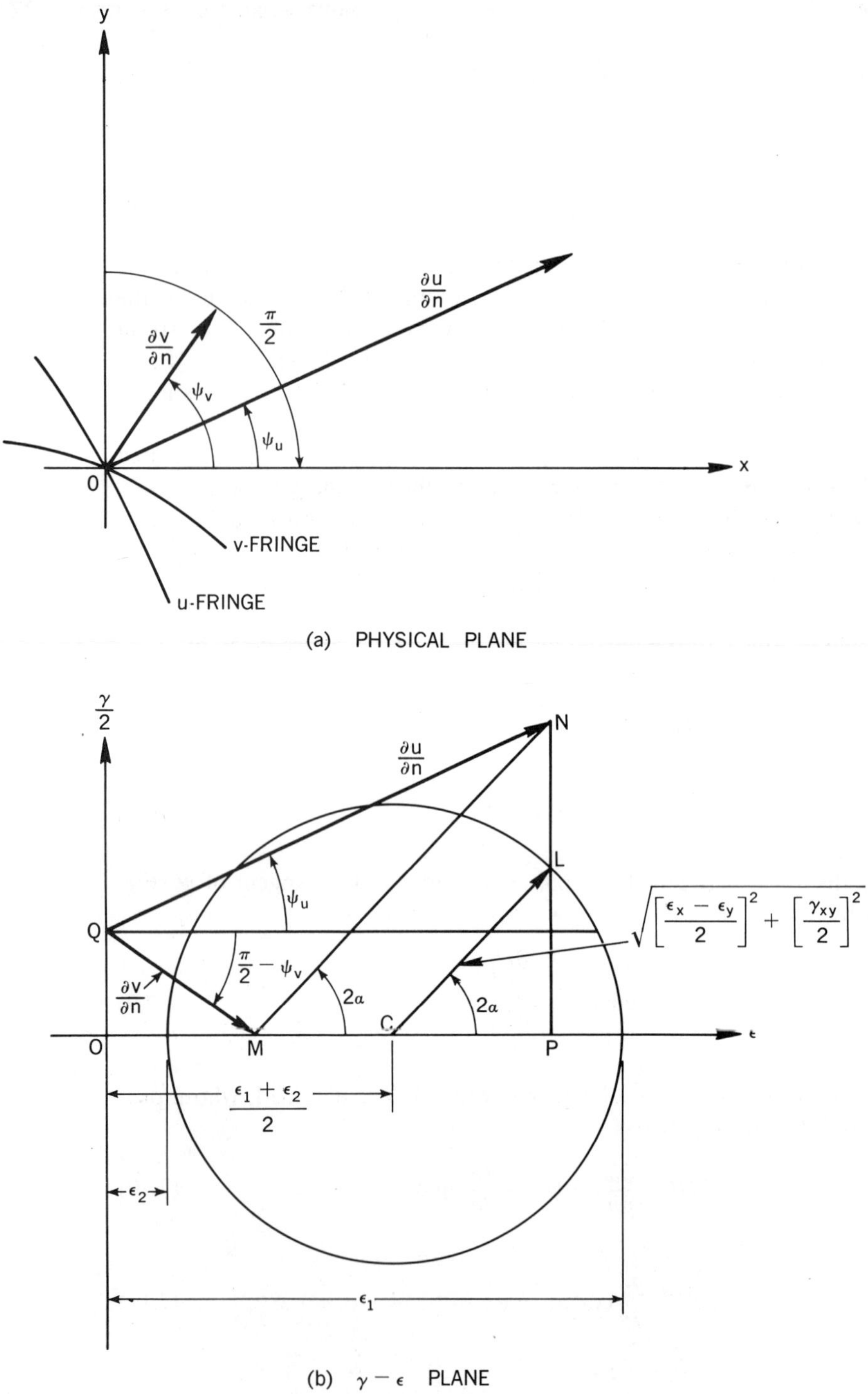

Figure 7.2. Mohr's-circle construction to determine strains using fringe gradients.

then

$$MP = \epsilon_x - \epsilon_y \tag{7.13}$$

The projection NP is equal to

$$NP = \frac{\partial u}{\partial n} \sin \psi_u + \frac{\partial v}{\partial n} \cos \psi_v = \gamma_{xy} \tag{7.14}$$

Point C, which is located half way between M and P is the center of Mohr's circle. The radius of the circle is obtained by joining point C with point L located half way between P and N. The origin of coordinates is the projection of the pole Q on the line MC.

7.5. Examples of Strain Determination

7.5.1. At Axes of Symmetry, by Graphical Differentiation

The study of a disk under diametral loading was performed to illustrate the application of the moire method. A $\frac{1}{2}$-in.-thick sheet of urethane rubber (Hysol 8705) was cast. A 300-lpi grating was photoprinted on the rubber using

Figure 7.3. Disk under diametral compression (horizontal displacements u).

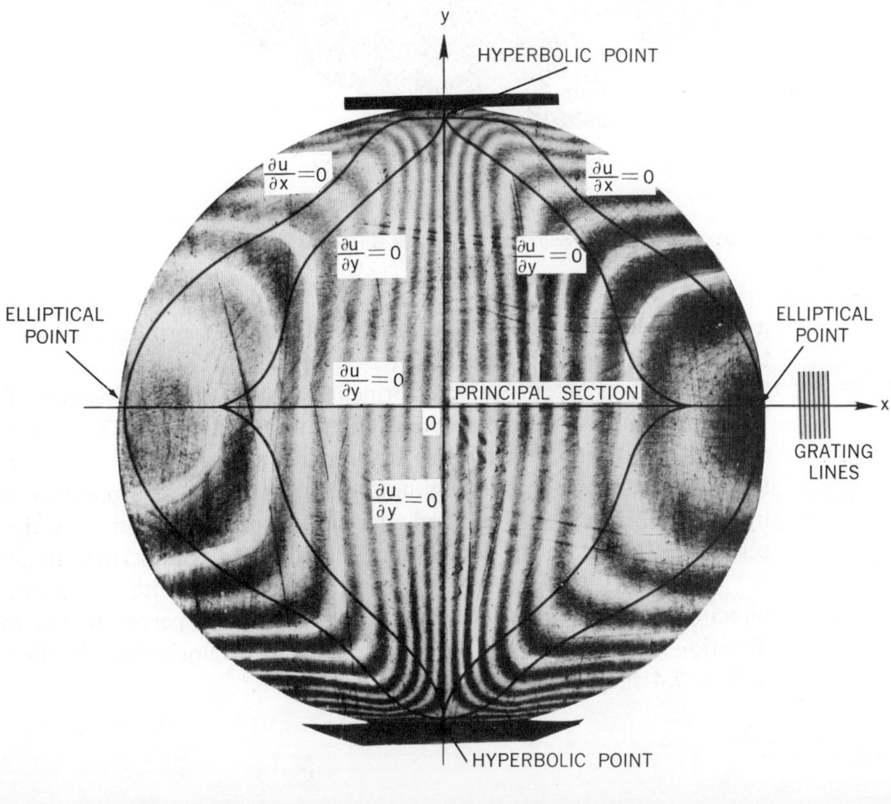

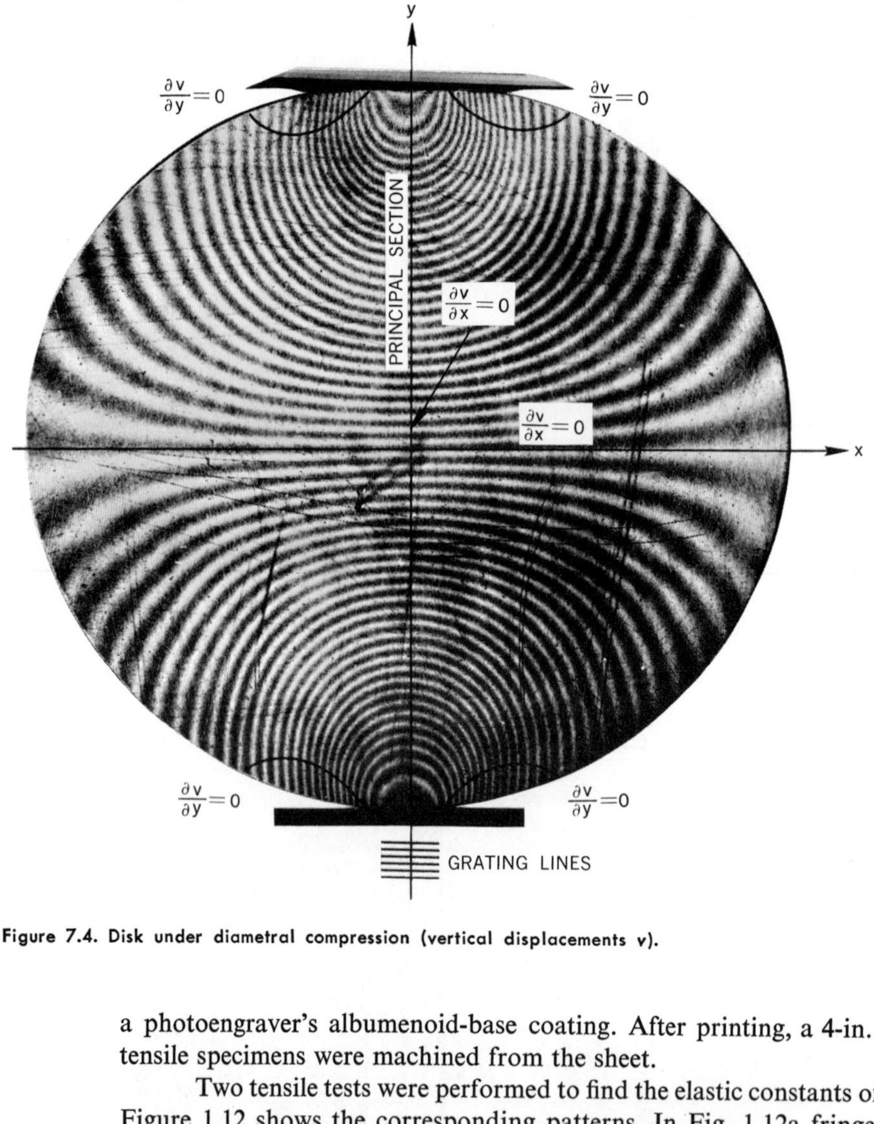

Figure 7.4. Disk under diametral compression (vertical displacements v).

a photoengraver's albumenoid-base coating. After printing, a 4-in. disk and two tensile specimens were machined from the sheet.

Two tensile tests were performed to find the elastic constants of the material. Figure 1.12 shows the corresponding patterns. In Fig. 1.12a fringes are the loci of points with equal transverse displacement, and in Fig. 1.12b fringes are the loci of points with equal longitudinal displacement. The patterns were obtained by the double-exposure method, the unloaded and loaded conditions being successively exposed on the same film. (Technique to be described in Sec. 15.2.2.)

The disk was loaded by means of flat plates, with the grating lines in the direction of the load (Fig. 7.3), and with the lines perpendicular to the loading direction (Fig. 7.4). In Fig. 7.3 moire fringes give horizontal displacements u; and in Fig. 7.4 they give vertical displacements v.

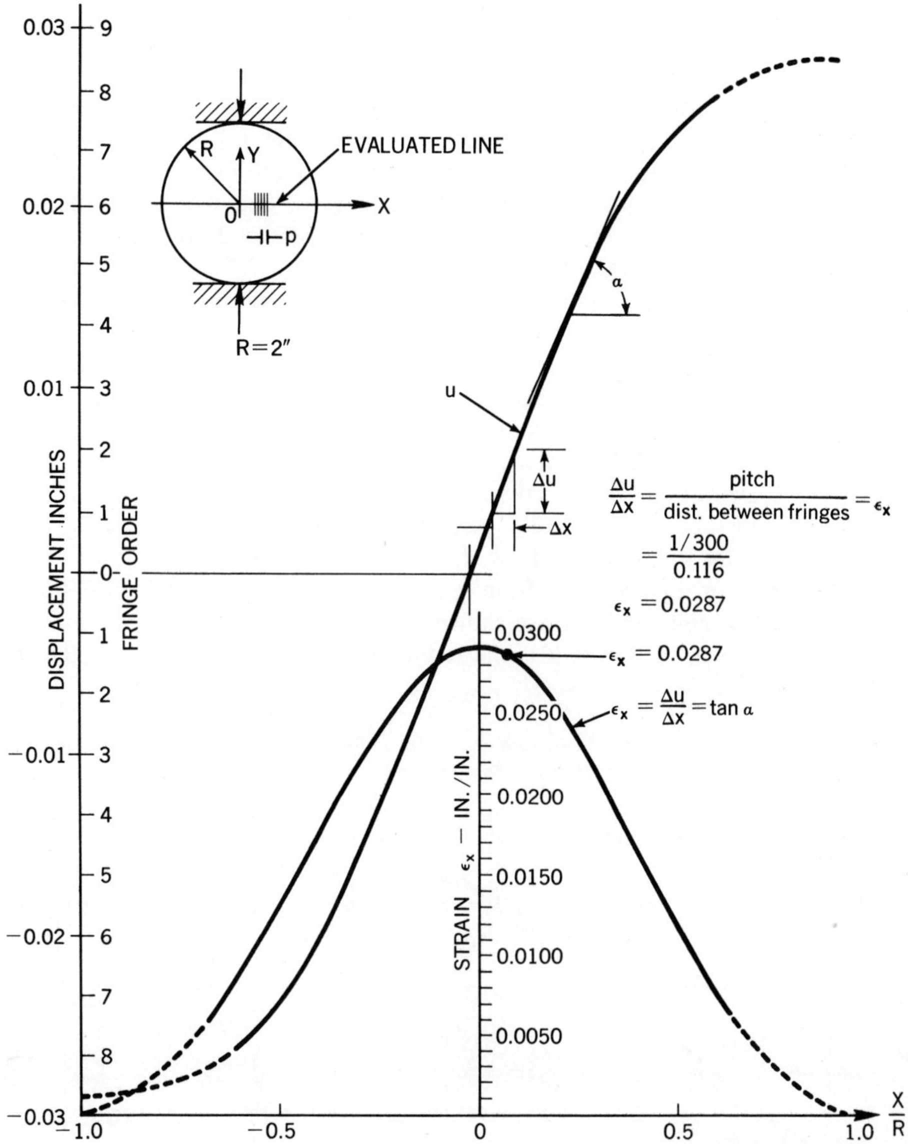

Figure 7.5. Horizontal displacements (u-field) and horizontal strains along the horizontal axis of a rubber disk subjected to diametral load, obtained from a moire pattern.

The loci of points at which the derivatives in the direction of the principal and secondary sections are equal to zero are indicated in Figs. 7.3 and 7.4. On the horizontal diameter (Fig. 7.3) the displacements increase in the positive x direction and, therefore, all of the central region enclosed by the line $\partial u/\partial x = 0$ has positive derivatives. In Fig. 7.4 the displacements in the vertical diameter increase in the negative y direction and, therefore, the derivatives in the direction of the principal section are negative throughout the model, except for small regions near the boundaries where the sign changes. The lines of zero derivative intersect the x- and y-axes at points called singular points. Two types of singular points are indicated in Fig. 7.3.

In Fig. 7.5 the u-displacements and the strains ϵ_x in the horizontal direction are shown for points at the horizontal axis of the disk. The procedure of graphical differentiation is also shown. Although the graphical differentiation method is general, in this case because of symmetry, $\partial v/\partial x = 0$ and the strain is obtained directly from one derivative.

To obtain ϵ_y along the horizontal axis, ten similar but shorter curves were drawn of the fringe values from the v-field plotted in the y-direction at discrete values of x. The strain perpendicular to the horizontal axis, $\epsilon_y = \partial v/\partial y$ was obtained from each of the short curves. Hooke's law was used to obtain stresses $\sigma_1 = \sigma_x$ and $\sigma_2 = \sigma_y$ from ϵ_x and ϵ_y. In Fig. 7.6, the principal stresses along the horizontal diameter are shown in dimensionless form. The ratio x/R is used as abscissa, in which R is the radius of the disk. The ratios of the stresses to the

Figure 7.6. Principal stresses on the horizontal axis of a disk (Hysol 8705) under diametral compression, obtained using the moire method.

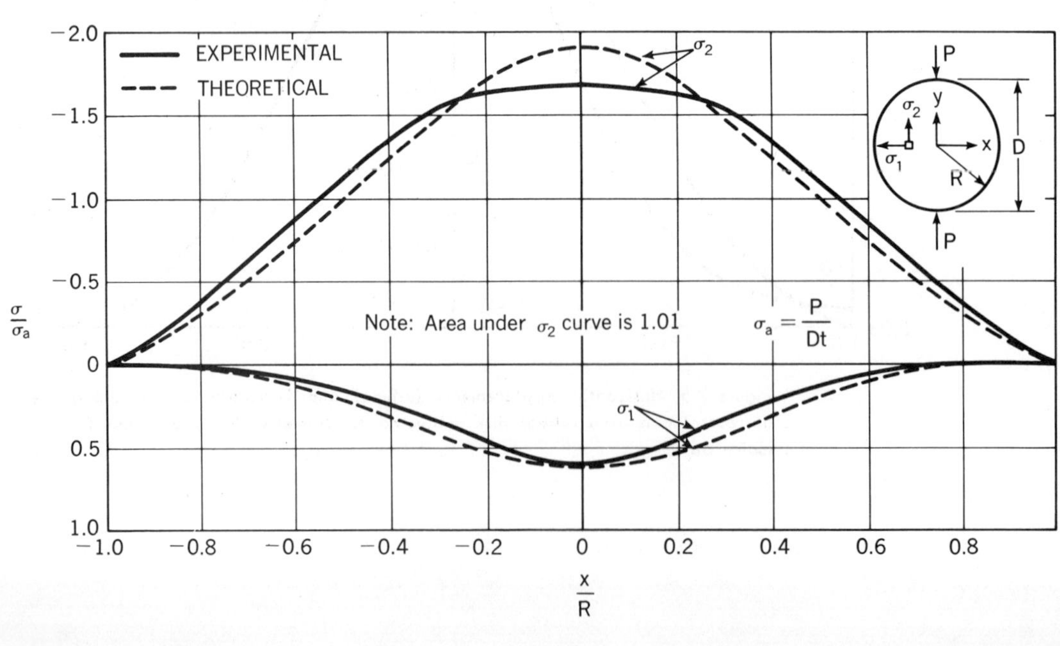

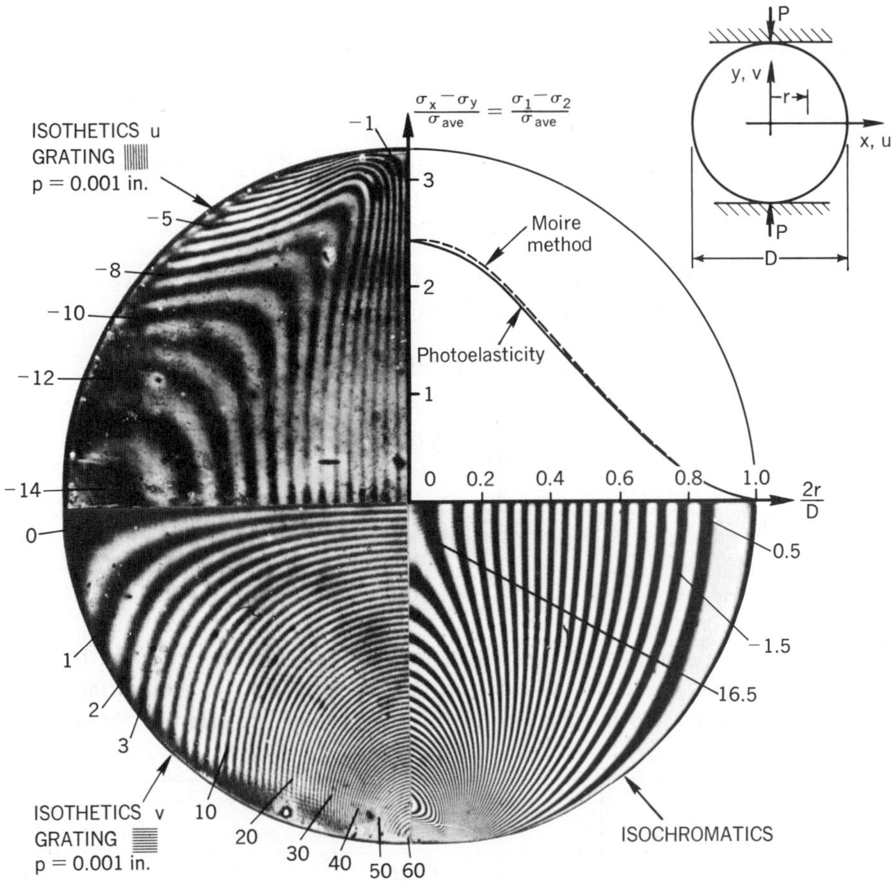

Figure 7.7. Moire patterns, isochromatics, and normalized principal stress differences on the horizontal axis for a disk loaded diametrically.

average stress $\sigma_{ave} = P/Dt$ are used as ordinates. The diameter of the disk and its thickness are denoted by D and t, respectively.

Differences with the theoretical solution can be seen. Figure 7.7 shows a similar model printed with a 1000-lpi grating and much less deformed than the previous one. Quadrants of the u- and v-fields are shown with a quadrant of the corresponding isochromatic-fringe pattern. Principal stresses were obtained as described above and the values subtracted to compare with the isochromatic values. The agreement between the results obtained from both experimental methods is good, indicating that the difference with theory in the first model was due to the different boundary conditions at the point of load application.

7.5.2. *At Generic Points, from Fringe Gradients*

For a point with coordinates $x = 0.2r$ and $y = 0.5r$, the gradients of displacement and the associated angles obtained directly from Figs. 7.3 and 7.4

(Eulerian description) are

$$\frac{\partial u}{\partial n} = 0.0262 \qquad \frac{\partial v}{\partial n} = -0.0509$$

$$\psi_u = 5^\circ \qquad \psi_v = 118^\circ$$

Applying Eqs. (7.7) and (7.8), the partial derivatives can be obtained from the measured gradients and fringe angles.

$$\frac{\partial u}{\partial x} = 0.0262 \cos 5^\circ \qquad = 0.0261$$

$$\frac{\partial u}{\partial y} = 0.0262 \sin 5^\circ \qquad = 0.0023$$

$$\frac{\partial v}{\partial x} = -0.0509 \cos 118^\circ = 0.0239$$

$$\frac{\partial v}{\partial y} = -0.0509 \sin 118^\circ = -0.0449$$

The second powers of the cross-derivatives are seen to be small enough so that the effect of rotation can be neglected in computing the direct strains. However, both exact and elementary values are given for comparison.

	Elementary Eqs. (4.25) and (4.26)	Exact Eqs. (4.8) and (4.12)
ϵ_x^E	0.0261	0.0258
ϵ_y^E	−0.0449	−0.0449
γ_{xy}^E	0.0262	0.0268

The deformation is large enough to require the Eulerian description to be used.

From Fig. 7.8, a similar analysis gives

$$\epsilon_{45^\circ}^E = -0.0175$$

This information could have also been obtained using the graphical differentiation of u- and v-isothetics, as explained in Sec. 7.2, but this method would have required the plotting of fringes in the neighborhood of the point of interest. Alternatively, with less precision, the distance between fringes, at the point of interest, in the directions parallel and perpendicular to the grating lines, could have been measured.

Several methods of computing principal strains and principal directions could be used with the data. The rosette equations could be used with ϵ_x, ϵ_y, and γ_{xy}. An alternative procedure would be to use ϵ_{45°. The principal strains are obtained here using the Mohr's-circle approach described in Sec. 7.4, with the values of $\partial U_i/\partial n$ and ψ_i.

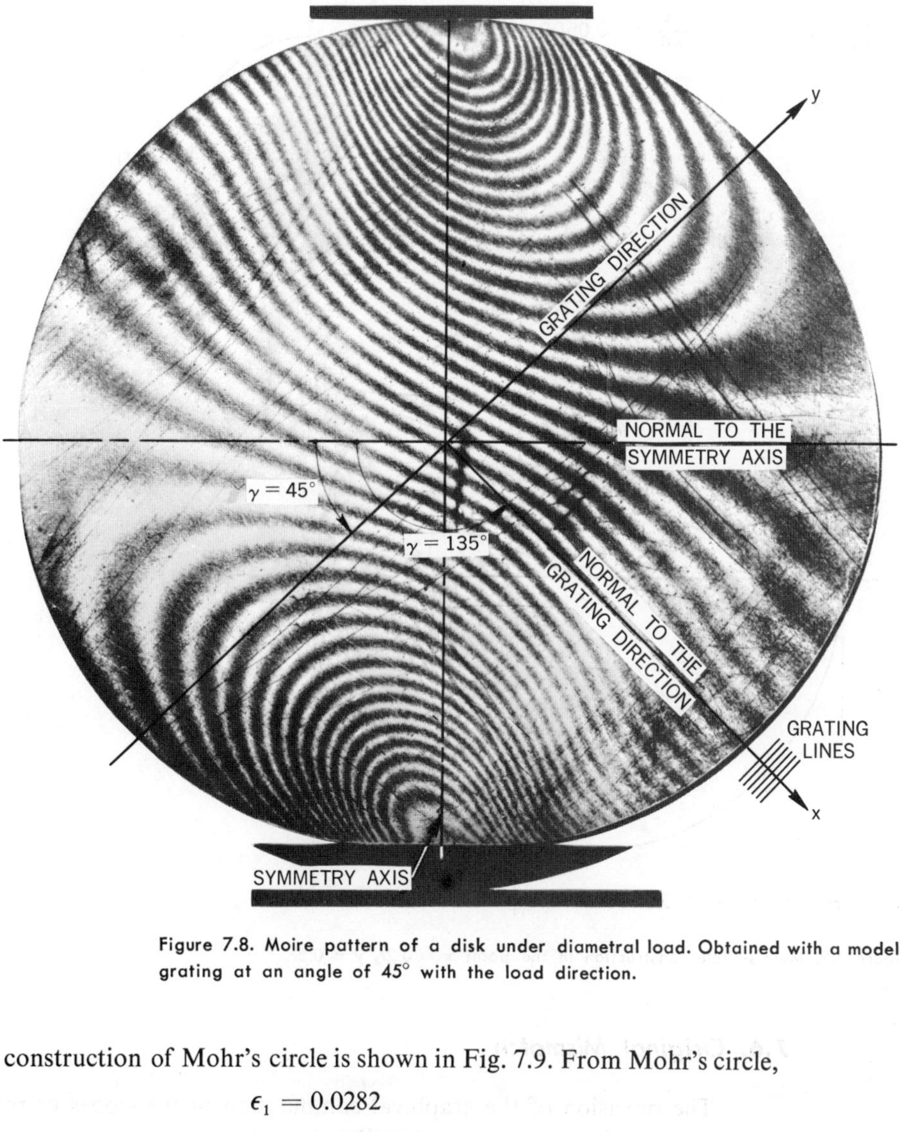

Figure 7.8. Moire pattern of a disk under diametral load. Obtained with a model grating at an angle of 45° with the load direction.

The construction of Mohr's circle is shown in Fig. 7.9. From Mohr's circle,

$$\epsilon_1 = 0.0282$$

$$\epsilon_2 = -0.0462$$

$$\alpha = 10°$$

The rigid rotation of the point is

$$\omega = \frac{1}{2}\left(\frac{\partial v}{\partial x} - \frac{\partial u}{\partial y}\right)$$

$$= \tfrac{1}{2}(0.0239 - 0.0023)$$

$$= 0.0108 \text{ radian}$$

which is about 37′ of rotation, counterclockwise.

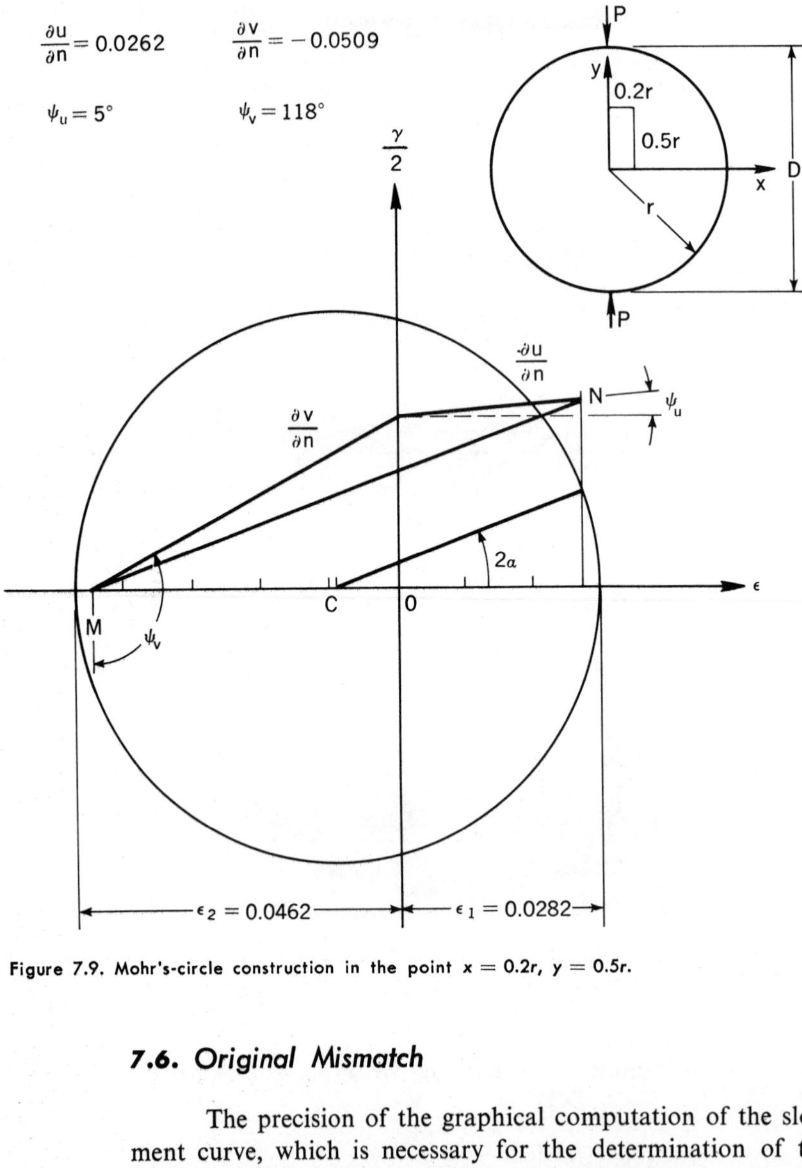

Figure 7.9. Mohr's-circle construction in the point $x = 0.2r$, $y = 0.5r$.

7.6. Original Mismatch

The precision of the graphical computation of the slopes of the displacement curve, which is necessary for the determination of the strains, depends on the number of points defining the differentiated curve. In the moire method a large number of points requires a large number of fringes. For a given strain the number of points depends on the grating pitch. If the grating pitch cannot be changed, an alternative procedure of increasing the number of points is to have an initial pattern and to work with the change of the fringe spacing.

In the following the advantages of grating "mismatch" will be discussed. It is assumed that the strains are small and that the moire pattern has been referred to the initial or undeformed state.

Calling ϵ_m the initial fictitious strain corresponding to the pitch mismatch and p the master-grating pitch, the fringe spacing is

$$\delta_m = \frac{p}{|\epsilon_m|} \tag{7.15}$$

If ϵ_d is the strain produced by the applied deformation, the corresponding fringe spacing is

$$\delta_d = \frac{p}{|\epsilon_d|} \tag{7.16}$$

The fringe spacing resulting from the superposition of ϵ_m and ϵ_d is

$$\delta_f = \frac{p}{|\epsilon_m| \pm |\epsilon_d|} \tag{7.17}$$

where the plus sign applies to the case $\epsilon_d\epsilon_m > 0$, and the minus sign to the case $\epsilon_d\epsilon_m < 0$.

To increase the precision it is desired that

$$\delta_f < \delta_d \tag{7.18}$$

This is achieved by any

$$|\epsilon_m| > 0 \tag{7.19}$$

when $\epsilon_m\epsilon_d > 0$.

When $\epsilon_m\epsilon_d < 0$, inequality (Eq. 7.18) is satisfied if

$$|\epsilon_m| > 2|\epsilon_d| \tag{7.20}$$

It is then clear that a mismatch of the same sign will always produce more fringes and so will always be advantageous; and that a mismatch of opposite sign, to be advantageous, must be at least twice the magnitude of the strain to be analyzed.

To increase the number of moire fringes it is also possible to introduce a rotation of the master grating relative to the specimen grating. This corresponds to a rigid-body motion and is not associated with initial strains.

7.7. Determination of the Derivatives of the Displacement by Shifting a Deformed Grating

In Chapter 2 it was shown that the order of the fringes of a moire pattern is the difference between the orders of the lines of any two superposed gratings (within some broad limits of pitch and rotation and provided the order numbers are chosen in the proper direction). If two copies of the same deformed grating (with no master) are superposed, but shifted a small amount so that the images do not line up, then a moire will be produced corresponding to the difference in orders of the pairs of points that happen to coincide in the shifted arrangement. It will be shown that the contour lines of the partial derivatives of the displacements can be obtained directly as moire fringes of shifted deformed gratings.

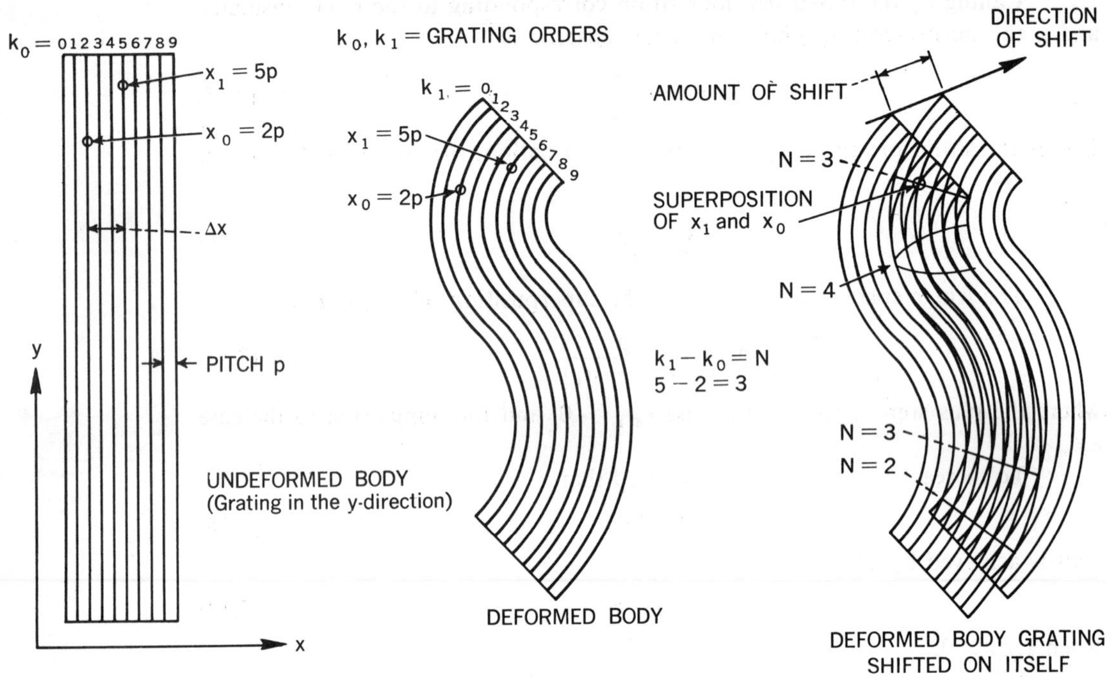

Figure 7.10. Moire pattern produced by the superposition of two copies of a deformed grating shifted one with respect to the other.

This method will therefore automatically produce the result obtained by graphical differentiation in Sec. 7.2.

7.7.1. *Distance Between Two Points in the Undeformed Body (Given by Moire Fringes)*

Consider a deformed grating (Fig. 7.10) produced from an originally undeformed system of straight equidistant lines with pitch p and initially lying parallel to the y-coordinate in an x-y Cartesian frame of reference. After deformation, the lines may no longer be straight, equidistant, or parallel. However, they will still represent the position the points had in the undeformed state; that is, the grating lines of order 1 can be said to have had a position $x = p$ in the undeformed state, and the grating line of order 2 had a position $x = 2p$, etc.

Suppose now that two copies of this grating are superposed, one slightly shifted with respect to other in any arbitrary direction. A moire pattern will be produced as shown in Fig. 7.10. The order of a fringe of the moire pattern can be defined as the difference between the orders of any pair of points (one point from each copy of the superposed deformed grating) that make the fringe. Suppose the two points in the undeformed body had positions x_1 and x_0. (Nothing need be said here about their y-position.) Call the associated grating orders k_1 and k_0 ($x_1 = k_1 p$, $x_0 = k_0 p$). The moire-fringe order N at the point of coincidence is then given in terms of grating orders as

$$N = k_1 - k_0 \tag{7.21}$$

Since

$$k_1 = \frac{x_1}{p} \quad \text{and} \quad k_0 = \frac{x_0}{p}$$

then

$$N = \frac{x_1}{p} - \frac{x_0}{p}$$

and

$$Np = x_1 - x_0$$

This happens to be the distance Δx between the points, in the x-direction before deformation; therefore,

$$Np = \Delta x \tag{7.22}$$

It is evident, then, that the fringe order of the moire pattern produced by a deformed grating (that originally lay parallel to the y-coordinate), superposed and shifted on itself in any direction, gives the x-component of the distance between two points in the undeformed grating—the two points which coincide after the shift.

7.7.2. Distance Between Two Points in the Deformed Body (Given by the Magnitude of a Shift of the Image of the Body)

A second Cartesian coordinate system (x', y'), referring to the deformed body, is introduced here. In Fig. 7.11 the (x', y') system is shown in coincidence with the x-y system. This is a common procedure when the moire method is used; however, in general, the undeformed and deformed shapes of a body can be referred to two different coordinate systems, and the shift of one over the other can take place in any Cartesian direction.

Consider now that a shift of the deformed grating has taken place in either the x- or y-direction (as shown in Fig. 7.11).

If the shift is in the x-direction, then every pair of superposed points will be separated in the actual deformed body by the same shifted distance $\Delta x'$.

7.7.3. Ratio Between Distances Before and After Deformation

Dividing both members of Eq. (7.22) by $\Delta x'$, an expression is obtained relating the x-component of the distance between two points, before and after

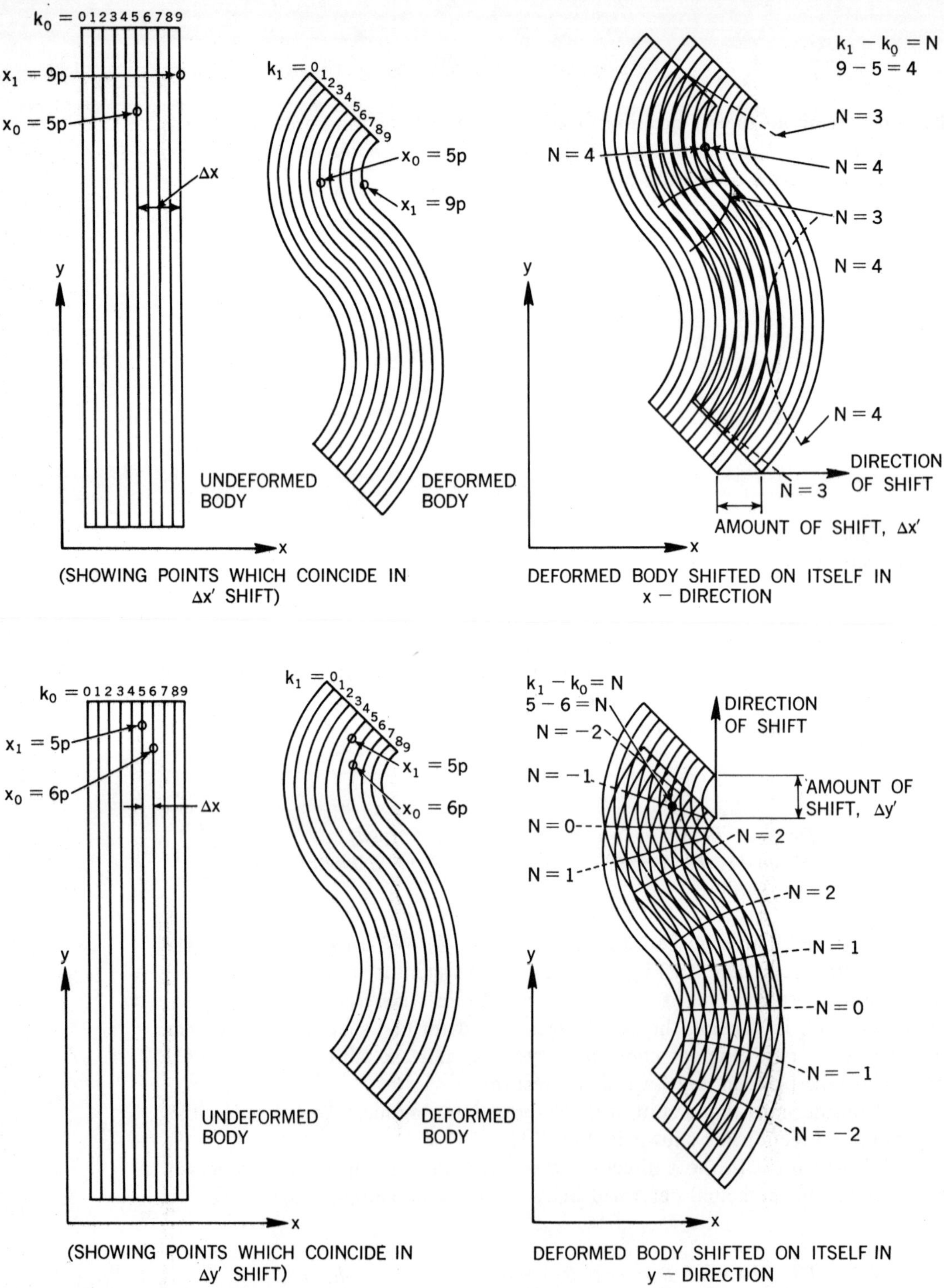

Figure 7.11. Moire patterns produced by superposing two copies of the deformed grating and shifting in the Cartesian directions.

deformation, to the fringe order:

$$\frac{\Delta x}{\Delta x'} = \frac{N_{x'}p}{\Delta x'}$$

Or, if the shift is in the y-direction:

$$\frac{\Delta x}{\Delta y'} = \frac{N_{y'}p}{\Delta y'}$$

The subscripts x' and y', added here to the fringe-order symbol N, indicate which one of the shifts $\Delta x'$ or $\Delta y'$ produced the fringe.

Two similar incremental ratios are obtained from a system of straight lines originally parallel to the x-axis, deformed and shifted in the x'- and y'-directions.

The four ratios so obtained are

$$\begin{aligned}\frac{\Delta x}{\Delta x'} &= \frac{N_{x'}^{x} p_x}{\Delta x'_x} \\ \frac{\Delta y}{\Delta y'} &= \frac{N_{y'}^{y} p_y}{\Delta y'_y} \\ \frac{\Delta x}{\Delta y'} &= \frac{N_{y'}^{x} p_x}{\Delta y'_x} \\ \frac{\Delta y}{\Delta x'} &= \frac{N_{x'}^{y} p_y}{\Delta x'_y}\end{aligned} \tag{7.23}$$

Superscripts x and y, added to the fringe-order symbol N, indicate that the fringe was obtained from gratings perpendicular to the x-coordinate and y-coordinate, respectively. The notation can be simplified by noting that the shift-to-pitch ratio is the number of fringes present along the length of the shift and calling this ratio N_s. The definition of N_s agrees with the general definition of N and gives the value of a fringe that would be produced uniformly throughout the field if an undeformed grating were shifted on itself. If a second numerical subscript is added to N_s to distinguish the four fields, the four shift-to-pitch ratios can be written as follows:

$$\frac{\Delta x'_x}{p_x} = N_{s1} \qquad \frac{\Delta y'_y}{p_y} = N_{s2} \qquad \frac{\Delta y'_x}{p_x} = N_{s3} \qquad \frac{\Delta x'_y}{p_y} = N_{s4} \tag{7.24}$$

The incremental ratios are then

$$\begin{aligned}\frac{\Delta x}{\Delta x'} &= \frac{N_{x'}^{x}}{N_{s1}} \\ \frac{\Delta y}{\Delta y'} &= \frac{N_{y'}^{y}}{N_{s2}} \\ \frac{\Delta x}{\Delta y'} &= \frac{N_{y'}^{x}}{N_{s3}} \\ \frac{\Delta y}{\Delta x'} &= \frac{N_{x'}^{y}}{N_{s4}}\end{aligned} \tag{7.25}$$

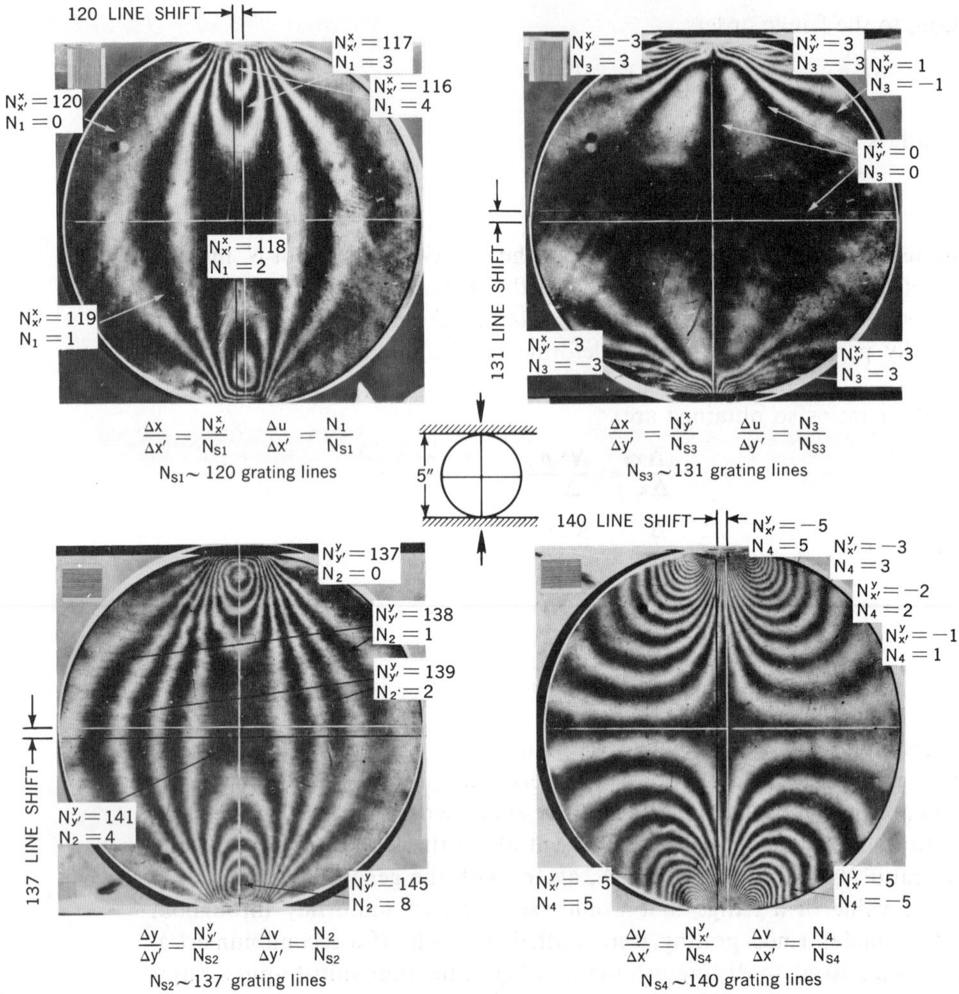

Figure 7.12. Four moire patterns from a 5-in. rubber disk subjected to diametral compression on the vertical axis. Each pattern was produced by shifting and superposing copies of a deformed grating, which, when undeformed, was made up of straight, equidistant lines with a pitch of 0.001 in. The upper two patterns were produced by vertical lines, shifted as shown. The lower two patterns were produced by horizontal lines, shifted as shown.

In many cases, both pitches are the same or all four shifts are the same. This further reduces the complexity of the notation.

If xy and $x'y'$ initially coincide and if the rotations of the fibers in the x- and y-directions are not too large, the ratios $\Delta x/\Delta x'$ and $\Delta y/\Delta y'$ are the ratios of initial length to final length of fibers in the x- and y-directions. The term *stretch* usually designates the ratio of final length to initial length of a fiber, so that for small rotations, the ratios are the reciprocals of the stretches.

Thus, the first of Eqs. (7.23) states that if a grating of lines $x = \text{const}$ with a pitch p_x is deformed and a moire pattern is produced by a $\Delta x'$ shift of the deformed grating, in the direction of the x'-axis, and if rotation is small, then the quantity $N^x_{x'} p_x / \Delta x'_x$ gives the reciprocal of the stretch of a point on the moire line of order $N^x_{x'}$. And if rotations are small, a moire fringe of order $N^x_{x'}$ is a line of equal stretch.

The shifts in Figs. 7.10 and 7.11 were made large enough to produce several fringes. If the grating were much finer, the shift could have been much smaller and the fringes would represent the incremental ratios over much shorter spans. In this respect, the illustrations are exaggerated and do not represent a typical analysis.

Figure 7.12 illustrates a typical analysis. The fringes for a set of four derivatives are shown for the case of a 5-in. rubber disk subjected to vertical diametral compression. The disk before deformation had a grating of 1000 equidistant straight lines per inch. The disk was loaded the same amount, first with the lines vertical and then with the lines horizontal. A photograph was made of both deformed gratings. Each photographic-film negative was copied and shifted on itself in the vertical and horizontal directions to produce the two

Figure 7.13. Loci of points exhibiting the same value of partial derivative $\partial v/\partial x$ (isoparagogic) in a circular ring subjected to diametral compression. All points in the same fringe have a derivative value of 0.0125 times the fringe order. The pattern was obtained by shifting a deformed grating on itself.

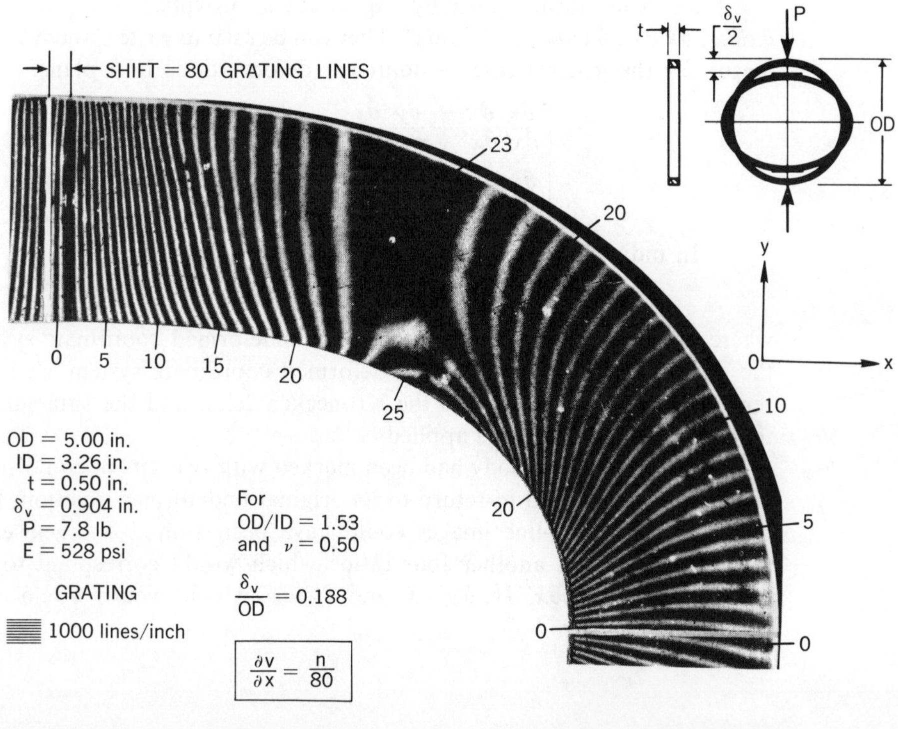

pairs of patterns shown. The shift is shown in terms of grating lines in each case. For an exact analysis, the shifts have to be determined to within a fraction of a grating line. This was not done in the illustration.

The accuracy of the method will be governed by the pitch of the grating, the amount of shift, and the magnitude of the deformation. Different combinations of the three features should be determined in each problem.

No "small-deformation" restriction has been introduced and the method is valid for large deformations. The one approximation that is required is associated with the shift. The quotients $\Delta x/\Delta x'$, etc. are the mean values over the shifted interval. That is, each fringe is produced by the association of two points in the body and will represent an average value between these two points.

Thus, the smaller the shift, the more precise the fringe positions will be located, but the less the response for a given deformation. On the other hand, the larger the deformation, the smaller the shift needed to produce the same response.

It is also noted that even if the deformations are small, as is probably true in the disk shown in Fig. 7.12, nonlinear analysis may be required, due to the magnitude of the cross-derivative shown in Fig. 7.12.

An application of this technique to the case of a thick ring diametrically loaded is shown in Fig. 7.13.

7.7.4. *Computation of Nonlinear Deformation Tensor*

The four ratios given by Eq. (7.25) approximate the partial derivatives $\partial x/\partial x'$, $\partial x/\partial y'$, $\partial y/\partial x'$, and $\partial y/\partial y'$. They can be used to write Cauchy's deformation tensor for the general case of nonlinear deformations in a plane,

$$\begin{bmatrix} \dfrac{\partial x}{\partial x'}\dfrac{\partial x}{\partial x'} + \dfrac{\partial y}{\partial x'}\dfrac{\partial y}{\partial x'} & \dfrac{\partial x}{\partial x'}\dfrac{\partial x}{\partial y'} + \dfrac{\partial y}{\partial x'}\dfrac{\partial y}{\partial y'} \\ \dfrac{\partial x}{\partial y'}\dfrac{\partial x}{\partial x'} + \dfrac{\partial y}{\partial y'}\dfrac{\partial y}{\partial x'} & \dfrac{\partial x}{\partial y'}\dfrac{\partial x}{\partial y'} + \dfrac{\partial y}{\partial y'}\dfrac{\partial y}{\partial y'} \end{bmatrix} \tag{7.26}$$

In indicial notation, this tensor can be represented by

$$c_{kl} = \delta_{KL} x^K_{,k} x^L_{,l} \tag{7.26a}$$

where the indexes K and L represent the undeformed coordinate system xy and the indexes k and l represent the deformed coordinate system $x'y'$, the comma represents differentiation, δ is the Kronecker delta, and the summation convention for repeated indexes is applied.

If the *deformed* body had been marked with two sets of equidistant straight lines and then allowed to return to its original undeformed position, the two sets of distorted grating-line images could have been shifted on themselves exactly as above to obtain another four ratios which would correspond to the partial derivatives $\partial x'/\partial x$, $\partial x'/\partial y$, $\partial y'/\partial x$, and $\partial y'/\partial y$, which in turn would give Green's deformation tensor—

$$C_{KL} = \delta_{kl} x^k_{,K} x^l_{,L} \tag{7.27}$$

—written here only in indicial form, with the same symbols as above.

Presented in this way the approach has even broader application, since the same two tensor fields—Cauchy's and Green's deformation tensors—can be represented in general coordinates as

$$\begin{aligned} c_{kl} &= G_{KL} x^L_{,k} x^L_{,l} \\ C_{KL} &= g_{kl} x^k_{,K} x^l_{,L} \end{aligned} \tag{7.28}$$

where the x's with capital-letter superscripts refer to any coordinate system before deformation and the x's with the lower-case subscripts refer to any coordinate system after deformation. The quantities G_{KL} and g_{kl} are the covariant metric tensors of the undeformed and deformed coordinate systems, respectively.

This would permit, for example, the determination of the deformation-tensor field from a set of circular and angular grating lines (polar coordinates), distorted, and then shifted in the x- and y-directions (Cartesian coordinates).

7.7.5. *Computation of the Nonlinear Strain and the Strain Tensors*

Equations (7.23) can be rewritten using the Cartesian displacement symbols u or v, defined as follows:

$$u_1 = x'_1 - x_1$$
$$u_0 = x'_0 - x_0$$

If Δu is defined as $u_1 - u_0$, then

$$\begin{aligned} \Delta u &= (x'_1 - x_1) - (x'_0 - x_0) \\ &= (x'_1 - x'_0) - (x_1 - x_0) \\ &= \Delta x' - \Delta x \end{aligned} \tag{7.29}$$

Likewise,

$$\Delta v = \Delta y' - \Delta y$$

Dividing Eqs. (7.29) by $\Delta x'$ and $\Delta y'$ and combining with Eqs. (7.23) gives

$$\frac{\Delta u}{\Delta x'} = \frac{\Delta x' - \Delta x}{\Delta x'} = 1 - \frac{\Delta x}{\Delta x'} = 1 - \frac{N^x_{x'} p_x}{\Delta x'_x} \tag{7.30a}$$

$$\frac{\Delta v}{\Delta y'} = \frac{\Delta y' - \Delta y}{\Delta y'} = 1 - \frac{N^y_{y'} p_y}{\Delta y'_y} \tag{7.30b}$$

The cross-ratios $\Delta u/\Delta y'$ and $\Delta u/\Delta x'$ take a somewhat different form. The substitution is as before,

$$\frac{\Delta u}{\Delta y'} = \frac{\Delta x' - \Delta x}{\Delta y'}$$

But here the $N^x_{y'}$ fringes were obtained for $\Delta x' = 0$, so

$$\frac{\Delta u}{\Delta y'} = -\frac{\Delta x}{\Delta y'} = -\frac{N^x_{y'} p_x}{\Delta y'_x} \tag{7.30c}$$

Likewise,

$$\frac{\Delta v}{\Delta x'} = -\frac{N^y_{x'} p_y}{\Delta x'_y} \tag{7.30d}$$

Again, the shift-to-pitch ratio (Eq. 7.24) can be used to simplify the expressions by substitution, as was done in Eq. (7.25). The expressions can be further simplified by assuming that the orders of the shifted grating lines are less than the orders of the unshifted set by the number of grating lines the grating was shifted (the shift-to-pitch ratio). From the definition of N, the orders $N^x_{x'}$ and $N^y_{y'}$ are then reduced by their N_s value, and all four sets of fringe orders are changed in sign. Thus, four new fringe-order symbols are created for the displacement derivatives: N_1, N_2, N_3, and N_4, and are defined as follows:

$$N_1 = -(N^x_{x'} - N_{s1})$$
$$N_2 = -(N^y_{y'} - N_{s2})$$
$$N_3 = -N^x_{y'}$$
$$N_4 = -N^y_{x'}$$

The incremental ratios are then

$$\begin{aligned} \frac{\Delta u}{\Delta x'} &= \frac{N_1}{N_{s1}} \\ \frac{\Delta v}{\Delta y'} &= \frac{N_2}{N_{s2}} \\ \frac{\Delta u}{\Delta y'} &= \frac{N_3}{N_{s3}} \\ \frac{\Delta v}{\Delta x'} &= \frac{N_4}{N_{s4}} \end{aligned} \tag{7.31}$$

The values N_{s1}, N_{s2}, N_{s3}, and N_{s4} are simply the fringe-calibration values found for each pattern by taking the ratio of shift to pitch as specified in Eq. (7.24). Values of N_1, N_2, N_3, and N_4 can be taken from $N^x_{x'}$, $N^y_{y'}$, $N^x_{y'}$, $N^y_{x'}$; or what is more practical in many cases is to order the fringes from some known part, for example, $N_1 = 0$ where $\Delta u/\Delta x' = 0$, etc.

Values of $N_1, N_2, N_3, N_4, N_{s1}, N_{s2}, N_{s3}$, and N_{s4} are all illustrated in Fig. 7.12.

The four ratios given by Eqs. (7.30) or (7.31) approximate the partial derivatives $\partial u/\partial x'$, $\partial v/\partial y'$, $\partial u/\partial y'$, and $\partial v/\partial x'$ associated with a grating placed on the undeformed specimen. These four partial derivatives can be used to compute the nonlinear Eulerian normal- and shear-strain components expressed as

$$\epsilon_{x'} = 1 - \sqrt{1 - 2\frac{\partial u}{\partial x'} + \left(\frac{\partial u}{\partial x'}\right)^2 + \left(\frac{\partial v}{\partial x'}\right)^2}$$

$$\epsilon_{y'} = 1 - \sqrt{1 - 2\frac{\partial v}{\partial y'} + \left(\frac{\partial u}{\partial y'}\right)^2 + \left(\frac{\partial v}{\partial y'}\right)^2} \tag{7.32}$$

$$\gamma_{x'y'} = \arcsin \frac{\dfrac{\partial u}{\partial y'} + \dfrac{\partial v}{\partial x'} - \left(\dfrac{\partial u}{\partial x'}\dfrac{\partial u}{\partial y'} + \dfrac{\partial v}{\partial x'}\dfrac{\partial v}{\partial y'}\right)}{(1 - \epsilon_{x'})(1 - \epsilon_{y'})}$$

The four partial derivatives can also be used to compute the Eulerian strain tensor defined as

$$\begin{bmatrix} \dfrac{\partial u}{\partial x'} - \dfrac{1}{2}\left[\left(\dfrac{\partial u}{\partial x'}\right)^2 + \left(\dfrac{\partial v}{\partial x'}\right)^2\right] & \dfrac{1}{2}\left[\dfrac{\partial u}{\partial y'} + \dfrac{\partial v}{\partial x'} - \left(\dfrac{\partial u}{\partial x'}\dfrac{\partial u}{\partial y'} + \dfrac{\partial v}{\partial x'}\dfrac{\partial v}{\partial y'}\right)\right] \\ \dfrac{1}{2}\left[\dfrac{\partial u}{\partial y'} + \dfrac{\partial v}{\partial x'} - \left(\dfrac{\partial u}{\partial x'}\dfrac{\partial u}{\partial y'} + \dfrac{\partial v}{\partial x'}\dfrac{\partial u}{\partial y'}\right)\right] & \dfrac{\partial v}{\partial y'} - \dfrac{1}{2}\left[\left(\dfrac{\partial v}{\partial y'}\right)^2 + \left(\dfrac{\partial u}{\partial y'}\right)^2\right] \end{bmatrix} \tag{7.33}$$

or, in indicial notation,

$$e_{kl} = \tfrac{1}{2}(u_{k,l} + u_{l,k} - u_{m,k}u^m_{,l}) \tag{7.33a}$$

The Eulerian strain tensor and the Cauchy deformation tensor are related as

$$e_{kl} = \tfrac{1}{2}(g_{kl} - c_{kl}) \tag{7.34}$$

Again, if the grating had been placed on a deformed specimen, ratios $\Delta u/\Delta x$, $\Delta v/\Delta y$, $\Delta u/\Delta y$, and $\Delta v/\Delta x$ would have been obtained, which would approximate $\partial u/\partial x$, $\partial v/\partial y$, $\partial u/\partial y$, and $\partial v/\partial x$. These derivatives give the nonlinear Lagrangian normal-strain and shear-strain components as

$$\epsilon_x = \sqrt{1 + 2\frac{\partial u}{\partial x} + \left(\frac{\partial u}{\partial x}\right)^2 + \left(\frac{\partial v}{\partial x}\right)^2} - 1$$

$$\epsilon_y = \sqrt{1 + 2\frac{\partial v}{\partial y} + \left(\frac{\partial u}{\partial y}\right)^2 + \left(\frac{\partial v}{\partial y}\right)^2} - 1 \tag{7.35}$$

$$\gamma_{xy} = \arcsin \frac{\dfrac{\partial u}{\partial y} + \dfrac{\partial v}{\partial x} + \left(\dfrac{\partial u}{\partial x}\right)\left(\dfrac{\partial u}{\partial y}\right) + \left(\dfrac{\partial v}{\partial x}\right)\left(\dfrac{\partial v}{\partial y}\right)}{(1 + \epsilon_x)(1 + \epsilon_y)}$$

The Lagrangian strain tensor, which is related to the Green deformation tensor, is given in indicial notation by

$$E_{KL} = \tfrac{1}{2}(C_{KL} - G_{KL}) = \tfrac{1}{2}(u_{K,L} + u_{L,K} + u_{M,K}u^M_{,K}) \tag{7.36}$$

and can be calculated from the displacement derivatives with respect to x and y obtained above.

7.7.6. Computation of Small Strains with Small Rotations

For the case of small strains and small rotations, the moire patterns give the conventional strains directly. Lagrangian and Eulerian values are approximately equal, and the corresponding strain-tensor terms all reduce to

$$\epsilon_x = \frac{\partial u}{\partial x}$$
$$\epsilon_y = \frac{\partial v}{\partial y} \qquad (7.37)$$
$$\gamma_{xy} = \frac{\partial u}{\partial y} + \frac{\partial v}{\partial x}$$

This is the same simplification used in Sec. 7.5, in which strains were obtained by graphical differentiation.

7.7.7. An Inherent Difficulty

In principle, the approach outlined in the previous section can be applied to any moire problem. However, there is an inherent difficulty in that approach, inasmuch as it is necessary to photograph directly the deformed grating. Dense gratings on specimens are difficult to photograph. The approach is suggested only for special problems, as where it would be difficult to use a master (say, if the model is in a furnace) or where it is easy to print or copy directly on the model (as with rubber models in light loading frames or on models with frozen deformations that can be annealed, such as the slices of three-dimensional models with frozen deformations). To avoid the difficulty, an alternative method is suggested: the moire-of-moire technique.

7.8. Determination of the Derivatives of the Displacement by Shifting Two Moire Patterns (Moire-of-Moire)

The same results obtained in the previous sections using deformed gratings can be obtained from the u-field and v-field moire patterns. A copy of each moire pattern is superposed on itself and shifted in the two Cartesian directions. The four resulting moire patterns correspond to the same partial derivatives discussed above. The moire-of-moire pattern in this case gives the difference in displacement —say, Δu if the u-field pattern is used—between two points—say, x_1 and x_0— which were superposed by a shift $\Delta x' = x'_1 - x'_0$. The moire-of-moire fringe is then

$$\Delta u = u_1 - u_0 = n_1 p - n_0 p = Np$$

where n_1 and n_0 now refer to the u-field fringe orders. Dividing by the shift $\Delta x'$,

$$\frac{\Delta u}{\Delta x'} = \frac{Np}{\Delta x'}$$

Here, N has the same meaning and value as the N_1 in Eq. (7.31). The same reasoning applies to the other incremental ratios. The moire-of-moire pattern has the disadvantage of retaining the original moire fringes in areas which have a low density of original fringes.

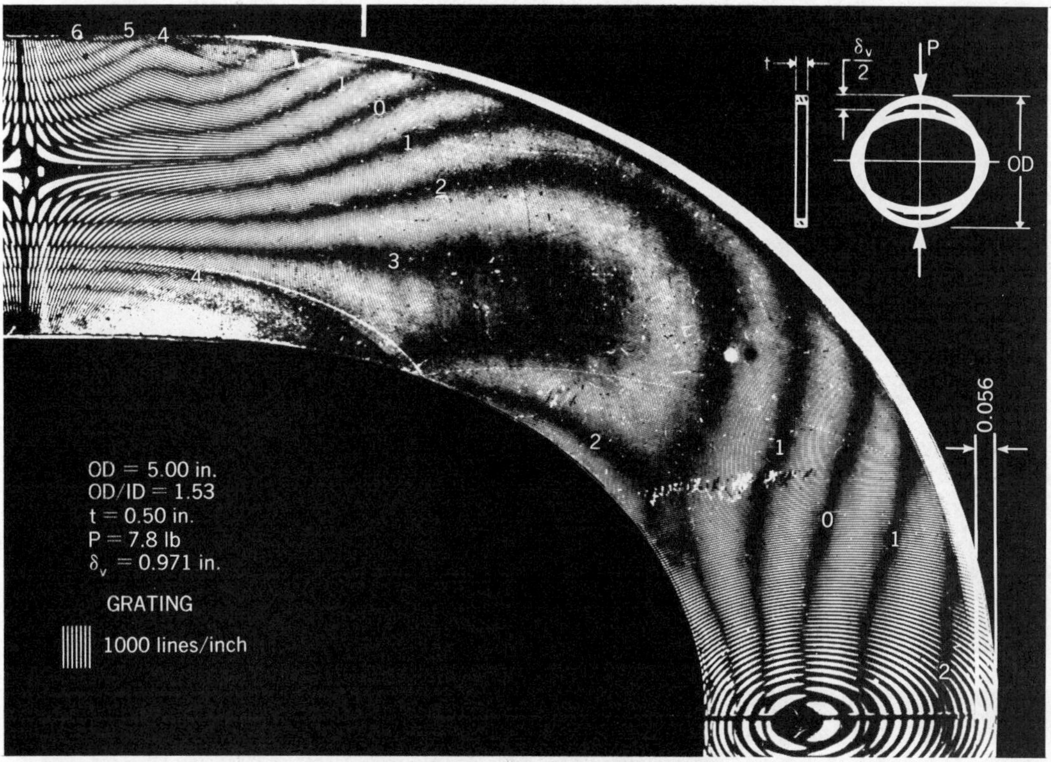

Figure 7.14. Loci of points exhibiting same value of partial derivative $\partial u/\partial x$ (isoparagogic) in a circular ring subjected to diametral compression. All points in the same fringe have a derivative value of 0.017 times the fringe order.

The initial fringe order in the u-field and v-field is often assigned in an arbitrary manner, since variation of the values in the field by a constant amount represents only a rigid-body translation, which is often unimportant; here, however, in the moire-of-moire or derivative field, the absolute value of N is significant. To order the fringes, it is best to find some point at which the same light fringe in the shifted and unshifted copies falls over itself. This will be a point of zero order ($N = 0$). The dark fringes will then be the half orders ($\ldots, -2\frac{1}{2}, -1\frac{1}{2}, -\frac{1}{2}, \frac{1}{2}, 1\frac{1}{2}, 2\frac{1}{2}, \ldots$).

Dark fringes of integral order can be obtained by shifting a negative copy on a positive copy. The point at which a dark line falls over its transparent image will give a fringe of zero order, and the surrounding fringes will be -3, -2, -1, 1, 2, 3, etc.

Illustrative examples of the four partial derivatives in a ring subjected to diametral compression are shown in Figs. 7.14 through 7.17, inclusive. These loci are called *isoparagogic*. Both sets of patterns shown in Fig. 7.13 and in Figs. 7.14 through 7.17 have been obtained by superposing a negative copy on a posi-

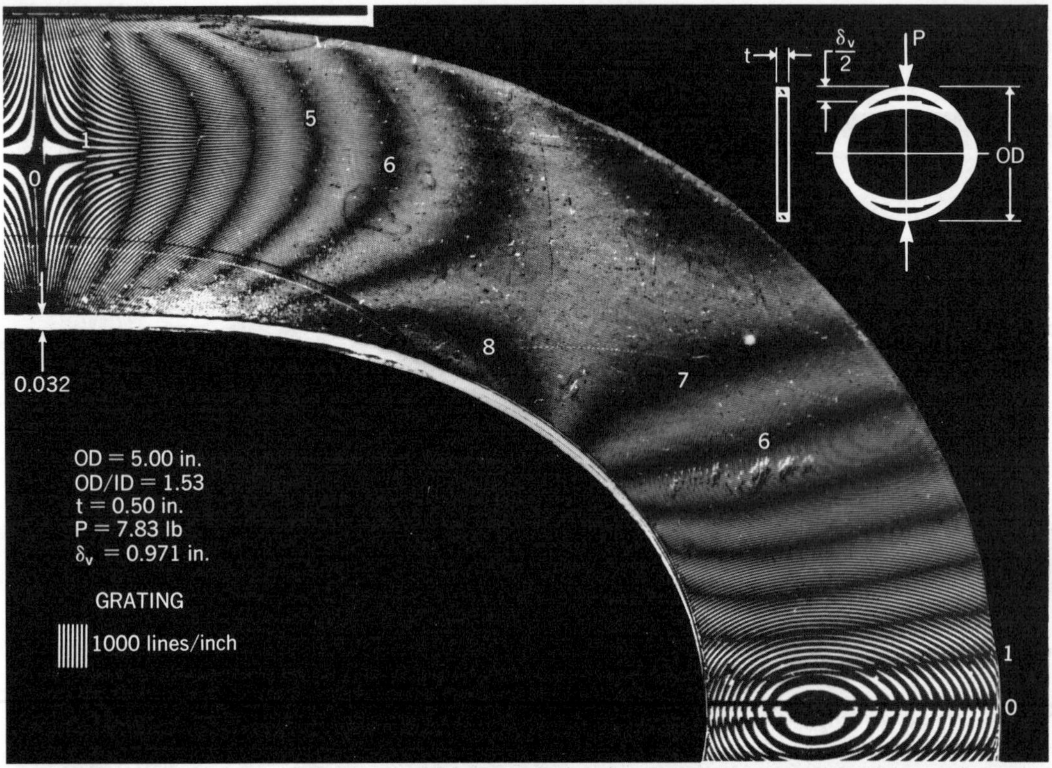
t
δ_v
2
P
OD
5
1
6
0
8
7
6
0.032
OD = 5.00 in.
OD/ID = 1.53
t = 0.50 in.
P = 7.83 lb
δ_v = 0.971 in.
GRATING
1000 lines/inch
1
0

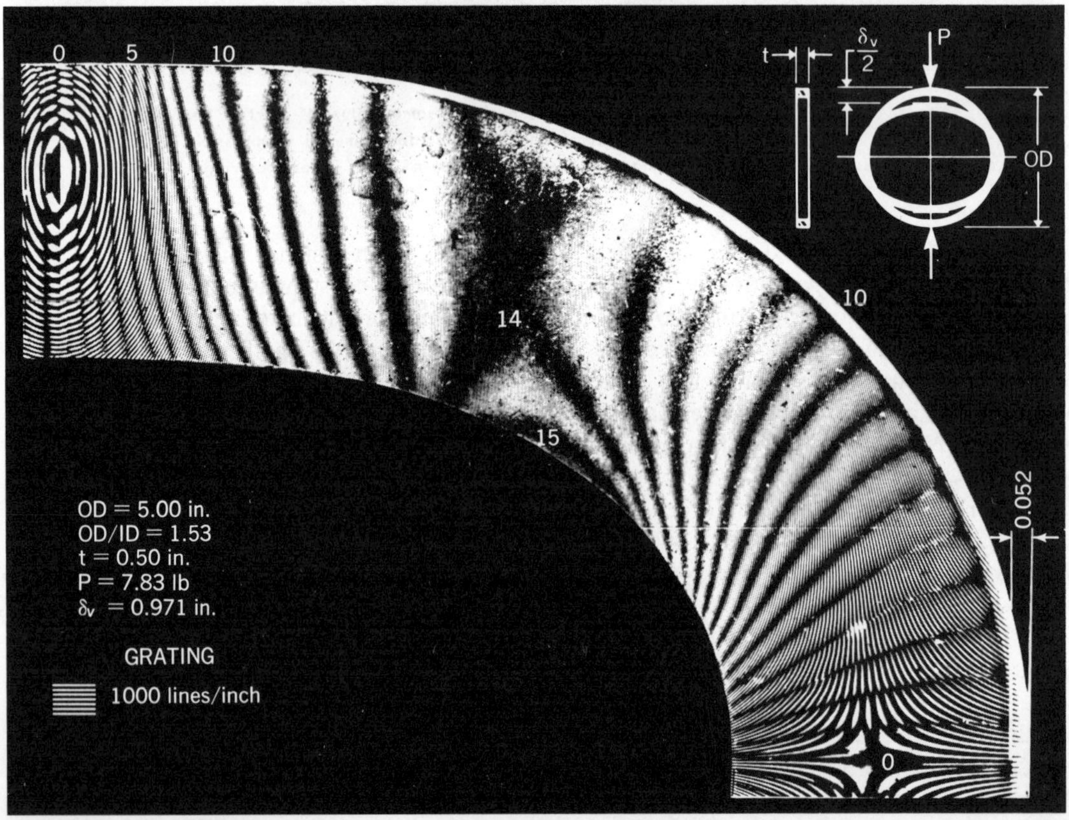
0
5
10
t
δ_v
2
P
OD
10
14
15
0.052
OD = 5.00 in.
OD/ID = 1.53
t = 0.50 in.
P = 7.83 lb
δ_v = 0.971 in.
GRATING
1000 lines/inch
0

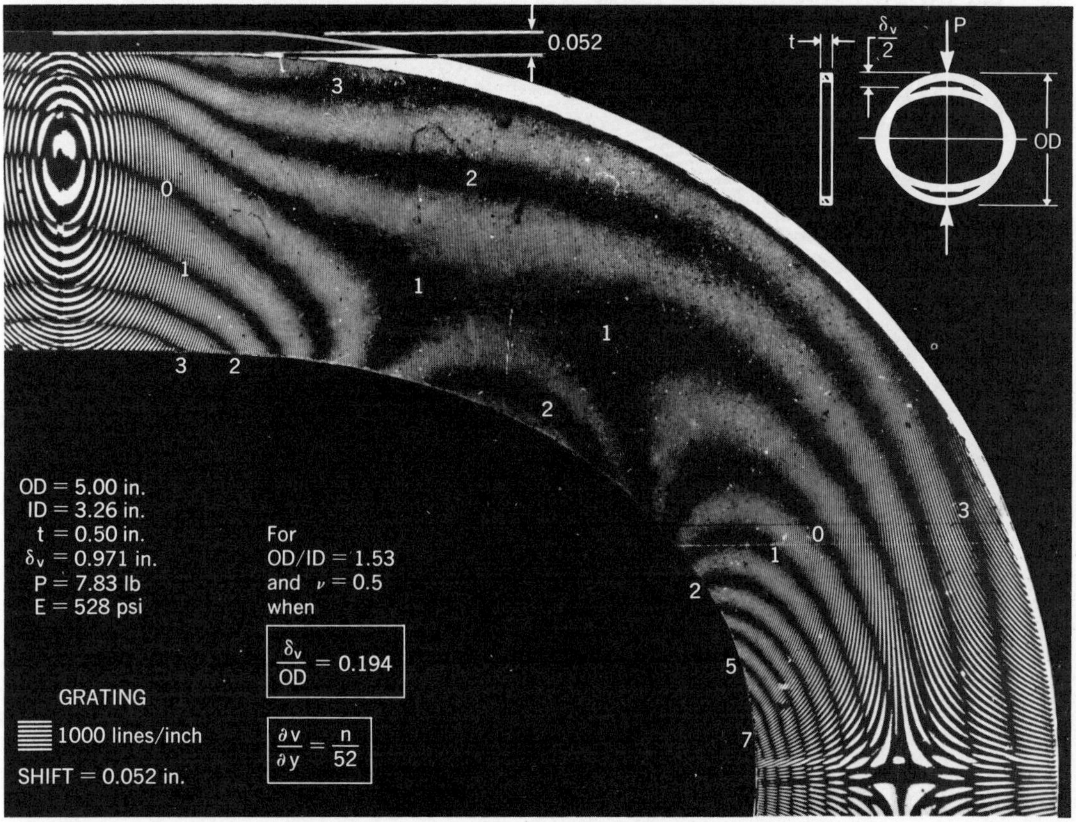

Figure 7.17. (Top) Loci of points exhibiting the same value of a partial derivative $\partial u/\partial x$ (isoparagogic) in a circular ring subjected to diametral compression. All points in the same fringe have a derivative value of 0.019 times the fringe order (large deformation).

Figure 7.15. (Upper left) Loci of points exhibiting the same value of partial derivative $\partial u/\partial y$ in a circular ring subjected to diametral compression. All points in the same fringe have a derivative value of 0.031 times the fringe order.

Figure 7.16. (Bottom left) Loci of points exhibiting the same value of partial derivative $\partial v/\partial r$ (isoparagogic) in a circular ring subjected to diametral compression. All points in the same fringe have a derivative value of 0.019 times the fringe order.

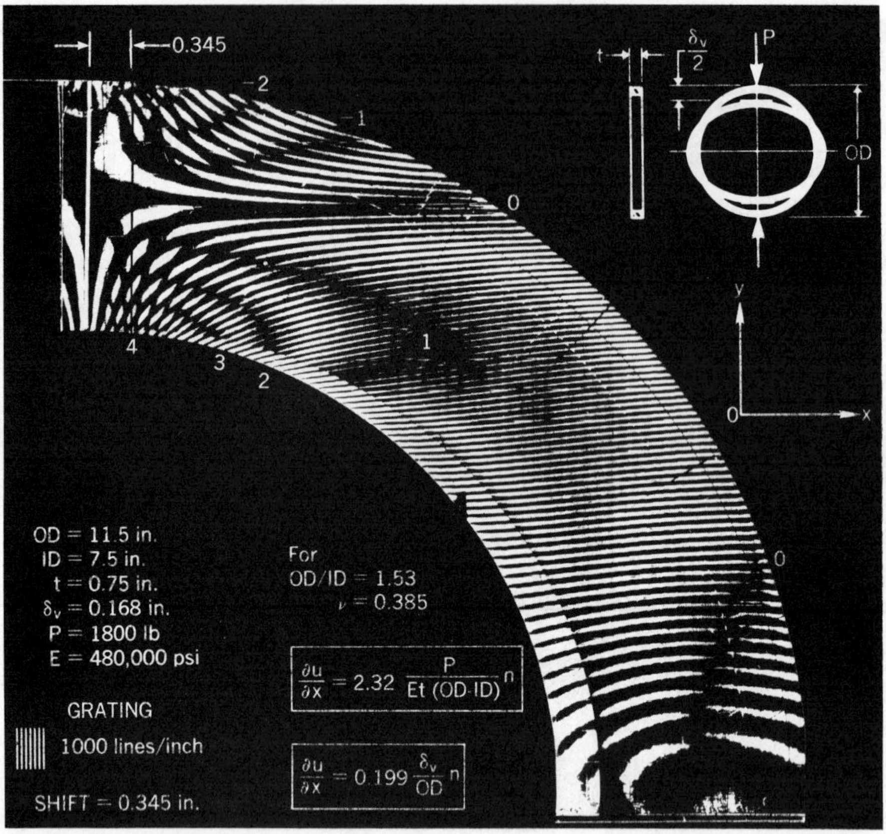

Figure 7.18. Loci of points exhibiting the same value of partial derivative $\partial u/\partial x$ (isoparagogic) in a circular ring subjected to diametral compression. All points in the same fringe have a derivative value of 0.00290 times the fringe order (integral order).

tive copy of the grating. The initial displacement fields had a very large number of isothetic fringes, which in turn permitted obtaining a large number of moire-of-moire fringes. When hard materials are used, the response is smaller. Figure 7.18 is a moire-of-moire pattern obtained by shifting negative and positive moire images from an epoxy ring. The maximum fringe order is only 4, in spite of a large shift of 0.345 in.

In cases such as this it may be worthwhile also to attempt to analyze the light fringes in the pattern, or even better to produce dark half-order fringes as suggested above, by superposing and shifting two negative (or two positive) images of the same pattern. The half-order pattern is shown in Fig. 7.19.

Basically, the light fringes of the whole-order fringe pattern will differ from the dark fringes of the half-order fringe pattern with respect to their continuity. Where the grating is visible the light fringes of the moire pattern are everywhere interrupted by the dark grating lines and, as such, are discontinuous.

Conversely, the dark fringes are continuous, since they are produced by alternate dark grating lines from specimen and master. This condition is more obvious in moire of moire. In either the whole- or half-order patterns the dark fringes are continuous and the light fringes are interrupted. When the fringes of both patterns are traced and represented together, a family like the one in Fig. 7.20 is obtained, which shows a sufficiently large number of loci of partial derivatives.

The generalized scales to apply the values of Figs. 7.18, 7.19, and 7.20 to rings of any size subjected to any load or displacement applied diametrically (provided $\nu = 0.385$) are

Figure 7.19. Loci of points exhibiting the same value of partial derivative $\partial u/\partial x$ (isoparagogic) in a circular ring subjected to diametral compression. All points in the same fringe have a derivative value of 0.00290 times the fringe order (half orders).

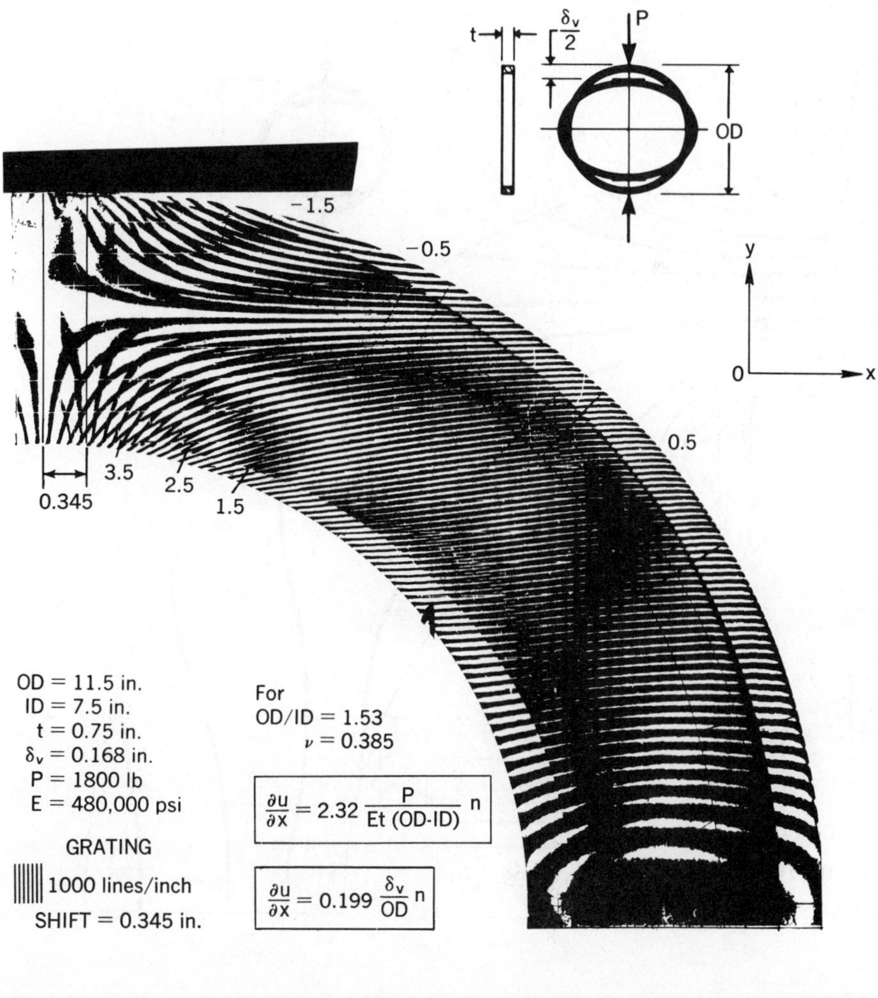

$$\frac{\partial u}{\partial x} = 2.32\frac{\mathrm{P}}{Et(\mathrm{OD}-\mathrm{ID})}n$$

$$\frac{\partial u}{\partial x} = 0.199\frac{\delta_v}{\mathrm{OD}}n$$

The same procedure applied to the determination of the cross-derivative gives the family of lines shown in Fig. 7.21 with the following generalized scale:

$$\frac{\partial v}{\partial x} = 3.98\frac{P}{Et(\mathrm{OD}-\mathrm{ID})}n$$

$$\frac{\partial v}{\partial x} = 0.341\frac{\delta_v}{\mathrm{OD}}n$$

Figure 7.20. Lines of constant $\partial u/\partial x$ obtained by shifting in the x-direction one u-field pattern with respect to another.

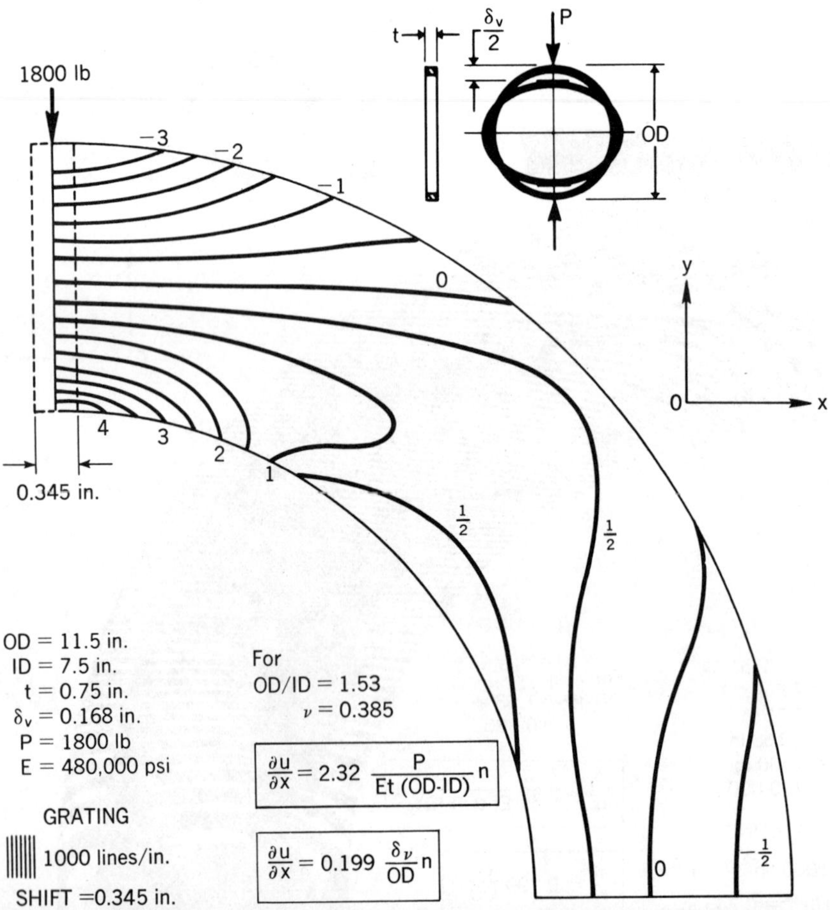

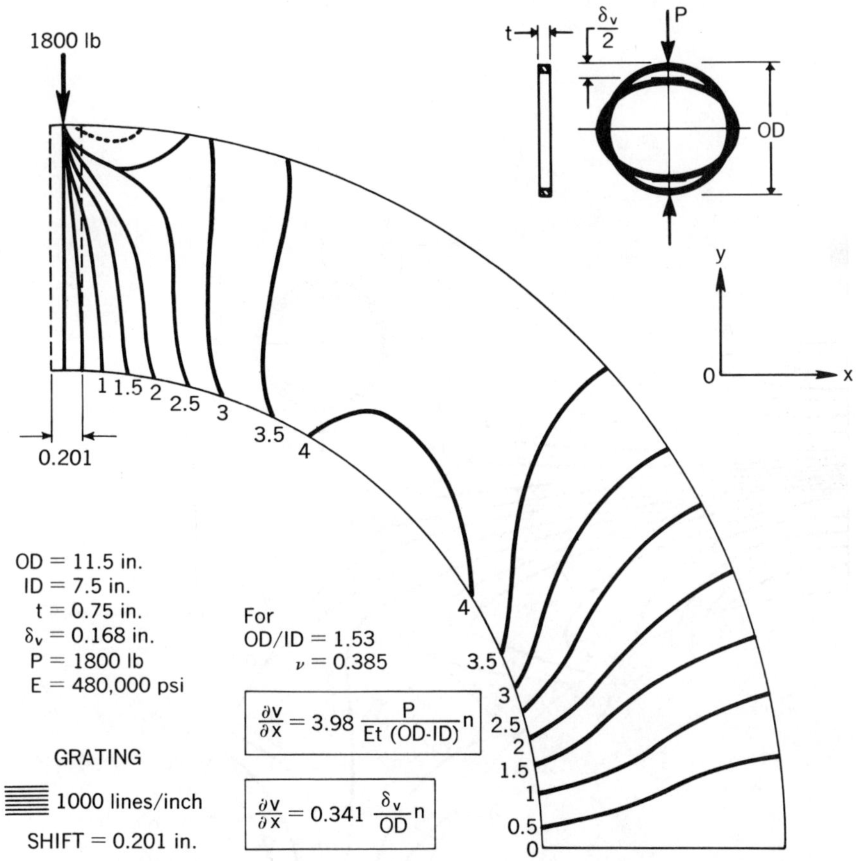

Figure 7-21. Lines of constant $\partial v/\partial x$ obtained by shifting in the x-direction one v-field pattern with respect to another.

The epoxy ring subjected to small levels of applied displacement is, in a first approximation, everywhere in a state of small strain and small rotation. Figure 7.20 can, therefore, be considered a first approximation to the solution of the direct strains (corresponding to Eq. 7.37). The addition and subtraction of the values in Fig. 7.21 and in the corresponding figure for the other cross-derivative will give the first approximation to the shear strains and rigid rotations, respectively. (This will be shown later.) To obtain more accurate values (corresponding to Eq. 7.35, etc.), the squares and products of the derivative have to be computed. This has been done and represented graphically in Figs. 7.22 and 7.23 for the two direct strains. Compare Fig. 7.20 ($\partial u/\partial x$) with Fig. 7.22 (ϵ_x). A computer was used for the evaluation, which was conducted only at a certain number of points of the field. The curves are affected by some error, therefore—in particular at the zone of high gradients. The general scales for these isotenics are

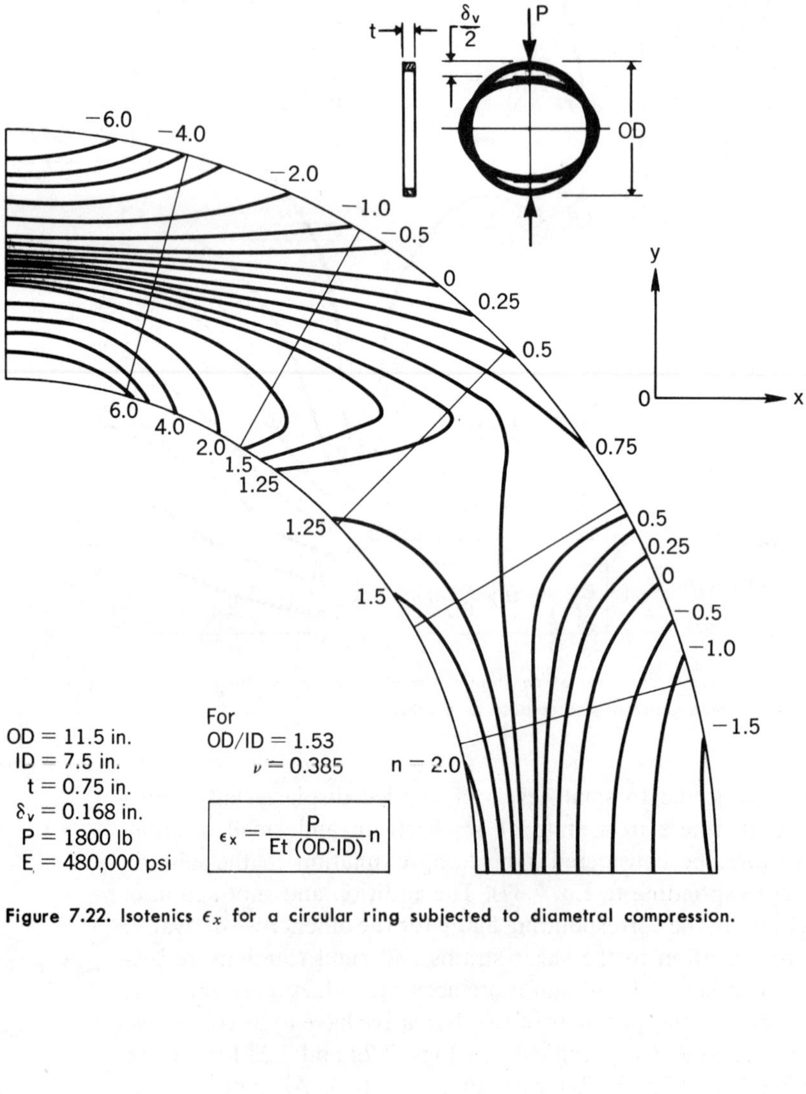

Figure 7.22. Isotenics ϵ_x for a circular ring subjected to diametral compression.

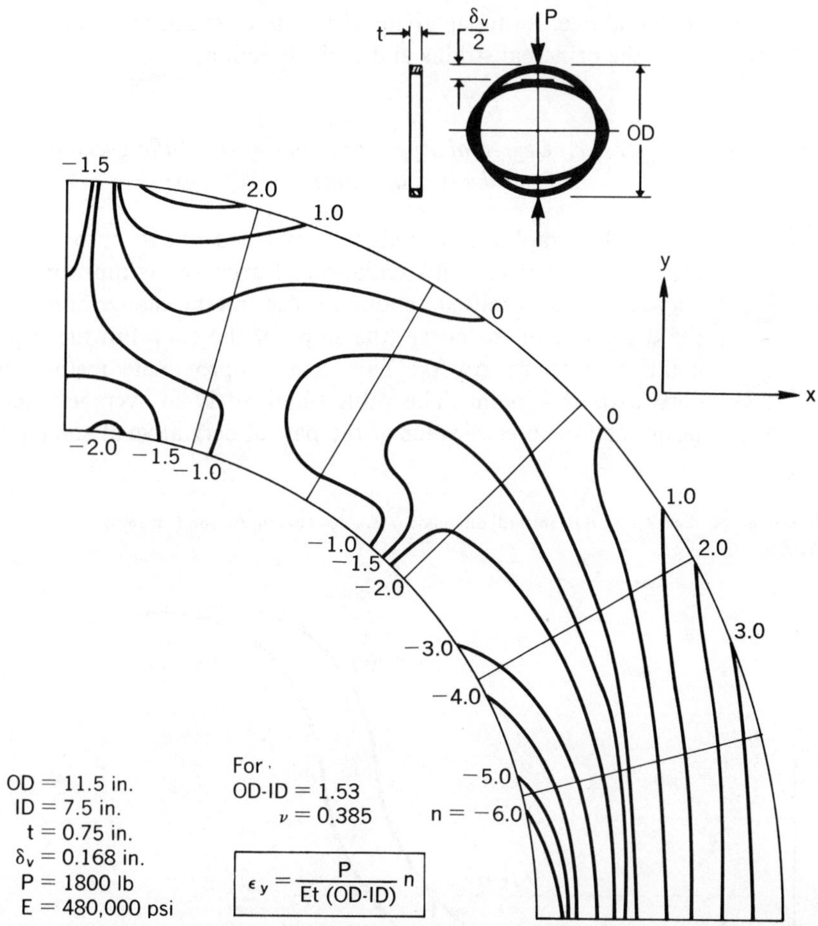

Figure 7.23. Isotenics ϵ_y for a circular ring subjected to diametral compression.

$$\epsilon_x = \frac{P}{Et(\text{OD} - \text{ID})}n \tag{7.38}$$

$$\epsilon_y = \frac{P}{Et(\text{OD} - \text{ID})}n \tag{7.39}$$

After the direct and shear strains have been determined by the method shown above, formulas giving the state of strain at a point can be used to determine the principal strains and their direction.

7.8.1. Degree of Approximation of the Differentiation Procedures Using Pattern Shiftings

The graphical procedures described above correspond to a well-known method of graphical differentiation of a curve, as illustrated in Fig. 7.24. If the curve $u = f(x)$ is shifted an amount Δx and the change of u (Δu) is measured, then the derivative of the curve (the slope of the curve) in the region at the measurement is given by $\Delta u/\Delta x$. This is an approximate method of determining the derivative at a point. The value obtained is an average value between the two points and the precise value of the partial derivative at some point in the interval.

Figure 7. 24. Curve $u = f(x)$ shifted an amount Δx to obtain Δu and from this $\Delta u/\Delta x$.

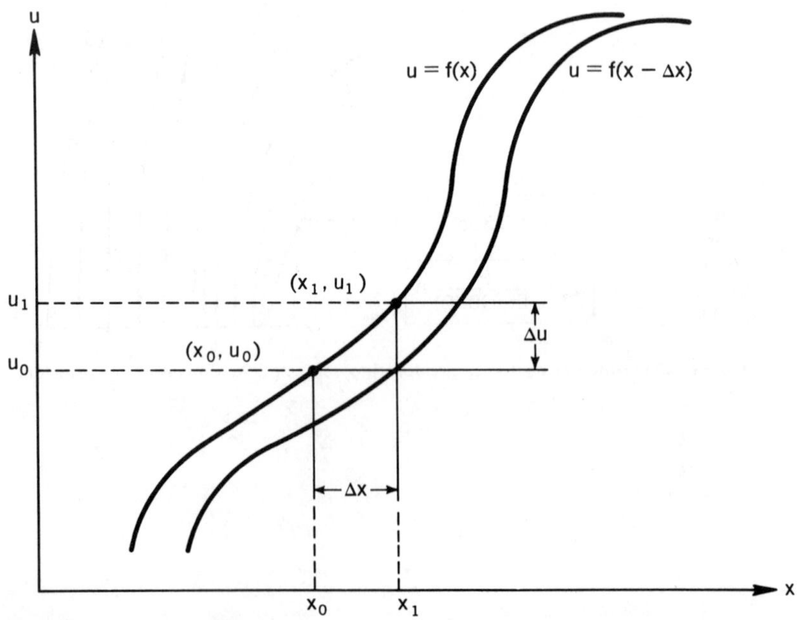

A Taylor-series expansion indicates the degree of approximation; thus

$$u_1 = u_0 + \Delta x \frac{\partial u_0}{\partial x} + \frac{\Delta x^2}{2!}\frac{\partial^2 u_0}{\partial x^2} + \frac{\Delta x^3}{3!}\frac{\partial^3 u_0}{\partial x^3} + \cdots$$

$$\frac{\Delta u}{\Delta x} = \frac{\partial u_0}{\partial x} + \frac{\Delta x}{2!}\frac{\partial^2 u_0}{\partial x^2} + \frac{\Delta x^2}{3!}\frac{\partial^3 u_0}{\partial x} + \cdots$$

If $\partial u_0/\partial x$ is constant—as in a homogeneous-strain field—or if the value of Δx is kept small by a very slight shift, then the approximation

$$\frac{\Delta u}{\Delta x} = \frac{\partial u}{\partial x}$$

is a good one, and the error can be computed from the series expansion.

For small strain, then, four combinations of the two Cartesian displacement fields superposed and shifted in the Cartesian directions will give approximate values of the Cartesian components of strain, as follows:

$$\begin{aligned}
\epsilon_x &= \frac{\partial u}{\partial x} \approx \frac{\Delta u}{\Delta x} = \frac{N_1}{N_{s1}} \\
\epsilon_y &= \frac{\partial v}{\partial y} \approx \frac{\Delta v}{\Delta y} = \frac{N_2}{N_{s2}} \\
\gamma_{xy} &= \frac{\partial u}{\partial y} + \frac{\partial v}{\partial x} \approx \frac{N_3}{N_{s3}} + \frac{N_4}{N_{s4}}
\end{aligned} \tag{7.40}$$

8

MOIRE USED TO DETERMINE STRAINS

Geometric approach

8.1. Introduction

The reader will recognize in the following discussions several of the procedures, explained in the previous chapters, in which strains were determined using the derivatives of the displacements. This could be expected, since actually the same phenomenon is observed from different points of view. Here, the basis of the reasoning will be the geometric considerations of Chapter 3.

8.2. Recording of the Data

In order to determine strains and rotations associated with two coordinate directions in a body, it will be convenient to have grating lines perpendicular to these directions. If the specimen is not identical when turned 90°, two specimens must be equipped with gratings; or, alternatively, two sets of grating lines can be placed on the same specimen and analyzed separately with a one-way master grating.

Individual measurements of the distance between two fringes can be obtained as shown in Chapter 3, and strains at the single point can be determined. However, these individual measurements cannot be very accurate, as is apparent from an examination of Figs. 7.3 and 7.4. Also, the point being examined may not be exactly in the middle between two fringes.

It is usually more convenient and accurate to determine the position of fringes along a line, including the point of interest, and plot the accumulated fringe distances from any convenient point, as in Fig. 8.1. A smooth curve is drawn through the plotted points, and the slope of this curve at any point gives fringes per inch, the reciprocal of which is the value in inches per fringe of δ_r or δ_s, as the case may be.

As is apparent from Figs. 7.3 and 7.4, even in a simplified case such as a disk it will not always be possible to measure both δ_r and δ_s at the same point

Figure 8.1. Determination of δ_r at all points along lines by means of a plot of accumulated fringe distance from any arbitrary beginning.

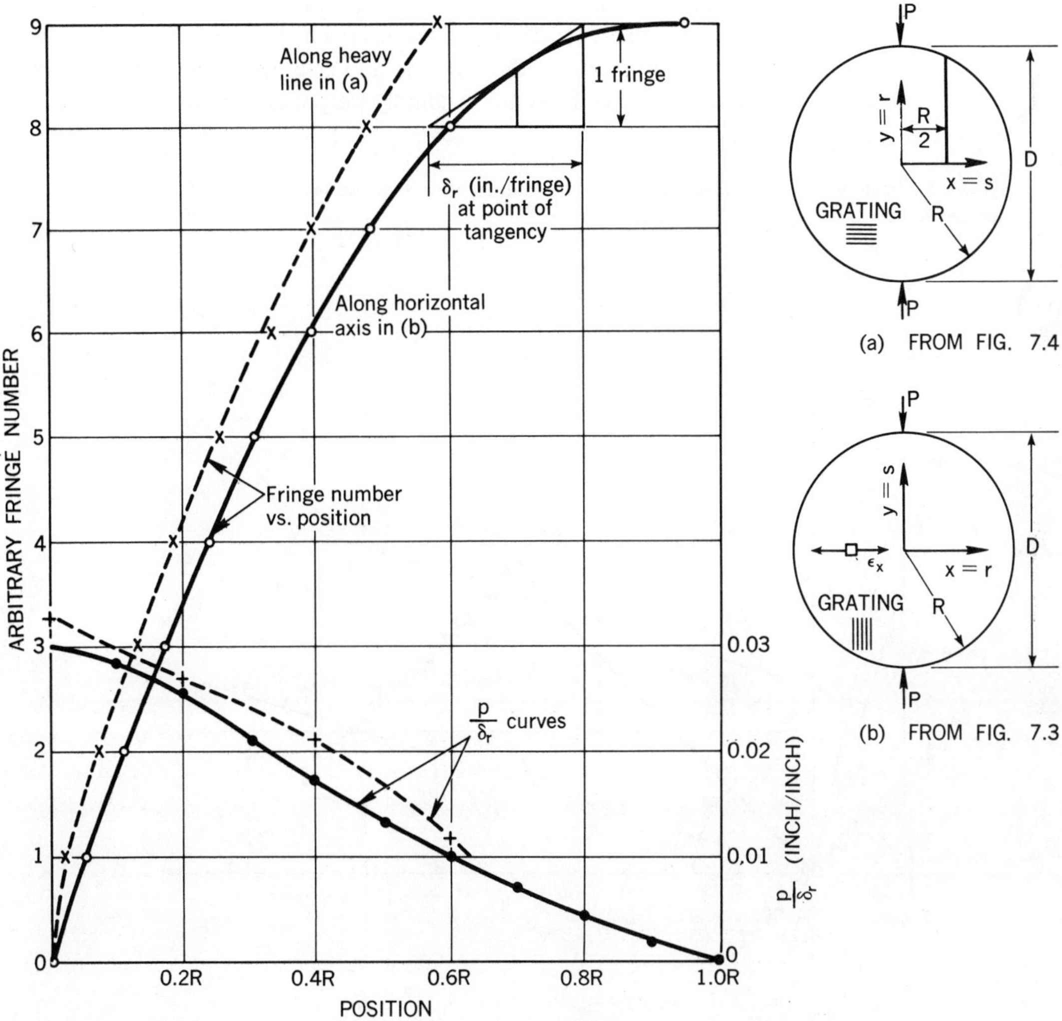

with acceptable precision; and at some points, neither can be measured. However, δ can always be measured, and values of $(1 - p/p_1)$ and θ can be computed from measurements of δ and can be used to supplement or check those obtained from measurements of δ_r and δ_s.

The measurements of ϕ will, in general, not be very accurate. However, since one can usually make a choice from among the measurements of δ, δ_r, and δ_s at any given point of interest, the sensitivity to errors in measurements of ϕ can be kept at a fairly low level for most fringe configurations.

8.3. Determination of Shear Strains

Shear strain, defined in Sec. 4.3, is the change in angle between two originally perpendicular lines. It will be computed from the rotation of two lines, one perpendicular to the grating of the u-family and one perpendicular to the grating of the v-family. These two lines are originally perpendicular to each other.

Figure 8.2. Determination of shear strain from the rotation of each of two mutually perpendicular lines.

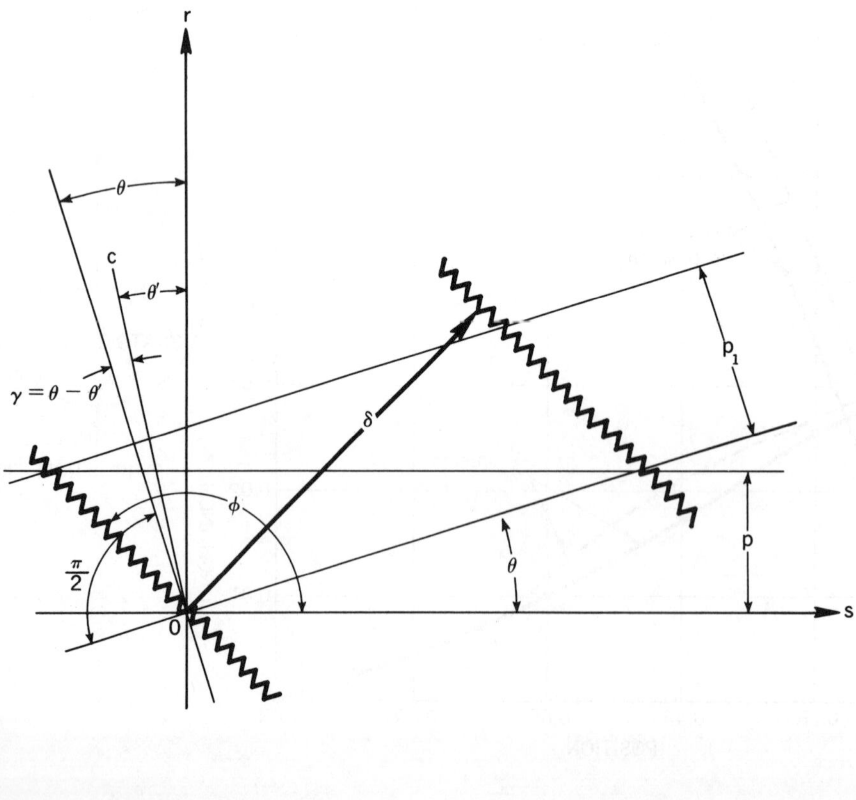

Figure 8.2 repeats Fig. 3.1 with a line $0C$ added to represent a grating line that in the undeformed body lies parallel to the r-axis. The rotation of the s-axis and r-axis can be obtained by measurements of ϕ and δ on two moire patterns associated with gratings perpendicular to each other, and by the determination of α_δ, α_s, or α_r from Figs. 3.2, 3.5, and 3.6.

The angle θ has previously been defined as the rotation of the s-axis. If a prime is used to denote all values corresponding to the moire pattern produced by the grating and the master initially parallel to the r-axis, then the rotations of the two directions can be computed in one of three alternative ways from Eqs. (3.13), (3.28), and (3.29). Thus,

$$\begin{aligned} \theta &= \arctan \alpha_\delta \frac{p}{\delta} = \arcsin \alpha_r \frac{p}{\delta_r} = \arcsin \alpha_s \frac{p}{\delta_s} \\ \theta' &= \arctan \alpha'_\delta \frac{p'}{\delta'} = \arcsin \alpha'_r \frac{p'}{\delta'_r} = \arcsin \alpha'_s \frac{p'}{\delta'_s} \end{aligned} \tag{8.1}$$

The shear strain γ is then

$$\gamma = \theta - \theta' \tag{8.2}$$

8.4. Determination of Direct Strains

The direct strains along the segment of line $0C$ in Fig. 8.2, which initially was parallel to the r-coordinate and had a length p, will be

$$\epsilon_r = \frac{\text{Final Length} - \text{Initial Length}}{\text{Final Length}} = \frac{\dfrac{p_1}{\cos(\theta - \theta')} - p}{\dfrac{p_1}{\cos(\theta - \theta')}} \tag{8.3}$$

Noting that $\gamma = \theta - \theta'$,

$$\epsilon_r = 1 - \frac{p}{p_1} \cos \gamma \tag{8.4a}$$

$$\epsilon_r = 1 + \frac{p}{p_1}(1 - 1 - \cos \gamma) \tag{8.4b}$$

$$\epsilon_r = \left(1 - \frac{p}{p_1}\right) + \frac{p}{p_1}(1 - \cos \gamma) \tag{8.4c}$$

From Eqs. (3.18), (3.19), (8.1), and (8.2),

$$\epsilon_r = \beta_\delta \frac{p}{\delta} + \left(1 - \beta_\delta \frac{p}{\delta}\right)\left[1 - \cos\left(\arctan \alpha_\delta \frac{p}{\delta} - \arctan \alpha'_\delta \frac{p'}{\delta'}\right)\right] \tag{8.5}$$

From Eqs. (3.18), (3.19), (8.1), (8.2), (3.30), and (3.31) a number of variations of this expression can be obtained; for example,

$$\epsilon_r = \beta_r \frac{p}{\delta_r} + \left(1 - \beta_r \frac{p}{\delta_r}\right)\left[1 - \cos\left(\arcsin \alpha_r \frac{p}{\delta_r} - \arcsin \alpha'_r \frac{p'}{\delta'_r}\right)\right]$$

$$\epsilon_r = \beta_s \frac{p}{\delta_s} + \left(1 - \beta_s \frac{p}{\delta_s}\right)\left[1 - \cos\left(\arcsin \alpha_s \frac{p}{\delta_s} - \arcsin \alpha'_s \frac{p'}{\delta'_s}\right)\right] \tag{8.6}$$

$$\epsilon_r = \beta_s \frac{p}{\delta_s} + \left(1 - \beta_s \frac{p}{\delta_s}\right)\left[1 - \cos\left(\arcsin \alpha_\delta \frac{p}{\delta} - \arcsin \alpha'_r \frac{p'}{\delta'_r}\right)\right]$$

8.5. *Simplified Expressions for the Direct Strain*

In many cases a shortened form of the equations for direct strain (Eq. 8.6) can be used. Two important simplifications will be discussed: the first, in which the entire second term on the right side of Eq. (8.6) is dropped; and the second, in which the entire second term is dropped and also β is made equal to one.

The first simplification implies that the shear strain is sufficiently small for the second term on the right side of Eq. (8.4c) to be neglected. Looking back at Eq. (8.4c), we can write the second term as follows:

$$\frac{p}{p_1}(1 - \cos\gamma) = \frac{p}{p_1}\left(2\sin^2\frac{\gamma}{2}\right)$$

For small values of γ,

$$2\sin^2\frac{\gamma}{2} \simeq \frac{\gamma^2}{2}$$

and the second term becomes

$$\frac{p}{p_1}\frac{\gamma^2}{2}$$

For small values of γ, this term can be neglected with respect to γ unless p/p_1 has extraordinary large values.

The shear strain cannot be larger than the maximum direct strain at any point. Thus, if the shear strain is small, then the error introduced in the direct strain by neglecting the second term will be negligible with respect to the maximum direct strain at that point. If the shear strain is less than 0.01 (about 35 minutes of arc), the error in neglecting the second term will be less than one per cent of the maximum direct strain.

The expression for direct strains neglecting shear strain is then written

$$\epsilon_r = \beta_\delta \frac{p}{\delta} = \beta_r \frac{p}{\delta_r} = \beta_s \frac{p}{\delta_s} \tag{8.7}$$

Now, this simplification will give the exact direct strain at the point if the r- and s-axes coincide with the principal directions at that point, regardless of the magnitude of the shear strains in other directions. The only shear strain that influences the computation of ϵ_r is that due to a change in angle between the r-axis and the s-axis.

The second simplification refers to the case where the shear strains can be neglected as above and the rotation also is small.

It is apparent from Figs. 3.3 and 3.7 that if ϕ is sufficiently close to 0° or 180° (depending also on the value of δ_r/p), the values of β and β_r are close to -1.0 or 1.0. This means that rotations θ are very small (Eq. 3.11). In many problems, sufficient accuracy may result from taking

$$\beta = \beta_r = \pm 1$$

and reducing Eq. (8.7) to

$$\epsilon_r = \pm \frac{p}{\delta} = \pm \frac{p}{\delta_r} \tag{8.8}$$

If p/δ_r is sufficiently small in comparison with unity (that is, if the strain ϵ_r is small), then p_1 will approach p and the value of ϵ_r will approximate the engineering definition of strain discussed below.

On the other hand, if in Fig. 3.5 ϕ is sufficiently close to 90° (depending on the value of δ_s/p), the value of α_s is close to 1.0 and sufficient accuracy may result from a reduction of Eq. (3.28) to

$$\sin\theta = \frac{p}{\delta_s} \tag{8.9}$$

Equations (8.6), (8.7), and (8.8) can be applied to both moire patterns to obtain direct strains in the two perpendicular directions. If the r-coordinate is first considered to coincide with the x-axis and then with the y-axis, the strains ϵ_x, ϵ_y, and γ_{xy} are obtained.

8.6. *Definition of the Strains Obtained Using the Geometric Approach*

Refer to Sec. 4.3 for a discussion of the definition of strain with regard to Eulerian and Lagrangian descriptions. The geometric approach (Eqs. 8.6, 8.7, and 8.8) will give strain values as indicated by the Eulerian formula $\epsilon^E = (l_f - l_i)/l_f$. Note, however, that the strain is evaluated at a line which is *initially* parallel to the coordinate direction and, therefore, the Eulerian strain is obtained in a Lagrangian description. If the Lagrangian strain is desired in a Lagrangian description, then the strain values obtained from the geometric approach can be reduced to Lagrangian strains by

$$\epsilon^L = \frac{\epsilon^E}{1 - \epsilon^E} \tag{8.10}$$

as indicated in Sec. 4.3. This is illustrated on Fig. 4.1.

8.7. Example of Strain Determination Using the Geometric Approach

The same rubber disk evaluated in Sec. 7.5, using the displacement-derivative approach, will be used here to illustrate the use of the graphs shown in Chapter 3.

Figure 8.1 is a plot of the positions of successive fringes along two representative lines where δ_r could be successfully measured in Figs. 7.3 and 7.4. On a separate scale in Fig. 8.1 are also plotted the values of the quantity p/δ_r. From the values of p/δ_r (also p/δ_s and p/δ from separate measurements in appropriate directions), the strains can be found as previously described.

From a series of measurements across the horizontal axis of Fig. 7.4, the value of ϵ_y^E was determined at a number of points. These values, combined with those for ϵ_x^E from Fig. 8.1, and with the modulus of elasticity (also determined by the moire method), allowed computation of σ_x and σ_y (σ_1 and σ_2, respectively, in this case), the results of which are shown in Fig. 7.6. Also shown in Fig. 7.6 are the results of a theoretical analysis of the disk when subjected to concentrated loads. As could be expected both approaches in this case give the same answer. Changes in pitch have also been determined using the geometric approach, for points of the vertical line shown in Fig. 8.1a. The values of $(1 - p/p_1)$ obtained from the nomograph in Fig. 3.7 all fell on the curve p/δ_r shown in Fig. 8.1, with the exception of the point at distance $0.6R$ from the horizontal axis which was 2% off the curve. Thus the coefficient β_r is near unity for all points on this line, indicating small values of rotation of the line.

The point analyzed for strain in Sec. 7.5.2 was again analyzed using the geometric approach. The nomographs in Figs. 3.2 and 3.3 were read as $\alpha_\delta = 0.08$; $\beta_\delta = 1.00$ for the v-field, and $\alpha_\delta = 0.45$; $\beta_\delta = 0.89$ for the u-field. Applying Eqs. 8.2 and 8.5 the strains were determined to be $\epsilon_x = 0.0265$, $\epsilon_y = -0.0450$, $\gamma_{xy} = 0.0249$.

To illustrate the application of the geometric approach to the analysis when large rotations are present, a point on the ring shown in Figs. 6.10 and 6.11 was selected. The point was located on the 45° diagonal, located midway between the outer and inner diameters. The measured values were $\delta = 0.051$ in. and $\phi = 96°$ (counterclockwise) in the u-field; and $\delta = 0.053$ in. and $\phi = 86°$ (counterclockwise) in the v-field. This gave $\alpha_\delta = 1.00$ and $\beta_\delta = 0.09$ in the u-field, and $\alpha_\delta = 0.99$, $\beta_\delta = 0.11$ in the v-field. The computed strains from Eqs. 8.2 and 8.5 are $\epsilon_x = 0.0017$, $\epsilon_y = -0.0020$, and $\gamma_{xy} = -0.0010$.

8.8. Comparison of the Geometric and Displacement-Derivative Approaches

The data and the first step of computation are the same for the geometric approach and the displacement-derivative approach. Thus, from the u-field, $\partial u/\partial x$, $\partial u/\partial y$, and $\partial u/\partial n$ are the same as p/δ_r, p/δ_s, and p/δ. And so for many cases the analysis will be identical, for example on axes of symmetry. The difference between the two approaches will develop when the computation of strain is affected by shears and rotations. In those cases both the equations and the final result may differ.

9

SENSITIVITY, PRECISION, AND ACCURACY

9.1. Introduction

The concept of *accuracy* is rather complex and not uniquely defined. It is sometimes related to a numerical statement of the *precision* of a measurement, sometimes only to a qualitative statement of that precision. In this chapter accuracy will be defined as including not only sensitivity and precision of measurements which account for random errors, but also the *systematic errors* of the method. Accuracy is frequently estimated by calibration or comparison of the observed measurements with "true" measurements—that is, with measurements taken by an instrument known to be more accurate than the instrument to be calibrated. Sometimes this calibration is compared with theoretical predictions.

It is helpful to distinguish two types of sensitivity, ultimate sensitivity and unit sensitivity, because depending on the instrument one or the other may be the most significant. *Ultimate sensitivity* refers to the smallest change of the measured quantity that will produce a detectable effect in the measuring instrument. *Unit sensitivity* is determined by the number of units of the measured quantity that are represented by one scale division. If observation with the naked eye is the determining factor, the unit sensitivity can be taken as about one-hundredth of an inch.

The ultimate sensitivity of moire measurements of displacement can approach the limits prescribed by the fluctuation phenomena of microphysics. More commonly, however, an image of the grating will be photographed at some stage in the measurement process and the ultimate sensitivity will be limited by the resolution of the optical imaging system.

Unit sensitivity is usually the significant factor in the case of measurements obtained by moire. An analysis of this quantity is given in the following three sections. Then the causes of error which can further reduce the attainable accuracy from the values estimated on the basis of unit sensitivity alone, will be considered. In the final sections certain modeling considerations which can be used to improve measurement accuracy are discussed.

A number of considerations will be presented here based on experience—in moire analysis and similar methods of experimental analysis—to help in the critical evaluation of moire results.

9.2. *Measurements in Moire Determinations*

It might be well to point out first that determinations of any field quantity, such as those found in moire, involve two measurements: the value of the fringe order and the value of a length giving the position of the point at which the fringe order occurs. The importance of this second measurement is sometimes neglected. Thus, for example, in photoelasticity, a fringe order may be determined to an accuracy of $\frac{1}{100}$ of a fringe; however, the scale used to determine the position of the point at which the birefringence occurs may be coarse and thus be the governing source of error. Despite this, highly sophisticated fringe-measuring devices are sometimes used with crude marks on photoelastic models, or applied to poorly defined boundaries.

The photoelastic example was introduced to emphasize the distinction between the two measurements. In moire the measurements are not quite as distinct, since one measurement defines a position and the second defines a displacement, both of which have the same dimensional units. Although the two measurements are less distinct in moire than in photoelasticity, they are more important, since the ratio of the displacement to the length giving the position is required.

9.3. *Sensitivity in Displacement Measurements*

Since moire is a means of measuring displacements, the sensitivity of moire measurements refers to a sensitivity to displacement measurement, or displacement sensitivity.[1] And since in moire analysis displacements are given by the fringe orders, the displacement sensitivity will depend on the determination of the position and order of fringes. These two determinations are sources of error in the displacement values.

The sensitivity of the measurement of the length that defines position will

[1] See A. J. Durelli, E. A. Phillips, and C. H. Tsao, *Introduction to the Theoretical and Experimental Analysis of Stress and Strain* (New York: McGraw-Hill Book Company, 1958), p. 242. In that book displacement sensitivity is called "deformation sensitivity."

depend on the measuring instrument. The best estimate with the naked eye and a good scale will be to about ± 0.01 in. If some sort of measuring microscope is used, measurements of lengths as small as 0.0001 in. can be made in some cases.

The sensitivity in the determination of the fringe order depends on the ability of some sensing device, often the naked eye, to specify the fringe order or fractional fringe order at some point. Usually the fringe width and the interfringe are in about the same ratio everywhere in a pattern. By judging the density of the shading of an individual fringe, the naked eye can probably estimate a whole fringe order to within about ± 0.05. In practice, the experimentalist brings some crosshair or indicator in line with the fringe and records the whole fringe value, say $n = 2$. The sensitivity is related to the fact that the order of the fringe under the crosshair cannot be determined exactly, but to within the range 1.95 to 2.05. Since this variation can be observed by moving the crosshair a specific amount, it is often thought of as an error of position and could be evaluated statistically. It is easier to think of this error as an error of fringe order when a sensing device like a photometer with a scale is used. Then the fringe-order error is seen as the error on the photometer scale and the position error is seen as the error in the measurement of the length that specifies the position and is associated with the length-measuring device.

These same concepts apply when a fractional fringe order is to be determined. In this case it is common to divide the field into equal intervals and to record fringe orders to within a fraction of a fringe at each division line. Here, the position error again depends on the linear measuring instrument, which is used either to set up the system of grid lines or to record the grid-line position. In this case the naked eye is probably less accurate than in estimating the whole-fringe centers, and an estimate of ± 0.1 fringe is probably realistic. The fringe represents 1 pitch increment. The displacement sensitivity DS can then be specified in equation form as

$$\text{DS} = \frac{p}{10} \tag{9.1}$$

To fix these ideas, consider a system with a pitch of 0.001 in. It is assumed that the measuring system used is a highly accurate traveling microscope moved by a precision-cut, calibrated screw that records the position with a degree of precision much greater than that obtained in sighting the fringes. This leaves the fringe-order observation as the dominant source of error. A fringe-position–displacement-position curve is plotted in Fig. 9.1. If the location on the x-axis of the fringe marked $n = 2$ is determined, its location can vary about 0.05 of the fringe spacing δ. If the value of the fringe order at $x = 0.25$ in. is determined, it can vary about 0.1 fringe. The displacement curve obtained by either method is about the same, the displacement sensitivity being from $\frac{1}{10}$ to $\frac{1}{20}$ of 0.001 inch or, in other words, 0.0001 in. to 0.00005 in. This is about the same displacement sensitivity as that of a very good dial gage. Note that, as with a dial gage, the displacement sensitivity is not related to the total displacement measured. The total displacement in either case could be $\frac{1}{10}$ in. or 10 in.

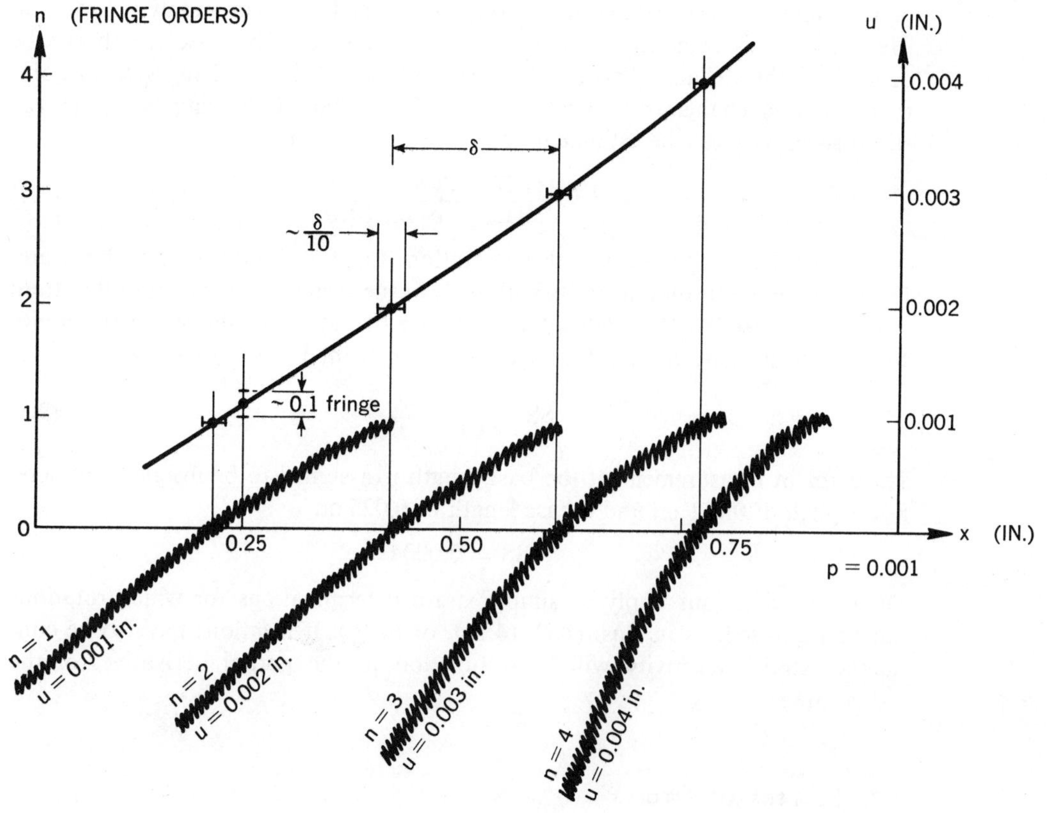

Figure 9.1. Precision in the determination of the position of the axis of a fringe in a pattern.

Techniques have been developed that determine displacements point-by-point using devices sensitive to the light intensity.[2] When procedures of this type are used, the displacement sensitivity will depend on the fraction of a fringe that a particular sensing instrument can determine.

9.4. Strain Sensitivity

The estimate of sensitivity of moire to strain is more complicated. Strain sensitivity SS can be defined as the displacement sensitivity DS divided by the base length BL. Base length in most strain gages, as in mechanical strain gages, is fixed for a particular gage. For a Huggenberger gage the base length is commonly 0.5 in.

[2] C. A. Sciammarella, "Techniques of Fringe Interpolation in Moire Patterns," *Proceedings of 2nd SESA International Congress on Experimental Mechanics*, Society for Experimental Stress Analysis, Westport, Conn. (1966).

or 1 in., and for electrical-resistance strain gages the base is commonly $\frac{1}{8}$, $\frac{1}{4}$, or $\frac{1}{2}$ in. The situation is somewhat different for moire fringes. The base length can be defined as the distance between fringes, and this distance changes from point to point, and changes also with the amount of load. Following Eq. (9.1), the strain sensitivity can be estimated as

$$\text{SS} = \frac{\text{DS}}{\text{BL}} = \frac{\frac{1}{10}p}{\delta} = \frac{1}{10}\frac{p}{\delta} \tag{9.2}$$

For the case of strain with no rotation (Eq. 1.7) the ratio p/δ is the strain. For this case the strain sensitivity will be $\frac{1}{10}$ of the strain, independent of the strain level. If the base length is defined as some fixed length, then the strain sensitivity will be specified by the pitch and the relation from Eq. (9.1) will be

$$\text{SS} = \frac{\text{DS}}{\text{BL}} = \frac{\frac{1}{10}p}{\text{BL}} \tag{9.3}$$

The error in measurement of the base length is assumed to be insignificant here. For a pitch of 0.001 in. and a base length of 0.025 in.,

$$\text{SS} = 0.0004$$

These considerations apply to simple strain determinations for which rotations can be neglected, as in Eqs. (1.7), (4.15), or (4.16). If rotations have to be considered, strain sensitivity will be a function of the partial derivatives of the displacement.

9.5. Sources of Error

It is practically impossible to catalog systematically all the factors that contribute to error in the physical measurement of a phenomenon such as moire. In any measurement the total or final error is the result of various sources of error, and the ones that are considered are those which have predominant influence in a particular problem. Two types of error can be studied: random errors which can be evaluated statistically, and systematic errors.

9.5.1. Random Errors

Random errors can be evaluated by analyzing statistically the values obtained when the same quantity is measured many times. The degree of reproducibility of the measured quantity determines the precision of the measurement.

It was pointed out in the previous sections that the concepts of displacement sensitivity and strain sensitivity are associated with the scale of the instrument, and that in the case of moire it is suggested that the scale will have units of approximately $\frac{1}{10}$ of a fringe. The precision of the measurements cannot be greater than this sensitivity, but the concept of precision encompasses reproducibility as well as sensitivity. Precision is usually measured by some index, such as the standard deviation. In the example shown in Fig. 9.1, if no special devices are used to increase precision, the approximate figure of 4×10^{-4} in. can be used as the standard deviation of the measured displacements u.

9.5.2. Pitch Error

Basic to all moire analysis, both for displacement and strain analysis, is the assumption of a constant known value of the pitches of both master and undeformed specimen gratings. If the two pitches are well matched but have a common error of 1 per cent, the error in displacement, assuming a density of 1000-lpi, will be only 0.00001 in., and on a 0.1-in. base length the error in strain will be 0.0001.

In the case of mismatch the pitch of master grating and the pitch of the undeformed specimen grating are different, but each has to be known, and errors in these figures will be reflected directly in any analysis. It is simple to obtain an overall estimate of any mismatch by overlaying master and undeformed specimen gratings and recording the total number of fringes produced in the field.

A serious systematic error will be that made by assuming a matched pair of gratings and neglecting mismatch, since a mismatch of 0.1 per cent will immediately show up as a large strain, 0.001. The chief precaution in moire analysis for overall variation in pitch (mismatch) is to record any initial fringes before loading and subtract out this "fictitious" displacement and strain in the analysis.

This same recommendation could be made for local variations in pitch. However, the variation may not be of sufficient magnitude to produce fringes in particular regions of the field.

It is estimated that the pitch lines may vary some $\pm\frac{1}{100}$ of the pitch spacing, so that the error in pitch of a 1000-lpi grating will be of the order of ± 0.00001 in. On a base length of 0.03 this would produce a fictitious strain as great as

$$\epsilon = \frac{\pm 0.00001}{0.03} = 0.0003$$

This, however, would require that the error in pitch be repeated over a number of grating lines in the region in order to effect the position of a moire fringe.

Essentially, this is the problem with periodic pitch error, which repeats itself systematically over a number of grating lines and produces herringbone effects when overlaid by a strain pattern.

9.5.3. Average Strain Error

The strain evaluated using the total displacement between two fringes is an *average strain* over the base length. From this point of view, the closer the

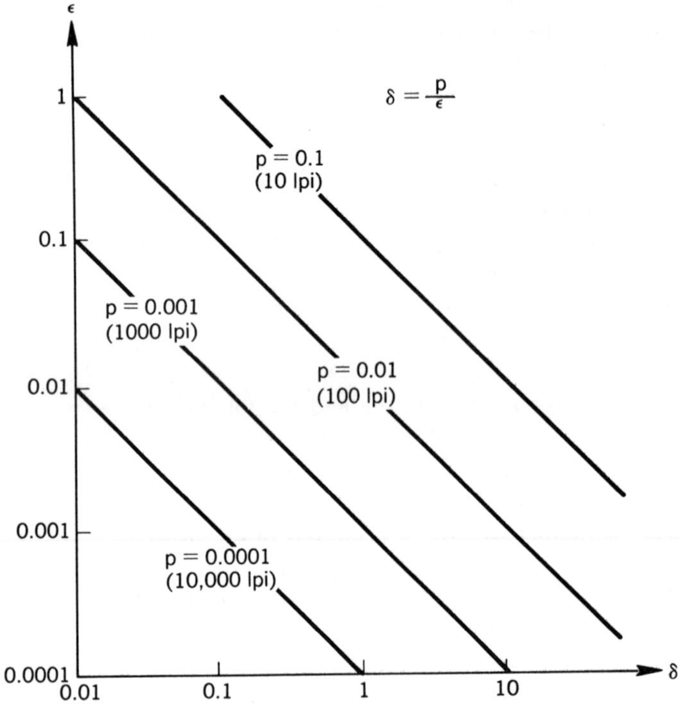

Figure 9.2. Relation of fringe spacing to strain and grating pitch for the simple case of no rotation and no shear. An estimate of the minimum measurable strain for a grating of no certain pitch can be obtained along the line of the given pitch by specifying the required base length as δ.

fringes, the more accurate will be the determinations, until they are so close that estimation to $\frac{1}{10}$ of a fringe becomes difficult.

Figure 9.2 is a plot of $\epsilon = p/\delta$ using p as a parameter. Crisp[3] suggests that these curves may actually be used as a guide to a minimum level of measurable strain. He first relabels the abscissa as a "gage length" (base length) and then argues that, if a given base length is required, when the fringe spacing is greater than the base length the response is too small to give strains for that base length. This concept implies that the determined strain is the average strain between two fringes. But it may also be a practical limit of measurable strain since, if the fringe spacing is greater than the gage length, the strain sensitivity begins to approach the order of the strain, as indicated above in Eq. 9.2.

[3] J. D. C. Crisp, "The Measurement of Plane Strains by a Photo-screen Method," *Proc. SESA*, Vol. 15, No. 1 (1957).

9.6. Variation of Material

Considerations on sensitivity, precision, and accuracy are important because of the limitations on the maximum allowable strain in many materials. Many engineering materials fail below 10 per cent strain, some below 1 per cent strain. Below the failure level, engineering materials generally have a proportional limit of strain which can be as low as 0.1 per cent. Thus, a great deal of strain analysis must be conducted below prescribed maximum allowable limits of strain. If the analysis is conducted on the prototype and the material is specified, then the maximum allowable limit of strain is usually fixed. Where the prototype can be simulated by a model, variation in the model material, especially for linear problems, can give a greater maximum allowable strain limit.

Mild steel has a proportional limit of about 0.1 per cent. Plastics such as epoxy have a proportional limit of about 1 per cent. Thus, for a linear analysis, an epoxy model will have 10 times greater response than the mild-steel prototype. Variation of material, then, is one way of increasing moire response.

9.7. Variation of Size

For a given material of a given geometric shape subjected to a given maximum allowable strain, multiplying the size of the specimen will multiply the displacements and therefore the moire fringes by a like amount. Thus, variation in size is a second way of increasing moire response.

If the available grating has a fixed size, increasing the size of the model beyond the grating size will, of course, reduce the region of analysis. But if only a small region need be analyzed, then the model size can be increased until the grating covers only the region of interest.

PART 3

SPECIAL METHODS

10

DETERMINATION OF THE VELOCITIES OF DISPLACEMENTS AND STRAINS

10.1. Introduction

The applications described in previous chapters deal with problems in which loads and associated strains are linearly related. This means that with increasing load the strains increase in the same proportion, and thus the distribution of strains, or ratio of strains to each other, throughout the body remains constant. However, there are a large number of problems in nonlinear elasticity and plasticity for which the solution of moire is also suited. In these problems the displacement field and the strain field depend on the load level. This dependence may be due to the fact that the material has nonlinear stress–strain relationship, or to the fact that the geometry of the loaded body changes appreciably in the process of deformation.

To help in the understanding of what follows, the concept of "metamorphosis" will be introduced.

Consider a system of lines drawn on the face of a urethane-rubber circular ring (Fig. 10.1). The ring is subjected to a diametral compression load P, associated with a vertical displacement δ_v of the point of application of the load. A superposed photograph of the undeformed and deformed shapes of the ring shows the metamorphosis which the ring goes through in the process of deformation. *Metamorphosis* refers to the process of transformation, but in particular will be used to specify a deformed shape.

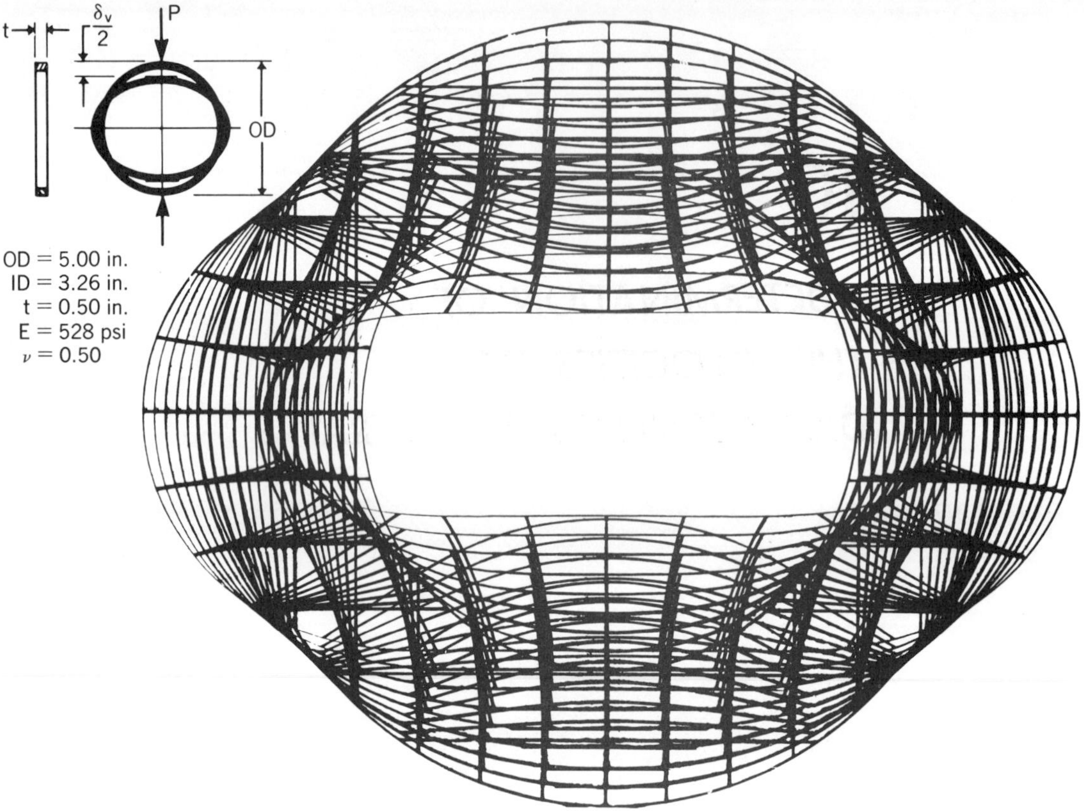

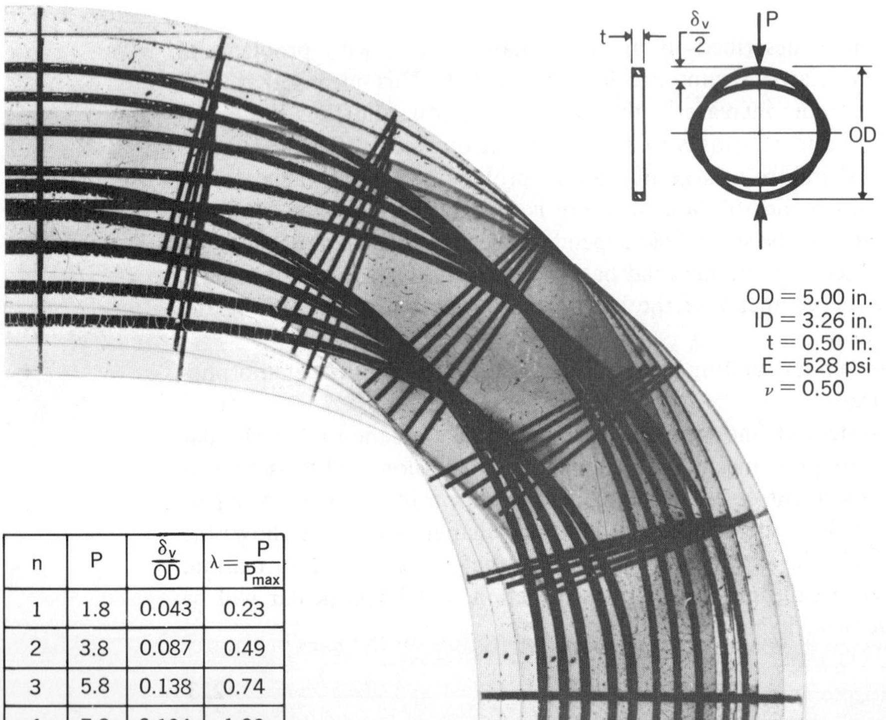

n	P	$\frac{\delta_v}{OD}$	$\lambda = \frac{P}{P_{max}}$
1	1.8	0.043	0.23
2	3.8	0.087	0.49
3	5.8	0.138	0.74
4	7.8	0.194	1.00

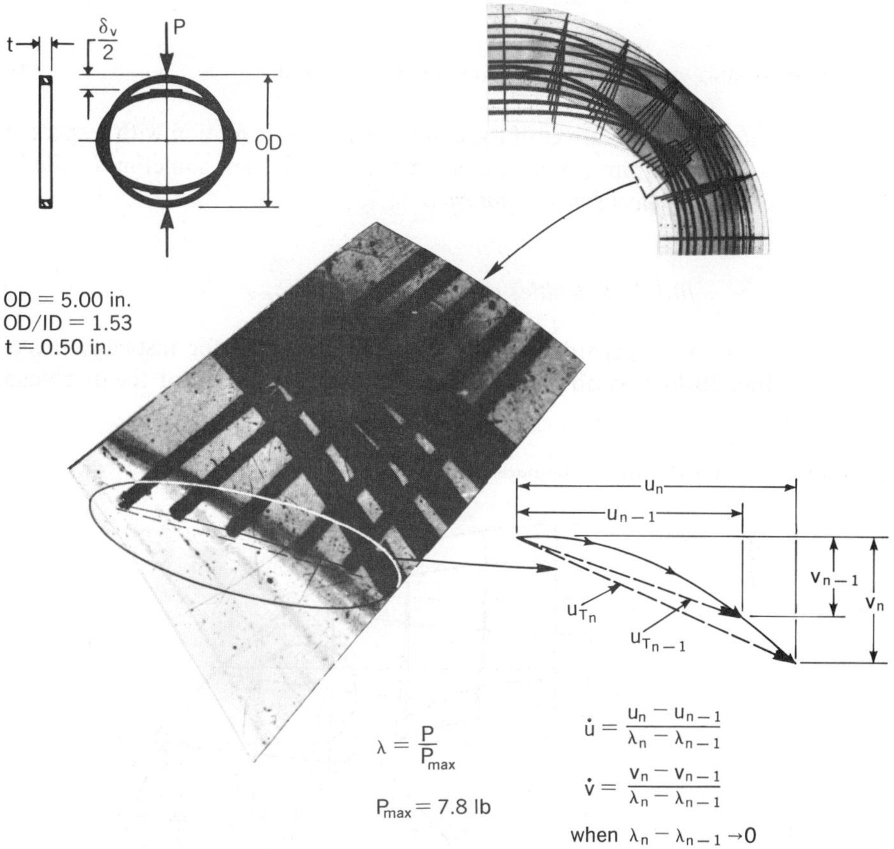

Figure 10.3. (Top) Displacement velocity as a function of load.

Figure 10.1. (Upper left) Metamorphosis of a circular ring subjected to diametral compression (Cartesian grid).

Figure 10.2. (Bottom left) Deformed shape of a ring subjected to four levels of diametral compression (metamorphosis of polar grid).

A detail of the metamorphosis at four levels of load when a polar grid is drawn on the face of the ring is shown in Fig. 10.2. The parameter λ_n indicates the level of loading, and is defined as

$$\lambda_n = \frac{P_n}{P_m}$$

where P_n is a given load level and P_m is the maximum load applied to the specimen.

10.1.1. Velocities of Displacement

The enlarged detail of the metamorphosis shown in Fig. 10.3 illustrates the case of nonlinear motion of a point and its projection on the axes of coordinates. The derivatives of the components of this motion with respect to the loading, or to some parameter related to the loading, are sometimes called the *velocities of the displacement components.*

10.1.2. Velocities of Strain

With regard to a single metamorphosis—for instance, the one shown in Fig. 10.4—it is obvious that the value and direction of the displacements change

Figure 10.4. Spatial derivatives of displacement.

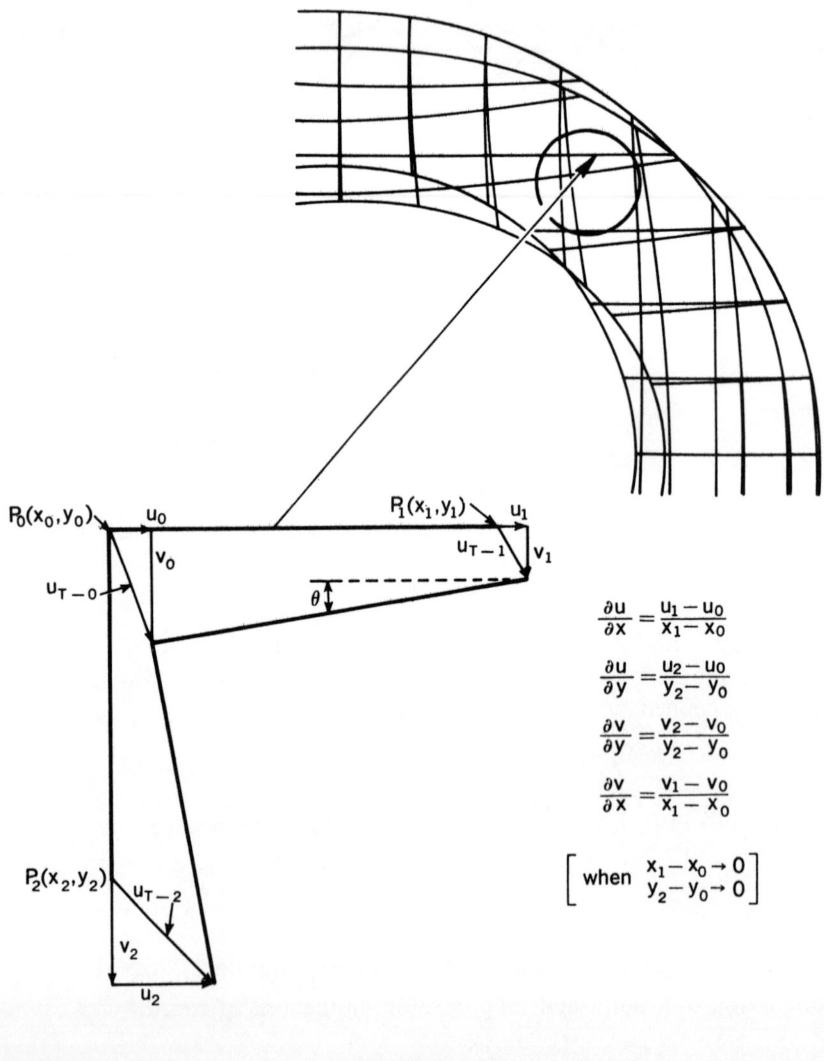

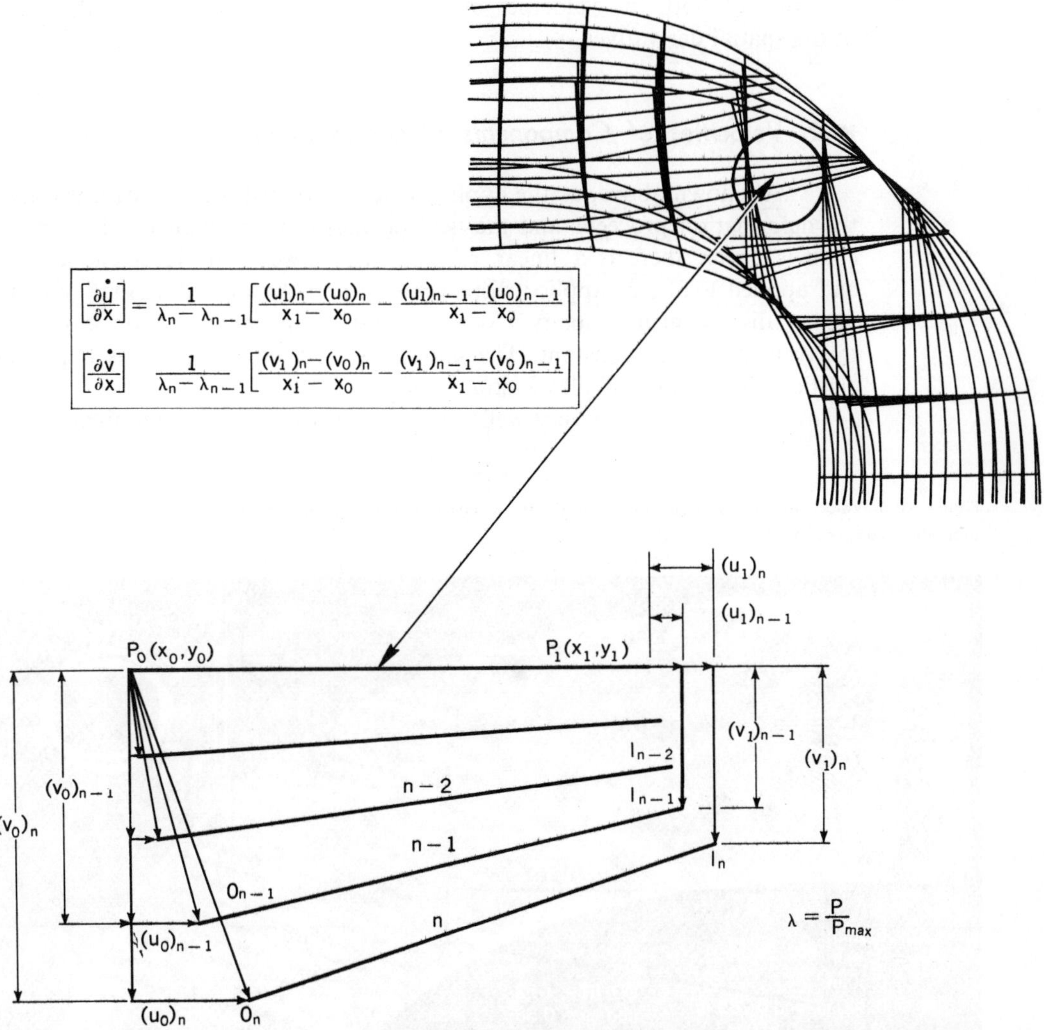

Figure 10.5. Velocities of spatial derivatives of displacement.

from point to point. This change in space (or as a function of the coordinates of the point) is expressed by the spatial partial derivatives of the displacement components. The determination of these spatial derivatives was discussed in Chapter 7.

It is easy to conceive that, as there is a change in displacement of a point as the deformation proceeds (displacement velocity), there is also a change in the spatial derivative as the deformation proceeds (*partial-derivative velocities*). These are illustrated in Fig. 10.5.

The object of this chapter will be to illustrate with an example the determination of both the velocities of displacement components and the velocities of the spatial derivatives.

10.2. Velocities of Components of Displacement

As shown above, as the applied load or applied displacement is increased, the deformation proceeds and the field of displacement changes. If the phenomenon is linear, there is a linear relationship between displacement at a point and applied loads, or applied displacement; and the rate of change of displacement ("displacement velocity") is a constant. If the phenomenon is not linear, this velocity is not constant. The knowledge of these velocities is important in elastic-finite-strain and plastic-strain analyses.

Moire effects are very well suited to the whole-field determination of dis-

Figure 10.6. Velocities of displacement (*u*-field) in a circular ring subjected to diametral compression ($u_{7.3} - u_{6.8}$).

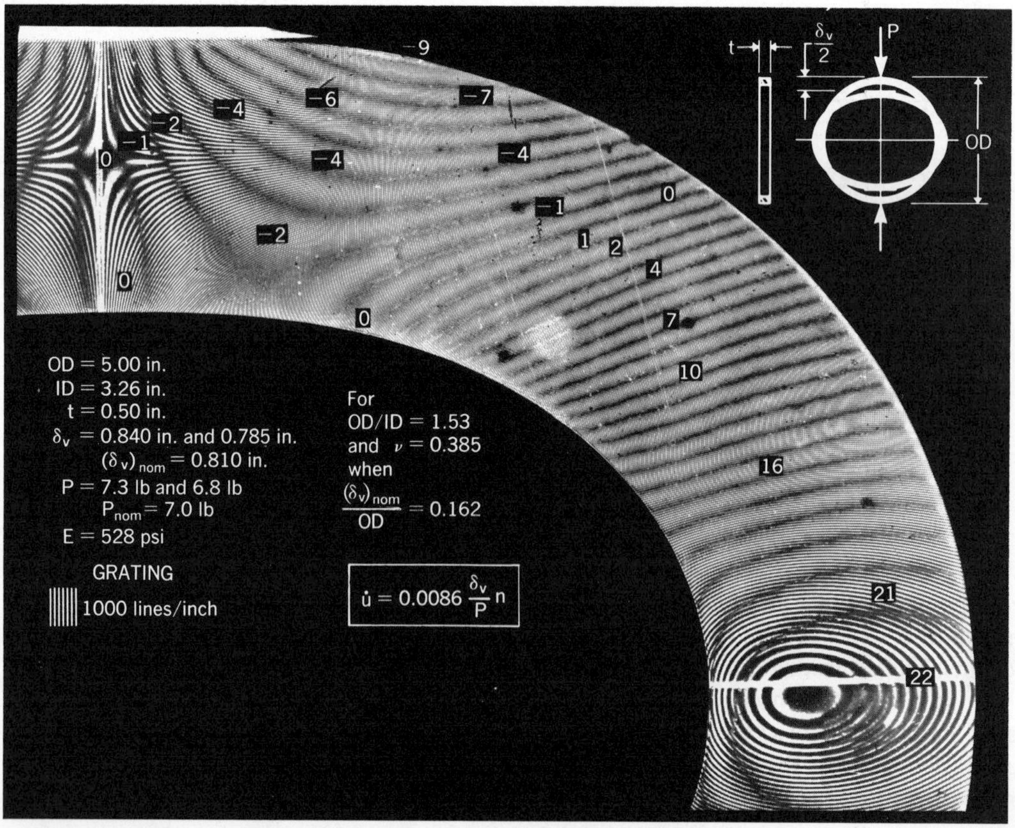

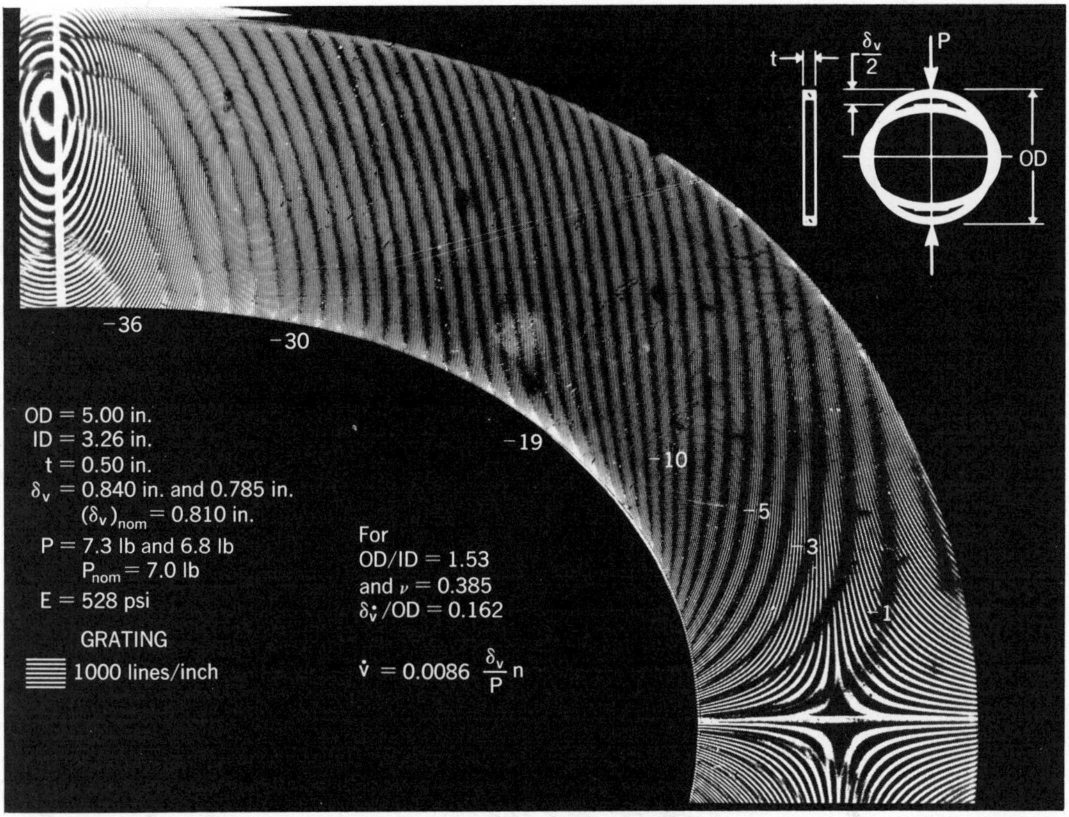

Figure 10.7. Velocities of displacement (v-field) in a circular ring subjected to diametral compression ($v_{7.3} - v_{6.8}$).

placement velocities. A moire pattern can be interpreted as the sum or difference of the two motifs producing the moire. In other words, if the lines of the two original gratings are ordered in the proper manner, the order of the moire fringe produced at one point is the sum or difference of the order of the lines of the gratings that intersect at that point. (See Chapter 2.)

In this case, the original gratings will be the moire isothetics corresponding to two levels of loads. Superimposing these two patterns, the moire-of-moire pattern at a point will be, in general, the difference between the order of the fringes of the two isothetics at the same point. Therefore, this value, divided by a parameter related to the increase in loading, will give the velocity of displacement.

Figures 10.6 and 10.7 are loci of points exhibiting the same velocity of u- and v-displacements, respectively, between load levels of 6.8 and 7.3 lb in the rubber ring. The phrase "isotachics of displacement component" (u or v) is suggested to identify these loci. The phrase "isotachics of isothetics" could be used as well.

These same patterns can also be obtained without the intermediate moire by superposing images of the deformed line gratings at the two levels of load using a procedure similar to the shifting method suggested in Sec. 7.7.

10.3. Velocities of Spatial Derivatives

It has been shown how the displacements change as the deformation proceeds and how the rate of change can be determined. The partial derivatives of the displacements also change as the deformation proceeds, and it is important in many cases to determine their rates of change or the velocities of the derivatives of displacements. When the rotations are small, this concept is sometimes called the *strain rate*.

It has also been shown how the velocities of displacements can be obtained photographically by exploiting the moire-of-moire effect (superposition of two isothetics obtained at two different levels of load). In principle, it is possible to

Figure 10.8. Isotachics of partial derivative $\partial \dot{u}/\partial x$ in a circular ring subjected to diametral compression (large deformation).

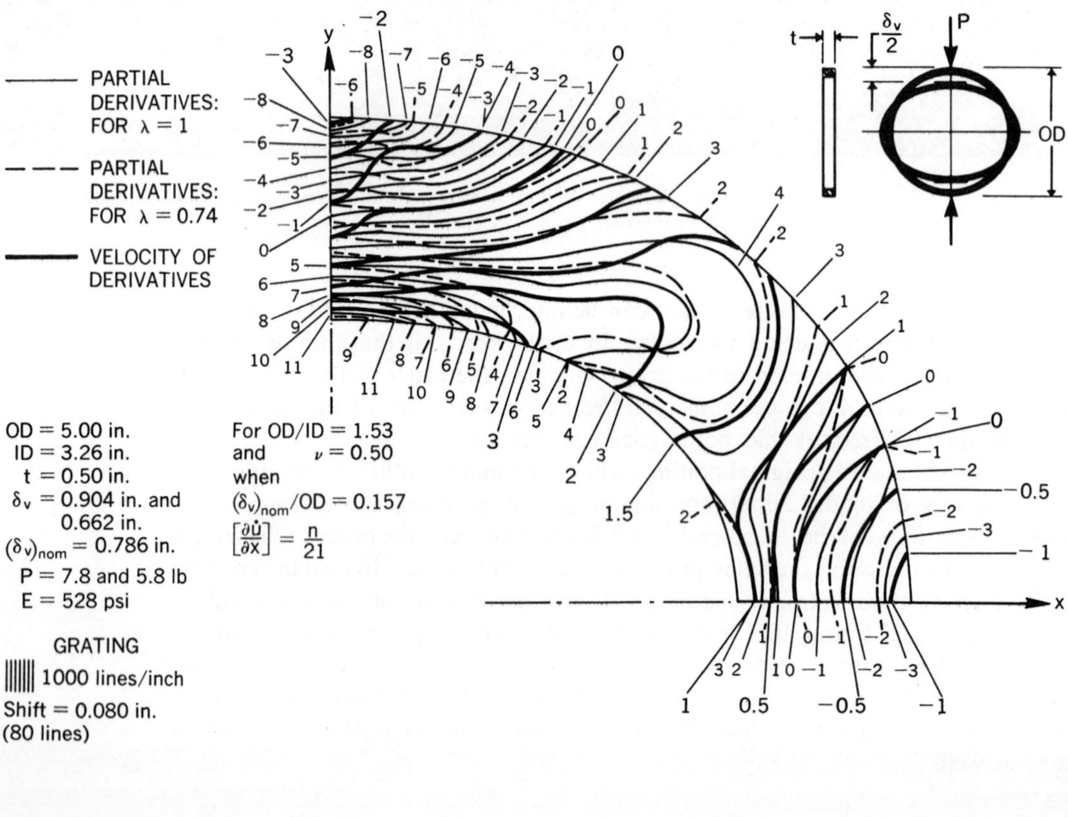

apply the same procedure to the loci of partial derivatives to obtain their velocities. From the point of view of the method this requires, however, another step: the moire of the moire-of-moire. This effect would be produced by the superposition of two patterns, loci of partials, obtained at two load levels. This method requires a high density of partial-derivative fringes and has not been accomplished yet. It is easy, however, to trace on the same scale two patterns of partial derivatives, obtained at two load levels, and to draw the family of lines corresponding to the difference in the parameters of the other two families, using the same procedure explained in Chapter 2. An example of this approach is shown for $(\partial u/\partial x)$ in Fig. 10.8. The phrases "isotachics of partial derivatives" or "isotachics of isoparagogics" are suggested to identify these loci.

Again, an alternative method, similar to the method suggested in Sec. 7.7, is available. Shifting two images of the isotachics of displacements (obtained as explained in Sec. 10.2) with respect to each other will give the isotachics of isoparagogics. Here again, the low orders of fringes will probably require connecting intersections with hand-drawn curves.

11

DETERMINATION OF ROTATIONS

11.1. *Introduction*

In this chapter two kinds of rotations will be considered: the rotation of a line and rigid rotation.

The rotation of a line is illustrated in Fig. 10.4, where it is indicated by the angle θ. The loci of points exhibiting the same value of rotation of specific lines going through the points are called *isogyros*. In the Eulerian description, the isogyro in the x-direction can be expressed by

$$\theta = \arcsin \frac{\partial v}{\partial x}\left(\frac{1}{1 - \epsilon_x}\right) \tag{11.1}$$

which for small ϵ_x becomes

$$\theta = \frac{\partial v}{\partial x} \tag{11.2}$$

The isogyro and cross-derivative of displacement can, therefore, be considered to approximate each other.

Rigid rotation is defined as the simultaneous rotation of the two sides of a right angle that remains right. At each point in a deformed body there are always at least three lines which remain at right angles, and in plane problems there are always two such lines in the plane. The loci of points exhibiting the same value of rigid rotation are called *isostrophics*. In the case of small strains, isostrophics are given by

$$\omega = \frac{1}{2}\left(\frac{\partial u}{\partial y} - \frac{\partial v}{\partial x}\right) \tag{11.3}$$

11.2. Determination of Isogyros

The angle θ of rotation of lines that have a given direction, can be obtained experimentally by rotating the moire master by the angle θ over the deformed grating (Figs. 11.1 and 11.2). Points at which the tangent to the moire fringes is parallel to the master-grating lines are connected by solid lines in the figures. These are points at which the deformed grating has rotated the same amount as the master grating. This property becomes obvious recalling that if a displacement is perpendicular to the common direction of the master and the specimen gratings, then the fringes are parallel to the direction of the gratings lines.

Figure 11.3 combines the lines shown in Figs. 11.1 and 11.2. Compare the θ_x values with the $\delta v/\delta x$ values obtained in Fig. 11.4.

Figure 11.1. Isogyros (u-field) in a circular ring subjected to diametral compression.

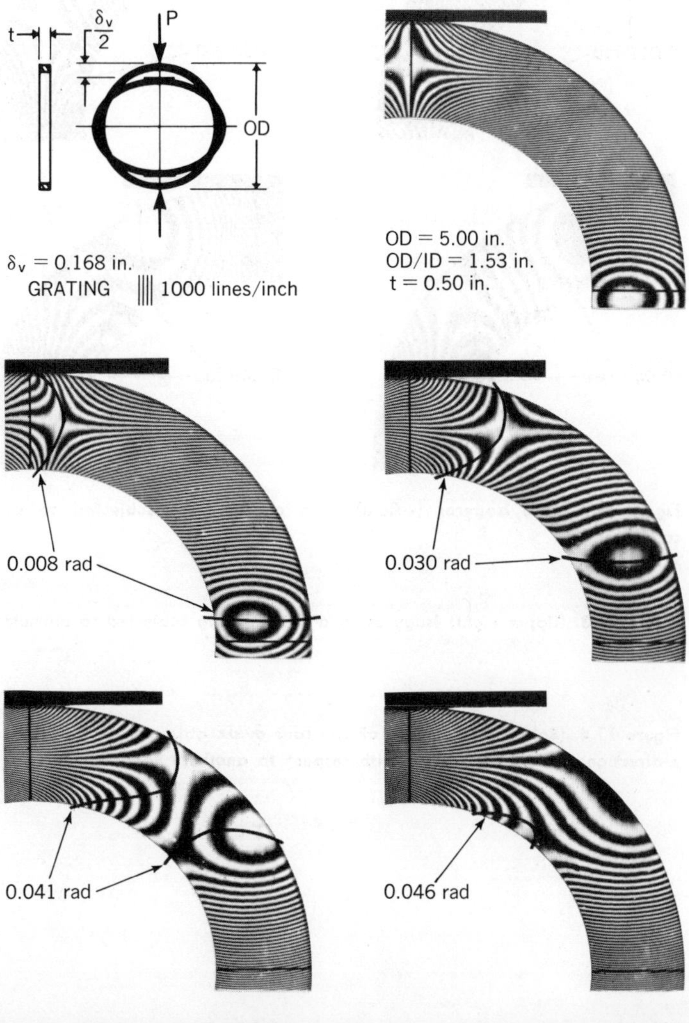

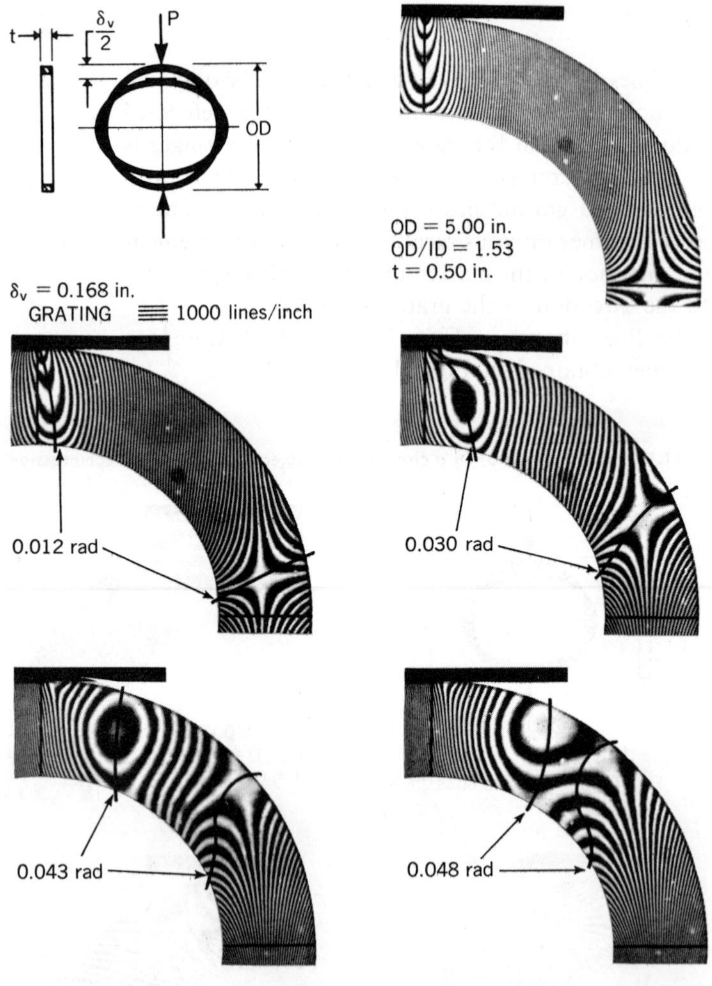

Figure 11.2. (Top) Isogyros (v-field) in a circular ring subjected to diametral compression.

Figure 11.3. (Upper right) Isogyros in a circular ring subjected to diametral compression.

Figure 11.4. (Bottom right) Lines of constant $\partial v/\partial x$ obtained by shifting in the x-direction one v-field pattern with respect to another.

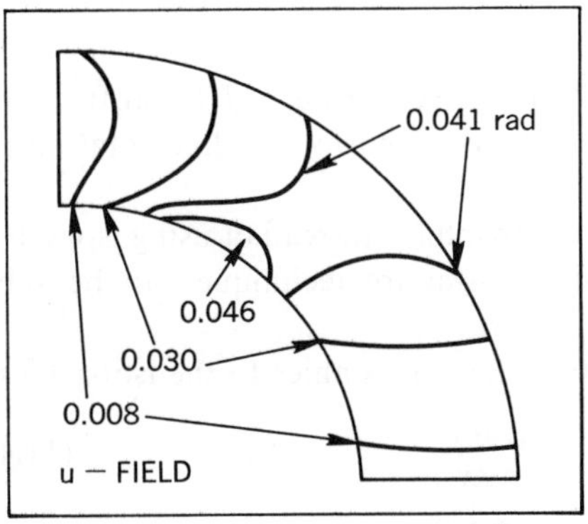

for OD/ID = 1.53 $\theta_y = 29.8 \dfrac{\delta_v}{OD} n$

OD = 5.00 in.
ID = 3.26 in.
$t = 0.50$ in.
$\delta_v = 0.168$ in.
$E = 528$ psi
$\nu = 0.50$

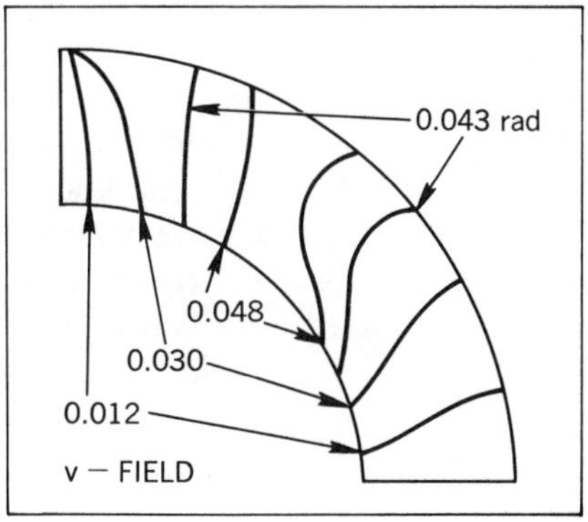

for OD/ID = 1.53 $\theta_x = 29.8 \dfrac{\delta_v}{OD} n$

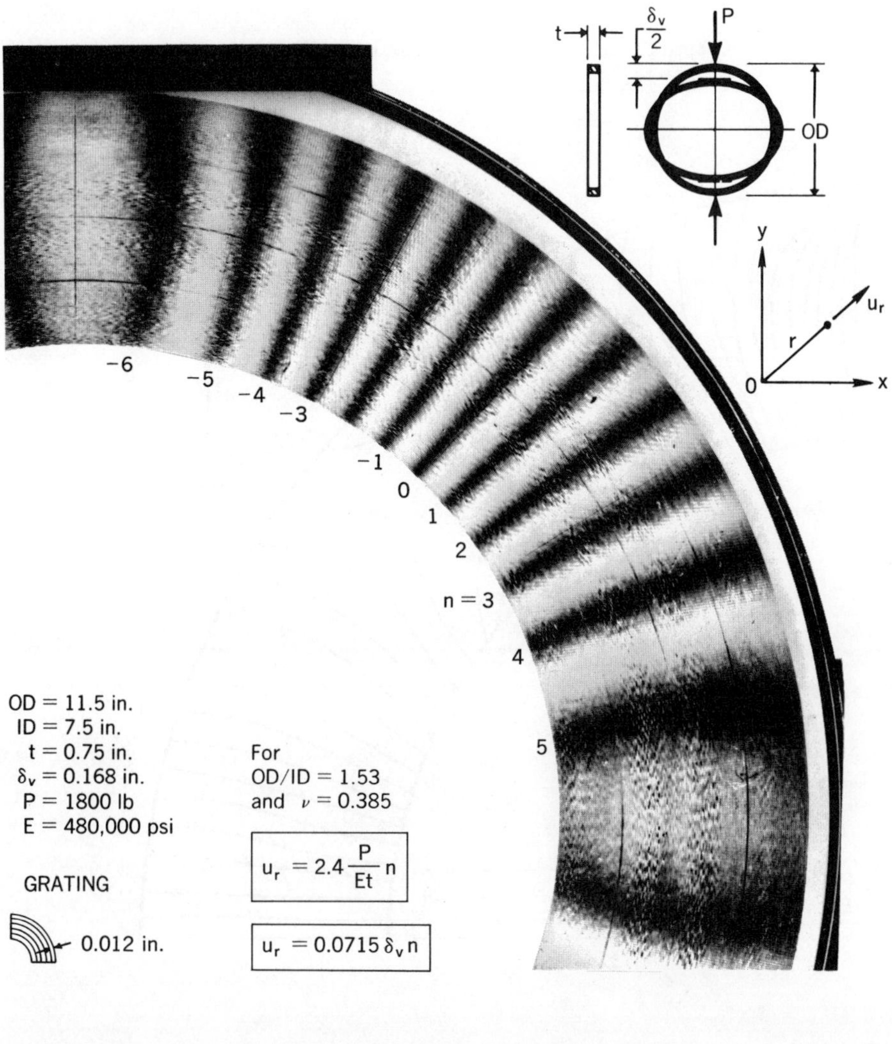

11.3. Determination of Isostrophics

Following the definition of rigid rotation given in Sec. 11.1, isostrophics can be determined by subtraction of the values obtained for the line rotation of two perpendicular lines.

There are other ways of determining isostrophics. Instead of using isogyros, the partial derivatives obtained by the moire-of-moire technique can be subtracted. This was done in Fig. 11.5.

An interesting property is that relating the isostrophics to the isopachics:

$$\frac{\partial(\sigma_x + \sigma_y)}{\partial y} = -E\frac{\partial \omega}{\partial x} \tag{11.4}$$

Equation (11.4) shows that the two families should be perpendicular to each other and, if the units are the same for both families, the distance between contour

Figure 11.5. Isostrophics in a circular ring subjected to diametral compression.

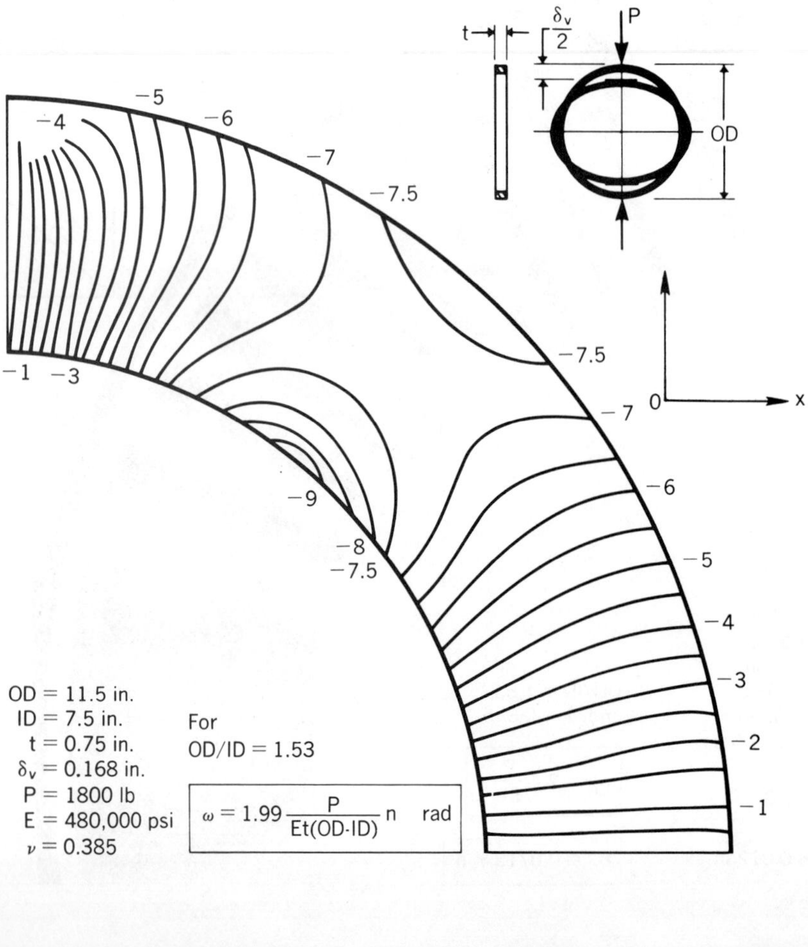

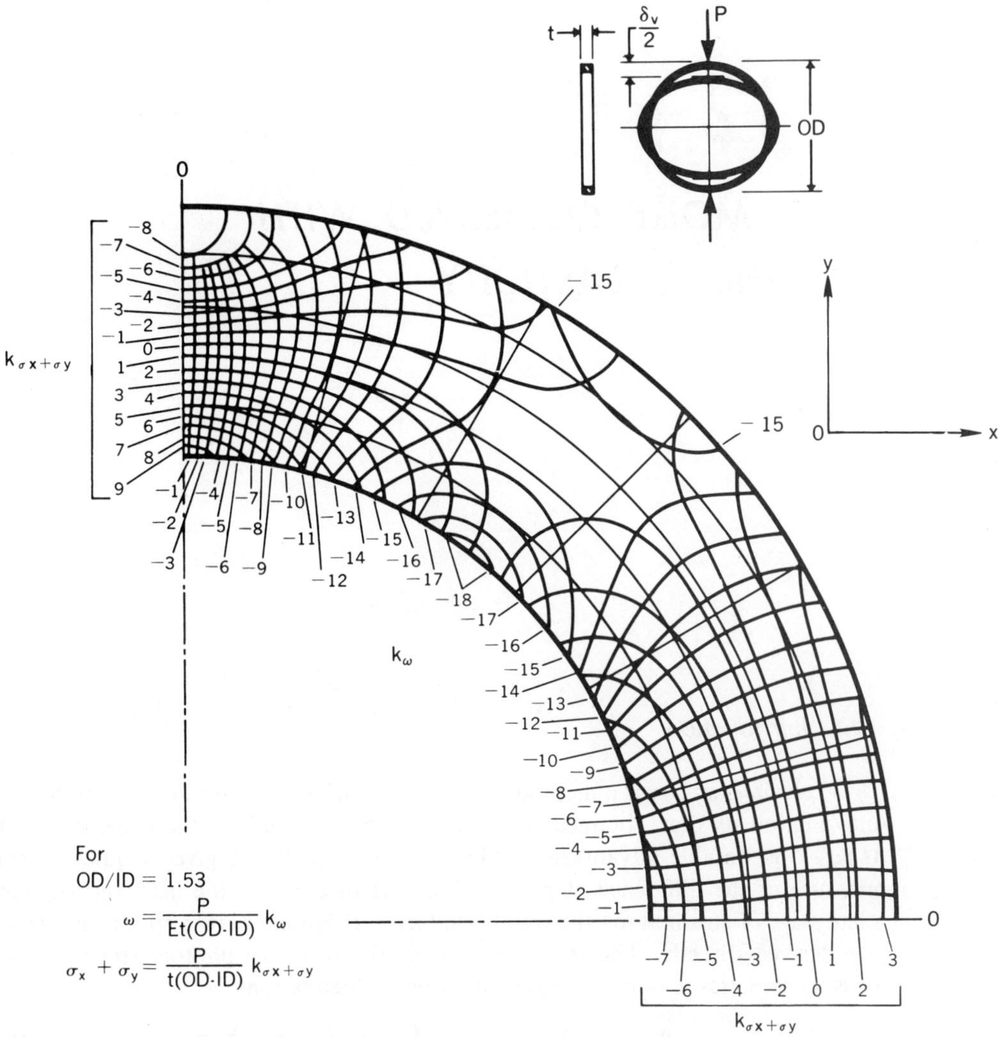

Figure 11.6. Isostrophics (obtained from moire) and isopachics (obtained from thickness measurements) in a circular ring subjected to diametral compression. The two families are related geometrically.

lines will be the same, so as to form a grid of approximately square elements. Figure 11.6 shows isostrophics (obtained from moire) superimposed on the isopachics (obtained from thickness measurements). The two families appear orthogonal over most of the field. No attempt has been made to adjust local inconsistencies.

12

MOIRE COMBINED WITH PHOTOELASTICITY

12.1. Introduction

Photoelasticity is a method of analysis which can be used to obtain the same information as that obtained with moire analysis. Each method has its advantages. One notable advantage of photoelasticity is that it gives a quick, precise measure of the difference of principal strains throughout the field and the complete state of strain along the free boundaries. If Hooke's law applies, the stresses are derived directly. The interpretation of the ordinary photoelastic pattern of of isochromatics is obtained from the following equations:

$$2\tau_{\max} = \sigma_1 - \sigma_2 = \frac{E}{1+v}(\epsilon_1 - \epsilon_2) = 2nF_\sigma \tag{12.1}$$

where n is the isochromatic-fringe order and F_σ is the model stress-fringe value (constant under usual conditions).

In cases where the individual stresses and strains are required within the field, much more elaborate analysis is necessary. There are a number of types of "adjunct" analyses that can be combined with the results obtained from isochromatics, to obtain all principal stresses; for example: (1) the use of other photoelastic patterns (*isoclinics*) with line-by-line integration of the equation of equilibrium; (2) measurement of the thickness changes in the model at each point, due to load, in order to obtain the sum of the principal stresses; (3) solution of Laplace's equation for the given boundary conditions, either theoretically (usually with numerical methods) or by some experimental analogy, to obtain the sum of the principal stresses; (4) absolute retardation; (5) oblique-incident methods.

Moire can, of course, be used alone to obtain the complete strain field in a plane and, from it, the complete stress field. However, there are times when it is advantageous to use moire analysis in conjunction with the easily obtained isochromatics of photoelasticity, rather than the other methods itemized above. The methods to be described in this chapter take advantage of the presence of relatively large strains in some linear elastic materials, which makes the measurements easier and increases their precision. Gratings can be printed or cemented to these photoelastic materials in the usual manner.

Combination of isochromatics and two families of isothetics and combination of isochromatics and one family of isothetics are the methods to be described in the following two sections. An application in which photoelasticity is used for stress analysis and moire for displacement measurements will also be shown.

The specimen to be analyzed can be made of a low-modulus, transparent material with birefringent properties, such as polyurethane rubber; both the isochromatics and the isothetics can be obtained at room temperature. Epoxies, when subjected to high temperature (of the order of 200°F) are also suitable materials. At a critical temperature that depends on the particular composition, epoxies are rubbery like polyurethanes. If an epoxy model is loaded at this critical temperature and cooled, both the birefringent response and the deformation are "frozen" or "locked in." The load can be removed at ambient temperature and the deformations and birefringence recorded. Analysis of frozen models is described in Sec. 12.4.

12.2. *Use of Two Moire Isothetics*

The sum of the two principal strains in the plane of a plate (in plane stress) is given by

$$\epsilon_1 + \epsilon_2 = \epsilon_x + \epsilon_y = \frac{1 - \nu}{E}(\sigma_x + \sigma_y) = \frac{1 - \nu}{E}(\sigma_1 + \sigma_2) \qquad (12.2)$$

The sum of the principal stresses can, therefore, be obtained from the determination of two normal strains in the plane of the plate. This information is available from the isothetic moire patterns. Two of these patterns are necessary for the general case—for instance, one corresponding to the u-displacements and the other corresponding to the v-displacements. From the differentiation of the displacements, the strains are obtained. This way of operating requires only the isochromatics (Fig. 12.1) and the isopachics (loci of points of equal change in thickness—Fig. 12.2) obtained from the two moire patterns (isothetics).

This method has been found particularly useful in dynamic analysis, and applications will be shown in Chapter 17. The method has been shown to be precise. The determination of the sum of the principal stresses from moire patterns can be done with little error. The method has the disadvantage, however, of requiring two isothetics such as the ones shown in Figs. 7.3 and 7.4. It should be noted that in analyzing the strains along a given line (say, in the x-direction), if one of the moire gratings is perpendicular to the line, the strains ϵ_x in the direction of that line can be obtained from a continuous differentiation of the displacement curve along the line. However, to obtain the orthogonal strain ϵ_y along the same line requires a more cumbersome process of analyzing the other moire pattern by a series of differentiations along several lines perpendicular to the given line. This situation brings into consideration the next method.

12.3. *Use of Isoclinics and One Moire Isothetic*

When the isoclinics have been obtained photoelastically (or the isostatics have been obtained by means of brittle coatings), the direction of the principal stresses is known, and only one moire pattern is required. Thus,

$$\begin{aligned}\epsilon_1 &= \epsilon_x - \frac{\gamma_{\max}}{2}\cos 2\theta + \frac{\gamma_{\max}}{2} \\ &= \epsilon_x + \gamma_{\max}\sin^2\theta \\ &= \epsilon_x + \frac{\tau_{\max}}{G}\sin^2\theta\end{aligned} \tag{12.3}$$

and similarly, for the other principal strain,

$$\epsilon_2 = \epsilon_x - \frac{\tau_{\max}}{G}\cos^2\theta \tag{12.4}$$

where θ is the angle between the direction of the lines in which ϵ_1 and ϵ_x act. Adding Eqs. (12.3) and (12.4), the sum of the principal strains is given by

$$\epsilon_1 + \epsilon_2 = 2\epsilon_x - \frac{\tau_{\max}}{G}\cos 2\theta \tag{12.5}$$

Substituting Eq. (12.5) into Eq. (12.2), the sum of the principal stresses is obtained as follows:

$$\frac{1}{2}(\sigma_1 + \sigma_2) = \frac{1}{1-\nu}[E\epsilon_x - (1+\nu)\tau_{\max}\cos 2\theta] \tag{12.6}$$

Combining the above with Eq. (12.1), the individual principal stress is obtained from

$$\begin{aligned}\sigma_1 &= \frac{1}{1-\nu}[E\epsilon_x + (1-\nu)\tau_{\max} - \tau_{\max}(1+\nu)\cos 2\theta] \\ &= \frac{1}{1-\nu}\{E\epsilon_x + \tau_{\max}[(1-\nu) - (1+\nu)\cos 2\theta]\}\end{aligned} \tag{12.7}$$

where ϵ_x is obtained from the isothetic, $\tau_{\max}$ is obtained from isochromatics, and

θ is obtained from isoclinics. For lines of symmetry $\theta = 0$, and this equation becomes

$$\sigma_1 = \frac{1}{1-\nu}(E\epsilon_x - 2\nu\tau_{\max}) \tag{12.8}$$

12.4. Use of Moire on Frozen Specimens

The illustrations in Figs. 7.3 and 7.4, among others shown in this book, correspond to moires produced on the surface of rubber specimens by means of gratings printed on an emulsion deposited on the surface of the rubber. Similarly, gratings can be printed on epoxy specimens before they are subjected to the

Figure 12.1. Isochromatics in a polyurethane rubber ring diametrically loaded.

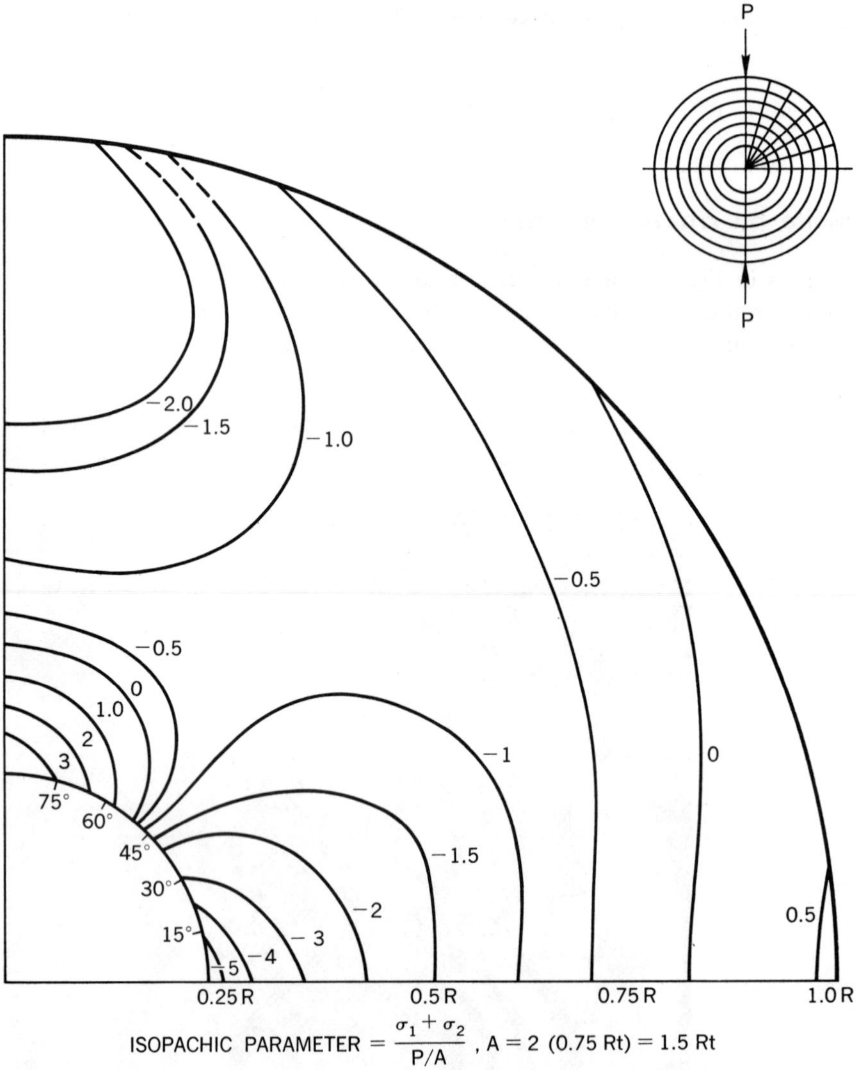

Figure 12.2. Isopachics in a thick ring subjected to diametral compression.

"freezing" cycle. This cycle will then "lock in" both the birefringence and the strains, with the strains producing the moire effect. Variation of the method consists in obtaining the isochromatics first and then printing the gratings on the deformed specimen. The moire is obtained in the process of annealing the "locked-in" pattern.

If the analyst does not have facilities to print the gratings on the surface

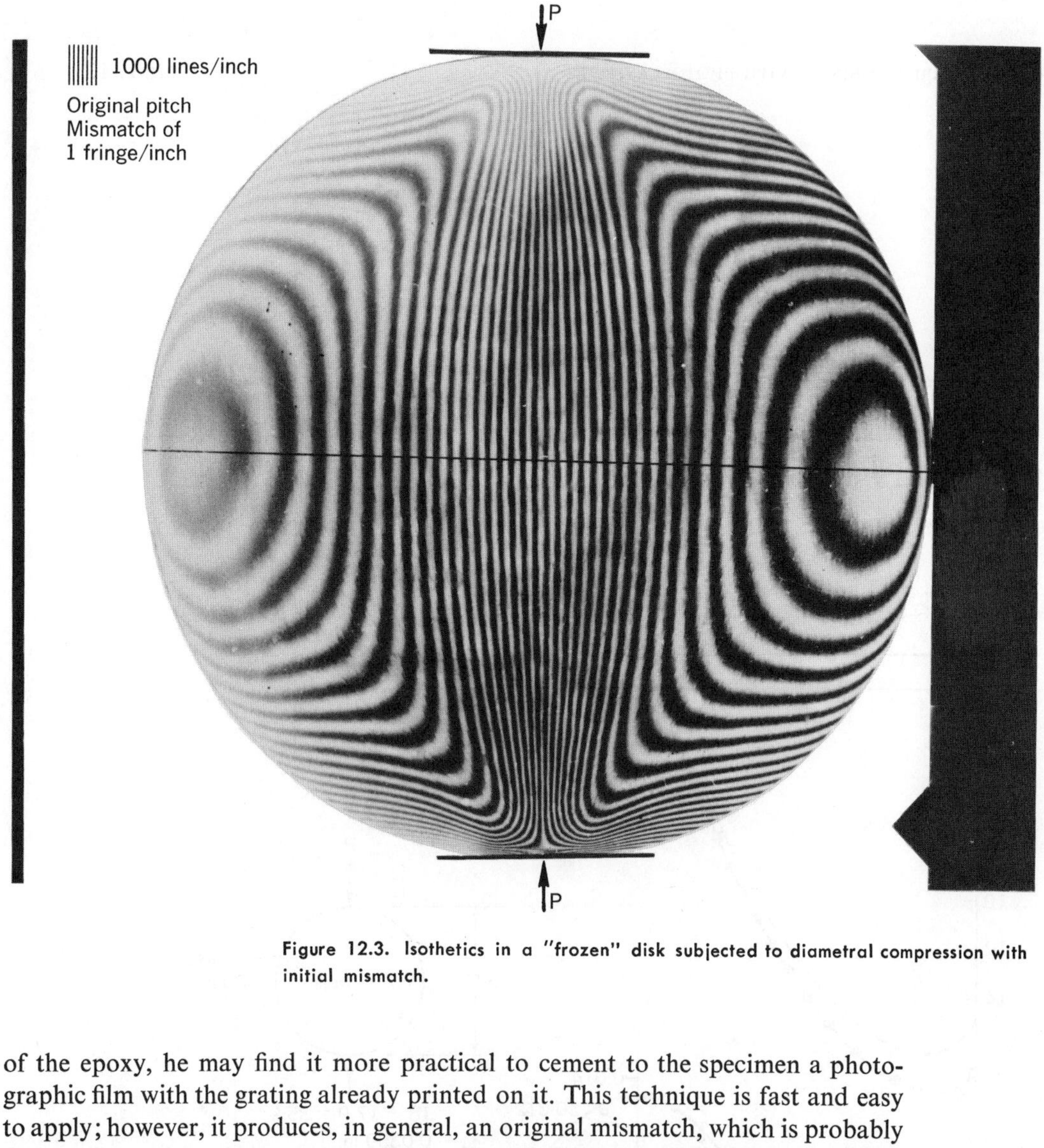

Figure 12.3. Isothetics in a "frozen" disk subjected to diametral compression with initial mismatch.

of the epoxy, he may find it more practical to cement to the specimen a photographic film with the grating already printed on it. This technique is fast and easy to apply; however, it produces, in general, an original mismatch, which is probably due to the curing of the cement. A photograph obtained using this technique is shown in Fig. 12.3. Evaluation of the pattern requires the deduction of the mismatch effect from the final effect, as shown in Fig. 12.4.

This method can be applied to frozen slices cut from three-dimensional specimens for photoelastic purposes, when the slices are thin and parallel to free boundaries, which is a case of approximately plane stress ("surface" slices). If the Poisson's ratio of the material used is appreciably lower than 0.5, it would also be practical to use in a similar way the generalized Hooke's law applied to internal slices of the frozen specimen. In this case two or three slices, depending on the state of stress, may be necessary to determine all the stresses.

Sometimes it may be convenient to use three-dimensional photoelasticity

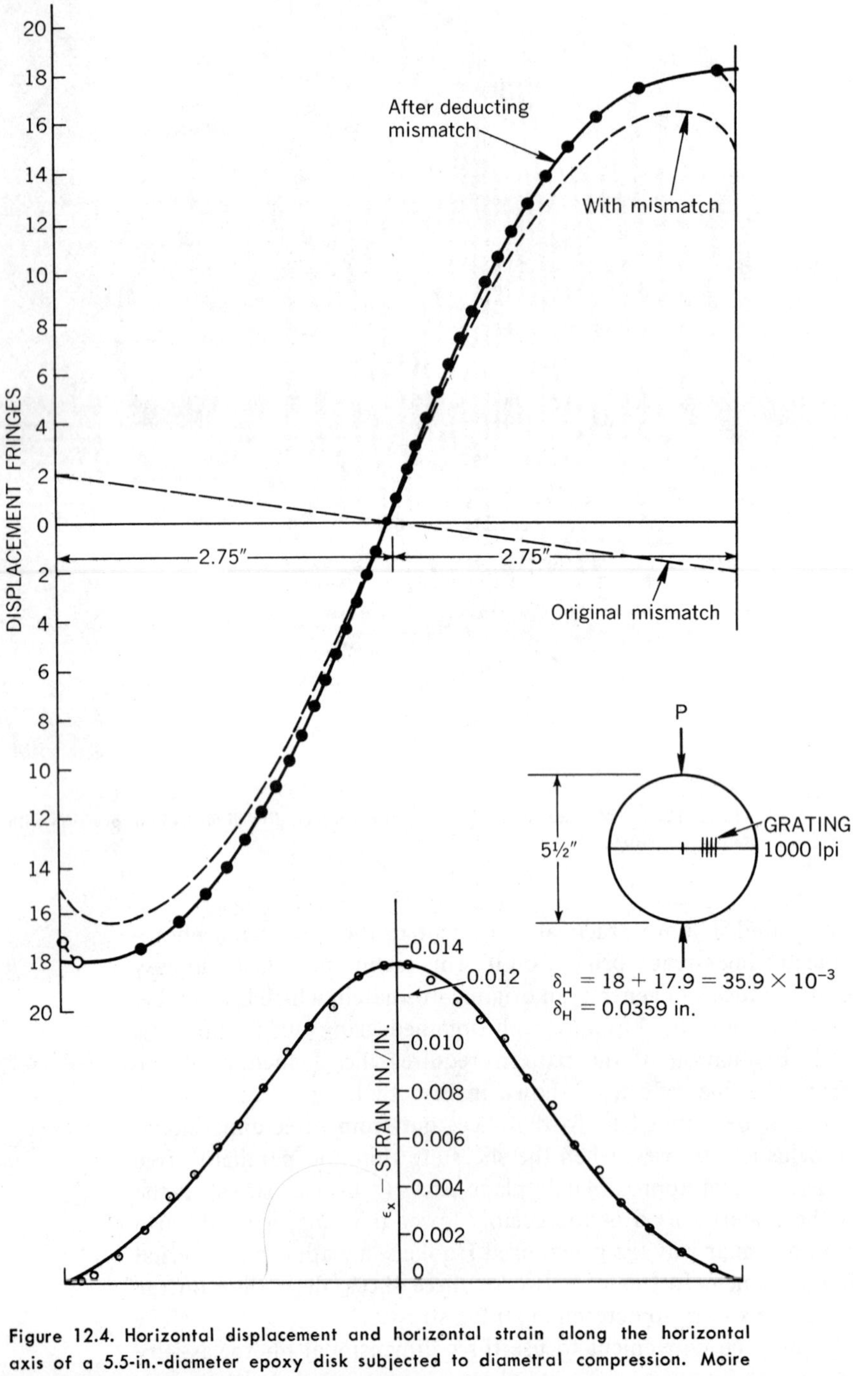

Figure 12.4. Horizontal displacement and horizontal strain along the horizontal axis of a 5.5-in.-diameter epoxy disk subjected to diametral compression. Moire analysis with mismatch of a frozen specimen.

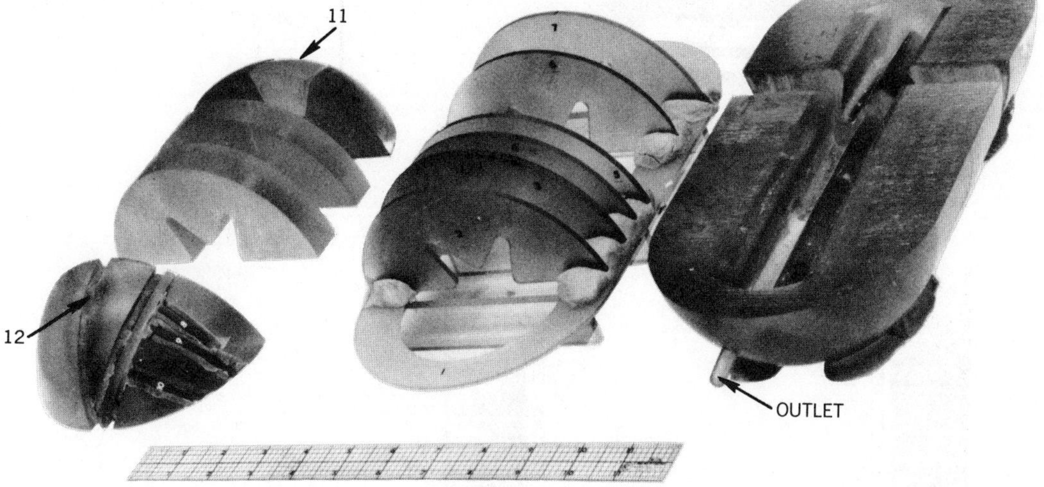

Figure 12.5. Epoxy model of a pressure vessel showing the star-shaped internal configuration and the slices removed for the analysis.

Figure 12.6. Isochromatics (light field) in a transverse slice ($t = \frac{1}{8}$ in.) of the pressure vessel.

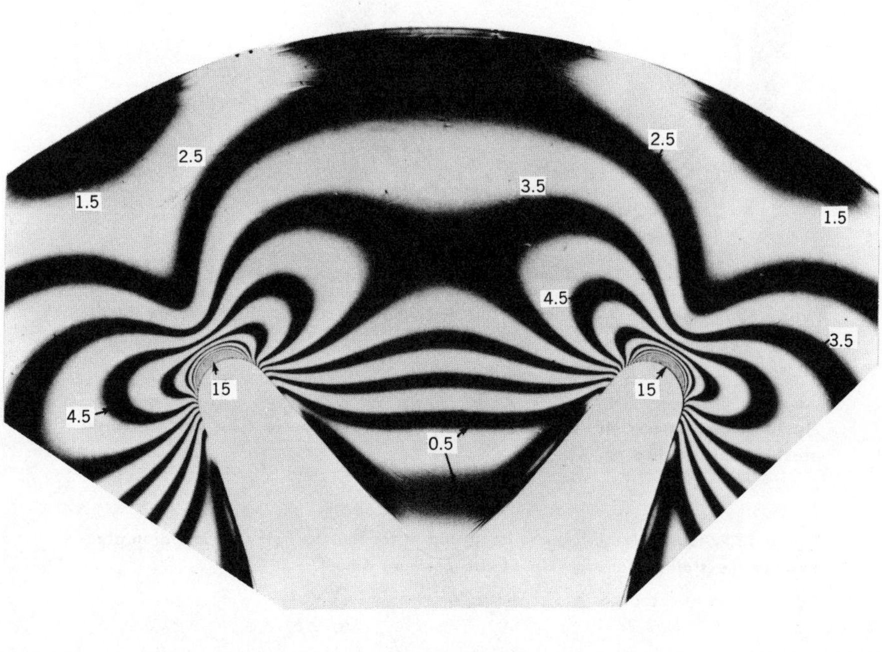

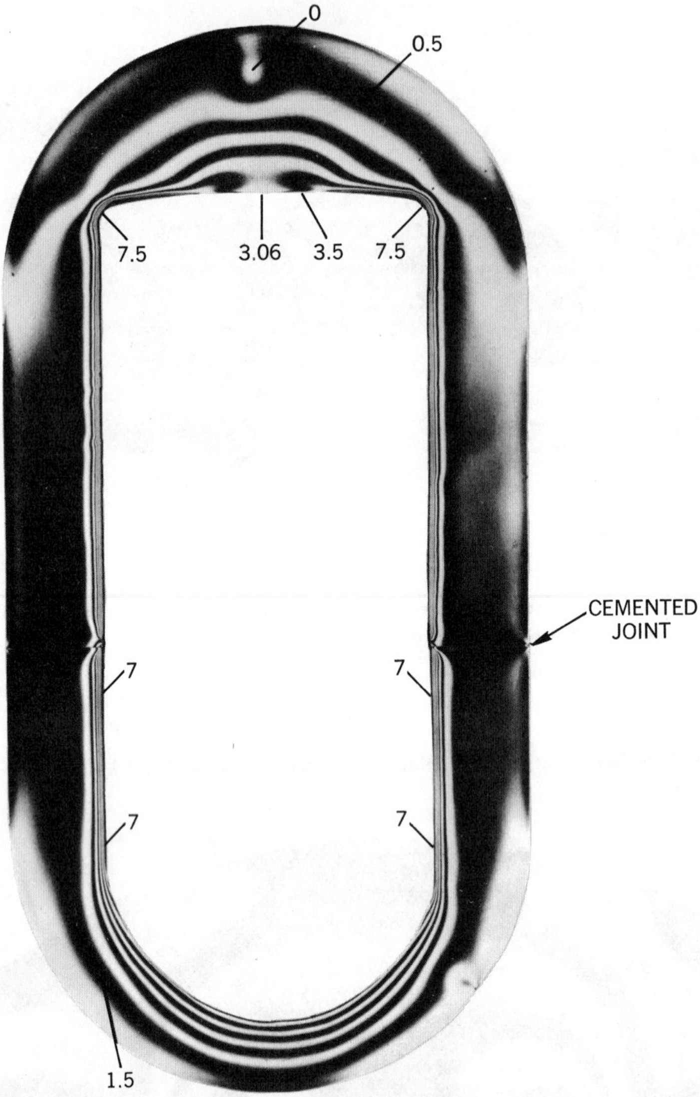

Figure 12.7. (Top) Isochromatics in the meridian slice of the pressure vessel ($t = \frac{1}{8}$ in.).

Figure 12.8. (Upper right) Isothetics *u* and *v* in a transverse cross-section of the pressure vessel (slice no. 4).

Figure 12.9. (Bottom right) Isothetics *u* and *v* for the top half of a meridian plane through the star perforation tip of the pressure vessel.

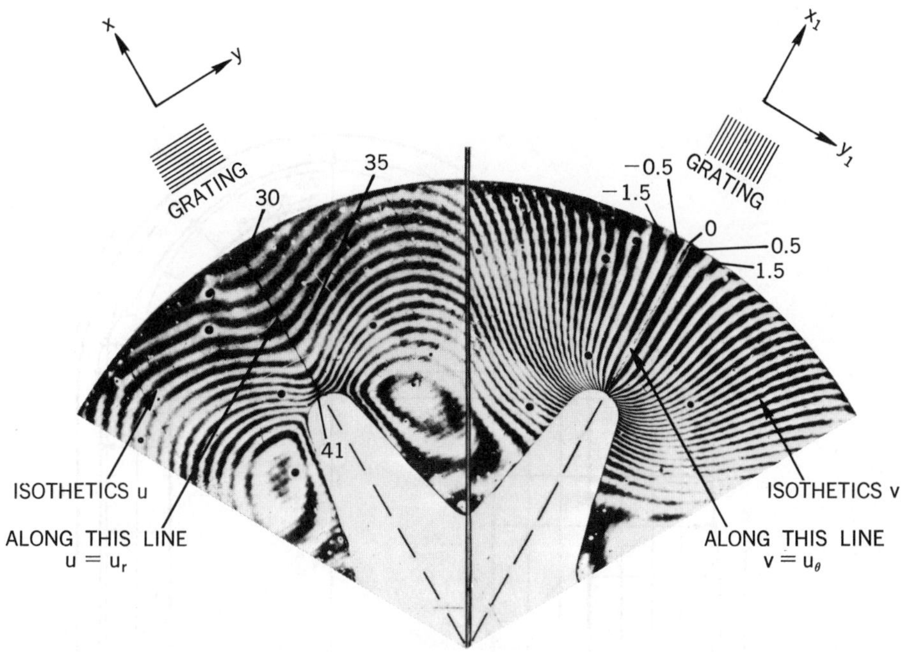

x
y
x_1
y_1
GRATING
GRATING
35
30
−0.5
−1.5
0
0.5
1.5
41
ISOTHETICS u
ALONG THIS LINE
u = u_r
ISOTHETICS v
ALONG THIS LINE
v = u_θ

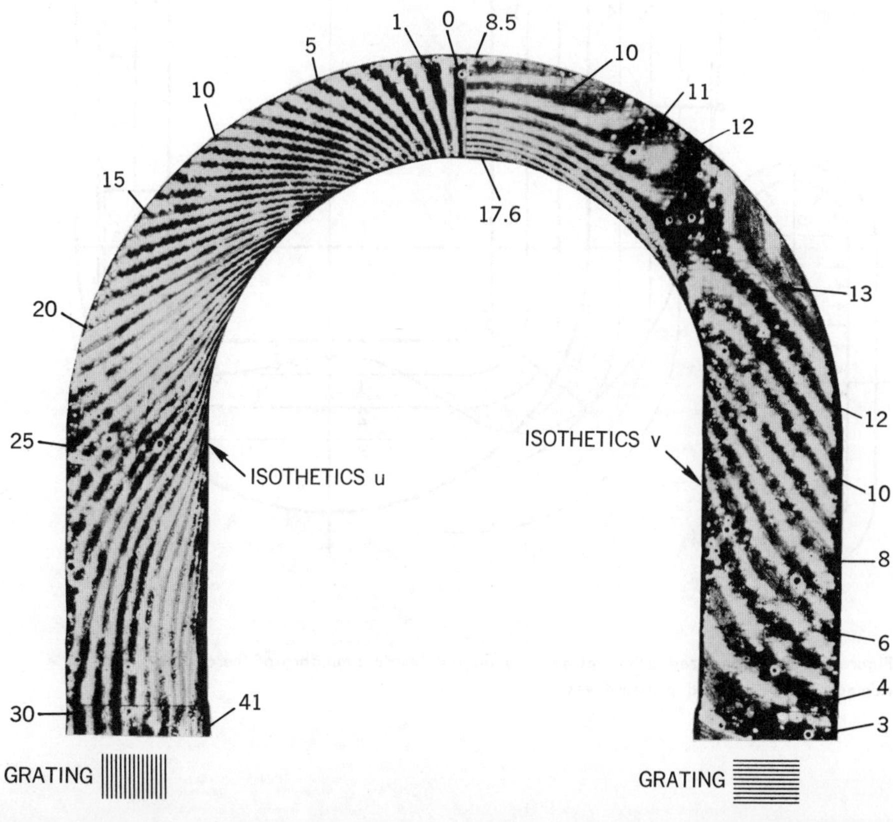

1
0
8.5
5
10
10
11
12
15
17.6
13
20
12
25
ISOTHETICS u
ISOTHETICS v
10
8
6
4
30
41
3
GRATING
GRATING

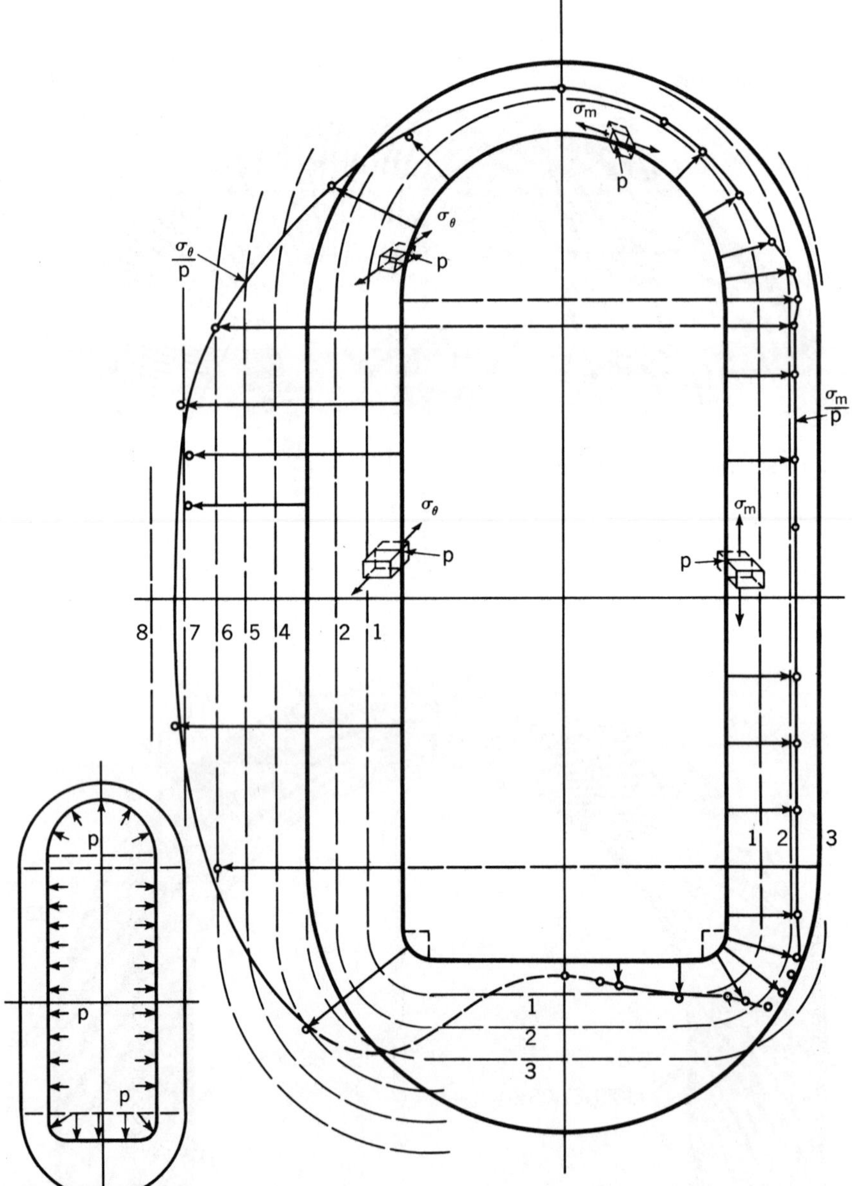

Figure 12.10. Normalized principal stresses on the inside boundary of the double-domed, star-perforated pressure vessel.

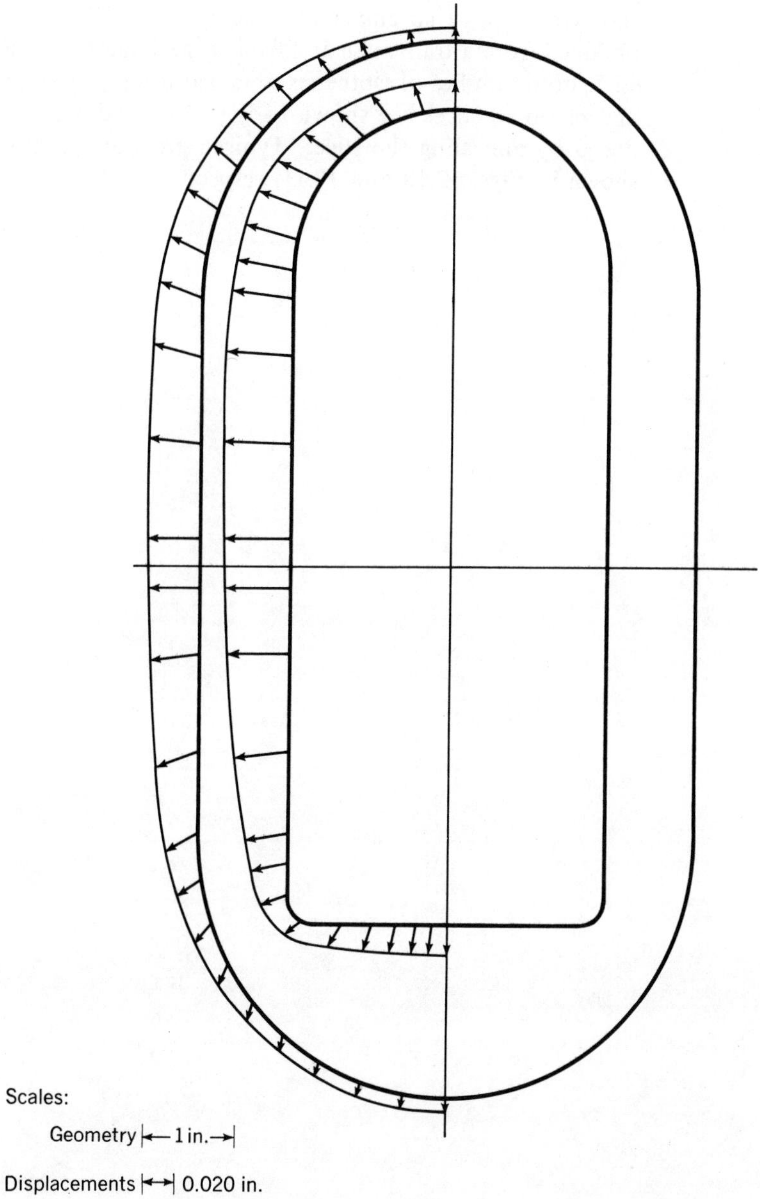

Figure 12.11. Boundary displacements of the pressure vessel in a meridian plane through the star perforation tip.

for the stress-and-strain analysis, and moire only to determine deformed shapes. This approach was followed in the study of stresses, strains, and displacements present in a thick-walled pressure vessel of very complicated shape. Figure 12.5 shows the epoxy specimen after being "frozen" and sliced. The isochromatics obtained from a transverse and from a meridian slice are shown in Figs. 12.6 and 12.7. Both families of isothetics obtained using 1000-lpi gratings in the transverse and meridian slices are shown in Figs. 12.8 and 12.9. The moire effect was produced by annealing the slices. Typical stress and displacement distributions are shown in Figs. 12.10 and 12.11, respectively.

13

OUT-OF-PLANE ANALYSIS

All the displacements and strains analyzed in previous chapters take place in a surface or plane (in-plane analysis). In general, displacement and strains occur in three directions at all points in a body, including the surface of the body. In this respect, it was noted in Sec. 1.7 that the displacement normal to a surface being analyzed using moire effects, can be a source of error if a gap is produced between the specimen and the master grating. Some particular characteristics of this gap were studied in Sec. 1.8.

The out-of-plane displacement can be analyzed independently from inplane displacement using one of several available moire techniques. In evaluating the results of in-plane analysis, treated in the previous chapters, an attempt was made to avoid the influence of out-of-plane displacements; the class of problems treated with this kind of analysis are problems in which out-of-plane displacements are small. The out-of-plane moire methods incorporate techniques to minimize the effect of in-plane displacements; the class of problems treated with this kind of analysis are problems in which out-of-plane displacement predominates.

13.1. Out-of-Plane Displacement Measurements

Out-of-plane displacements are mainly useful for the analysis of plane-stress problems. For these problems, the out-of-plane displacement, determined by moire methods, should be doubled and divided by the specimen thickness to get the strain in the direction normal to the specimen. This strain is assumed to be uniform through the thickness and (making the usual assumption of the linear theory of elasticity) is proportional to the sum of the pair of orthogonal in-plane strains at the point. The loci of points of constant change in thickness are called *isopachics*. The mathematical relations are

$$\frac{\Delta t}{t} = \epsilon_z = \frac{-\nu}{1-\nu}(\epsilon_x + \epsilon_y) \tag{13.1}$$

where t is the specimen thickness, Δt is the change in thickness, ϵ_z is the out-of-plane strain, ϵ_x, ϵ_y are an orthogonal pair of in-plane strains, and ν is Poisson's ratio of the material.

Out-of-plane displacements have previously been measured using an initially optically flat specimen, which when deformed would produce a change in the airgap between its surface and an undeformed master optical flat. Light sent perpendicular to the surface produces the well-known Newton's rings. This type of light interference gives directly the isopachics. Interferometers can be used to measure the change in thickness of a specimen by arranging the beam splitter and mirrors of the interferometer so that one beam passes through the specimen (usually twice) while the other does not. If the two beams are then superimposed, the interference pattern which they produce contains fringes proportional to lines of constant thickness. If the specimen were initially optically flat, then these fringes would also be proportional to the isopachics. This method and several applications are discussed in Chapter 20. More recently, moire obtained with a shadow technique was suggested to obtain isopachics.[1] In this case the specimen is coated with a matte finish to give a distinctive shadow, and a master grating is placed in front of it. Collimated light is directed at an angle to the master grating and the interference between the master grating and its shadow produces a moire pattern. When viewed normal to the master grating, the moire pattern can be analyzed to give the out-of-plane displacement and isopachics (Fig. 13.1). The fringe is produced by the master and a shadow of the master displaced in the plane of the grating an amount e. From geometry, the distance to the model w is given by $w = e/\tan i$, where i is the angle between the light beam and the normal to the grating. If the angle i is kept less than about 0.1 radian, $\tan i \approx i$, and $w \approx e/i$.

The distance e divided by the pitch p gives the number of grating lines between the entering and exiting ray, and this in turn specifies the fringe order; thus

$$n = \frac{e}{p}$$

or

$$n = \frac{w \tan i}{p}$$

[1] P. S. Theocaris, "Moire Fringes of Isopachics," *Journ. of Sc. Instruments*, Vol. 41, No. 3 (1964), pp. 133–138.

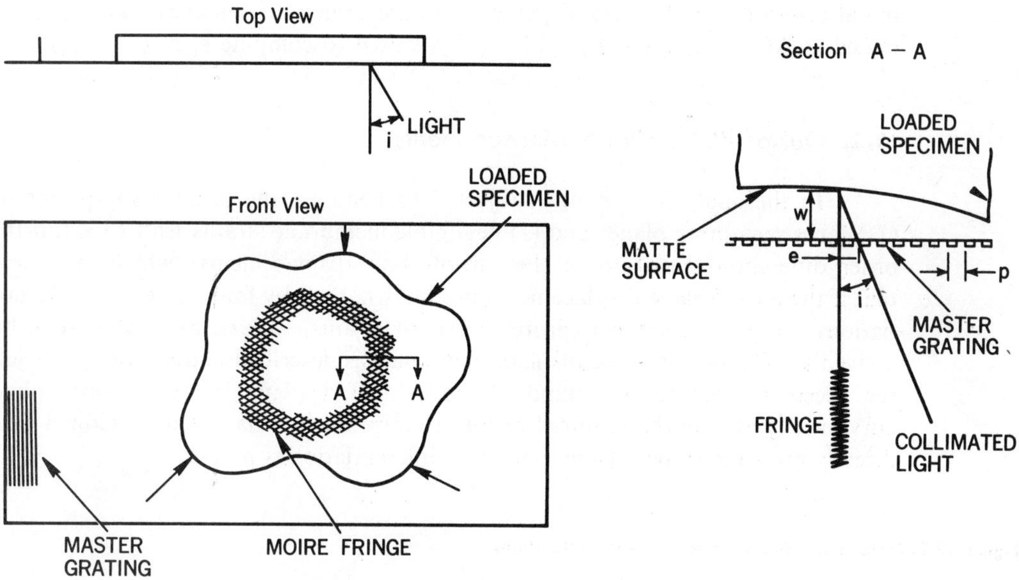

Figure 13.1. Method to obtain moire of out-of-plane displacements and isopachics.

The out-of-plane displacement is then given by

$$w = \frac{pn}{\tan i}$$

or, for small angles i,

$$w = \frac{pn}{i} \tag{13.2}$$

If the fringes are considered positive, then a positive value of w will represent the inward displacement of the surface. If any part of the surface moves outward, the master grating will have to be placed at the plane of the outermost displacement of the surface, and the fringe at that point will be a zero fringe. It may be necessary to measure the out-of-plane displacement independently at some point on the surface to determine the position of the initial undeformed plane. In other cases a point or region of the surface may have a known theoretical out-of-plane displacement value which can be used as a reference.

It may be convenient in some cases to fix the master with a small known gap between the undeformed plate and the master, and record an initial moire pattern. This initial pattern serves simultaneously as a datum plane and as a measure of the initial specimen surface if it is not flat. Subtraction of the initial from the final pattern provides the actual out-of-plane displacements. After the values of w have been adjusted to give the actual out-of-plane motion from the initial position, for the case of plane stress the change of thickness can be determined as $\Delta t = -2w$, and Eq. (13.1) can be used to compute ϵ_z or $(\epsilon_x + \epsilon_y)$.

13.2. Out-of-Plane Slope Measurements

In the analysis of plates subjected to bending, the usual assumptions of (1) planes remaining planes and (2) negligible membrane strains lead to a fourth-order differential equation of the out-of-plane displacements; which is to say that if the out-of-plane displacements are known, then by four successive differentiations—slopes, curvatures (proportional to strains), shears, and loads—can be arrived at. The out-of-plane displacement analysis described above would provide the necessary data for this kind of analysis of a plate. However, since plate curvature is usually the required result, a moire method has been developed that directly gives the slopes. Then, only one differentiation is necessary.

Figure 13.2. Method to obtain moire of bent plate slopes.

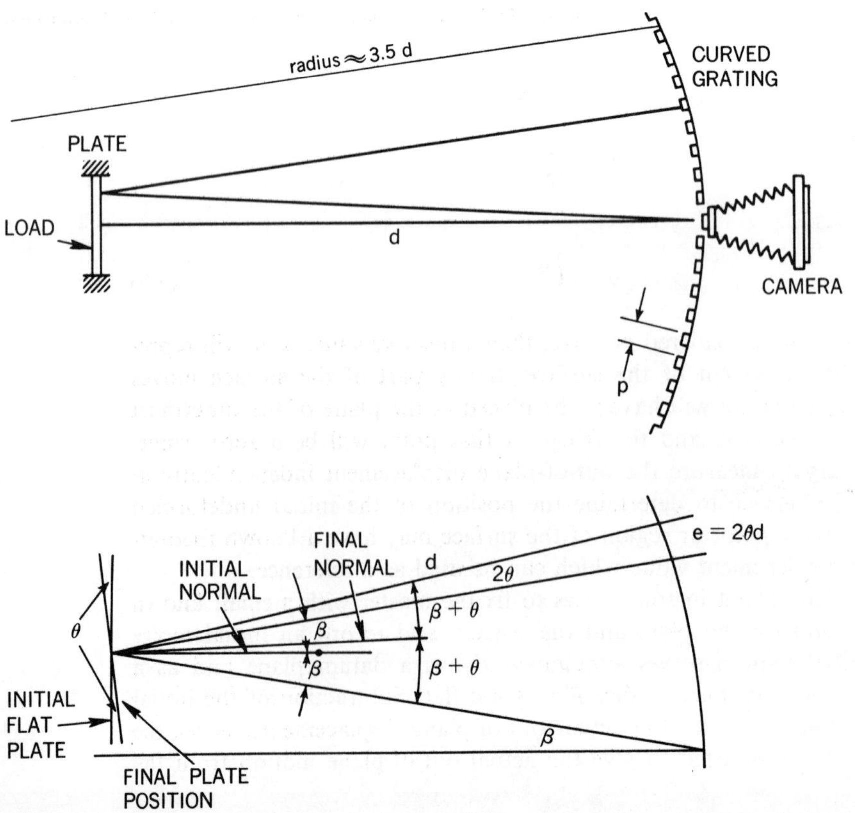

The method of measurement of slope can be thought of as similar to the method of measurement of out-of-plane displacement, except that the specimen surface is made reflective instead of matte. Since the reflective image of the grating on the plate does not depend on the angle of the incident light, collimated light is not necessary. In this case it is not possible to use the same arrangement as shown Fig. 13.1, since a reflected image arranged as shown in Fig. 13.1 would be affected by both the slope of the plate and the distance between master grating and specimen. To create a situation in which only the change in slope of the plate affects the image motion, the master grating is usually placed far from the specimen.

The oldest, simplest, and most common approach is that introduced by Ligtenberg[2] and shown in Fig. 13.2. A camera views the image of a large, slightly curved grating reflected by the plate. Moire is produced by the superposition of the grating images before and after loading. This superposition can be obtained by double-exposure photographically on the same film.

Again, $n = e/p$. From the properties of the triangles in Fig. 13.2, it can be shown that the vertical motion e is proportional to the distance from the plate surface to the grating, and to the angle of rotation of the plate surface $e = 2d\theta$, where d is the distance from the plate to the curved grating and θ is the change in angle between initial and final positions, in terms of radians.

Solving for θ,

$$n = \frac{e}{p} = \frac{2d}{p}\theta$$
$$\theta = \frac{np}{2d} \tag{13.3}$$

Note that the value of d must be large so that out-of-plane displacements change the value of d only slightly. Ligtenberg recommends that to limit the error in the relation $e = 2d\theta$ to less than $\frac{1}{2}$ per cent the drum holding the curved grating should have a radius of the order of $3.5d$ for a plate of width $0.4d$.

If the plate is iritially flat, the change in angle of the light ray is the inclination of the bent plate in the direction normal to the grating direction; this is the plate slope referred to above. Two moire patterns of these slopes are required to analyze the plate completely. Just as with the in-plane displacement fields, two orthogonal sets of curves can be obtained—say, θ_x and θ_y—by rotating the grating 90°. These are differentiated, just as the in-plane displacement patterns, to obtain curvatures. The shifting method (moire-of-moire technique) discussed in Sec. 7.8 or curve plotting along a line illustrated in Sec. 7.1 can be used. In either case four partial derivatives are obtained which represent curvature. The direct derivatives $\partial\theta_x/\partial x$ and $\partial\theta_y/\partial y$ give the curvature, which is the reciprocal of the radius of curvature in the associated direction. The cross-derivatives $\partial\theta_x/\partial y$

[2] F. K. Ligtenberg, "The Moire Method—A New Experimental Method for Determination of Moments in Small Slab Models," *Proc. SESA*, Vol. 12, No. 2 (1955).

and $\partial\theta_y/\partial x$ are equal one to the other and, in plate analysis, are termed twist. These values can be related directly to surface strains by the usual assumptions of plate theory and with the knowledge of the plate thickness. They also are used to compute plate moments directly. For small curvatures, the three independent derivatives can be solved as a symmetric tensor (Mohr's circle) for the principal curvatures in the plate. These topics are all covered in detail in texts on plate theory.

In a series of papers,[3] Duncan has developed a very closely related method for analysis of the slope of plates. An optical system is used to obtain the reflected image of a grid which is directly proportional to the plate slope. The method is more sensitive than the moire method described above, and thus is more suitable for shifting the slope-pattern image (Sec. 7.8) to obtain a moire pattern proportional to curvatures.

[3] J. P. Duncan, "Interferometry Applied to the Study of Elastic Flexure," *Proc. Inst. Mech. Eng.*, Vol. 176, No. 16 (1962), p. 379. J. P. Duncan and C. J. E. Brown, "Slope Contours in Flexed Elastic Plates by the Martinelli–Ronchi Technique," *Exp. Mech.*, Vol. 3, No. 12 (Dec. 1963) and "An Experimental Method for Recording Curvatures Contours in Flexed Elastic Plates," *Exp. Mech.*, Vol. 5, No. 1 (Jan. 1965). J. P. Duncan, "Grid and Moire Methods of Stress Analysis," Chapter 14 of *Stress Analysis* (eds. V. C. Zienkiewicz and A. S. Holister) (New York: John Wiley & Sons, Inc., 1965).

PART 4

TECHNIQUES

14

GRATINGS AND PRINTING OF GRATINGS

14.1. Introduction

Some information on the availability of gratings and on the techniques used for printing them will be useful. Most of the information in this chapter, however, is based on present technical development in the United States. Techniques improve continuously and new ways are found of making gratings and of printing them. It is also likely that in some countries the materials described may go by other names.

14.2. Coarse Gratings

Gratings of up to about 70 lpi can be obtained from commercial-art or drafting suppliers as "screens" or "shading" lines. They are usually available in sheets of 9 by 12 in. and $16\frac{3}{4}$ by $21\frac{3}{4}$ in.[1] These sheets are transparent and have adhesive backing, which makes very easy the attachment to the faces of plastics such as Plexiglas. Their density is sufficient to obtain fairly good response in deflection studies in frames, arches, etc., as illustrated in Chapter 6. These gratings are available, in several ratios of line width to line spacing and as one-way gratings or as cross-gratings ("dots"). They are inexpensive, the cost being of the order of one dollar per sheet.

Gratings made of parallel, radial, circumferential, elliptical, and logarithmic lines are available through some companies selling items of scientific

[1] One manufacturer is Para-tone, Inc., 512 W. Burlington Avenue, LaGrange, Ill. The sheets are sometimes called "Zip-A-Tone."

interest to highschool students.[2] These gratings are usually printed in transparent sheets $8\frac{1}{2}$ by 11 in.; their cost is of the order of 20 dollars. Examples of these gratings are shown in Figs. 1.3 and 1.4, and an illustration of the moire patterns that can be obtained is shown in Fig. P.7.

14.3. *Medium-Density Gratings*

Some gratings of medium density, up to about 200 lpi, can be manufactured using a milling machine. A good engraving material is Plexiglas. An example of a moire pattern obtained using polar gratings and manufactured using this procedure is shown in Fig. P.6. A good illustration of a Cartesian grating can be found in a paper by Knauss.[3] However, in spite of the fact that the density was 165 lpi, the author evaluated it only as a grid without producing moire.

14.4. *High-Density Gratings*

Gratings with densities of the order of 250 and 300 lpi are usually available from photoengravers. They are rectangular, and as large as 8 by 10 in. Gratings with densities of 1000 and 2000 lpi were, up to now, more difficult to obtain. In some cases relatively good quality gratings could be obtained for prices up to \$1000 for a grating of 6 by 6 in. with 1000 lpi, or 3 by 3 in. with 2000 lpi.[4] This situation has recently been substantially changed following the developments in England by Fidler and Nurse[5,6] and by Sayce[7] on the techniques of manufacturing high-density gratings. Fairly good quality gratings with densities of 1000 lpi can be obtained commercially[8] in the size 9 by 12 in. for about \$70 (£25) each. It is understood that gratings with densities 1 per cent above and below the nominal are also available for the same price. Metal film gratings on strippable metal backing, for bonding to specimens, are available in both one-way and crossed arrays. Sizes are as large as 4 by 4 in. with from 200 to 1000 lpi, and 1 by 1 in. with up to 2000 lpi. Corresponding masters on glass are also available from the same source.[9]

---

[2] One company is Edmund Scientific, 400 Edscorp Building, Barrington, N. J. 08007.

[3] W. G. Knauss, "Stresses Near a Crack in a Rubber Sheet," *Exp. Mech.*, Vol. 8, No. 4 (April 1968), pp. 177–181.

[4] The authors' sources were Max Levy and Company, Wayne Ave. and Berkley St., Philadelphia, Pa.; and Buckbee Mears Co., St. Paul. Minn.

[5] R. Fidler and P. Nurse, "A Method of Ruling Circular Diffraction Gratings and Their Use in the Moiré Technique of Strain Analysis," *Journ. of Strain Analysis*, Vol. 1, No. 2 (1966), pp. 160–164.

[6] R. Fidler, and P. Nurse, "Further Developments in the Moiré Technique of Strain Analysis," Central Electricity Research Laboratories, Leatherhead, England, RD/L/R 1479 (Sept. 8, 1967).

[7] L. A. Sayce, "Automatic Measurement," Ministry of Technology, National Engineering Laboratory, Penguin Technology Survey (1967).

[8] F. E. Brown and Co. Ltd., Empire Works, Cork Lane, Glen Parva, Leicester, England.

[9] Photoelastic, Inc. 67 Lincoln Highway, Malvern, Pa. 19355.

14.5. *Printing*

There is no difficulty in transferring the coarse gratings mentioned in Sec. 14.2, as they are available in sheets with adhesive backing. The method of printing the other kinds of gratings will depend on their density, on the type of material on which they will be printed, and on the testing environment (for instance, special techniques are required for tests at high temperature). In what follows, some of the methods used will be described.

14.5.1. Photosensitive Coating Based on Van Dyke Solution

The preparation and printing with Van Dyke solution is rather critical. The authors have found that, when carefully printed, this coating produced a high-contrast grating on polyurethane-rubber specimens. When this solution is used, the coating adheres perfectly to rubber up to strains of the order of 100 per cent, and to temperatures of around 250°F. It has also been used frequently to coat epoxy specimens. A detailed account will be given of the procedure for coating rubber and epoxy specimens with a mixture containing Van Dyke solution and for printing gratings on them.

The photosensitive mixture is made of three components: Van Dyke solution, silver-nitrate solution, and Polyco emulsion.

(a) Preparation of Van Dyke Solution

1. Distilled Water	100 cc	Heat to 180°F and stir until dissolved (5–10 min).
2. Ferric Oxalate	22.0 g	
3. Ferric Ammonium Oxalate	4.2 g	Continue stirring and add to ferric oxalate solution (5–10 min).
4. Citric Acid	7.5 g	
5. Urea	6.0 g	
6. Diethylene Glycol	8.3 cc	Allow to cool to 120°F, and add to the mixture of items 1 to 5 (5–10 min).
7. Isopropanol	4.2 cc	
8. Cooper's Glue	1.0 g	Slowly add water to glue and stir until mixed. Heat as required to dissolve (5–10 min).
9. Distilled Water	12.5 cc	

Items 8 and 9 are optional and are not required for most applications. If items 8 and 9 are used, they are added slowly to the mixture of items 1 to 7 stirring until mixed with the mixture. The purpose of using Cooper's glue is to increase adhesion, but frequently no adverse effects have been found if it is not used. Store the solution in a brown bottle under darkroom conditions.

(b) Preparation of Silver-Nitrate Solution

1. Distilled Water	100.0 cc	Add silver nitrate to water and stir until dissolved.
2. Silver Nitrate	27.6 g	

Store the solution in a brown bottle.

(c) Preparation of Photosensitive Mixture

1. Van Dyke Solution .... 80 cc
2. Silver-Nitrate Solution .... 20 cc
3. Polyco 2719 Emulsion (Borden Co.)............100 cc

First, add the silver-nitrate solution slowly to the Van Dyke solution and stir until mixed. Then, slowly add the Polyco 2719 emulsion to the addition of both solutions and stir until mixed.

The above solutions are mixed at room temperature under normal incandescent-lighting conditions, and after approximately 1 hr the mixture is filtered through several layers of cheesecloth and standard-grade, semicrimped rapid-filter paper. After filtering, the mixture is to be stored in a brown bottle in a darkroom.

The shelf life of the coating mixture is 1–2 months. Shelf life can be increased to about 6 months if the mixture is stored at about 40–50°F.

(d) Procedure of Application of Photosensitive Mixture to Rubber or Epoxy Surfaces

(1) *Surface cleaning*. Clean the surface with methanol for a rubber model, or with acetone for an epoxy model. Allow to dry for 5–10 min. Use incandescent lights only.

(2) *Applying the coating to the surface*. Place the rubber model on a spin table and spin at 30–40 rpm. Coat the surface with the mixture while spinning and allow it to spin-dry for a half-hour at room temperature.

Place the epoxy model on a spin table. First, coat the surface with Polyco emulsion diluted 1:1 with water, and spin-dry. If model surface is rough, this coating is important. Allow the surface to dry for approximately 2 hr at 110°F. Replace the model on spin table and apply the coating mixture while spinning at 30–40 rpm. Allow the surface to spin-dry for a half-hour at room temperature.

Coating may also be applied with an airgun (20–25 psi). In this case it is important to apply the coating in layers and turn the model after each layer to insure a uniform coating. For small models, the coating may be spread uniformly by dragging a tissue paper over the surface of the specimen. The coating should be applied uniformly, but thick enough to insure full coverage. It should not be too thick to become opaque.

(3) *Drying*. Dry the surfaces of the rubber model by keeping it in an oven at 140°F for 1–2 hr. Dry the surfaces of the epoxy model by keeping it in an oven at 110°F for 3–4 hr.

(4) *Cooling*. Remove the model from oven and allow it to cool to room temperature. Keep the models at room temperature for approximately 3 hr.

Coating on rubber models may crack if they are bent before exposure. Epoxy models should be cooled from 110°F at a slow rate to prevent steep thermal gradient.

(5) *Printing of gratings.* Expose the coating to a carbon arc lamp through a reference array in a vacuum frame for 30–60 min, keeping the distance between the coating and light source at 20 in. Instead of a carbon arc lamp, a battery of 18-in.-long black lights can also be used. Rubber models should be subjected to approximately a 1–2-in. vacuum. Epoxy models should be subjected to a 20–26-in. vacuum. Due to long exposure time, the temperature of the model increases to 90–100°F. The surface should be cooled for approximately 20 min to room temperature before developing. Printing on rubber models is also done by laying the model on the reference array on a glass table and lighting from below.

(6) *Developing and fixing.* Develop the exposed model in water for 10–20 sec at 70–75°F, using on orthochromatic safelight in a darkroom. Keep the model in rapid-fix solution for 15–30 sec at 70–75°F.

Wash the model in running water, bathe 5–10 min at 70–75°F. Air-dry at room temperature for 1–2 hr.

(7) *Coating cleaning and removal.* The coating on the model can be cleaned using a mild soap-and-water solution at 70–80°F, with a fine-grain cellulose photographic sponge or similar material.

The coating can be removed by cleaning it with acetone and the alcohols.

(8) *Photographing.* Paraffin oil can be used as a medium of contact for photographic purposes.

14.5.2. Photosensitive Coatings Made with Diazo Compounds

The authors have not found the use of diazo compounds as effective as the use of Van Dyke solution for printing on polyurethane-rubber models. However, diazo compounds have been used in the strain analysis of solid propellants.[10]

A very thin layer of white latex paint is applied first, with an artist's airbrush; this provides contrast for the grating. Next, 100 cc of coating solution is prepared by using the following formula:

1. Glycol	3.0 cc
2. Thiourea	4.5 g
3. Citric Acid	4.0 g (Stabilizer)
4. Resorcinol	0.3 g
5. Zinc Chloride	4.5 g

[10] W. D. Hart, "Moire Method for Measurements of Strains in Solid Propellants," Bulletin of Third Meeting of Working Group on Mech. Behavior Chemical Propulsion Information Agency, Vol. I, CPIA No. 61U (October 1964), pp. 545–560.

6. 1-3-6 Coupler[11] 2.0 g (Universal Developer of Black and Blue Lines)
7. Diamethyl Diazo[12] 1.0 g (Diazo Salt)
8. 3 RDZ Salt 1.0 g (Diazo Salt)

First, a coating of the solution is applied to the surface of the specimen. Next, the specimen is placed in a vacuum printing frame in contact with the grating. Illumination with ultraviolet light for approximately 1 min will destroy the diazo salt which is not covered by the lines of the grating. Development of the unexposed portion of the surface is accomplished by exposing it to ammonia vapor.

14.5.3. Photosensitive Coatings Made Using Photoengravers' Processes

Several photoengravers' processes have been used successfully to deposit coatings on metal surfaces. Detailed instructions on the procedure for applying these coatings can be found in commercially available catalogs.[13]

In the application described in Chapter 18, the photosensitive coating deposited on the aluminum plate is an albumenoid-base coating used by photoengravers. The polyurethane-rubber specimen used for the application described in Chapter 17 is coated with a polyvinyl-alcohol enamel which is also used by photoengravers.[14]

Several investigators have successfully used the techniques developed to print circuits. These materials are also available commercially,[15] and the techniques to apply them are described in catalogs. Some of these materials resist high temperature and have been used for thermal-strain determinations.[16] Detailed instructions on the use of Kodak's Resifax are given by Dantu.[17]

14.5.4. Stripping Films

The use of stripping films[18] has been found convenient for some applications. The grating is printed on these films before they are applied to the surface

---

[11] 1-3-6 Coupler is the trade name for material supplied by the Driazo Corp. Typical coupling agents are (a) 2-napthol-3, 6 disulfonic acid and (b) 2,3 dihydroxynaphthalene-6 sulfonic acid.

[12] DiMethyl Diazo and 3 RDZ Salt are trade names for materials received from Driazo Corporation. Typical diazo compounds are (a) 2 diazo-1-hydroxynapthalene-6 sulfonic acid and (b) diazotized N, N dialkyl-p-phenylene diamine.

[13] See, for instance, "Pitman Special Processes," Harold M. Pitman Co., 51st Avenue and 33rd St., Chicago Ill.: Black Print Process (p. 5), Blue Print Process (p. 8), and Cold Top Enamel (p. 12).

[14] The trade name is Gaco.

[15] See "Kodak Photosensitve Resists for Industry," Publication P-7, Sales Service Division, Eastman Kodak Co., Rochester, N. Y.

[16] C. A. Sciammarella and B. E. Ross, "Thermal Stresses in Cylinders by the Moire Method," *Exp. Mech.*, Vol. 4, No. 10 (October 1964), pp. 289–296.

[17] P. Dantu, "Extension of the Moire Method to Thermal Problems," *Exp. Mech.*, Vol. 4, No. 3 (March 1964), pp. 64–69.

[18] Kodalith stripping film is manufactured by Eastman Kodak.

of the specimen. Epoxy cement can be used as an adhesive for fixing the films onto the surface of the specimen.

Before the film is cemented to the specimen, it is recommended to apply a thin layer of a white paint to the gelatine side of the film, to increase the contrast. As used by Dantu, this paint is a mixture of a monomer, a catalyst, and an epoxy cement. After the paint is dried, it is recommended to roughen its surface lightly with a fine sandpaper. After the adhesive on the specimen surface is cured, the back of the film is removed mechanically, leaving only the gelatine and a thin layer of cellulose nitrate. It is claimed that this layer could be removed by using a solution of 50 per cent acetone and 50 per cent methyl alcohol.

14.5.5. Photosensitive Coatings on Plexiglas Surfaces

For the particular case of Plexiglas, plates are commercially available with a photosensitive coating ready to be printed. They are manufactured by Kodak[19] and yield a grating with very good contrast.

14.5.6. High Temperature Gratings

Printing on metals or other materials for high temperature analysis requires special techniques. One approach is to begin as mentioned above with a photosensitive resist. When a grating of this material has been printed on metal, then the process can be completed by etching the uncoated dots or lines using an acid bath, spray or electric etching. Another suggested approach is to apply a metallic grid by vacuum deposit in a vapor bath.[20]

14.6. Natural Gratings

The gratings described in Secs. 14.2 to 14.4 could be called "artificial" gratings in the sense that they have been especially prepared to be printed on surfaces and produce moire. In some cases the gratings are part of the natural phenomenon to be studied, and could be called "natural" gratings. An example of this situation was shown in Fig. 1.25 in which photoelastic isochromatics are used as gratings to produce a moire effect (in that case the isotachics of isochromatics). Another example are the isothetics obtained by moire of artificial gratings, but are used as gratings to obtain moire-of-moire effects, either isotachics of isothetics as shown in Fig. 10.6 or isoparagogic of isothetics as shown in Fig. 7.16.

---

[19] Photoplast plates are available in sizes of 8 by 10 in. (thickness: 0.060 in., 0.130 in., and 0.250 in.) Larger sizes up to 30 by 40 in. are available on special orders. (Eastman Kodak Co., Rochester, N. Y.)

[20] J. W. Neudecker, R. G. Lawson, and S. D. Stoddard, "Moire-Fringe Techniques at Extremely High Temperatures," presented at SESA 1969 Fall Meeting, Houston, Texas.

Another example should be mentioned here, which is important, and will be described in more detail in Chapter 20. The idea has already been used by Mesmer.[21] When light is reflected by the front surface of a transparent plate, as well as by the back surface of the plate, an interference may be produced, associated with the thickness of the plate. Fringes are then loci of points exhibiting the same value of thickness of the plate and are called *isobathics*. When the plate is loaded, its thickness changes because of Poisson's effect. This change in thickness is associated with the sum of the principal stresses at the point. It can be obtained by subtraction of the isobathics of zero loading, from the isobathics under load. If the number of isobathics is sufficiently dense, this operation can be done by moire, using the parametric properties explained in Chapter 2. In this case the gratings producing the moire effect are the natural isobathics of the plate. If a piece of Plexiglas is conveniently selected so that the plate has a wedge shape, the isobathics will look like straight parallel lines. Otherwise they have complicated shapes.

Obviously, this kind of "natural" grating does not have to be prepared or printed, and the photographic techniques used to record these gratings are usually the same ones used to record the moire fringes. Examples will be found in Chapter 20.

Phase Gratings. These are a class of gratings which transmit light over their whole surface. The grating changes the phase of the transmitted light by a periodic variation in thickness or index of refraction. For this reason they are called *phase gratings*, as opposed to amplitude gratings that block the light with periodic dark lines. Some gratings have the combined characteristics of phase and amplitude gratings. Except in this section, the gratings referred to are amplitude gratings.

The classic example of a phase grating is a glass plate furrowed with prallel grooves spaced at a given pitch. To the unaided eye such a grating looks clear, but when viewed through a spatial filtering system the grating lines become visible. When combined with a second phase or amplitude grating, this glass plate can produce moire patterns similar to those obtained from two amplitude gratings. Furrowed gratings can have upward to 30,000 lpi and are very suitable for diffraction work.

In the past, phase gratings were either ruled glass plates or molded replicas of the ruled glass plates. They were seldom used for moire analysis. By exposing and developing a nondarkening-type emulsion [22], or by bleaching a darkened emulsion, phase gratings are now produced on film from amplitude gratings.

---

[21] G. Mesmer, "The Interference Screen Method for Isopachic Pattern (Moiré Method)," *Proc. S.E.S.A.*, Vol. XIII., No 2, pp. 21–26.

[22] See discussion by Fu-pen Chiang of a paper by G. S. Holister and H.R. Luxmoore, "Production of High Density Moire Gratings," *Experimental Mechanics*, Vol. 8, No. 5 (May 1968—paper) and Vol. 9, No. 6 (June 1969—discussion).

15

PHOTOGRAPHY OF GRATINGS AND FRINGES

15.1. Introduction

Several techniques are used to produce and photograph moire patterns. Some of these have been suggested or implied in various parts of the text. Here, a more detailed and organized description will be presented.

15.2. Position of the Master Grating

The master grating used to produce a moire pattern can have several positions with respect to the specimen grating. The following are some of the most commonly used.

15.2.1. Master and Specimen Gratings in Contact with Each Other

The most commonly used method to obtain moire fringes is to photograph the master and specimen gratings in contact with each other. The fringes are then visible to the naked eye and can be photographed easily. Dense gratings are not usually visible on the photographic record. To use this technique, the spaces of the master grating must be transparent. The master grating can be rigid (usually printed on glass or Plexiglas) or flexible, as when printed on photographic film. Frequently, there is an advantage in using a flexible master and making it adhere to the surface of the specimen by means of a layer of light paraffin oil. This permits the production of moire from surfaces which are not absolutely flat, since the master grating will follow, to some extent, the displacement of the face in a direction perpendicular to its plane, but will offer no resistance to motions in the plane. Most of the photographs shown in this book have been obtained using this method.

The fringe pattern can be photographed with an ordinary camera, or by contact printing, as described elsewhere.[1]

15.2.2. Photography of Deformed and Undeformed Gratings

The simplest method of producing a fringe pattern is to photograph, before loading, the grating printed on the specimen (for practical reasons this is often better done when the specimen is subjected to a small load), then to load the specimen and photograph the grating a second time. The two photographic images are used to produce the moire pattern.

The method has some disadvantages: (1) It cannot be used if the original density of the gratings is higher than about 1000 lpi or if the density of the photographed images of them is more than 700 lpi. This is due, first, to the resolving power of the lens and, second, to the fact that the greater the density, the more critical the focusing becomes, since the grating lines are photographed rather than the fringes. (2) The fringes are seen only after development of the film. (3) No pattern is produced by those regions of the specimen near the boundaries that have been displaced outward, because two grating images are necessary to produce the moire, and in those regions only the deformed specimen grating is recorded. This point is of little importance in small-deflection problems, but may be important in large-deflection problems.

The fringe pattern may be obtained by exposing the undeformed and deformed gratings on the same film (double-exposure) or by photographing separately the images before and after loading and then superposing the two films.

(a) *Double-Exposure.* Double-exposure requires a somewhat longer exposure time for the first image than for the second, the difference depending on the characteristics of the film. It may be necessary to find by trial and error a suitable exposure time for each exposure using this method. Figures 7.3, 7.4, and 7.8 were obtained by double-exposure.

(b) *Separate Photography of Deformed and Undeformed Gratings.* The advantage of using a separate photograph of the undeformed grating as a master is that the magnification of the camera setup is maintained. A variation of the approach is to photograph only the deformed grating and use an independent master. This is possible when the magnification of the deformed grating is fixed

[1] A. J. Durelli, V. J. Parks, and L. Ferrer, "Photography Without Lenses," *Exp. Mech.*, Vol. 8, No. 3 (March 1968), pp. 142–144.

at unity or can be recorded on the photograph. Then any independent master of suitable pitch can be used with the photographic image of the deformed grating.

The spatial derivatives shown in Figs. 7.12 and 7.13 required no master and were obtained by photographing only the grating.

If an event produces nonreproducible deformations, as is the case in some dynamic loadings, a cross-grating on the specimen can be photographed this way and analyzed for both isothetic fields one after the other with a one-way master.

15.2.3. Master Grating on Ground Glass of Camera

The master grating can be located on the ground glass of the camera. The image of the specimen grating then produces a moire pattern with the master grating. Like the method described in Sec. 15.2.2, this method does not require that the master be kept against the specimen, which may be an advantage in some thermal studies. The positioning of the master grating, however, is critical and the errors associated with mismatch discussed in Sec. 9.6.1 must be considered. This technique is useful when fringes are shifted to increase response, as will be described in the next chapter.

This method, or the method described in Sec. 15.2.2, is the method used for three-dimensional applications, when the specimen grating is embedded inside a transparent body. It requires, in general, a sturdy camera of the type used in the graphic arts (commonly called a *copy camera*), or one specially designed for moire analysis.[2]

15.3. Moire-of-Moire

Time and spatial derivatives are obtained using moire-of-moire, or second-order moire, as described in Chapter 7. This is done by superposing two films on which the first-order moire has been recorded. The second-order moire is visible to the naked eye, can be adjusted by displacement of one film with respect to the other, and can be photographed either on a contact-printing frame or in an enlarger. Second-order moire patterns shown in Figs. 7.14 through 7.19 have been obtained using this procedure. These examples are all spatial derivatives. The shift and pitch are used to relate the moire-of-moire fringe order to the spatial derivatives of displacement.

Similar analysis can be conducted to obtain time derivatives of displacements, the velocity field, as described in Chapter 10. For velocities, two displacement-component patterns of different loads are superposed. The loads should be near the same value. The moire-of-moire effect gives the difference of the displacement component throughout the field for the given choice of increment of load (or time, as it is sometimes specified). The superposed patterns must be

[2] For instance, Model 102 made by Photolastic Inc., 178 Lincoln Highway, Malvern, Pa. 19355.

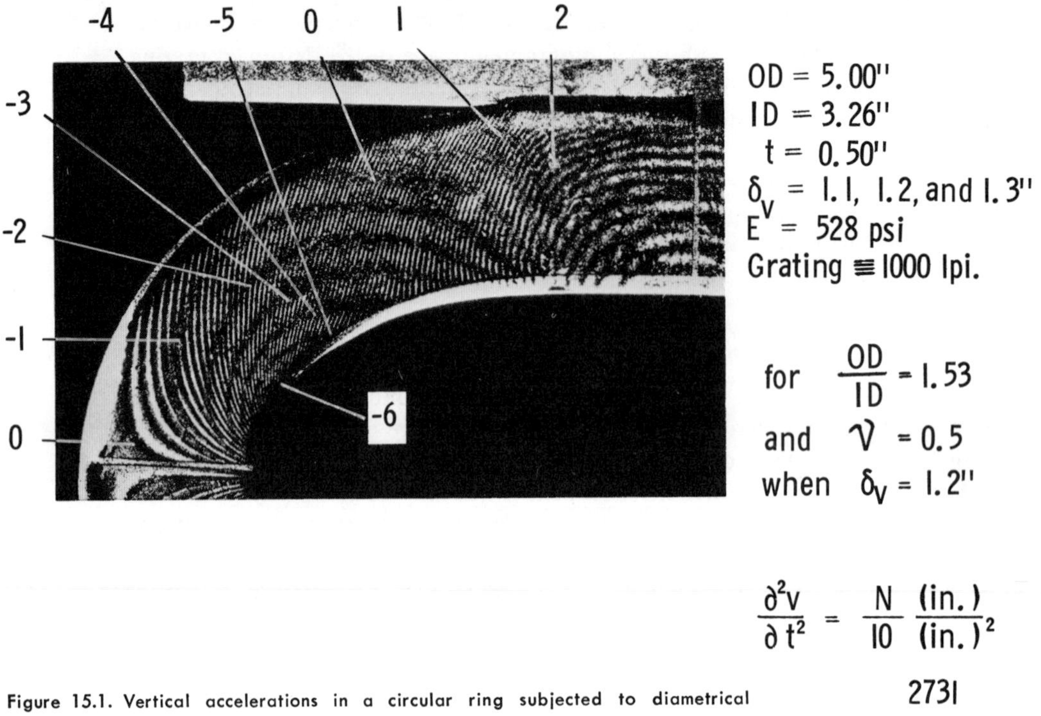

Figure 15.1. Vertical accelerations in a circular ring subjected to diametrical compression.

aligned (no rotation) and should not be translated one with respect to the other. Some judgment must be used to superpose points in the field which can be assumed not to have moved. It may be necessary to decide on a fixed reference point outside the field of the specimen. This point in some cases may be an arbitrary datum, as the intersection of two lines of symmetry, on the specimen. In any case, the velocities obtained will be referred to the chosen reference point. Figures 10.6 and 10.7 illustrate this type of application.

Figure 15.1 shows a moire-of-moire pattern of accelerations obtained by combining two velocity patterns. Each of the velocity patterns was obtained by superposing deformed gratings obtained at two consecutive levels of load, so that the velocities in this case were recorded as a first-order moire. Three load levels were used; the deformed gratings from the first two levels were used to produce one velocity pattern, and the deformed gratings from the second and third levels were used to obtain the second velocity pattern.

The vertical displacement of the ring is used as a load parameter, and the values 1.1, 1.2, and 1.3 correspond to inches of applied displacement. The vertical velocity of any point of the ring is then specified in the isotachic pattern as

$$\frac{\partial v}{\partial t} = \frac{0.001n}{0.1} = 0.01n \text{ in./in.}$$

where n is the fringe order.

The vertical acceleration at 1.2-in. deflection is given for any point in the ring by:

$$\frac{\partial^2 v}{\partial t^2} = \frac{0.001}{(0.1)^2} N = 0.1N \text{ in./(in.)}^2$$

where N is the fringe order in the loci of the acceleration pattern.

The units are inches of vertical displacement at the point, per inch of applied displacement, per inch of applied displacement.

15.4. *Third-Order Moire*

Techniques to obtain third-order moires are not yet completely developed. One figure in Chapter 10 (Fig. 10.8) shows an example of velocities of spatial derivatives $\partial u^2/\partial x \partial t$ obtained by hand. In principle, the same techniques followed to obtain second-order moire could be used, but the first-order moire, which was used as the master grating for the second-order moire, and which is still in the photograph, produces a parasitic effect which confuses the desired pattern. Methods of "cleaning," or filtering out, this first-order moire are being developed using the diffraction properties of light (see Sec. 1.8). An example of a filtered third-order moire of strain gradients $\partial^2 v/\partial y^2$ is shown in Fig. 15.2. It may be worth noting that third-order moire would also allow determination of accelerations $\partial^2 v/\partial t^2$.

Figure 15.2. Strain gradient pattern of $\partial^2 v/\partial^2 y$ represented by a filtered third-order moire pattern.

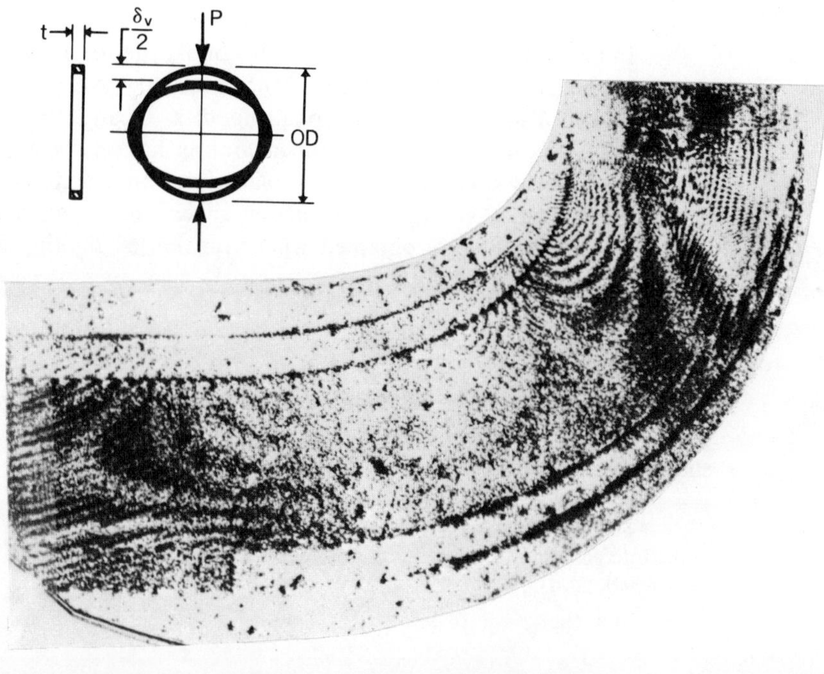

15.5. Other Combinations

Moire patterns can be obtained from a reference grating and its shadow, as when isopachics, or out-of-plane displacements of a plate, are desired, as discussed in Chapter 13.

Moire patterns can also be obtained from the reflection of a reference grating before and after loading to determine the slopes of plates in bending, also as explained in Chapter 13. Here again, the moire-of-moire effect can be used to obtain the spatial derivatives of the pattern. This gives the plate curvature.

15.6. Nonphotographic Recording

Throughout this book, the emphasis has been on recording the moire patterns by photographing. However, photographing is not always necessary. In some cases, measurements can be made directly on the specimen to analyze displacements and even strains. Occasionally, simple devices are designed with two gratings and a scale and mounted directly on the specimen to indicate local strain.

Light-sensitive devices can be used to record the moire fringe at a point, along a line, or throughout a field. Such devices have been used to increase sensitivity by recording the variation of light intensity continuously along a line. Such analysis is, of course, also possible on the photographic image.[3]

15.7. Opaque Specimens

In most of the tests mentioned in this book, the specimen grating is mounted on a transparent body. However, this is not necessary. Nearly all of the techniques described can also be applied to opaque bodies. Opaque bodies must be front-lighted; this requires that the specimen grating have sufficient contrast between the lines and spaces. On metal bodies, black lines on a polished surface are often sufficient; on other surfaces, a white or silver undercoat is often applied. In general, better results are obtained when transmitted lighting is used.

15.8. Moire of Isochromatics

Moire patterns corresponding to derivatives in space or derivatives in time (rate effects) can be obtained from a variety of patterns. For example, Fig. 15.3 shows an isochromatic pattern shifted on itself to give fringes corresponding to

---

[3] See especially B. E. Ross, C. A. Sciammarella, and D. Sturgeon, "Basic Optical Law in the Interpretation of Moire Patterns Applied to the Analysis of Strain, Part 2," *Exp. Mech.* (June 1965), pp. 161–166.

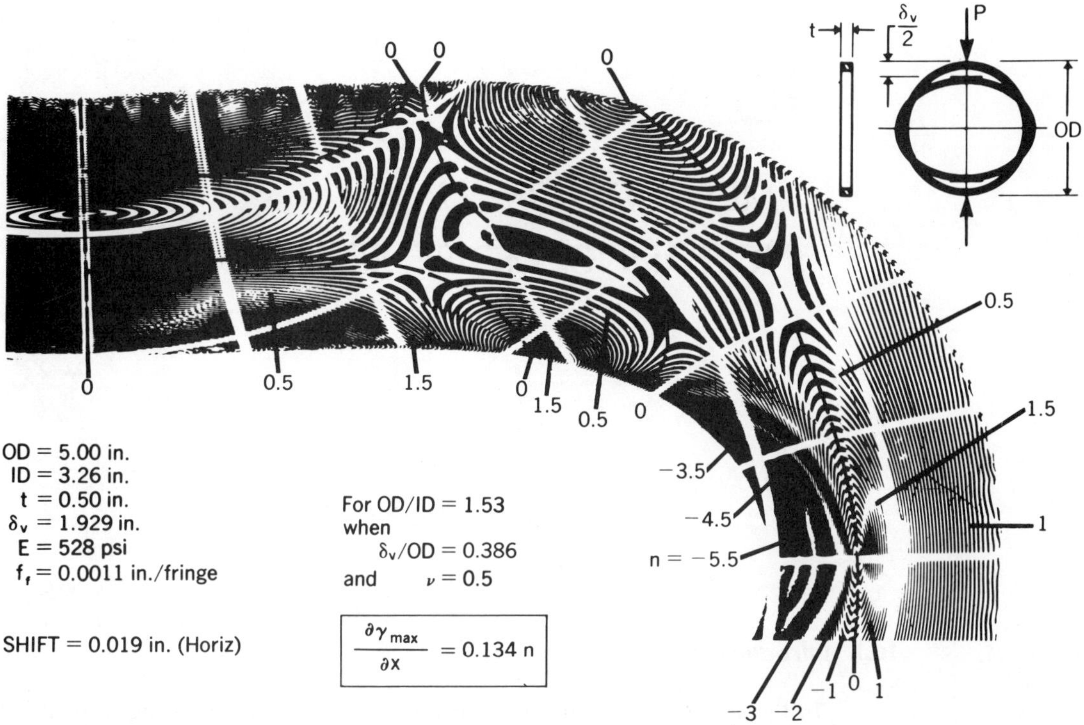

Figure 15.3. Loci of points exhibiting the same value of partial derivative $\partial\gamma_{max}/\partial x$ in a circular ring subjected to diametral compression. All points in the same fringe have a derivative value of 0.134 (1/in.) times the fringe order (large deformation).

a constant derivative of maximum shears (isochromatics) in the direction of shift. Figure 1.25 shows two isochromatics from two consecutive levels of loads superposed to give the fringes corresponding to a constant rate of change of maximum shear with respect to load.

The moire patterns were obtained by contact printing the two superposed isochromatics.

15.9. Moire of Absolute Retardation Patterns

There are some very specialized applications of moire effects that require unique techniques for their recording. An example is the determination of isobathics and related loci (isopachics and their isotachics and isoparagogics) in static and dynamic problems. Description of the details of the procedure to be used in this case will be found in Chapter 20.

16

METHODS OF INCREASING MOIRE RESPONSE

16.1. Introduction

In the study of elasticity problems, a serious drawback of moire methods is low sensitivity. Sensitivity can be increased by increasing the line density of the grating, but practical difficulties have limited this density to about 1000 lpi. In some cases it may be practicable to use 2000-lpi gratings, but applications can seldom be found in which higher-density gratings are used. Several techniques of interpolating or multiplying fringes have been suggested, among which are the continuous light-intensity displacement relation of Sciammarella[1] and the optical method of Sciammarella and Lurowitz.[2] The former method requires a microdensitometer, while the latter needs a six-piece optical bench. Dantu,[3] Durelli and Sciammarella,[4] Chiang,[5] and Post[6] suggest pitch and rotation mis-

---

[1] C. A. Sciammarella, "Basic Optical Law in the Interpretation of Moire Patterns Applied to the Analysis of Strains," *Exp. Mech.*, Vol. 5, No. 5, Part I (May 1965).

[2] C. A. Sciammarella and N. Lurowitz, "Multiplication and Interpolation of Moire Fringe Orders by Purely Optical Techniques," *Journ. Appl. Mech.*, Vol. 34, No. 2 (June 1967), pp. 425–430.

[3] P. Dantu, "Utilisation des Réseaux pour L'Étude des Déformations," Pub. No. 57–6, Lab. Central des Ponts et Chaussées, Paris (1957).

[4] A. J. Durelli and C. A. Sciammarella, "Elastoplastic Stress and Strain Distribution in a Finite Plate with a Circular Hole Subjected to Uni-dimensional Load," *Journ. Appl. Mech.*, Vol. 30, No. 1 (March 1963), pp. 115–121.

[5] Fu-pen Chiang, "A Method to Increase the Accuracy of Moire Method," *Journ. Eng. Mech. Div., Proc. ASCE*, Vol. 91, No. EM-1 (Feb. 1965), pp. 137–149.

[6] D. Post, "Moire Grid Analyzer Method for Stress Analysis," *Exp. Mech.*, Vol. 5, No. 11 (Nov. 1965), pp. 366–377.

match to increase the number of fringes. Following Guild,[7] Sciammarella,[8] Low,[9] and Post,[10] each suggest multiplying the number of fringes using diffraction gratings, as summarized in Sec. 1.8.

A simple mechanical way of interpolating and multiplying moire fringes by fringe shifting is presented here. This method does not require elaborate optical or mechanical devices. The precise control of fringe motion is achieved through a moire vernier. The method gives whole-field interpolation and multiplication of fringes with gratings of density up to 1000 lpi.

The concept of fringe shifting was also used by Sampson and Campbell[11] to obtain strains in solid propellant using 100- and 200-lpi gratings. Sampson and Campbell shifted the specimen and recorded fringes on the master grating, point by point.

16.2. Mechanism of Fringe Shifting

Fringe positions are functions of the intersection points of the two gratings, as shown in Fig. 16.1. In this figure, the master grating has been shifted one-quarter of its pitch in the direction normal to the lines; the fringe positions have changed accordingly. The same amount of movement of the fringes can be obtained if the master grating is held fixed and a proper shift is given to the specimen grating. However, this, in general, is less practical. If the shifting is done through the movement of the specimen grating, and the fringe recorded with respect to the master grating, a whole new system of analysis has to be developed to obtain the strains. It can be seen in Fig. 16.1 that a ratio α can be defined such that

$$\frac{EF}{CF} = \frac{BC}{AC} = \alpha \tag{16.1}$$

This simple relation is used in the design of the "moire vernier" described later.

In Fig. 16.1, the fringe spacing δ is not changed by the shifting of the master grating because the specimen grating is uniform, representing a state of homogeneous strain. In general, when strain changes from point to point, the fringe spacings as well as orientations will be changed by the shifting of the master grating, as shown in Fig. 16.2. This isothetic moire has been obtained from a disk subjected to diametral compression. The master grating has 300 lpi and a linear "tensile" mismatch. The two halves of the pattern differ in that

---

[7] J. Guild, *The Interference System of Crossed Diffraction Gratings* (Oxford: Clarendon Press, 1956).

[8] See Footnote 2, this chapter.

[9] I. A. B. Low, "The Moire Method Using Diffraction Gratings," presented to the SESA, Detroit (May 1966).

[10] D. Post, "Analysis of Moiré Fringe Multiplication Phenomena," *Applied Optics* (Nov. 1967). Also "New Optical Methods of Moiré Fringe Multiplication," *Exp. Mech.*, Vol. 8, No. 2 (Feb. 1968).

[11] R. C. Sampson and D. M. Campbell, "The Grid-Shift Technique for Moire Analysis of Strain in Solid Propellants," *Exp. Mech.*, Vol. 7, No. 11 (Nov. 1967), pp. 449–457.

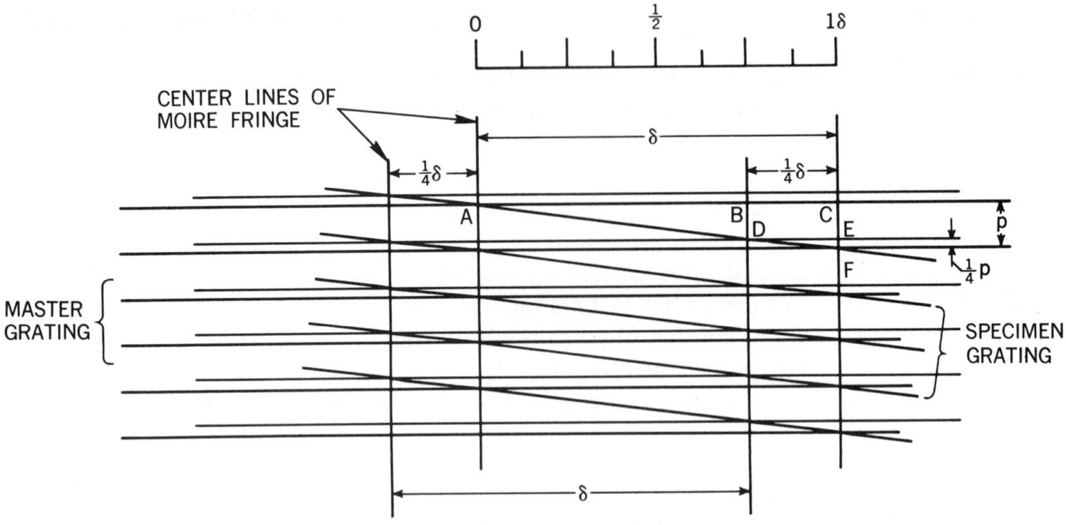

Figure 16.1. Mechanism of moire shifting. The fraction of a pitch that the master grating shifted is equal to the fraction of the fringe spacing that the fringes shifted.

Figure 16.2. Isothetic moire in a disk under diametral compression. The master gratings on the two halves differ in position by one-half pitch normal to the grating lines.

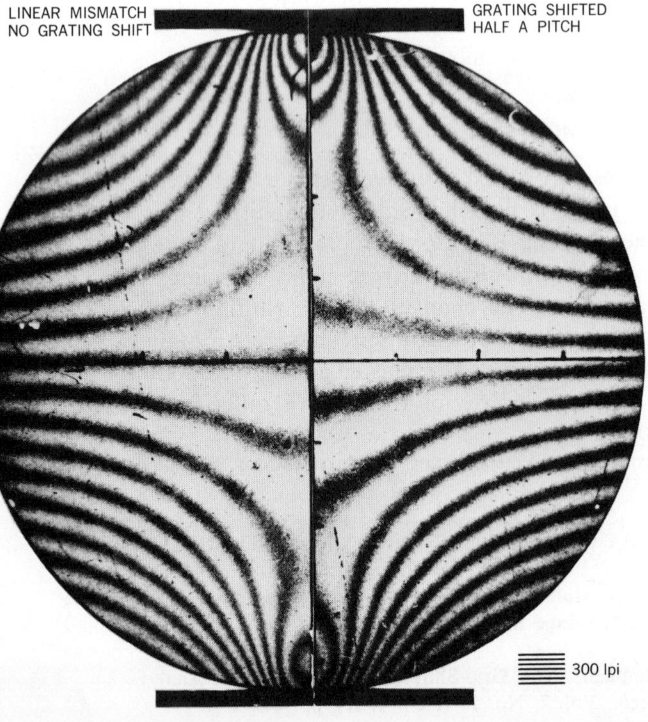

between pictures the master grating was shifted one-half of its pitch in the direction normal to the lines. Both fringe spacings and orientations are changed.

16.3. Fringe Interpolation

Points belonging to two immediate isothetic fringes have a difference in the value of the component of displacement (in the direction normal to the lines of the master grating) equal to 1 pitch. Hence, the isothetic pattern gives discreet information at the center of each fringe, throughout the field. What is happening between the fringes is not known. But, for the same deformation of a specimen, finer gratings produce more fringes.

It is then obvious from Fig. 16.2 that if the fringes are shifted from their original positions to the spacings between them by steps and the patterns so obtained are superimposed, a result is obtained which is equivalent to using finer gratings, and an increase in sensitivity will have been achieved.

Figure 16.3. Plotting of a displacement curve from shifted moire patterns.

p = pitch of master grating

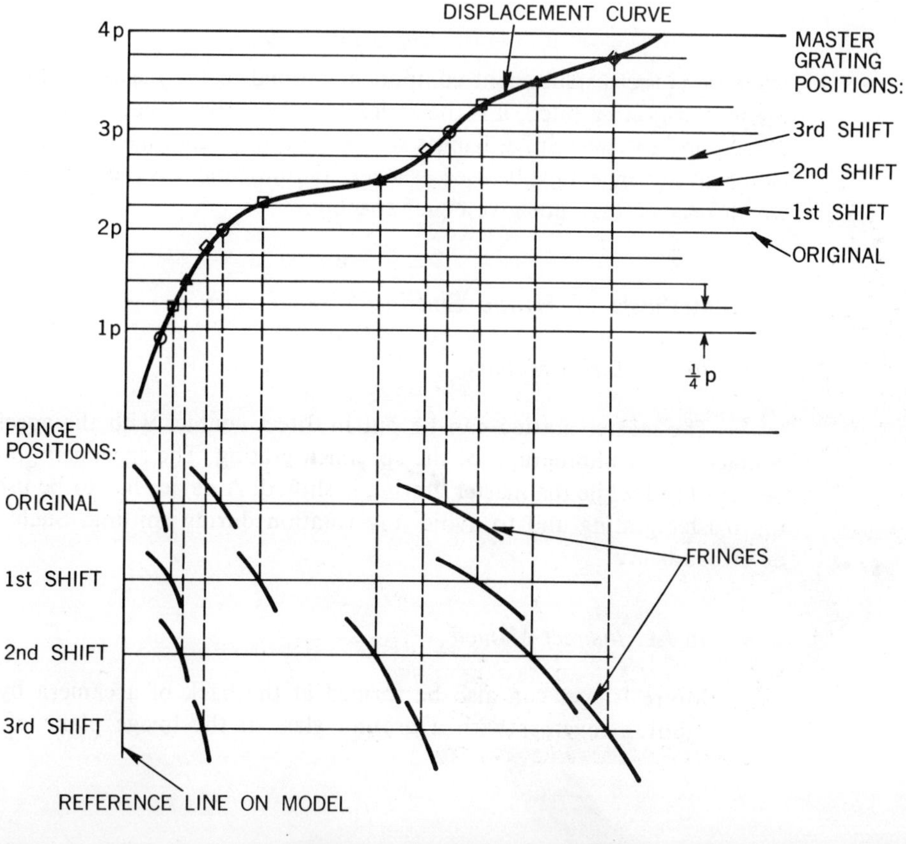

An illustrative sketch of the shifting method is shown in Fig. 16.3. Four sets of fringes are shown as a result of the master grating being shifted three times in the direction normal to the lines with the amount of shifting being one-quarter of the pitch p each time. The fringe positions along one particular section are then plotted at their corresponding grating positions. It is obvious that the same displacement curve, composed of the same points, would have been obtained if a grating with four times as many lines per inch as the original grating had been used to produce the moire pattern. Thus, the sensitivity of the method is increased four times as a result of three equal fringe shifts. Sometimes it may be found more convenient to shift the master grating an arbitrary portion of a pitch, provided the exact position of the grating can be measured. It is easy to see from Fig. 16.3 that as long as there are enough points on the curve, a correct displacement curve can be obtained.

The maximum sensitivity of the moire method that can be obtained by fringe shifting depends on the number of times that distinguishable shiftings can be made using the line density of the original gratin. It was found that with a grating of 1000 lpi, at least five distinguishable shifts can be made. This is equivalent to saying that the same results obtained using gratings of 5000 lpi can be achieved with a 1000-lpi grating.

The actual displacement components can be obtained graphically, as shown in Fig. 16.3, or from the expression

$$u_i = p(n + \alpha) \tag{16.2}$$

where u_i is the displacement component normal to the master-grating lines, p is the master-grating pitch, n is the series of consecutive integer fringe orders which correspond to consecutive fringes of one setting of specimen and master grating (some given fringe must be used as a zero datum), and α is the fraction, Eq. (16.1), which may or may not give equal spacing.

16.4. Methods of Fringe Shifting

16.4.1. Direct Method

The master grating can be put in direct contact with the specimen or in contact with a photograph of the specimen grating. The specimen grating has to be kept fixed while the master grating is shifted. A device has to be used to guide the master grating and to avoid any rotation during shifting. Such a device is described below.

16.4.2. Indirect Method

Moire fringes can also be formed at the back of a camera by placing a master grating together with a ground glass at the image plane of a camera.

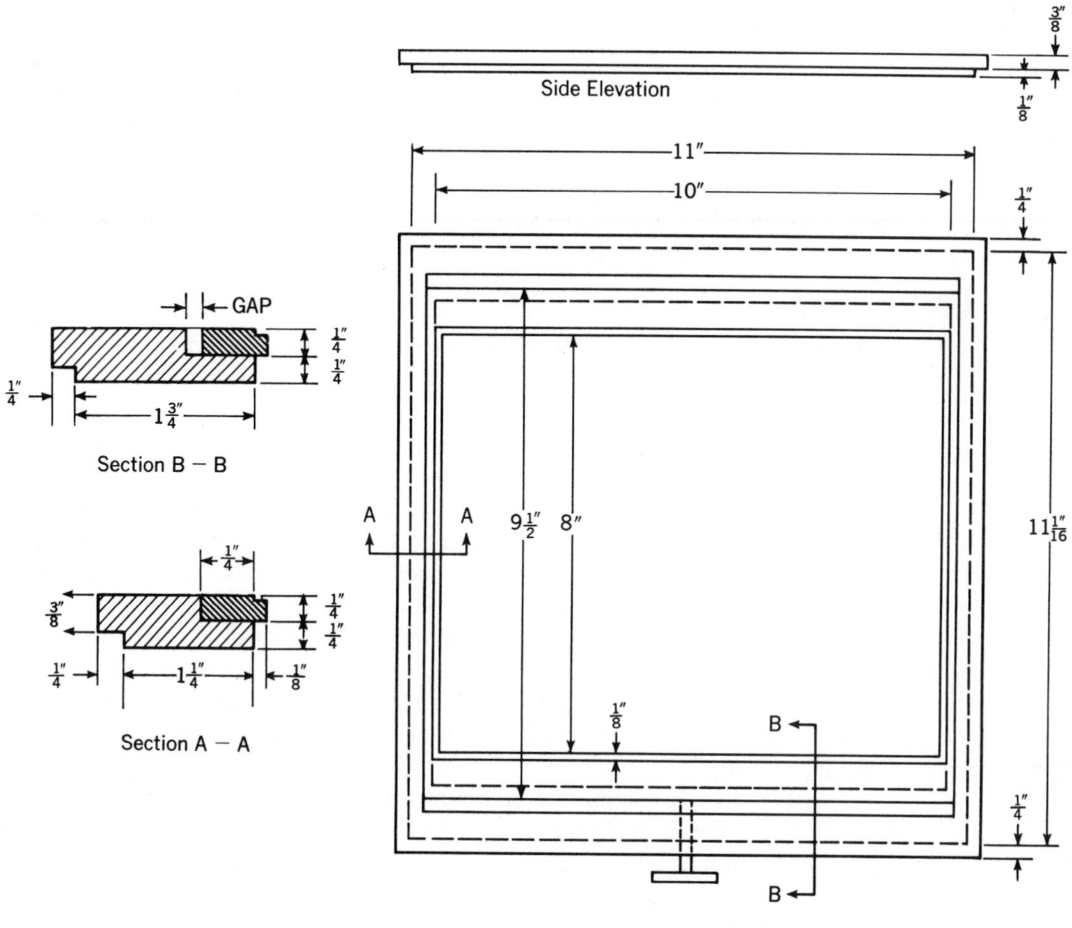

Figure 16.4. Frame designed for moire fringe shifting.

The moire is formed when the specimen grating is brought to thc image plane through a lens. The following two methods of fringe shifting can be used for this fringe-forming technique.

(a) *Camera-Back Shifting*. Obviously, if the master grating mounted on the camera back can be shifted in its plane, the moire pattern will also be shifted. A camera back designed for this purpose is shown in Fig. 16.4. It consists of two frames: the outer frame fits into the camera back; the inner frame, on which the master grating is mounted, can slide inside the outer frame without rotation. A screw with fine threads is used to push the inner frame. The device should be mounted on a revolving camera back (rotatable in the image plane) to allow alignment. The direction of the grating lines can be inclined with respect to the direction of motion on the frame so that a larger travel distance than the pitch of the grating will be required to shift the fringes and give a corresponding increase in precision.

(b) *Lens Shifting.* Some copy cameras are provided with a mechanism that permits shifting the lens in its plane along two perpendicular directions. If moire fringes are formed on the back of a camera of this kind, fringe shifting can be achieved by moving the lens in a direction perpendicular to the master-grating lines. This, in effect, moves the image of the specimen grating along this direction. The patterns shown in Fig. 16.2 were obtained by lens shifting.

16.5. Moire Vernier for the Measurement of the Shifting Distance of the Master Grating

For gratings of 1000 lpi, movements of the order of 0.0001 in. have to be measured in order to determine the grating position after every shift. A simple way to measure the traveled distance is to use a reference grating which has a line density either equal to or slightly different from that of the master grating. This reference grating is called a *moire vernier* because it functions similar to a vernier. The reference grating is placed in the plane of the specimen grating so that both will be covered by the same master grating either directly or at the back of a camera. When superposition is achieved, two moire patterns will be formed, one belonging to the specimen, the other to the moire vernier. The pattern of the moire vernier will be, in general, a pattern of straight lines produced by mismatch, rotation, or both.

If Fig. 16.1 is assumed to be a moire vernier, then from Eq. (16.1) it can be seen that if the master grating is shifted in the direction normal to the lines a distance equal to αp, the fringes will move a distance equal to $\alpha\delta$ in the direction normal to the fringes. Essentially, the pitch of a master grating is magnified by the moire effect to the spacing of fringes. If the distance between any two fringes is divided into equally small intervals, it can be used to determine the distance that the master has traveled by the location of an index fringe. This concept is demonstrated in Fig. 16.1, in which the fringe spacing δ is divided into eight equal parts. It is obvious that the larger the index-fringe spacing, the greater the accuracy. The fringe spacing is controlled by the rotation and mismatch that the moire vernier has with the master grating. The less the rotation and mismatch, the larger the spacing. However, fringes will become too wide if the fringe spacing is very large. Another way of increasing the magnification is to place the scale at a skew angle with respect to the fringes so that an equivalent large spacing can be obtained.

16.6. Comparison of Different Shifting Methods

The advantage of the direct-shifting method is better contrast.

The advantage of camera-back-shifting method is its adaptibility to most cameras available in a laboratory. In applying this method, the moire vernier does not have to be in the specimen plane—it can be placed in the image plane. Another advantage of this method is that rotation and mismatch can be easily imposed. If the camera has a lens that can be moved in the lens plane, then the third method—lens shifting—is the easiest to use. In other respects, this method is identical to the camera-back-shifting method.

16.7. Fringe Multiplication

As presented above, fringe shifting allows the recording of fringes between those obtained with the conventional moire pattern. The intermediate fringes can be assigned a displacement value according to the ratio of shift to pitch. This process is called *interpolation* here.

Also mentioned above is the idea that the shifts could be equally spaced over one pitch, and could fraction the pitch equally ($\frac{1}{2}$, $\frac{1}{3}$, $\frac{1}{4}$, etc.). If such patterns are obtained, it would be desirable to superpose a set of these fringe patterns. Superposing a set of fringes obtained by such fractional fringe shifts is called *multiplication* here. It is obvious that in order to produce multiple patterns, the individual fringe has to be either very thin or very wide (since superposition can be made using light fringes also). Techniques are presented below to sharpen or broaden fringes. Two techniques to produce multiple patterns by fringe shifting are then given.

16.7.1. Fringe Sharpening and Widening

In general, gratings have dark and light lines of equal width. Moire fringes thus obtained have the best contrast and dark and light fringes are of the same width. Moire patterns of thin dark fringes are produced when one grating with thin lines and broad spacing is superimposed on another grating with broad lines and thin spacing; moire patterns of broad dark fringes are produced by two gratings both having broad lines and thin spacing.[12] Special treatment of a 1: 1 grating is necessary to achieve the widening or sharpening effect. If a contact photographic plate of a grating is made by overexposure, the resulting grating has wider lines and thinner spacings. If a third contact plate is made from the second grating with normal exposure, the line-spacing–width ratio will be reversed. If this two-step cycle is repeated several times, gratings with very wide or very thin lines can be made. In practice, it is better to print a grating of wide lines and thin spacing on the specimen. In this case, patterns with either wide dark fringes or thin dark fringes will be produced, depending on whether the master grating has wide or thin lines.

Kodak high-resolution plates are recommended for making copies of the

[12] D. Post, "Sharpening and Multiplication of Moire Fringes," *Exp. Mech.*, Vol. 7, No. 4 (April 1967).

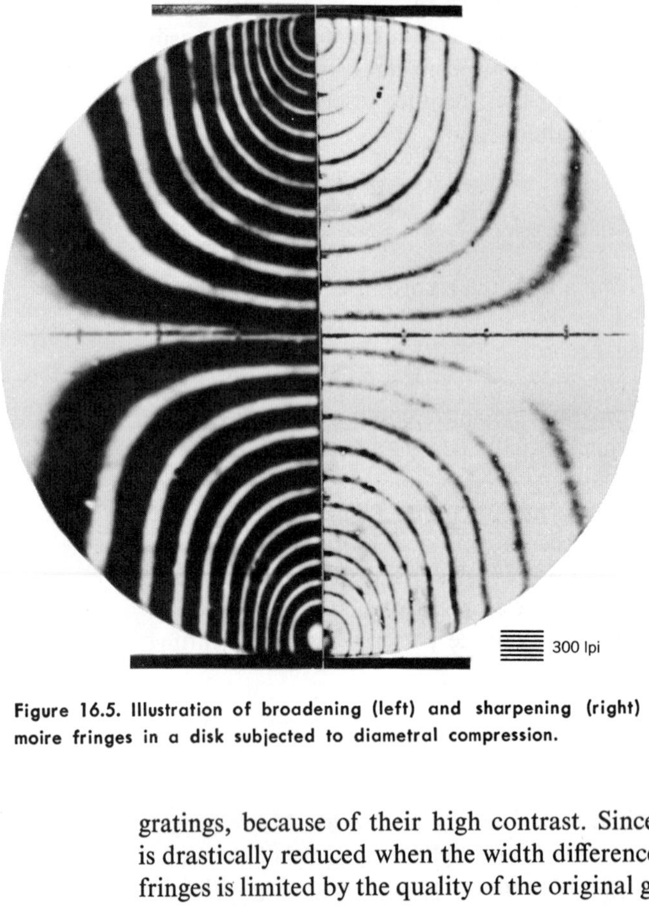

Figure 16.5. Illustration of broadening (left) and sharpening (right) of isothetic moire fringes in a disk subjected to diametral compression.

gratings, because of their high contrast. Since the contrast of the moire pattern is drastically reduced when the width difference is large, sharpening or broadening fringes is limited by the quality of the original grating, the photographic plate used, and the line density of the gratings. The finer the grating, the less the fringe can be sharpened or broadened without losing too much contrast. An example of fringe sharpening and broadening is shown in Fig. 16.5. The specimen is a disk under diametral compression with wide grating lines. The grating used has a line density of 300 lpi. The picture is a composite of thick moire fringes produced by a wide-line master grating and thin moire fringes produced by a thin-line master grating. The specimen and master gratings were in direct contact in each case.

16.7.2. Fringe Multiplication by Multiple-Exposure

This technique of fringe multiplication consists of exposing onto one film several patterns, each of them obtained after incremental shifting. It is necessary to have broad dark fringes so that, on the negative, only places corresponding to thin light fringes are exposed to the light. A difficulty in applying this technique is to divide properly the length of time given to each exposure in order to obtain

Figure 16.6. Moire fringe multiplication of two obtained by multiple exposure.

a uniform field. Also, it is difficult to obtain wide dark fringes with acceptable contrast. Furthermore, pictures made by multiple-exposure have inherently less contrast than single-exposure pictures. Nevertheless, with proper care, reasonably good pictures can be obtained with this technique. An example is given in Fig. 16.6, in which a fringe multiplication of 2 is achieved. The isothetics are taken from a disk subjected to diametral compression, and the grating used has 300 lpi. The specimen and master gratings were in direct contact. The first exposure took two-fifths of the total exposure time, and the second took three-fifths. The film used was Kodalith Ortho, Type 3. Judging from the crowding of the fringes, a multiplication of 3 is probably the limit that can be achieved by this method.

16.7.3. Fringe Multiplication by Superposition

In applying this technique, pictures of moire patterns at different stages of shifting are taken individually and then superimposed to yield the multiplication effect. In principle, either thin or wide fringes can be used in this technique. If the pattern is composed of broad dark fringes, superposition can be achieved

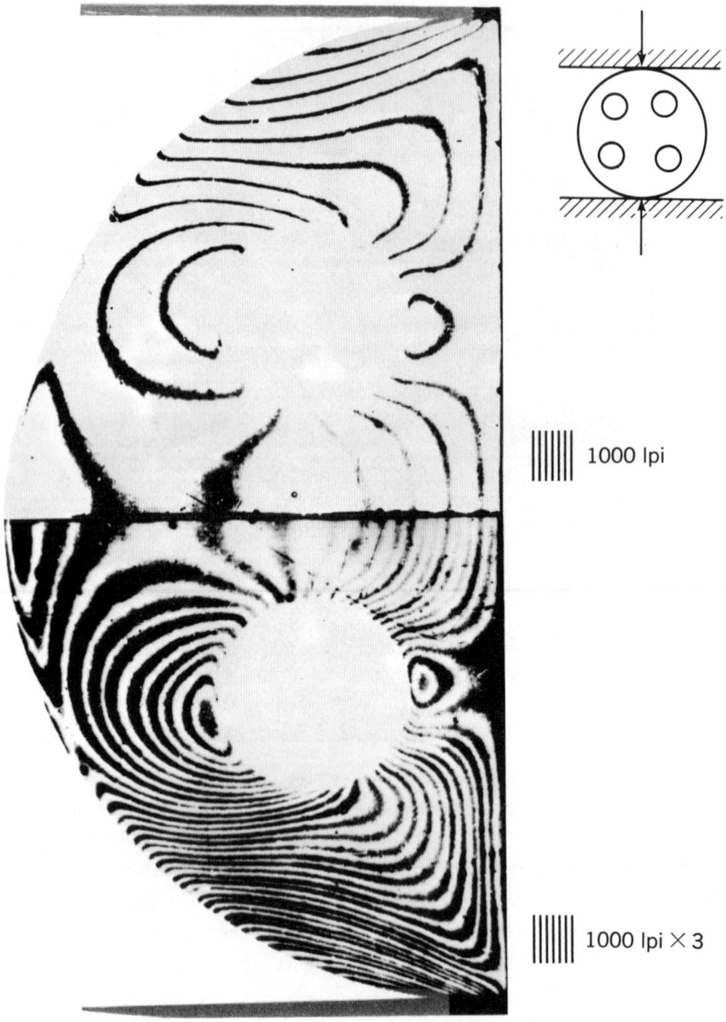

Figure 16.7. Isothetics in a perforated disk (pattern orders multiplied by three).

directly on the negative because it is made of thin dark fringe. On the other hand, if the pattern is composed of thin dark fringes, the negatives have to be transferred into positives in order to be superposed. Although the latter method requires one more step, better results are obtained this way. The difficulty of the former method stems from the fact that broad dark fringes are difficult to obtain without sacrificing contrast. The multiplied pattern shown in Fig. 16.7 is the pattern that would have been obtained if a grating of 3000 lpi had been used. The direct-shifting method was used with the frame shown in Fig. 16.4. Difficulties were encountered when attempts were made to achieve multiplication higher than 3. It was found that when shifted moire patterns were superposed, portions of the fringes were incorrectly spaced. The degree of this misalignment increases as the order of multiplication increases. This is because higher-order multiplication means higher sensitivity. It may be due to any of the following reasons: (1) local friction between specimen and master gratings, (2) rotation (in-plane or out-of-plane or both) of the master grating in the process of shifting, or (3) error in pitch uniformity of the master grating. Further work is needed to eliminate this difficulty and achieve higher-order fringe multiplication.

16.8. Comparison with Other Methods

The choice of method to increase moire sensitivity will be dictated to a large extent by the particular problem to be solved and the available equipment. The shifting method presented here is simple to use. The method is essentially mechanical and may, in some cases, be less precise than the optical methods mentioned in the introduction. Both the shifting method and the various optical methods differ from mismatch and rotation methods in that mismatch and rotation are methods of *addition* of fringes rather than *multiplication* of fringes. These and similar considerations will enter into the comparison of methods and influence the selection of the most suitable method.

PART 5

APPLICATIONS

17

DETERMINATION OF TRANSIENT STRESS-AND-STRAIN DISTRIBUTIONS

17.1. Introduction

A number of experimental methods have been used in the past to determine transient stress-and-strain distributions. Each method has advantages and limitations, depending on the particular problem under study. Electrical-resistance strain gages have been used by numerous investigators to determine transient strains on the surface of metal bars and plates. Gages of this type theoretically have a very high frequency response; however, the adhesive used with the gages will not always satisfactorily transmit high-frequency strains to the gage. These gages also suffer from the limitation that they give an integrated value of strain over a finite area. They are, however, relatively sturdy and easy to use. A second type of unit used for transient measurements utilizes the principle of a condenser. A portion of the surface of the specimen serves as one plate of the unit and, with proper instrumentation, the movements of the surface can be measured. These units can be made with a high sensitivity and with a very high frequency response. They suffer from the limitation that they can be utilized only for average surface measurements. A third method used by some investigators makes use of a piezoelectric crystal such as barium titanate. These units can be made extremely sensitive.

Optical methods have also been used by numerous investigators in studying the transient stress-and-strain problem. These include direct visual observation of a ruled grid, optical-interference techniques, and the well-known photoelastic method. The photoelastic method has the advantage of providing full-field information for the two-dimensional specimens in which the method is applicable. The method suffers from the limitation that auxiliary information is required for separation of the principal stresses. The method is further limited by the requirement of a transparent birefringent material and by the fact that the photoelastic effect observed is an integral effect; therefore, no information is obtained regarding the distribution of stress along the path of light.

In recent years, dynamic photoelastic methods using low-modulus materials[1] have been developed to the point where accurate dynamic boundary-stress distributions can be determined in certain two-dimensional problems.[2,3] In order to provide the additional information needed to separate the principal stresses at interior points of the model, embedded-grid techniques were developed for determining dynamic-strain fields. The measurements were difficult to make, since the quantities to be measured were extremely small. The method remains, however, as one of the very few practical suggestions for the determination of transient stress distributions in three-dimensional components. The moire method in its current state of development can give in certain cases accurate two-dimensional-surface strain distributions for both static and dynamic loadings. In the following sections, an application to the solution of dynamic problems will be shown.

17.2. Preparation of Specimens

The models for the tests reported here were machined from a sheet of low-modulus urethane rubber known commercially as Hysol 8705. A photographic contact print on glass of an original 1000-lpi ruling was available for use in preparing the master grating and the specimen gratings.

The master grating was prepared by making a contact print on film. The specimen gratings were prepared by contact printing using a polyvinyl-alcohol enamel known commercially as Gaco as the photosensitive material.[4] Normally, this material requires baking at 400°F as a part of its fixing process. With the

---

[1] J. W. Dally, W. F. Riley, and A. J. Durelli, "A Photoelastic Approach to Transient Stress Problems Employing Low-Modulus Materials," *Journ. of Applied Mechanics*, Vol. 26, *Trans. ASME*, Series E, Vol. 81 (1959), pp. 613–620.

[2] J. W. Dally, A. J. Durelli, and W. F. Riley, "Photoelastic Study of Stress Wave Propagation in Large Plates," *Proc. SESA*, Vol. 17 (1960), pp. 33–50.

[3] A. J. Durelli and W. F. Riley, "Stress Distribution on the Boundary of a Circular Hole in a Large Plate During Passage of a Stress Pulse of Long Duration," *Journ. of Applied Mechanics*, Vol. 28, *Trans. ASME*, Series E, Vol. 83 (June 1961), pp. 245–251.

[4] Later-developed, more efficient photosensitive coatings are described in Chapter 14.

soft-rubber materials, only partial processing was employed, and the lines were blue rather than the black normally obtained with full processing. Extremely careful washing was also required, since the photographic image was not completely fixed.

In order to obtain uniform contact between the rubber model and the master grating during exposure of the photosensitive material, a vacuum-printing frame was used. The pressure caused the rubber to deform slightly and, as a result, exactly 1000 lpi were not obtained. This resulted in a slight initial mismatch between the master grating and the model grating. The effects of this mismatch will be explained later.

17.3. Application to Determination of Transient Stresses at a Point in a Semi-Infinite Plate

A problem of considerable interest was the development of methods for determining the two principal stresses and their directions at any point in a semi-infinite plate when the plate was subjected to a time-dependent loading at a point on one edge.[5] One phase of this problem can be approached by using photoelasticity methods, which give values for the difference between the principal stresses at each point in the field. The auxiliary information required for determining the individual values for the two principal stresses can be provided by two moire-fringe patterns with grating lines perpendicular to each other.

The plate selected for use in the program was 0.37 by 12 by 12 in. The point of interest for stress determinations was located 4 in. from the load point along a radial line 30° from the centerline of the plate. The location of the point was determined by considerations not pertinent to the use of the method, and similar determinations could be made for any other point in the plate.

The loading of the model was accomplished by dropping a weight (108 g) from a height of 16 in. on a hard-plastic striker which was permanently positioned on the model. (The reasons for using the striker will be discussed later.) A sketch of the model is shown in Fig. 17.1.

A complete photographic record of the photoelastic-fringe patterns produced by the impact was obtained by using a 16mm Fastax camera, operating at 6900 frames per second. Some selected frames from a typical record are shown in Fig. 17.2.

A similar record of the moire-fringe patterns was obtained with the same camera operating at 6400 fps. Typical frames from this type of record are shown in Fig. 17.3. In these tests, the master grating was rigidly fastened to the frame

[5] The subject matter of this chapter has also been treated in W. F. Riley and A. J. Durelli, "Application of Moire Methods to the Determination of Transient Stress and Strain Distributions," *Journ. of Applied Mechanics*, Vol. 29 (March 1962), pp. 23–29.

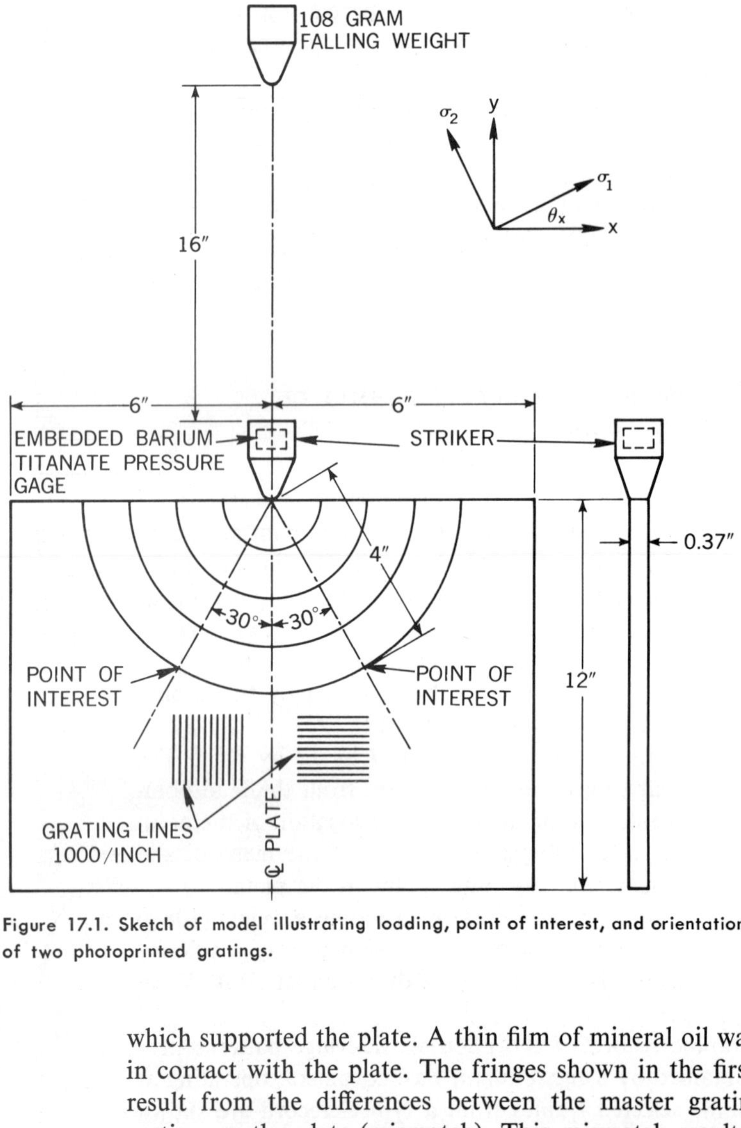

Figure 17.1. Sketch of model illustrating loading, point of interest, and orientation of two photoprinted gratings.

which supported the plate. A thin film of mineral oil was used to keep the master in contact with the plate. The fringes shown in the first few frames in Fig. 17.3 result from the differences between the master grating and the photoprinted grating on the plate (mismatch). This mismatch resulted from the printing procedure explained previously, and the apparent residual strains which it indicates must be added to or subtracted from the strains produced by the loading.

Curves showing the fringe order and the two strains at the specific point of interest obtained from such records are presented in Fig. 17.4. Curves showing the modulus of elasticity and the stress-fringe value of the model material as a function of strain rate are shown in Fig. 17.5. These data were then used in the following equations to obtain the principal stresses and their directions at the

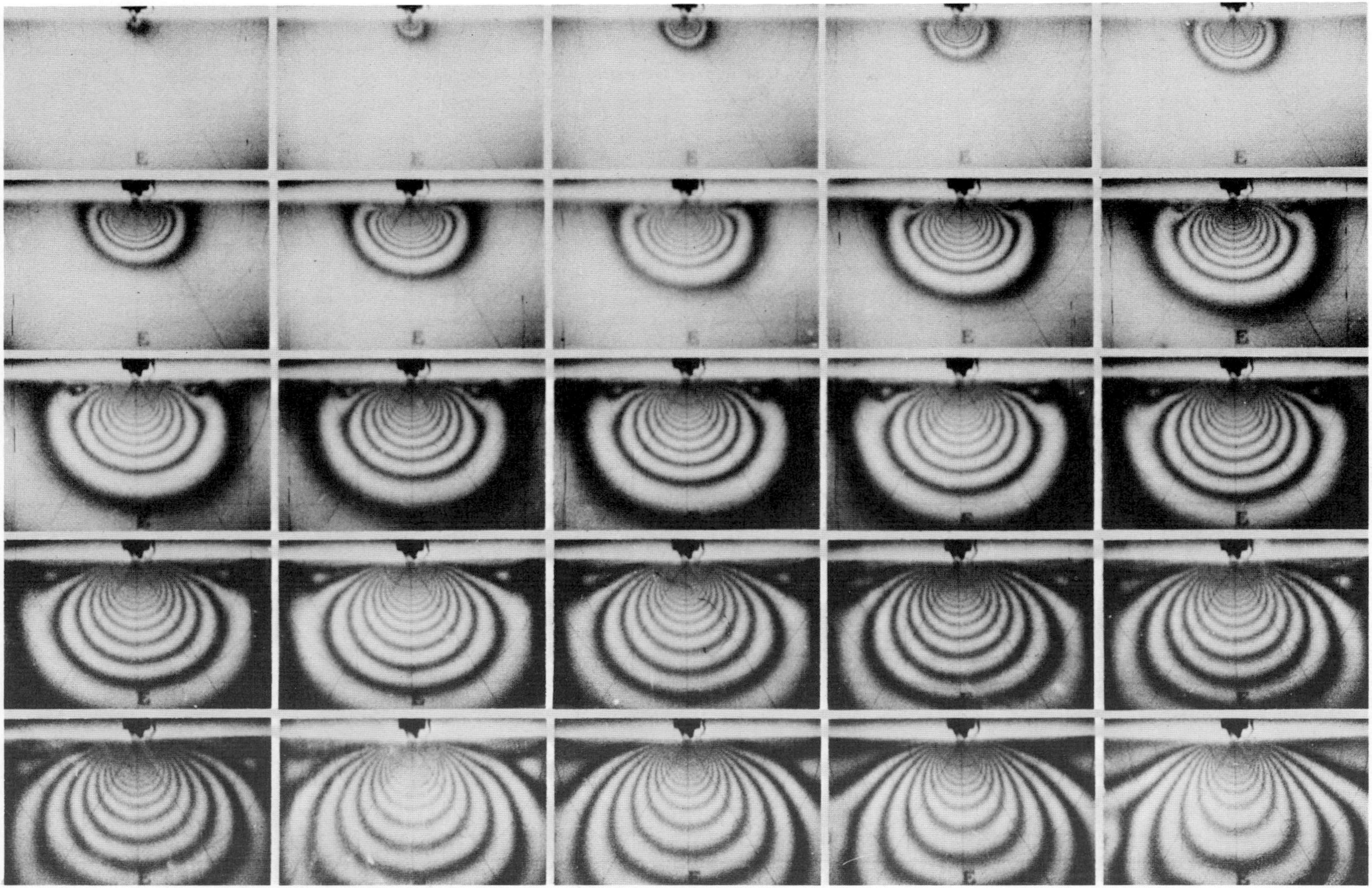

Figure 17.2. Selected frames from a typical photoelastic fringe pattern record obtained using a Fastax camera operating at 6900 frames per second.

Figure 17.3. Selected frames from typical moire fringe pattern record obtained using a Fastax camera operating at 6400 frames per second.

selected point. Poisson's ratio, which is also required in the following equations, has a value of 0.46 for the model material and is independent of strain rate:

$$\sigma_1 - \sigma_2 = \frac{2nf_\sigma}{t}$$

$$\sigma_1 + \sigma_2 = \frac{E}{1-\nu}(\epsilon_x + \epsilon_y)$$

$$\epsilon_1 - \epsilon_2 = \frac{2(1+\nu)}{E}\frac{nf_\sigma}{t}$$

$$\cos 2\theta_x = \frac{\epsilon_x - \epsilon_y}{\epsilon_1 - \epsilon_2}$$

where σ_1 and σ_2 are the principal stresses, ϵ_1 and ϵ_2 are the principal strains, ϵ_x and ϵ_y are the normal strains in the x- and y-directions, respectively, n is the photoelastic-fringe order at a point of interest,[6] f_σ is the stress-fringe value of material in shear, t is the thickness of the model, E is the modulus of elasticity, ν is Poisson's ratio, and θ_x is the direction of σ_1 measured with respect to the x-axis.

[6] Photoelasticity-fringe orders and moire-fringe orders are represented by the same symbol n. It is believed that no confusion will develop from this fact.

Figure 17.4. Fringe order, horizontal strain ϵ_x, and vertical strain ϵ_y at point of interest as a function of time after impact.

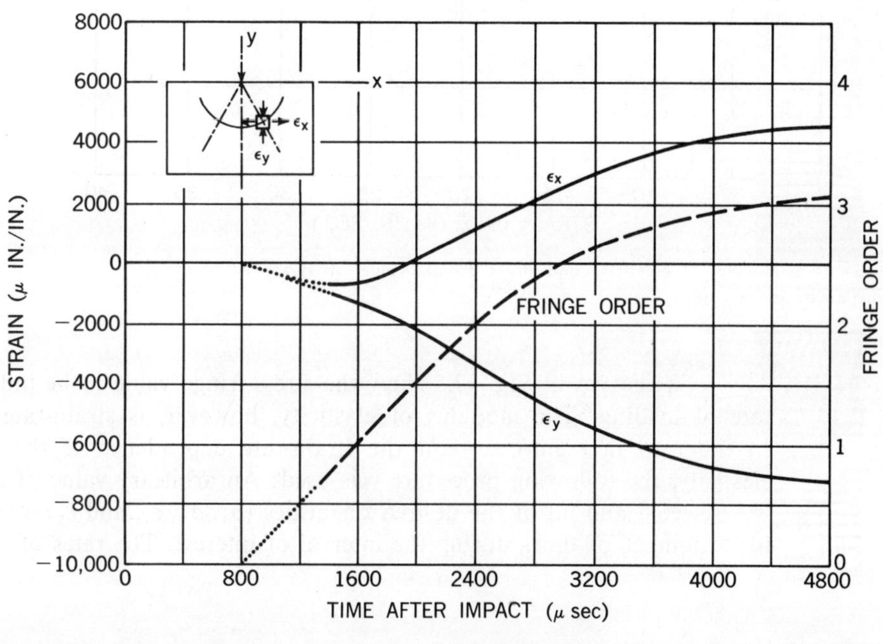

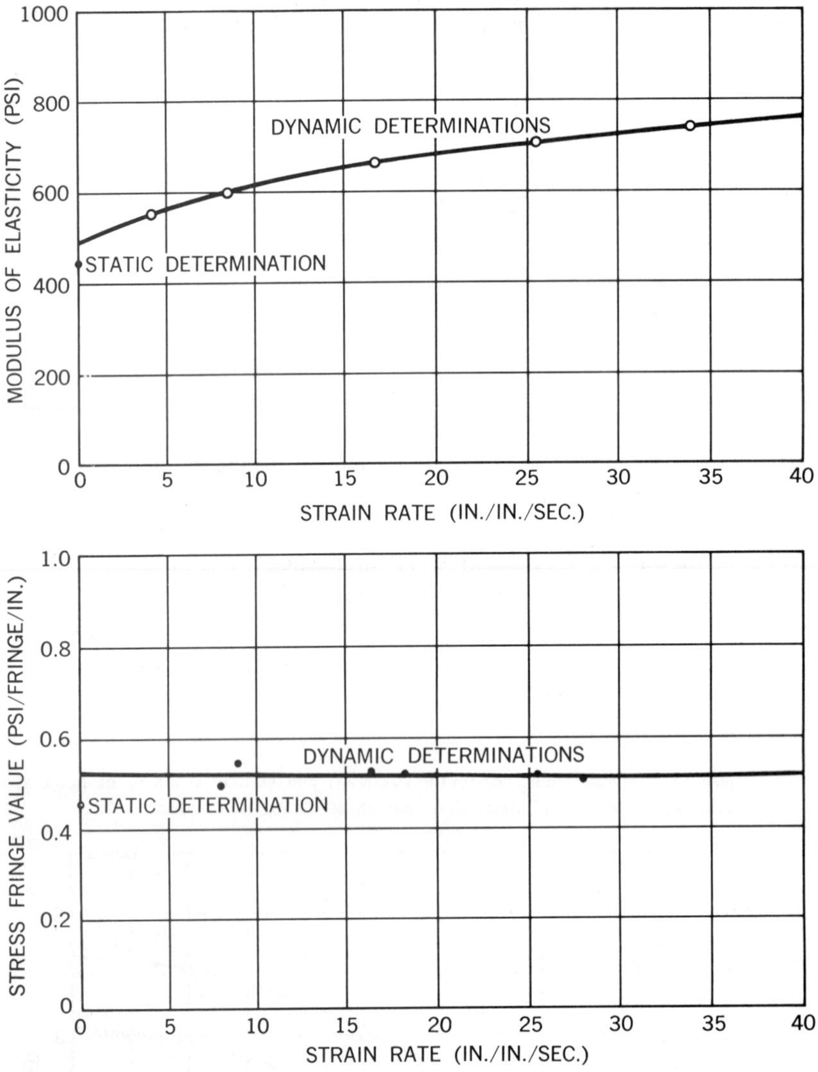

Figure 17.5. Modulus of elasticity and stress fringe value of Hysol 8705 as a function of strain rate.

The curves in Fig. 17.5 show the stress-fringe value to be independent of rate of loading. The modulus of elasticity, however, is strain-rate–dependent. In order to take into account the strain-rate dependence of the modulus of elasticity, the following procedure was used: An arbitrary value of the modulus was selected and all of the desired quantities (σ_1, σ_2, ϵ_1, and ϵ_2) were computed for a number of times during the interval of interest. The rates of change of ϵ_1

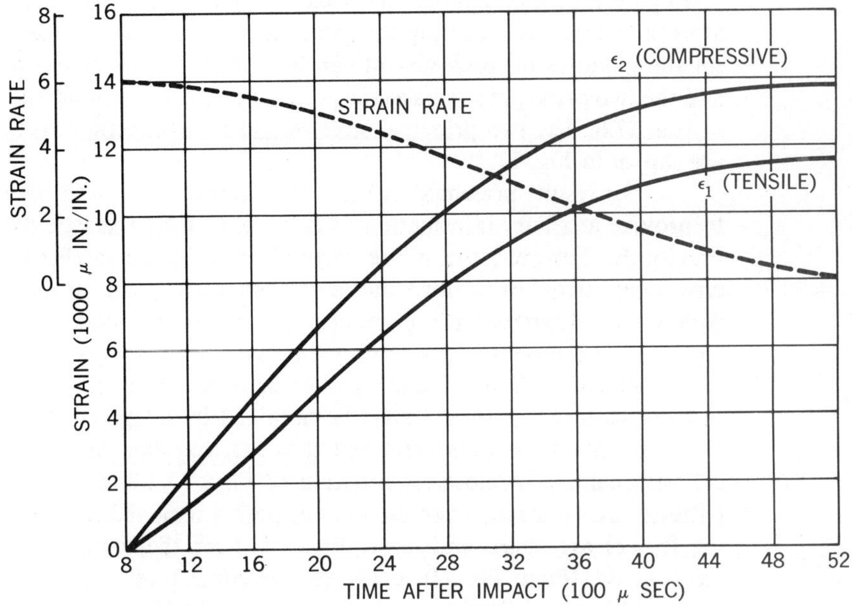

Figure 17.6. Strain rate and two principal strains at point of interest as a function of time after impact.

Figure 17.7. Two principal stresses and their orientation at point of interest, as a function of time after impact.

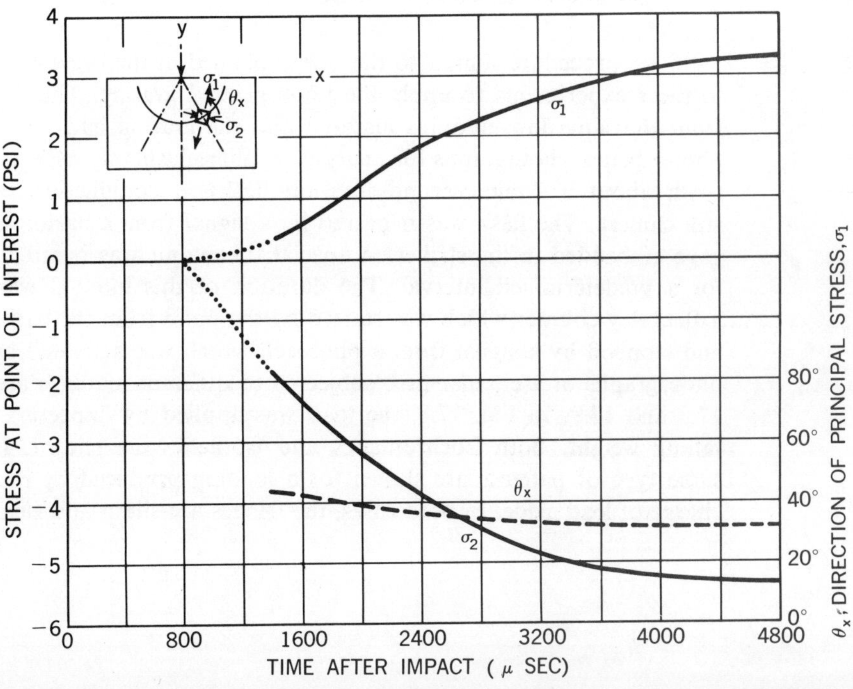

and ϵ_2 were then computed and used to obtain a better estimate of the modulus at each instant. By repeating this procedure three or four times, the correct value for the modulus for each instant can be obtained. Curves showing the strain rate and the two principal strains at the point of interest as a function of time are shown in Fig. 17.6. The two principal stresses and their directions as a function of time are shown in Fig. 17.7.

The results presented indicate that moire fringes can be utilized effectively to provide auxiliary information to complement photoelastic data. As mentioned previously, however, the moire method is self-sufficient, and by obtaining the cross-derivatives, or a third moire pattern at 45°, the problem could be completely solved without the photoelastic data. If the model has an axis of symmetry, as in the present case, two moire patterns can be recorded simultaneously. If no such axis exists, the loading must be of a reproducible nature, since the two or three patterns in general must be obtained in independent loadings.

In all of the tests described thus far, the data have been recorded using a Fastax camera. A close examination of Figs. 17.2 and 17.3 indicates that these patterns are satisfactory for use only at points removed from the load area where the fringes are sharp and clear. In areas near the load, the fringes cannot be resolved on 16mm film. In order to overcome these limitations, the techniques described in the following section were developed.

17.4. *Application to Determination of Transient Strains in a Circular Disk*

A procedure similar to the one explained in the previous section was used in these experiments to apply the photoprinted grating. The models were made from the same low-modulus material and the grating had 1000 lpi. In order to obtain better photographs for analysis, a microflash was used as the light source (with about a 2-microsecond-duration flash) in conjunction with a 5-by-7-in. still camera. The flash was triggered by a signal from a barium-titanate pressure gage embedded in the striker. A special timer unit was based to delay the flash for a predetermined interval. The duration of this interval was measured with a Berkeley counter which was started by the pulse from the barium-titanate gage and stopped by a signal from a photocell which was activated by the flash. Two photographs of a circular disk subjected to different loadings are shown in Figs. 17.8 and 17.9. In Fig. 17.8, the load was applied by impacting the disk with a falling weight. Both isochromatics and isothetics are shown. In Fig. 17.9 the same type of patterns are shown for a loading produced by detonating a small charge of lead azide. In both cases, the fringes are sharp and clear over the entire

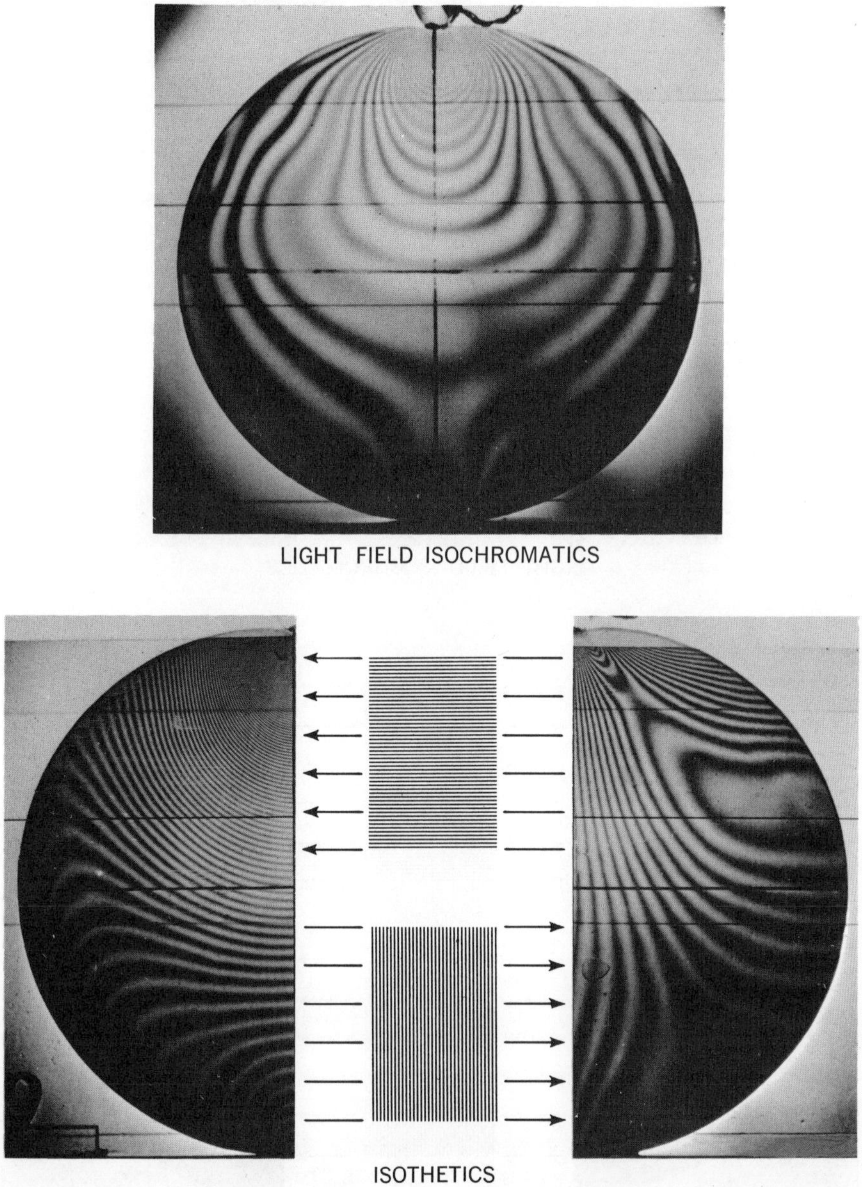

Figure 17.8. Microflash photographs of the photoelastic fringe pattern and moire fringe patterns in a 5-in. diameter disk 1805 μsec after it was impacted with a 30-g weight falling from a height of 30 in.

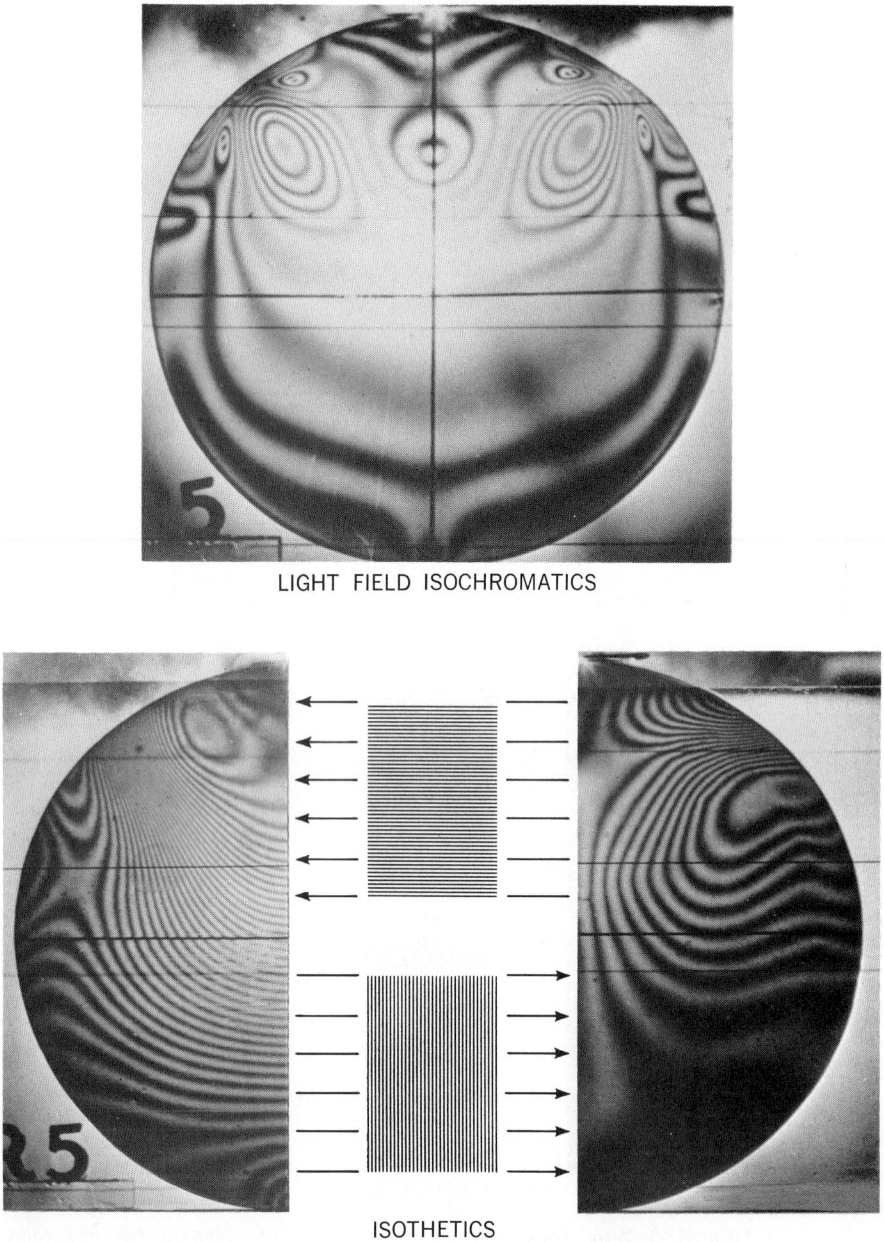

Figure 17.9. Microflash photographs of the photoelastic fringe pattern and moire fringe patterns in a 5-in. diameter disk 900 μsec after a 70-mg charge of lead azide was detonated on the boundary.

surface of the disk. The interpretation of such patterns requires that a number of photographs be taken over relatively short intervals in order for the strain rates to be determined. This imposes no special difficulties, except that the loading must be reproducible.

18

ELASTOPLASTIC STRESS AND STRAIN DISTRIBUTION IN A FINITE PLATE WITH A CIRCULAR HOLE SUBJECTED TO UNIDIMENSIONAL LOAD

18.1. Introduction

The problem of determining the distribution of strains and stresses in a plate with a circular hole under unidimensional load has been theoretically and experimentally studied by several authors. The value of the maximum strain and the distribution along the transverse axis can be found in Griffith,[1] Box,[2] and Miller.[3] A study of the strain distribution in a region surrounding the hole has been presented in Merril[4] and Vinckier and Dechaene.[5] The stress distribution on the transverse axis of a plate made of a perfectly plastic material has

[1] G. E. Griffith, "Experimental Investigation of the Effects of Plastic Flow in a Tension Panel with a Circular Hole," NACA TN 1705 (1948).

[2] W. A. Box. "The Effect of Plastic Strains on the Stress Concentrations," *Proc. SESA*, Vol. 8, No. 2 (1951), pp. 99–110.

[3] J. A. Miller, "Improved Photo Grid Techniques for the Determination of Strain Over Short Gage Lengths," *Proc. SESA*, Vol. 10, No. 1 (1952), pp. 29–34.

[4] P. S. Merril, "Photodot Investigation of the Plastic Strain Pattern in a Unidirectionally Loaded Flat Sheet with a Hole," *Proc. SESA*, Vol. 18, No. 2 (1961), pp. 73–80.

[5] A. Vinckier and R. Dechaene, "Use of the Moiré Effect to Measure Plastic Strains," *Jrnl. of Basic Engineering*, *Trans. ASME*, Vol. 82, Series D (1960), pp. 426–434.

been theoretically studied in Stowell[6] and in Budiansky and Vidensek,[7] and photoelastically in Thompson.[8]

The object of the investigation described in this chapter to analyze the distribution of stresses and strains on the transverse axis of a finite plate as the plastic deformation increases and, for one of the loading stages, to study the strain distribution in a region surrounding the hole. To improve the precision of the moire method, an initial moire-fringe pattern was introduced.[9]

18.2. *The Specimen*

An aluminum plate (Fig. 18.1) 24-in. long, 8-in. wide, and $\frac{1}{8}$-in. thick was used in the test. A circular hole 1.25 in. in diameter was drilled at the center. A grating of 300 lpi with lines parallel and perpendicular to the axial load was photoengraved on one face of the model. Since the strains to be determined were small, the mismatch method was used (see Sec. 7.6). Two different master gratings were prepared in order to take pictures alternatively of the moire patterns in the longitudinal and transverse directions. The strains in the longitudinal directions are tensile, whereas those in the transverse direction are compressive. Following the considerations noted, master-grating mismatches with the same signs as the corresponding strains were used. For the longitudinal direction, the mismatch used was $+1.8$ per cent; for the transverse direction, -1.16 per cent. The specimen was loaded in steps; at each step, pictures were taken of the moire patterns in the longitudinal and transverse directions after superimposing the master gratings. An initial load of 6500 lb was applied, and then the load was increased to 8000 lb; after that, the load was increased in steps of 2000 lb, up to the maximum of 18,500 lb. Strain values were obtained by graphic differentiation of the displacement curves and by subtracting from these values those of the initial mismatches. All the results were referred to the undeformed or initial shape of the model.

18.3. *Results of Test*

A tension test on several samples of the aluminum used in the plate was conducted in order to obtain the stress–strain relationship. The result is given in Fig. 18.2 in a dimensionless form. The ordinates are the ratio σ/σ^*; the stress

6 E. Z. Stowell, "Stress and Strain Concentration at a Circular Hole in an Infinite Plate," NACA TN 2073 (1950).

7 B. Budiansky and J. Vidensek, "Analysis of the Stress in the Plastic Range Around a Circular Hole in a Plate Subjected to Uniaxial Tension," NACA TN 3452 (1955).

8 R. A. Thompson, "Studies in Photoelasticity," doctoral thesis, Illinois Institute of Technology (January 1958).

9 This problem has also been studied in A. J. Durelli and C. A. Sciammarella, "Elastoplastic Stress and Strain Distribution in a Finite Plate with a Circular Hole Subjected to a Unidimensional Load," *Journ. of Applied Mechanics*, Vol. 30, No. 1 (March 1963), pp. 115–121.

σ^*, corresponding to the flat portion of the stress–strain curve, is called *yield stress.* The value of σ^* is 21,700 psi. As abscissas, the ratio $\epsilon E/\sigma^*$ is plotted, where $E = 10{,}000{,}000$ psi is the elastic modulus.

In order to represent the results of the test, it is convenient to have a dimensionless parameter characterizing the load level at each stage. The parameter

$$\lambda = \frac{P_{act}}{P_{max}} \tag{18.1}$$

has been chosen to express the load level, where P_{act} is the actual load of the plate and P_{max} is the maximum load that the plate would be able to take, assuming

Figure 18.1. Aluminum plate (24 in. × 8 in.) with a central circular hole (1.25 in.), subjected to uniformly distributed axial tensile load. Moire fringes for (a) v-family, (b) u-family.

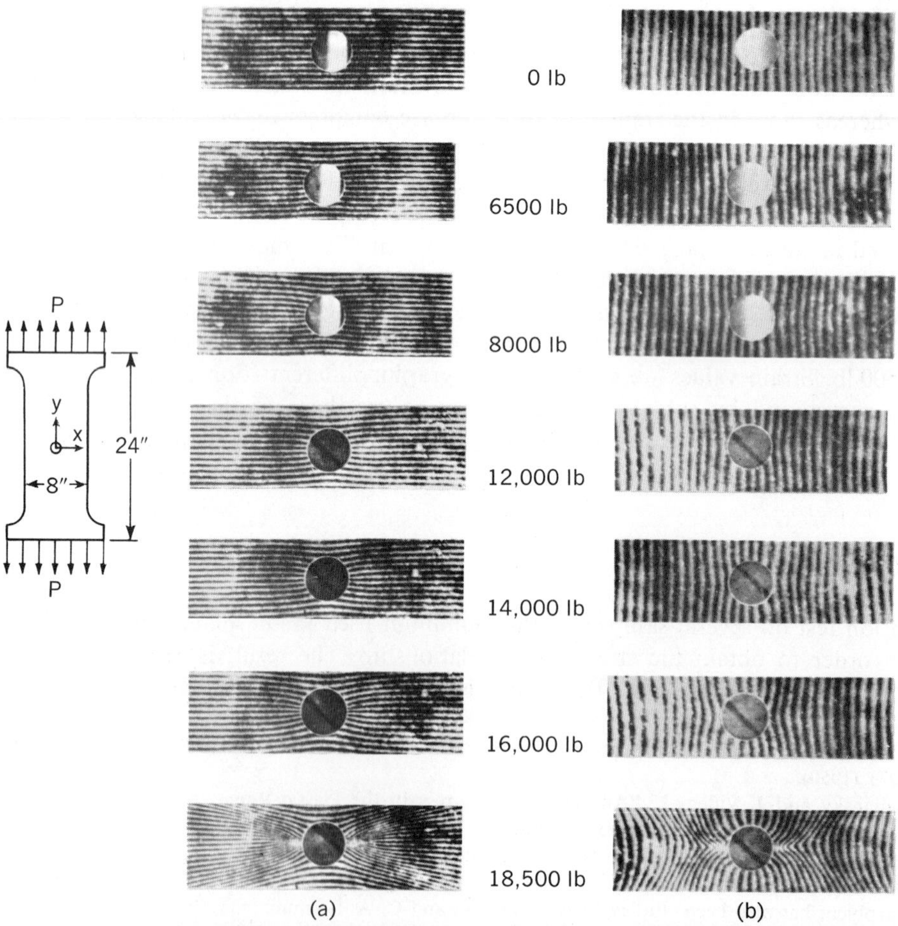

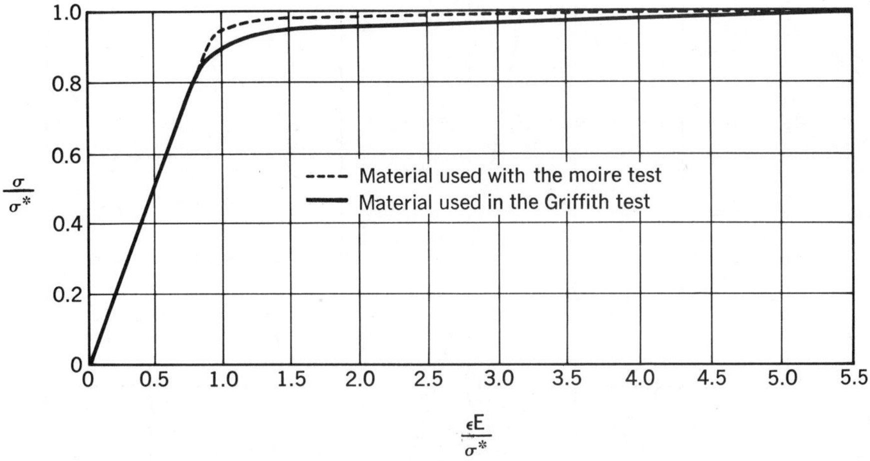

Figure 18.2. Dimensionless stress-strain curve of the aluminum employed in the test of a plate with a circular hole.

that it has no circular hole. If b is the width of the plate and h the thickness, Eq. (18.1) may be written as

$$\lambda = \frac{\sigma_{\text{avg}} bh}{\sigma^* bh} \tag{18.2}$$

Therefore, λ is equal to the ratio of the average stress on the gross area to the yield stress of the material.

A complete study of the deformations for all the loading steps was made for the axis of symmetry. The results are represented in dimensionless form in Fig. 18.3. The abscissas are dimensionless coordinates x/r, where r is the radius of the hole. The ordinates are dimensionless ratios $\epsilon E/\sigma^*$.

The maximum load taken by the plate was $P_{\text{ult}} = 18{,}500$ lb. This load corresponds to an average stress in the full section equal to

$$\sigma_{\text{avg}} \frac{18{,}500}{\frac{1}{8} \times 8} = 18{,}500 \text{ psi} \tag{18 3}$$

For this value of the load,

$$\lambda = \frac{18{,}500}{21{,}700} = 0.85 \tag{18.4}$$

The study of the deformations was carried out up to $\lambda = 0.82$, the value of λ at which the necking of the section started to form at the edge of the hole.

To check the results obtained with the moire method, comparison was made for the elastic case ($P = 6500$ lb) with the theoretical results obtained by

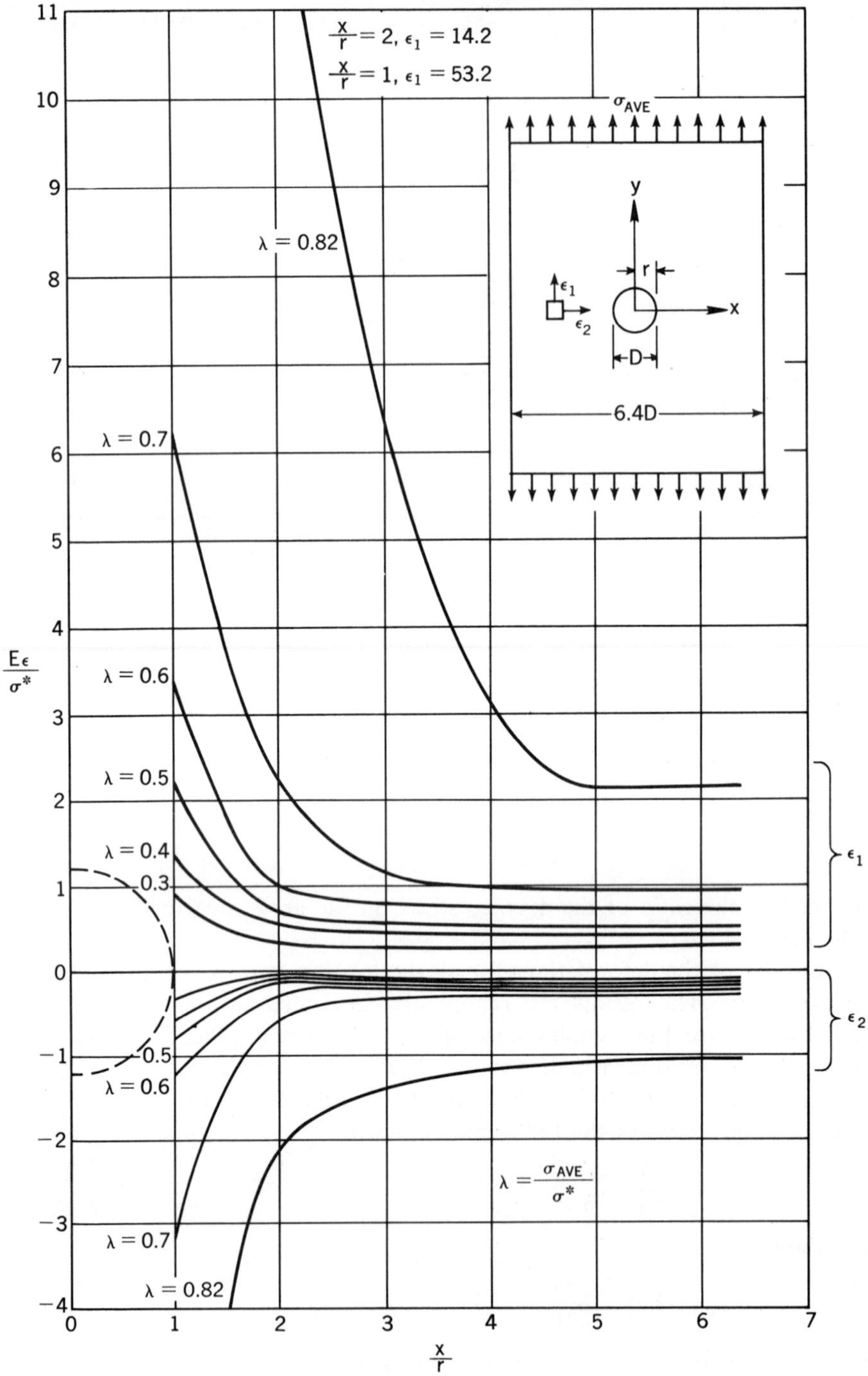

Figure 18.3. Plate with a circular hole. Principal strains on the transverse axis for different levels of plastic deformation.

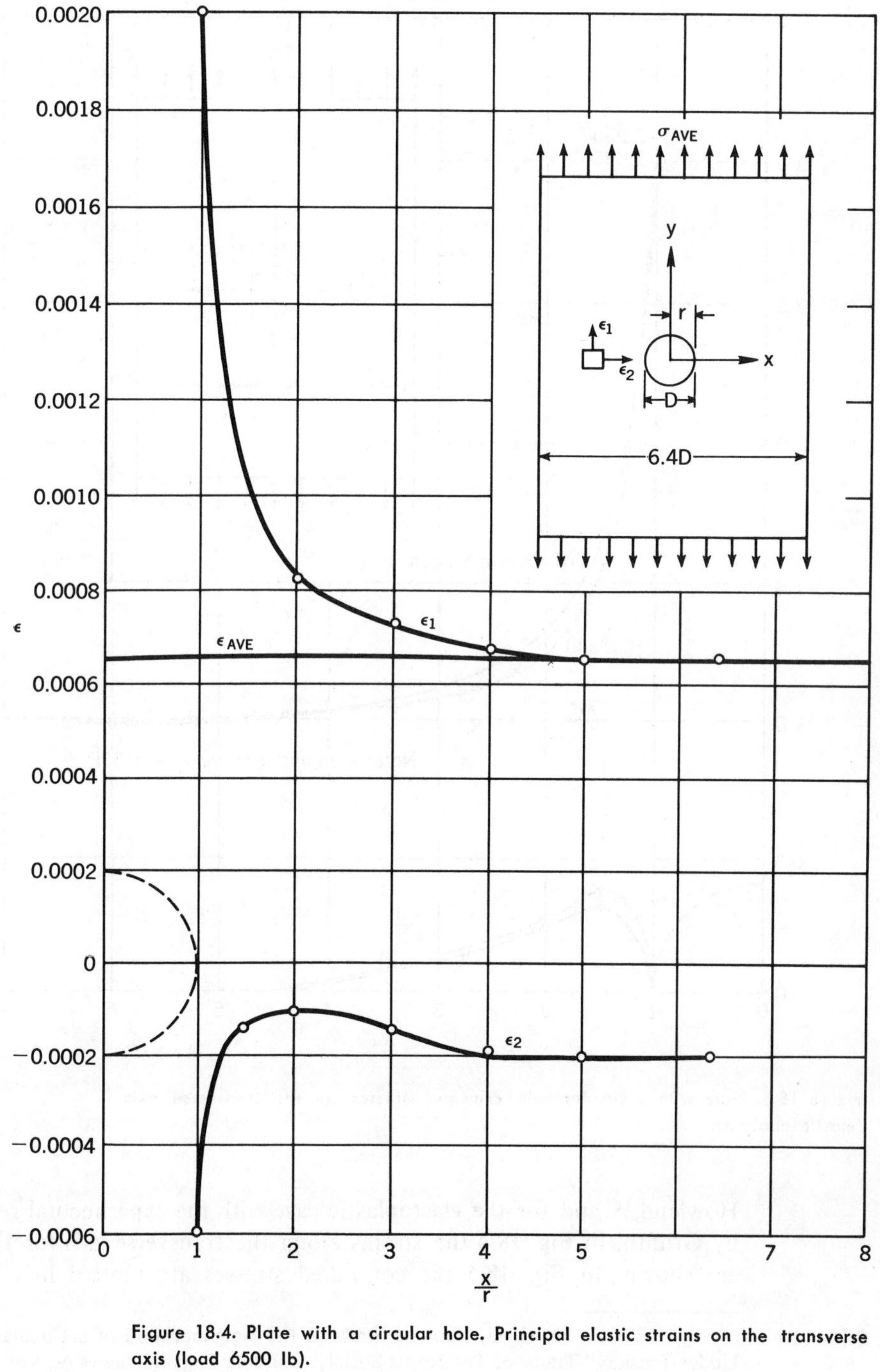

Figure 18.4. Plate with a circular hole. Principal elastic strains on the transverse axis (load 6500 lb).

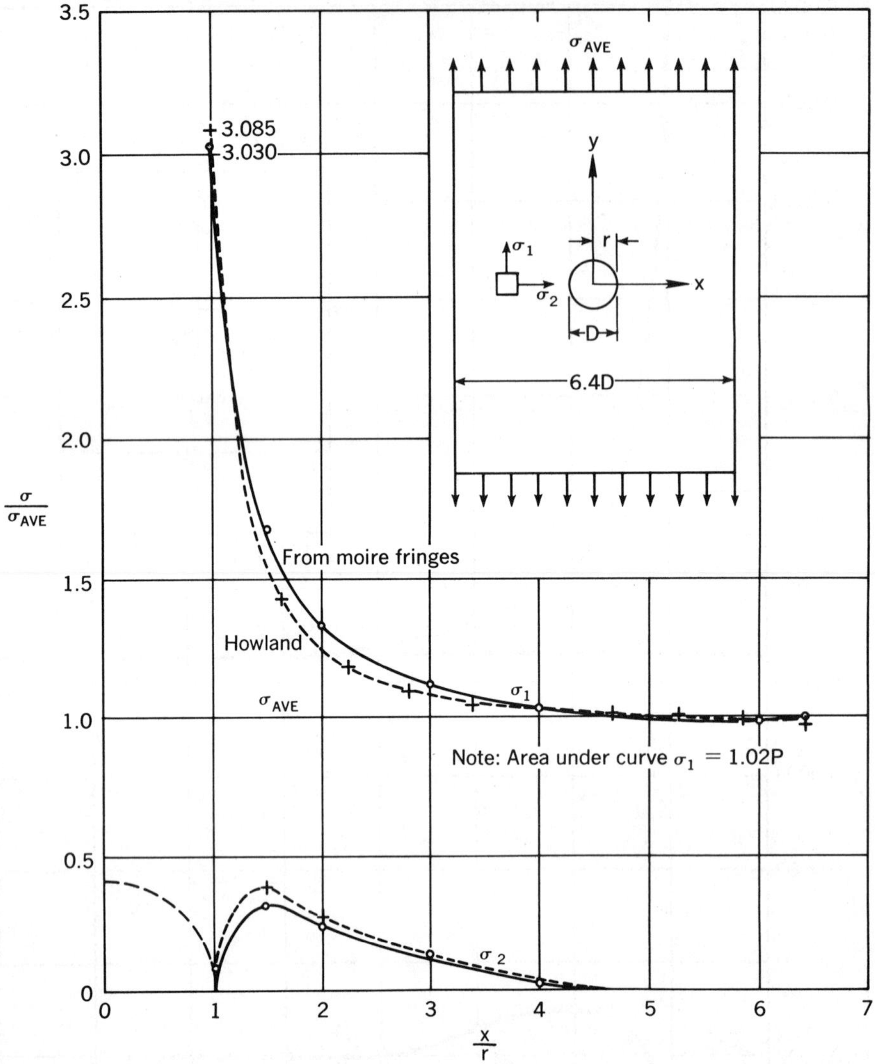

Figure 18.5. Plate with a circular hole. Principal stresses on the transverse axis (elastic analysis).

Howland,[10] and for the elastoplastic case with the experimental results obtained by Griffith. In Fig. 18.4 the strains along the transverse axis for the elastic case are shown. In Fig. 18.5 the computed stresses are plotted in a dimensionless

[10] R. C. J. Howland, "On the Stress in the Neighborhood of a Circular Hole in a Strip Under Tension," Trans. of The Royal Society, London, England, Series A., Vol. 229 (1929–1930), pp. 49–86.

form. The agreement of experimental and theoretical values is satisfactory. The maximum predicted strain at the boundary of the hole is 3.08 and the measured value is 3.03. The results obtained by A. Vinckier and R. Dechaene, (Footnote 5, this chapter), using the moire method, cannot be compared with the results of the test reported here because the ranges of deformations and the materials used are different.

For a rigorous comparison of the moire results with those of Griffith, it is necessary to point out that, to compare results in the plasticity field, the following conditions must be fulfilled: (1) The geometry must be similar, (2) the yield condition for both materials must be the same, (3) the stress–strain curves expressed in dimensionless form should be identical, and (4) Poisson's ratio must be the same.

Figure 18.6. Plate with a circular hole. Comparison of the moire test results with Griffith's results (for slightly different material and geometry).

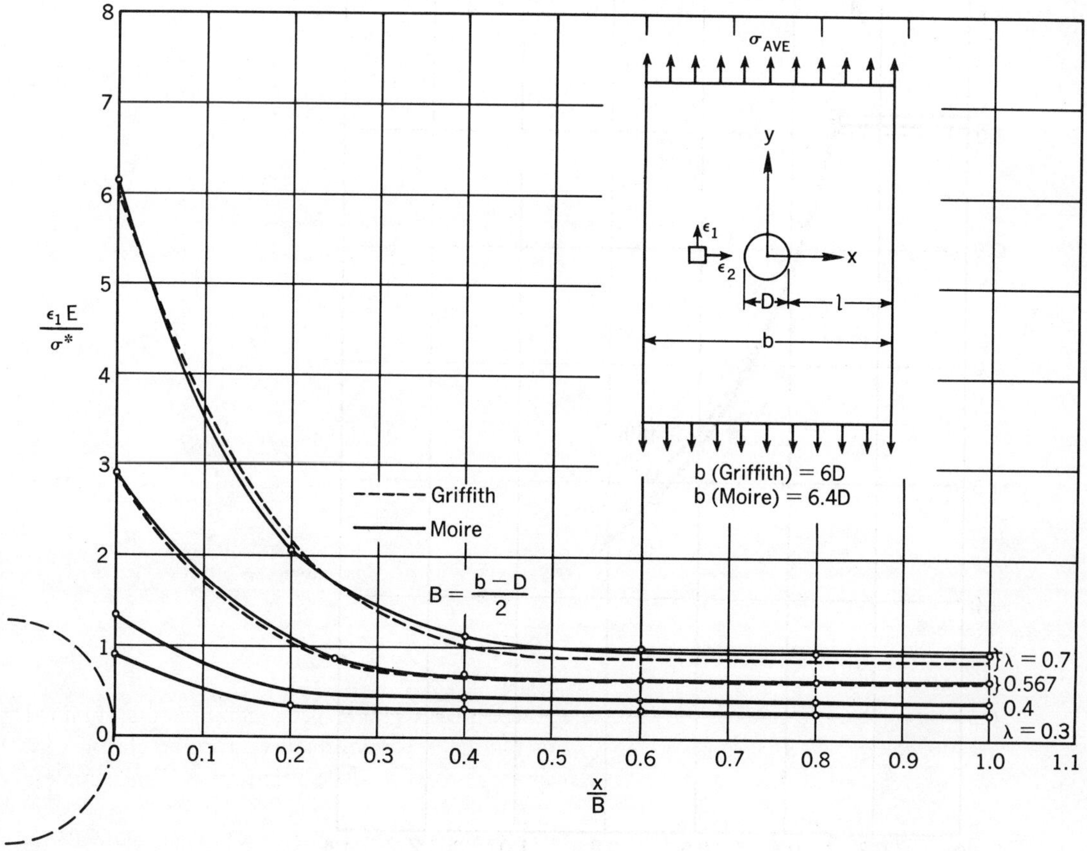

In Griffith's test and in this test, the material used was aluminum. Therefore, conditions (2) and (4) are fulfilled. From Fig. 18.3, it may be seen that neither condition (3) nor condition (1) is exactly satisfied, because the ratio b/D, where b is the width of the plate and D the diameter of the hole, is equal to 6 for the Griffith test and equal to 6.4 in the present test.

A comparison has been established, however, with the aim of providing criteria for the evaluation of the results obtained. This comparison is shown in Fig. 18.6. Abscissas are the dimensionless coordinates x/B, where

Figure 18.7. Plate with a circular hole. Stress concentration factor as a function of the degree of plastic deformation λ.

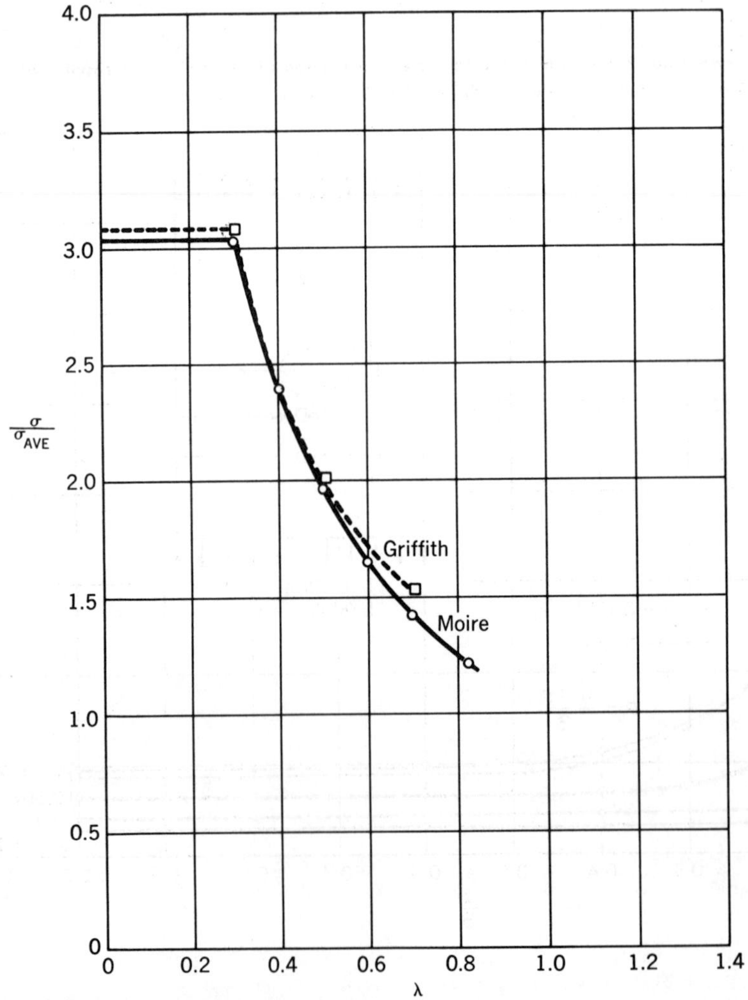

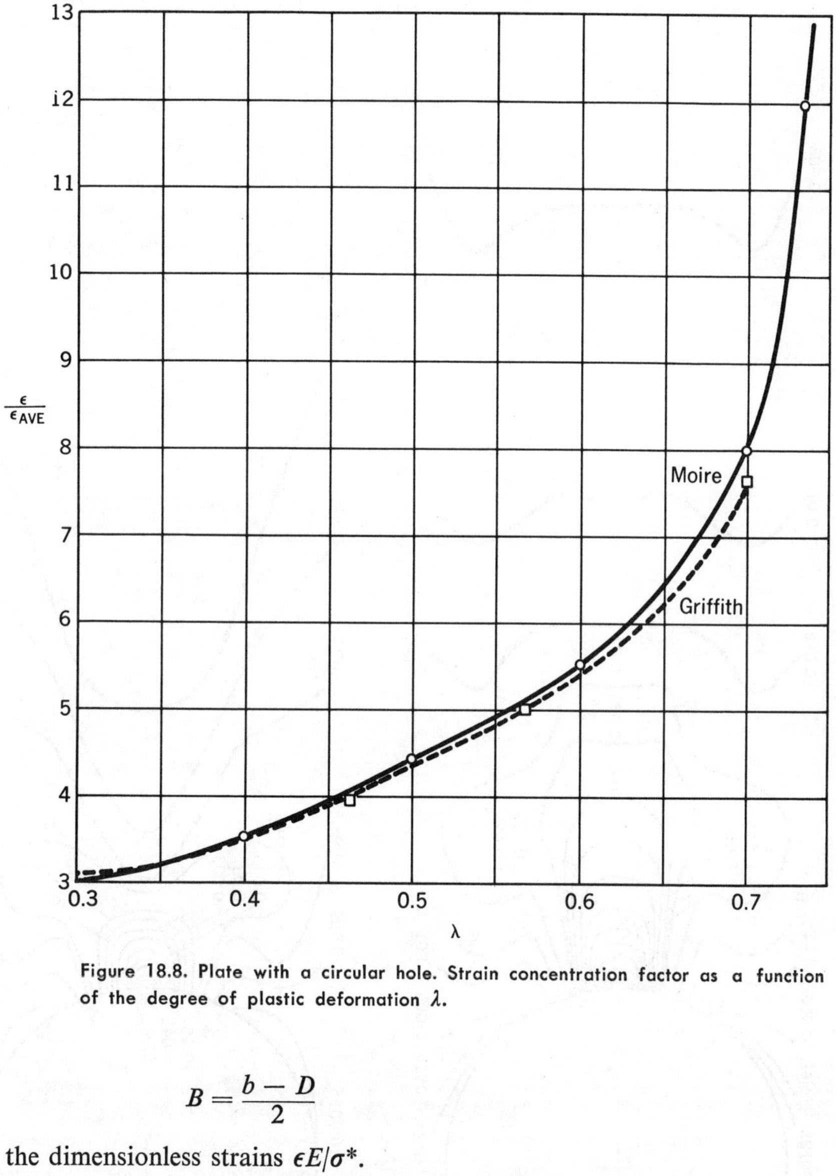

Figure 18.8. Plate with a circular hole. Strain concentration factor as a function of the degree of plastic deformation λ.

$$B = \frac{b - D}{2}$$

and ordinates are the dimensionless strains $\epsilon E/\sigma^*$.

For $\lambda = 0.3$, for $\lambda = 0.4$, and for $\lambda = 0.567$, the results of both tests were practically coincident. For $\lambda = 0.7$, the difference is larger, as should be expected. Whereas, in the present test, the complete section yielded for $\lambda = 0.7$, in the Griffith test part of the section remained elastic. The change of strain- and stress-concentration factors as functions of λ can also be compared with Griffith's results (Figs. 18.7 and 18.8). The strain-concentration factors obtained by Griffith are somewhat smaller than those obtained by the moire method. This should be

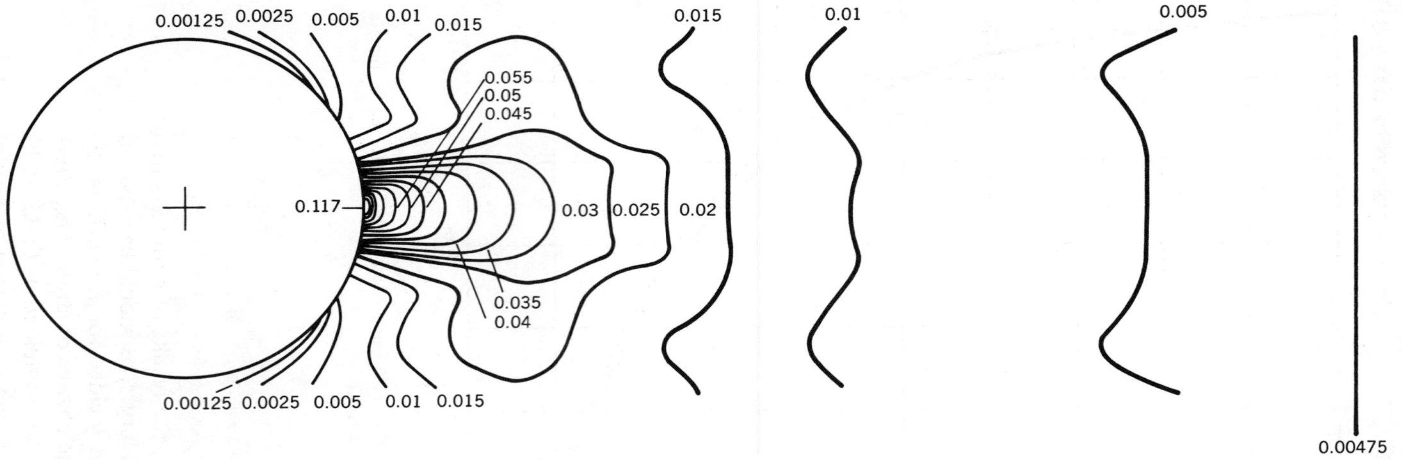

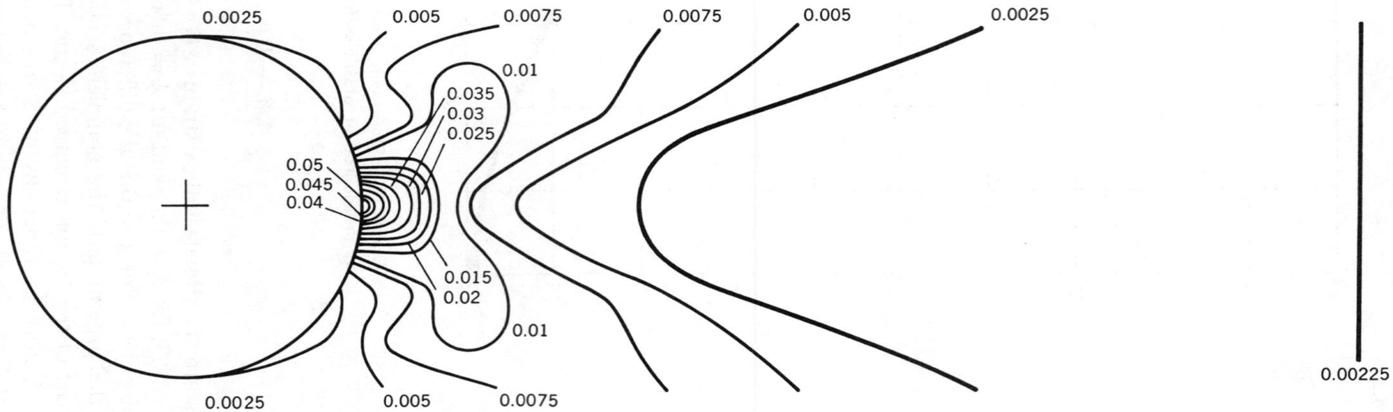

Figure 18.9. (Top) Plate with a circular hole. Family of the isotenic lines giving the ϵ_1 strains for the degree of plastic deformation $\lambda = 0.82$.

Figure 18.10. (Bottom) Plate with a circular hole. Family of the isotenic lines giving the ϵ_2 strains for the degree of plastic deformation $\lambda = 0.82$.

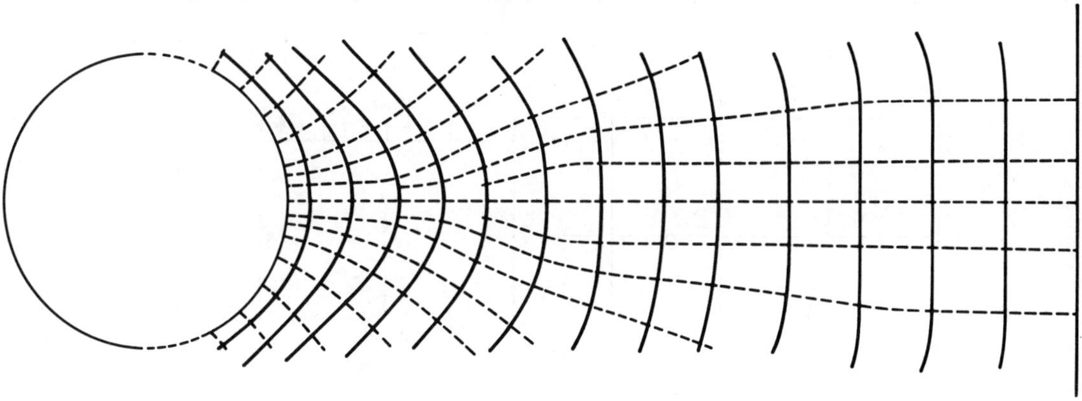

Figure 18.11. Plate with a circular hole. Isostatics for the degree of plastic deformation $\lambda = 0.82$.

expected, if it is taken into account that the values obtained by Griffith are averaged over the base of the extensometer used.

The complete field of strains in a region bounded by the hole—the external boundary of the plate, and two lines tangent to the hole and parallel to the transverse axis—was determined for $\lambda = 0.82$. For this purpose, a net of sections parallel respectively to the longitudinal and transverse axes was formed. In the neighborhood of the hole, a finer net was used; at each point of the net, the direct derivatives and the cross-derivatives in the x- and y-directions were computed. From these data, principal strains and principal directions were calculated. The results are presented as families of isotenic lines in Figs. 18.9 and 18.10. Isostatic lines are shown in Fig. 18.11.

18.4. Computations of Stresses on Transverse Axis

When the yield stress is reached, the instantaneous deviator of strains is proportional to the deviator of stresses, and the stress–strain relationship has a differential form. The significant values are the velocities of strain, rather than the strains themselves. To obtain the curves of velocities of strain, curves of $\epsilon = f(\lambda)$ for $x/r = 1, 2, 3 \ldots 6$ were plotted.

From these curves, by graphic differentiation with respect to the parameter λ, the curves $\dot{\epsilon} = f'(\lambda)$ were obtained. In Figs. 18.12 and 18.13, these curves are represented.

To compute the stresses, the Prandlt–Reuss stress–strain relationships were used. For purposes of computation, a convenient form of these equations for the case of the strain-hardening material was derived. A strain-hardening material was considered, in spite of the fact that the aluminum used has a well-

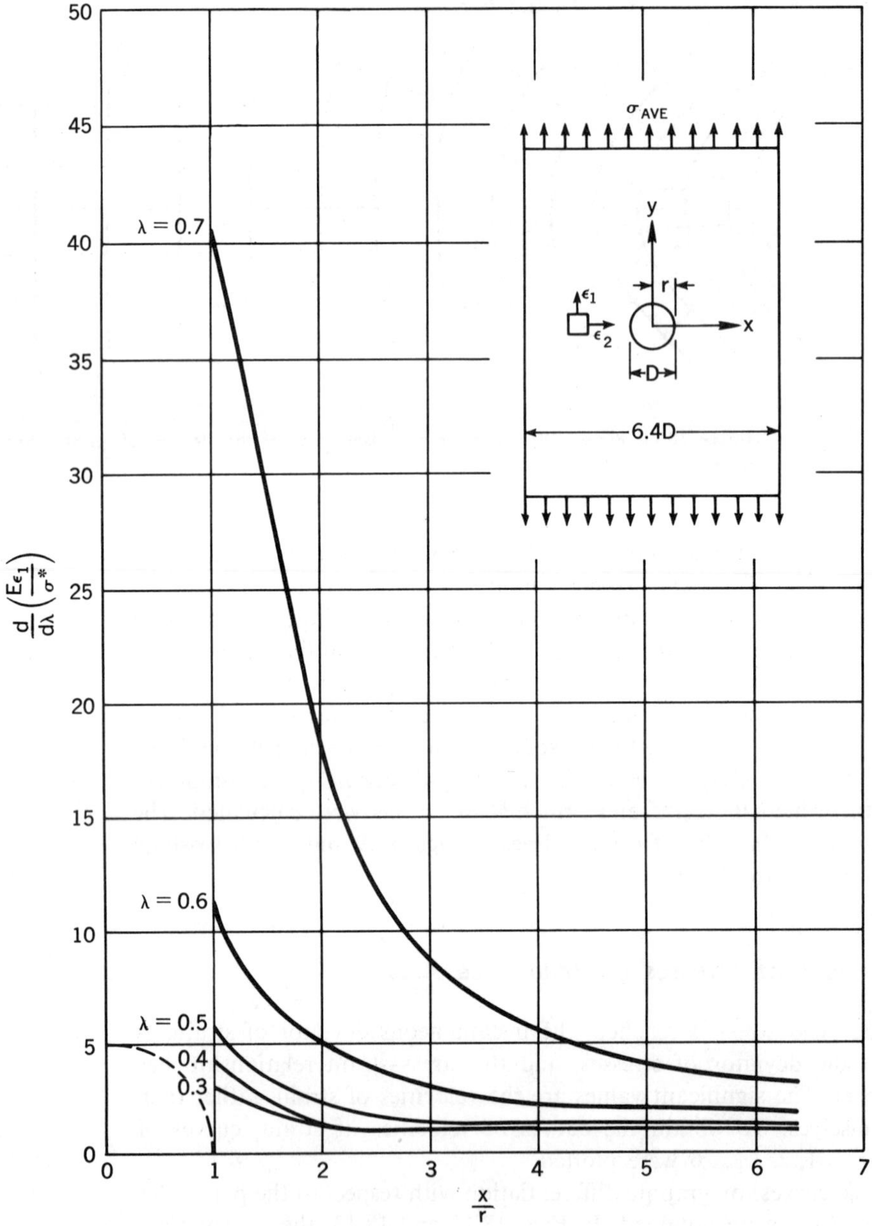

Figure 18.12. Plate with a circular hole. Dimensionless vertical velocities of strain along the transverse axis.

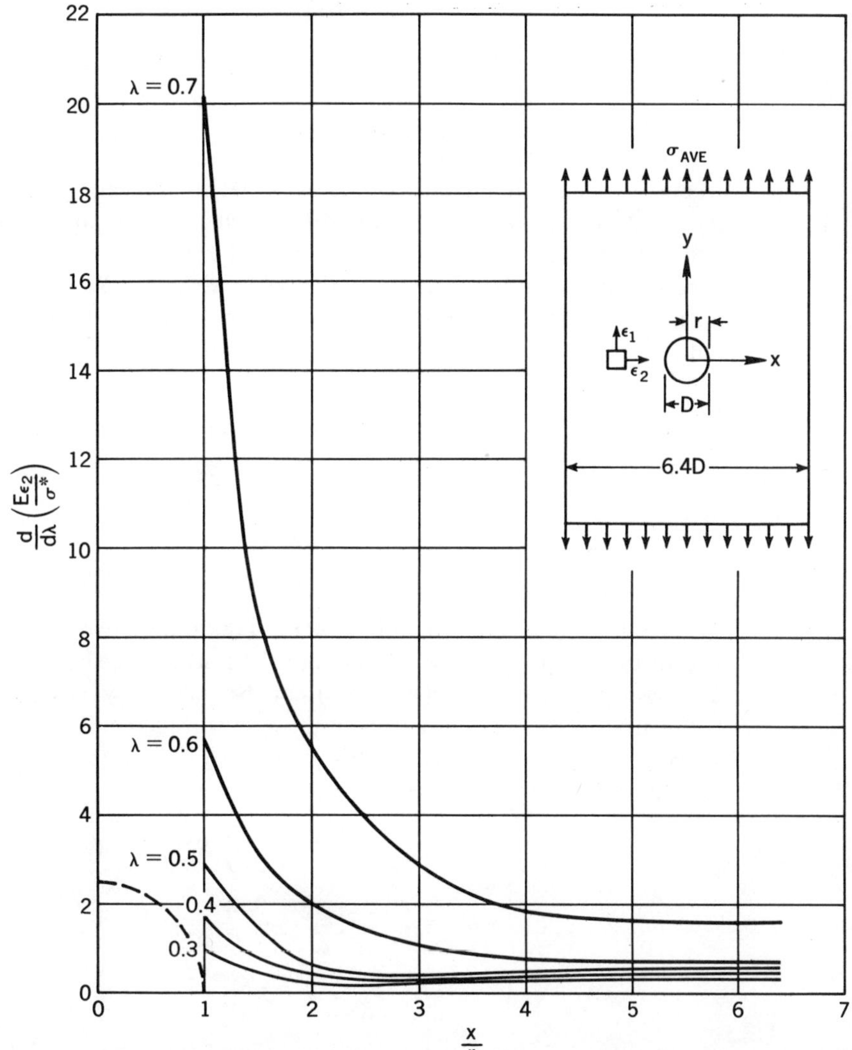

Figure 18.13. Plate with a circular hole. Dimensionless horizontal velocities of strain along the transverse axis.

defined yield stress, to avoid the use of an ideal stress–strain relationship composed of two straight lines.

Assuming that the rate of plastic work is large compared to the rate of elastic work, the Prandtl–Reuss equations take the following form:

$$\dot{S}_1 = 2G\left(\dot{e}'_1 - \frac{3}{2}\frac{\dot{W}}{\bar{\sigma}^2}S_1\right)$$

$$\dot{S}_2 = 2G\left(\dot{e}'_2 - \frac{3}{2}\frac{\dot{W}}{\bar{\sigma}^2}S_2\right)$$
$$\dot{S}_3 = 2G\left(\dot{e}'_3 - \frac{3}{2}\frac{\dot{W}}{\bar{\sigma}^2}S_3\right) \tag{18.5}$$
$$\dot{S} = 3K\dot{e}$$

where $\dot{S}_1$, $\dot{S}_2$, $\dot{S}_3$ are the rates of the principal components of the deviatoric-stress tensor; $\dot{e}'_1$, $\dot{e}'_2$, and $\dot{e}'_3$ are the rates of the principal components of the deviatoric strain tensor; $\bar{\sigma}$ is the effective stress; $\dot{W}$ is the rate at which the stresses do work in connection with the change of shape ($\dot{W} = \dot{e}'_1 S_1 + \dot{e}'_2 S_2 + \dot{e}'_3 S_3$); S is the mean normal stress; $\dot{e}$ is the mean normal strain rate; and G and K are the shear and bulk modulus, respectively. Since a two-dimensional stress field is being considered,

$$S_1 = \sigma_1 - S$$
$$S = \frac{\sigma_1 + \sigma_2}{3} \tag{18.6}$$
$$S_3 = -S$$
$$\dot{S}_3 = -\dot{S}$$

Of the first three equations of Eq. (18.5), only two are independent. Only two of them (for example, the first and the last) are needed to obtain a solution. The system of equations (Eq. 18.5) can be solved by an iteration procedure, if the strain rates are known, as in the case considered. To perform the iteration process, it is convenient to write the system of equations (Eq. 18.5) in the following way:

$$S_1 = \int_{\lambda_1}^{\lambda_2} 2G\left[\dot{e}'_1 - \frac{3}{2}\frac{\dot{W}}{\bar{\sigma}^2}S_1\right] d\lambda + C_1$$
$$S_3 = \int_{\lambda_1}^{\lambda_2} 2G\left[\dot{e}'_3 - \frac{3}{2}\frac{\dot{W}}{\bar{\sigma}^2}S_3\right] d\lambda + C_2 \tag{18.7}$$
$$\dot{e} = -\frac{\dot{S}_3}{3K}$$

where C_1 and C_2 are constants to be obtained from the initial conditions.

It was assumed that the material starts to yield when a strain equal to 0.0015 is reached in pure tension. This corresponds approximately to the point where the stress–strain curve departs from the straight line. Following the strain history at a point, the value of λ associated with yielding at that point was determined. After this value of λ was obtained, Eqs. (18.7) were used. For the first approximation in the iteration process, values of σ_1 and σ_2 were assumed, from which values of S_1 and S_3 were computed from eqs. 18.6. If the assumed values were correct, a graphical integration in the corresponding interval (λ_1, λ_2) would give the assumed values. In general, some difference was obtained. Once the first approximation was computed, a second approximation was calculated by replacing

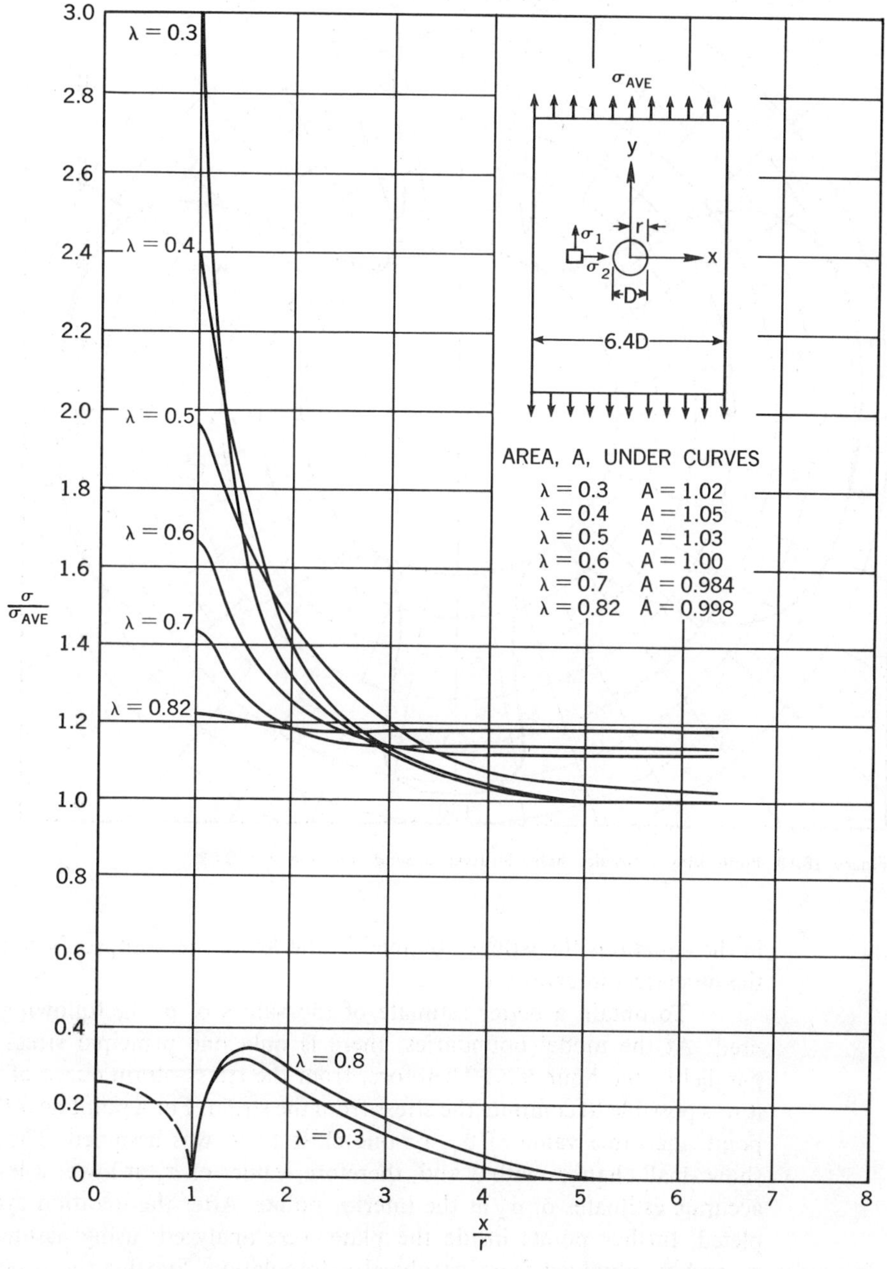

Figure 18.14. Plate with a circular hole. Plastic stress distribution along the transverse axis.

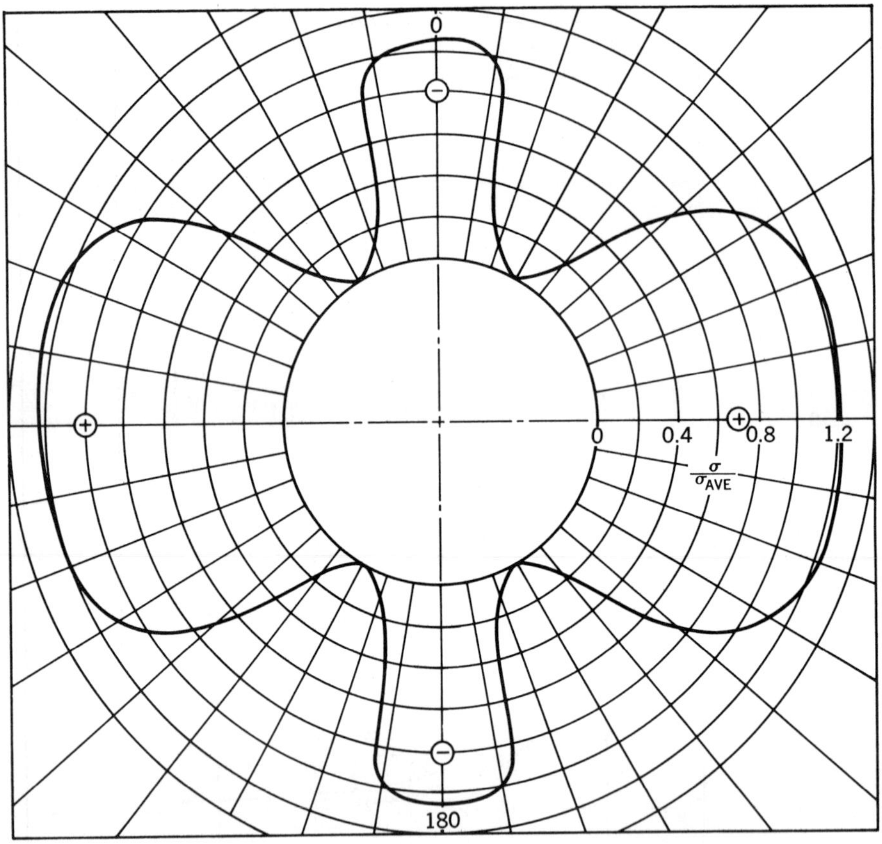

Figure 18.15. Plate with a circular hole. Stresses around the hole $\lambda = 0.82$.

in the equations the values obtained in the first cycle. Simpson's rule was used in the numerical integration.

To obtain a better estimate of the values of σ, the following process was used: At the model boundaries, there is only one principal stress σ_1, which is parallel to the boundary. Therefore, from the stress–strain curve of the material, it was possible to compute the stress from the strain. For a point near this boundary point, the same value of σ_1, or one close to it, was assumed. The values of σ_2 show small changes with λ and, therefore, values of σ_2 at lower λ levels provided accurate estimates of σ_2 at the interior points. After the iteration cycle was completed, further points inside the plate were analyzed, using assumed values of σ_1 and σ_2 obtained from graphical extrapolation. Starting the process from the two boundaries, a sufficient number of points can be determined to use graphical interpolation in order to obtain at the remaining points the values of σ_1 and σ_2 required to initiate the iteration cycle.

The results obtained are plotted in Fig. 18.14. The integrated areas of the σ_1 curves check the equilibrium with small errors. Although σ_1 tends to become uniform, σ_2 changes almost in proportion to the average stress. In order to make the plot clearer, only the values of σ_2 for the extreme values of $\lambda = 0.3$ and $\lambda = 0.82$ have been represented. There is no other means of judging the results obtained than the equilibrium check. Stowell[11] has applied a deformation-theory solution of the infinite plate with a hole. This solution does not satisfy the compatibility conditions. Budiansky and Vidensek[12] have also solved this problem, employing a deformation theory, but with a more rigorous theoretical basis. Thompson[13] has studied the stress distribution in a finite plate with a hole using photoelasticity. The results obtained correspond to another geometry and to a different type of material and therefore cannot be compared with this test. The values of the stress at the edge of the hole, in a dimensionless form, are plotted in Fig. 18.15. It is interesting to note that the boundary has the point of zero stress at the same position as in the elastic case.

---

[11] E. Z. Stowell, "Stress and Strain Concentration at a Circular Hole in an Infinite Plate," NACA TN 2073 (1950).

[12] Budiansky and Vidensek, *op. cit.*

[13] Thompson, *op. cit.*

19

DETERMINATION OF FINITE STRAINS IN A CIRCULAR RING SUBJECTED TO DIAMETRAL COMPRESSION

19.1. Introduction

Strain distributions in bodies subjected to large deformations have been determined in very few cases, among which the somewhat simple problems solved by Rivlin[1] should be mentioned. The theoretical solution of this kind of problem presents great difficulties and, as pointed out by Sokolnikoff,[2] the principle of superposition of effects, in general, is not applicable. Recent developments in experimental methods, on the other hand, permit a relatively easy solution of at least the two-dimensional problems. A first attempt at the experimental solution of one of these problems was made by one of the authors, who analyzed the strains on the horizontal diameter of a circular disk subjected to large diametral compression.[3] It was shown in that paper that it is possible to use photoelasticity and thickness measurements to analyze large strains. In this chapter, three methods

---

1 R. S. Rivlin, "Large Elastic Deformations," Chapter 10 in *Rheology* (ed. F. R. Eirich) (New York: Academic Press, Inc., 1966).

2 I. S. Sokolnikoff, *Mathematical Theory of Elasticity*, 2nd ed. (New York: McGraw-Hill Book Company, 1956), p. 33.

3 A. J. Durelli and A. P. Mulzet, "Large Strain Analysis and Stresses in Linear Materials," *Journ. of the Eng. Mech. Div., Proc. ASCE*, Vol. 91 (June 1965), pp. 65–91.

are evaluated: grids, moire, and photoelasticity, with particular emphasis on the moire application. The methods are applied to the determination of strains at selected points in a thick ring diametrically loaded. The problem of the determination of large strains in such a ring has not been solved previously; however, nonlinear theoretical solutions are available for thin rings.[4,5] The experimental results presented here were obtained using a ring with a thickness of 0.35 of its outer radius. The material of the ring behaved in an approximately linear manner up to about 25 per cent conventional strain. However, in order to simplify the conversion from large strain to stresses, notice was taken of a phenomenological linearity in the material to even higher strains. It was noted and reported elsewhere[6] that, for some natural and synthetic rubber, the incremental stress–strain ratio was constant throughout the range of strains present in the ring. (The maximum at the boundary is about 40 per cent.) As natural strain is a total or summation of incremental strain, a corresponding definition of "natural stress" is used to allow the total stress to be computed from the total strain.[7]

19.2. Model and Loading

A ring of polyurethane rubber (Hysol 4485) having an outer diameter of 5 in., an inner diameter of 3.26 in., and $\frac{1}{2}$-in. thick was used. The ring was compressed diametrically between two platens. The load was initially applied at the two ends of the vertical diameter of the ring, and with increasing load the contact area increased as the top and bottom of the ring flattened out against the platens. Both load and displacement in the vertical direction were measured and are given in Fig. 19.1. A maximum vertical displacement of 2 in. was applied on the ring. The vertical displacement in dimensionless form is represented by λ (where λ is the ratio of vertical displacement to outer diameter), so that λ_{max} is 0.4.

From observation of Fig. 19.1, it is obvious that rings of this configuration, made of linear materials, can be compressed one-third of their original diameter and still have a linear relationship between load and diametral contraction.

19.3. Definition of Strain

From Table 4.1 the following three definitions of strain are transcribed:

1. Lagrangian strain ϵ^L, defined as the change in length of a fiber divided

---

[4] H. H. Pan, "Non-Linear Deformation of a Flexible Ring," *Quart. Journ. Mech. and Appl. Math.*, Vol. 15, Pt. 4 (1962), pp. 401–412.

[5] M. Schoenburg and N. H. Bleich, "Viscoelastic Rings with Finite Deflections," Columbia Univ., ONR Tech Report No. 39 (April 1966).

[6] V. J. Parks and A. J. Durelli, "Natural Stress," *Int. Journ. of Non-Linear Mech.* (March 1969).

[7] This problem has also been dealt with in A. J. Durelli, V. J. Parks, and T. L. Chen, "Stress and Finite-Strain Analysis of a Circular Ring under Diametral Compression," *Exp. Mech.*, Vol. 9, No. 5 (May 1969), pp. 210–214.

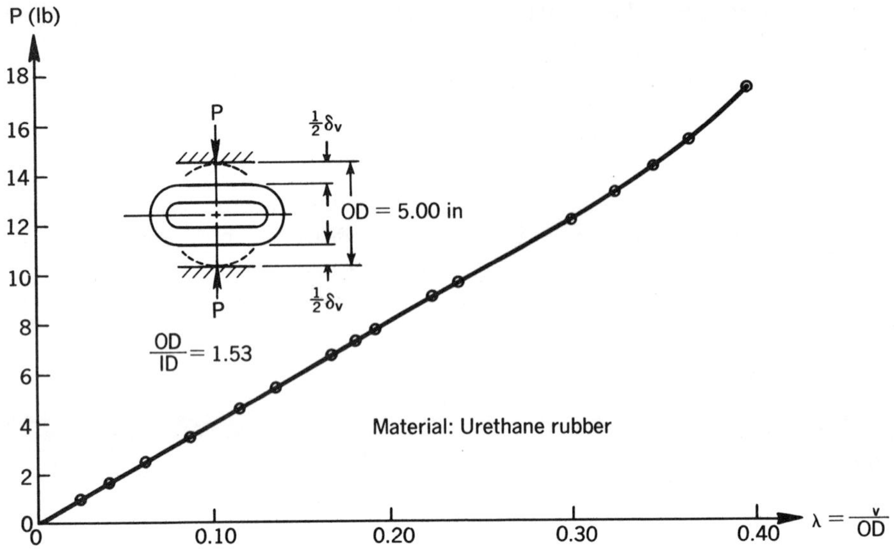

Figure 19.1. Relationship between applied load and applied displacement in a circular ring subjected to diametral compression.

by its initial length l_i (commonly associated with a coordinate system on the undeformed body). In symbolic form this is

$$\epsilon^L = \lim_{l_i \to 0} \frac{l_f - l_i}{l_i} \tag{19.1}$$

2. Eulerian strain ϵ^E, defined as the change in length of a fiber divided by its final length l_f (commonly associated with a coordinate system on the deformed body). In symbolic form this is

$$\epsilon^E = \lim_{l_f \to 0} \frac{l_f - l_i}{l_f} \tag{19.2}$$

3. Natural strain $\bar{\epsilon}$, most generally defined by the following mathematical expression

$$\bar{\epsilon} = \lim_{l \to 0} \lim_{\Delta l \to 0} \sum_{l_i}^{l_f} \frac{\Delta l}{l} \tag{19.3}$$

where l is the length of the fiber at any load level and Δl is the change in length due to a change of load at that load level. If the total loading which carries l from l_i to l_f is subdivided in such a manner that all the increments Δl approach zero, then the sum of the ratios $\Delta l/l$ will approach $\bar{\epsilon}$ and will be equal to $\bar{\epsilon}$ in the limit (may be associated with a coordinate system in either the undeformed or deformed body).

Each of the three definitions of strain specifies exactly the geometry of the deformation and, provided they refer to the same fiber as indicated in Sec. 4.3, each strain so defined has a fixed relationship with strain defined in the other two ways, as follows:

$$\bar{\epsilon} = \ln(1 + \epsilon^L) \tag{19.4}$$

$$\epsilon^L = \frac{\epsilon^E}{1 - \epsilon^E} \tag{19.5}$$

$$\bar{\epsilon} = \ln\left(\frac{1}{1 - \epsilon^E}\right) \tag{19.6}$$

However, each of the three definitions is particularly suited to one of the three methods of measurement, and all of them were used in the analyses described below.

The grid and moire measurements were made over finite base lengths and, therefore, give average values, over the base lengths.

19.4. *Methods of Measurement*

19.4.1. Grids

A Carteisan grid was marked on the undeformed ring. The distance between the lines of the grid was approximately 0.41 in. By measuring the change in distance between the lines, the average strain could be determined. This procedure suggests the use of the Lagrangian definition of strain; however, since both initial and final lengths were measured, the Eulerian strain can also be easily computed. With the grid method it is advantageous to use the Lagrangian definition of strain primarily because the Cartesian grid is put on the undeformed body and, as such, the strains computed along the lines of the grid are most easily identified with their undeformed position. If Eulerian strains are of particular interest, it would be possible to put the Cartesian grid on the deformed body and unload.

Figure 19.2 shows the grid on the ring with no load and at 8 load levels. Average strains between grid lines were computed using the finite form of Eq. (19.1). Figures 19.3, 19.4, and 19.5 give the distribution of the strains along the inner boundary (ϵ_t^L), a selected horizontal line ϵ_x^L, and a selected vertical line ϵ_y^L, for $\lambda = 0.1$, 0.2, 0.3, and 0.4. The curves were obtained by plotting the strains at the midpoint of each interval of the grid and passing a smooth curve through them. The strains tangential to the boundary were obtained from length measurements of segments of the boundary defined by the intersection of the boundary with the grid lines.

19.4.2. Moire

A grating of densely spaced parallel lines was marked on the undeformed body. The interference or isothetic pattern, produced by the distorted grating of

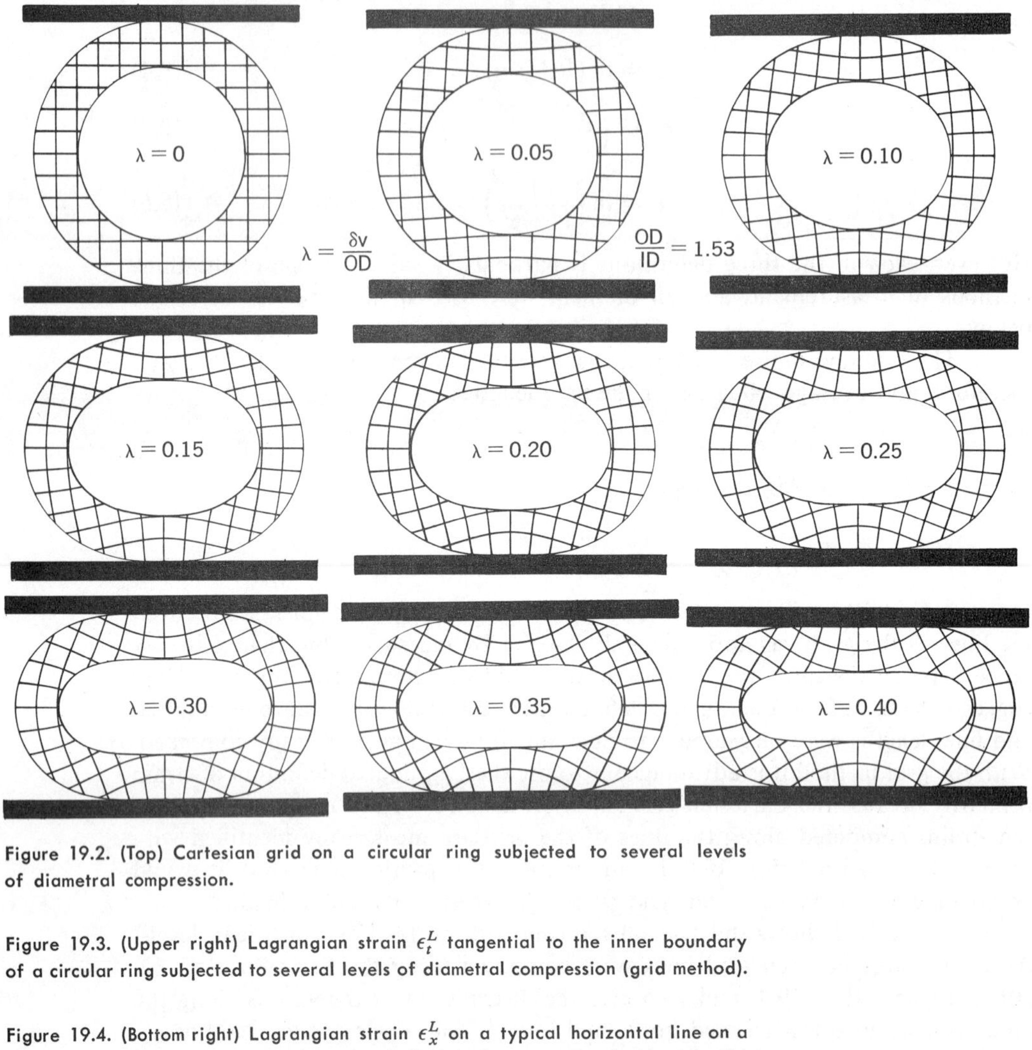

Figure 19.2. (Top) Cartesian grid on a circular ring subjected to several levels of diametral compression.

Figure 19.3. (Upper right) Lagrangian strain ϵ_t^L tangential to the inner boundary of a circular ring subjected to several levels of diametral compression (grid method).

Figure 19.4. (Bottom right) Lagrangian strain ϵ_x^L on a typical horizontal line on a circular ring subjected to several levels of diametral compression (grid method).

the deformed body and an undistorted master grating (both of which have the same pitch and orientation before deformation), gives for all points in the field the component of the displacement of the points along a direction perpendicular to the master-grating lines.

At each level of load, two isothetic patterns were obtained from two systems of 1000-lpi grating lines, which on the undeformed body were in the vertical and horizontal directions, respectively. Figure 19.6 is an example of the moire patterns obtained at $\lambda = 0.089$. These and the similar pairs of isothetics

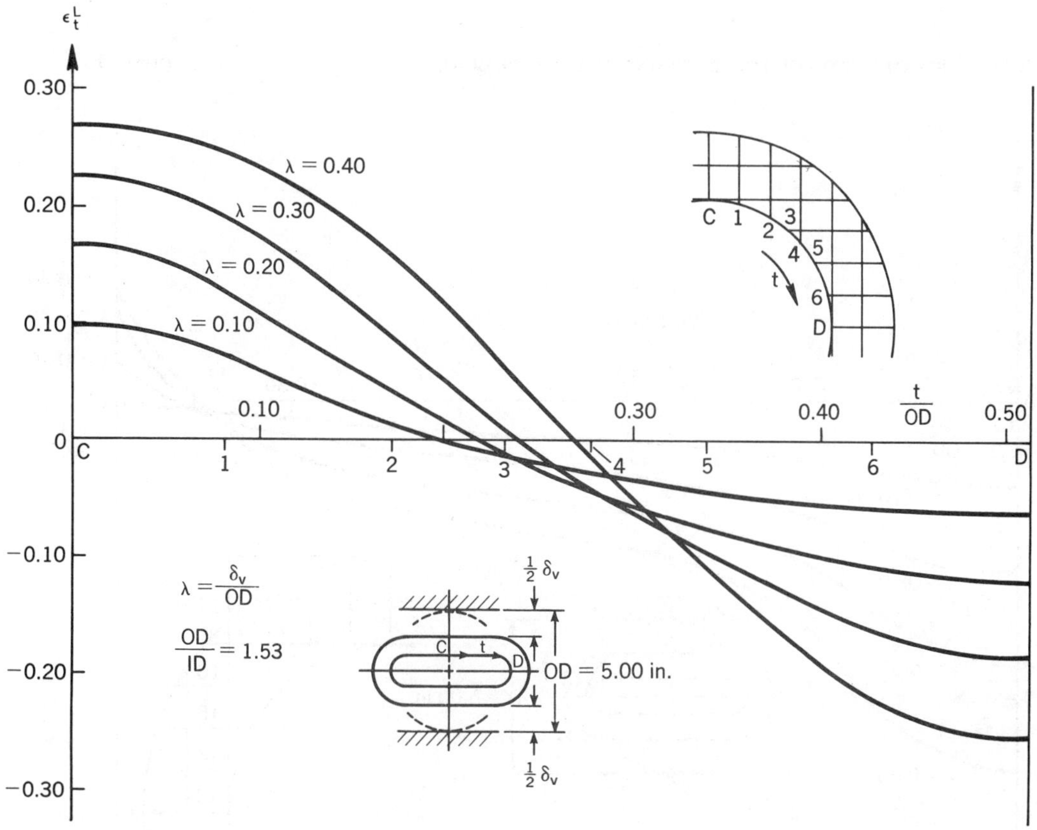

ϵ_t^L
0.30
0.20
0.10
0
−0.10
−0.20
−0.30
λ = 0.40
λ = 0.30
λ = 0.20
λ = 0.10
0.10
0.30
0.40
0.50
t/OD
C
1
2
3
4
5
6
D
t
λ = δv/OD
OD/ID = 1.53
½ δv
OD = 5.00 in.

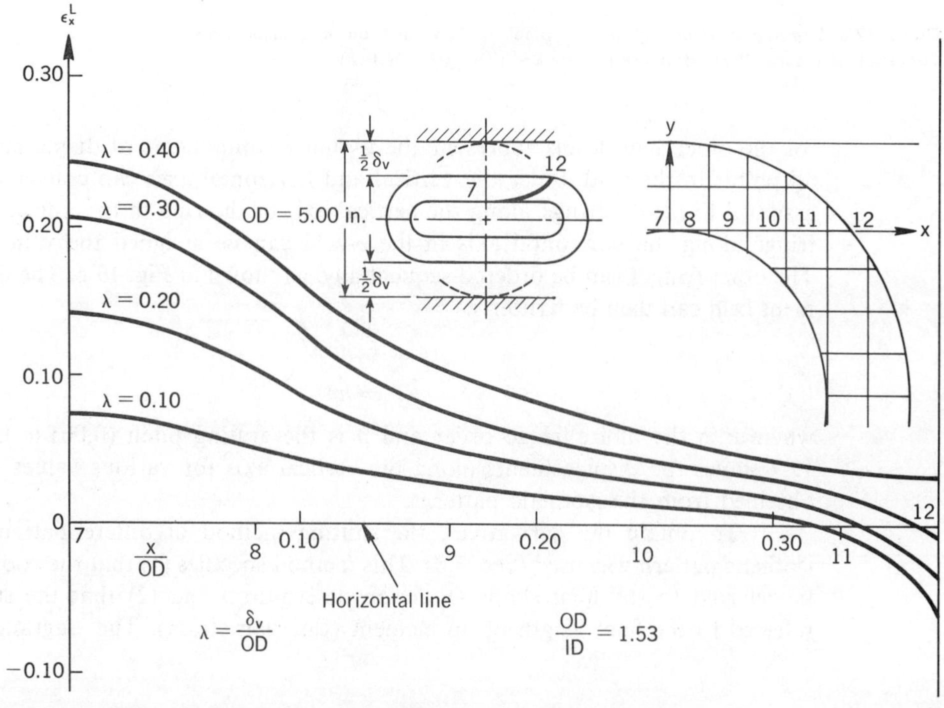

ϵ_x^L
0.30
0.20
0.10
0
−0.10
λ = 0.40
λ = 0.30
λ = 0.20
λ = 0.10
½ δv
OD = 5.00 in.
7
12
y
x
7
8
9
10
11
12
x/OD
0.10
0.20
0.30
Horizontal line
λ = δv/OD
OD/ID = 1.53

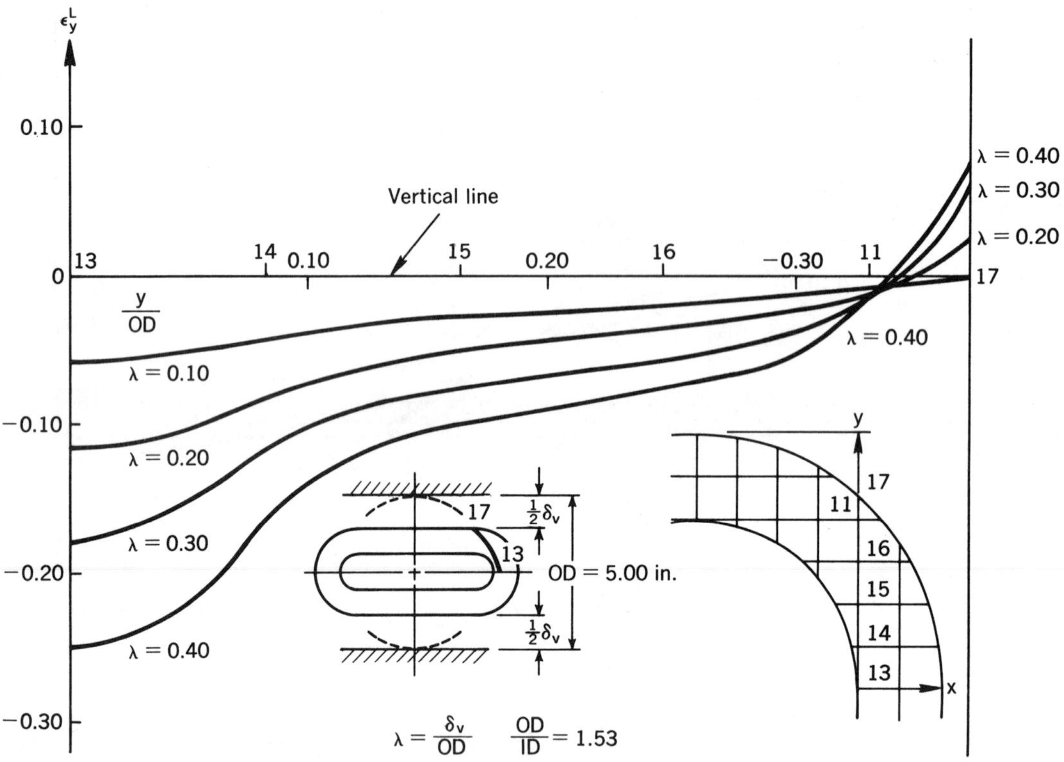

Figure 19.5. Lagrangian strain ϵ_y^L on a typical vertical line on a circular ring subjected to several lines of diametral compression (grid method).

for the other load levels represent the u- and v-components of displacement for all points in the field. Since the vertical and horizontal axes can conveniently be assumed fixed, the fringe along the vertical axis of the ring in the u-field and the fringe along the horizontal axis in the v-field can be assigned the value $n = 0$. The other fringes can be ordered sequentially, as shown in Fig. 19.6. The displacement field can then be written as

$$\begin{aligned} u &= np \\ v &= np \end{aligned} \tag{19.7}$$

where n is the moire-fringe order and p is the grating pitch (0.001 in.). Figure 19.7 shows the displacements along the vertical axis for various values of λ, as obtained from the isothetic patterns.

To obtain the derivatives, the shifting method of differentiation of the isothetic pattern was used (Sec. 7.8). This method specifies (1) that the coordinates be referred to the final shape (Eulerian description) and (2) that the strain be referred to the final length of an element (Eulerian strain). The Lagrangian description could also be used if a grating were marked on the deformed body, the body then unloaded, and the differentiation done by shifting the isothetics as before. Two partial derivatives of each displacement component can be obtained by shifting each isothetic pattern on its duplicate copy, in the two orthogonal directions. The four shifted patterns for 7 load levels are shown in Fig. 19.8. If the fringe orders in these moire-of-moire patterns are given the symbol N, then the

Figure 19.6. Illustrative example of u- and v-field isothetic pattern of a circular ring subjected to diametral compression.

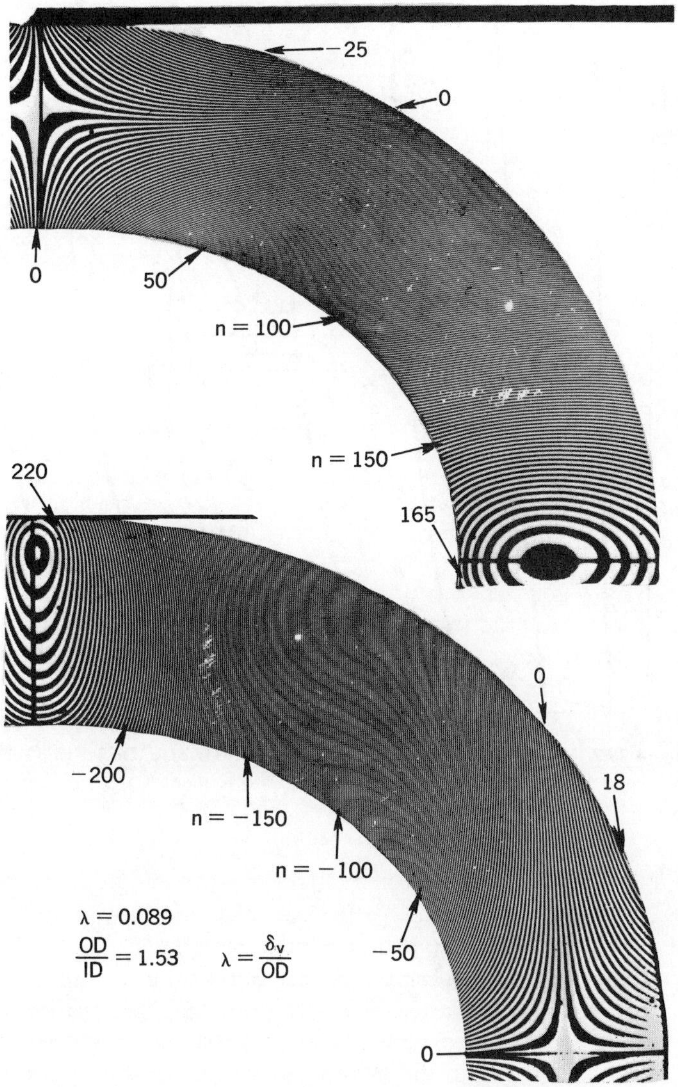

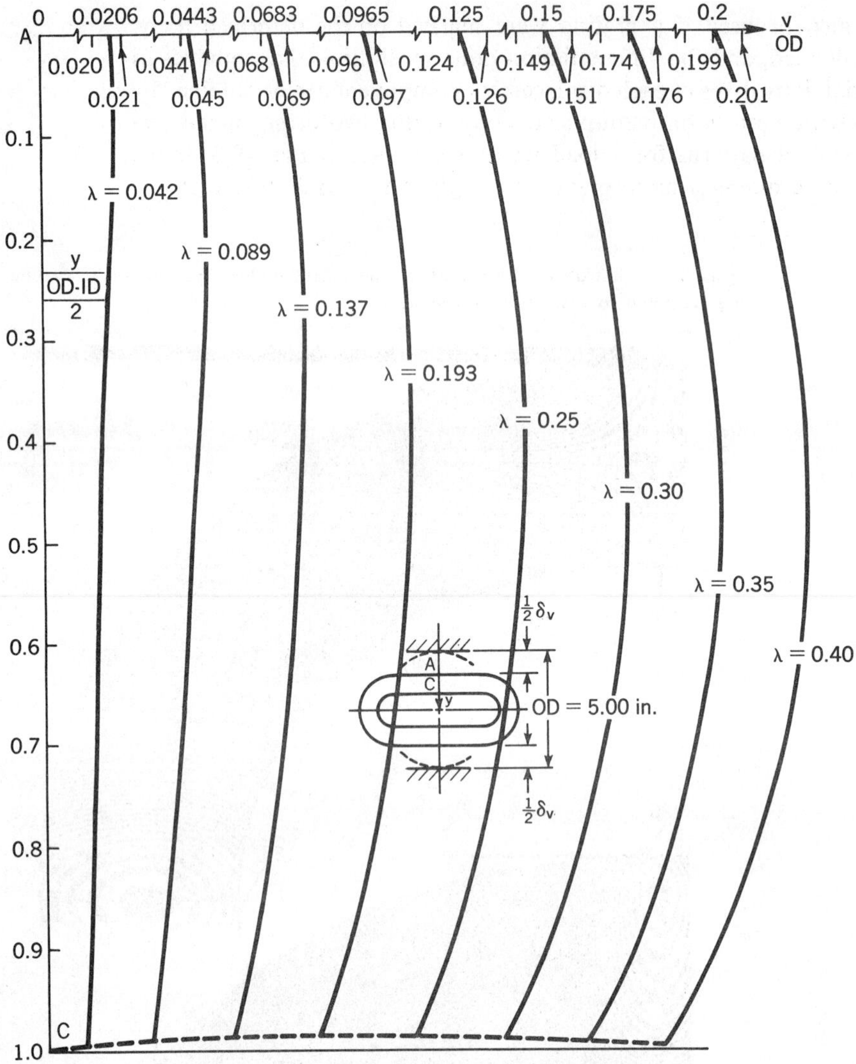

Figure 19.7. Vertical displacement of the vertical axis of a circular ring subjected to several levels of diametral compression (moire method).

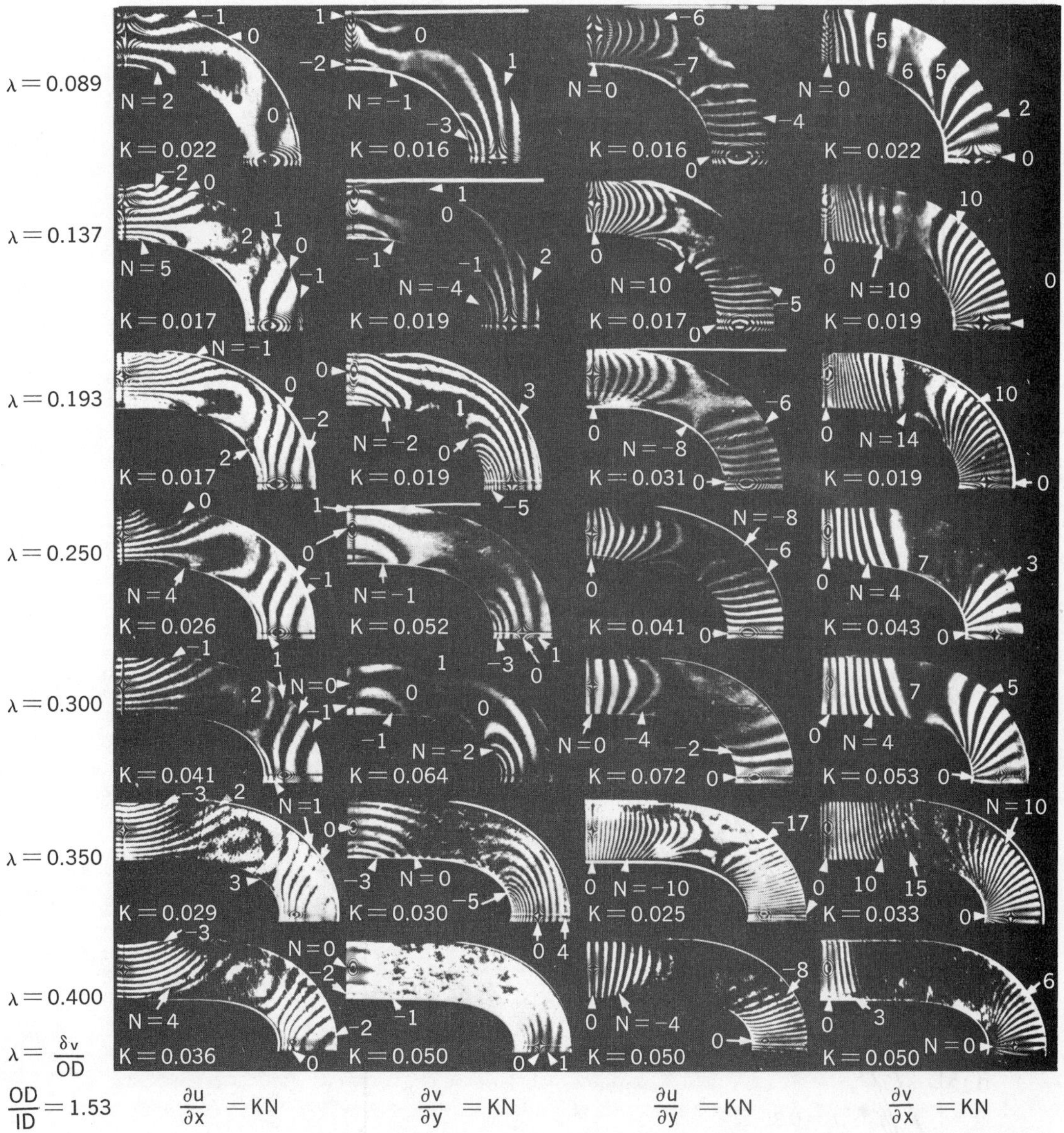

Figure 19.8. The four isoparagogics in a circular ring subjected to several levels of diametral compression (moire method).

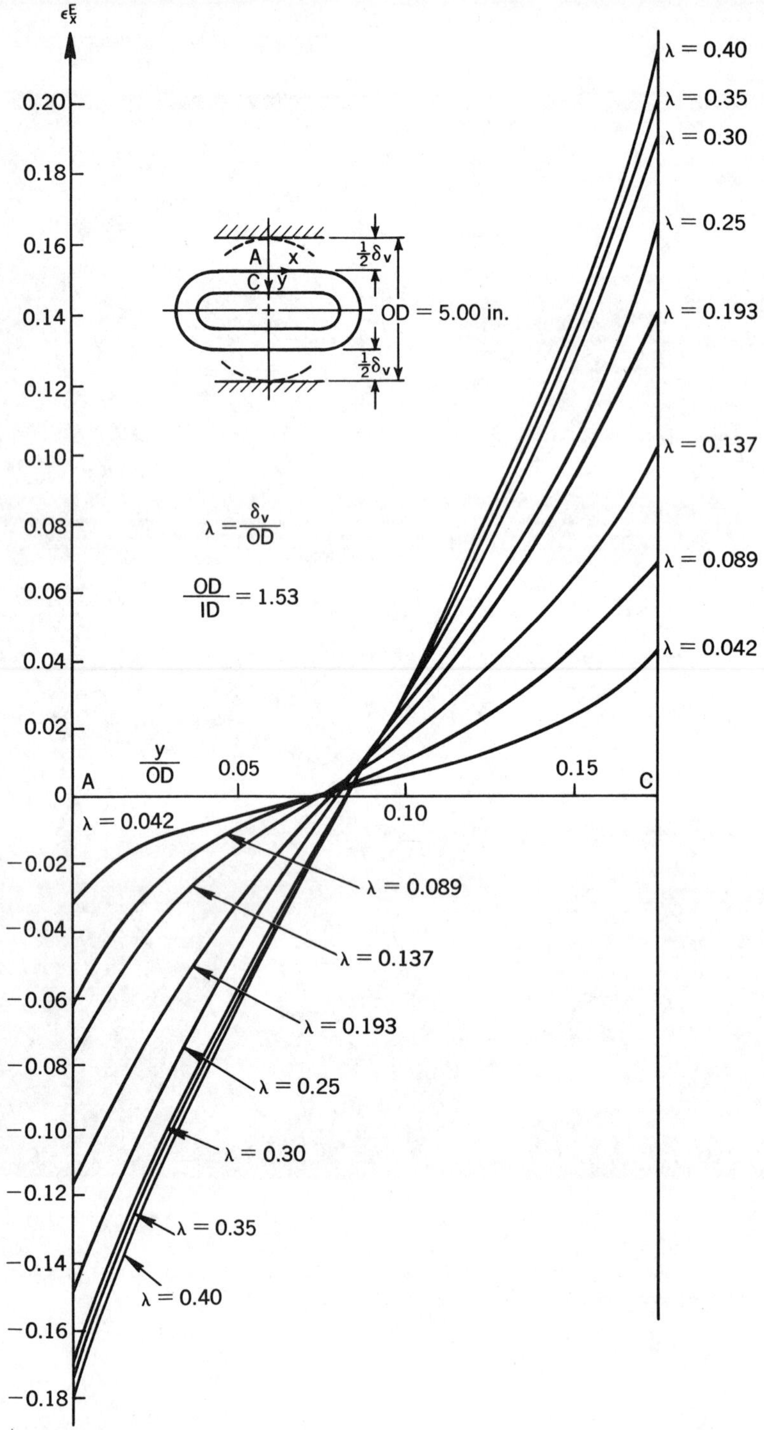

Figure 19.9. Eulerian horizontal strain ϵ_x^E on the vertical axis of a circular ring subjected to several levels of diametral compression (moire method).

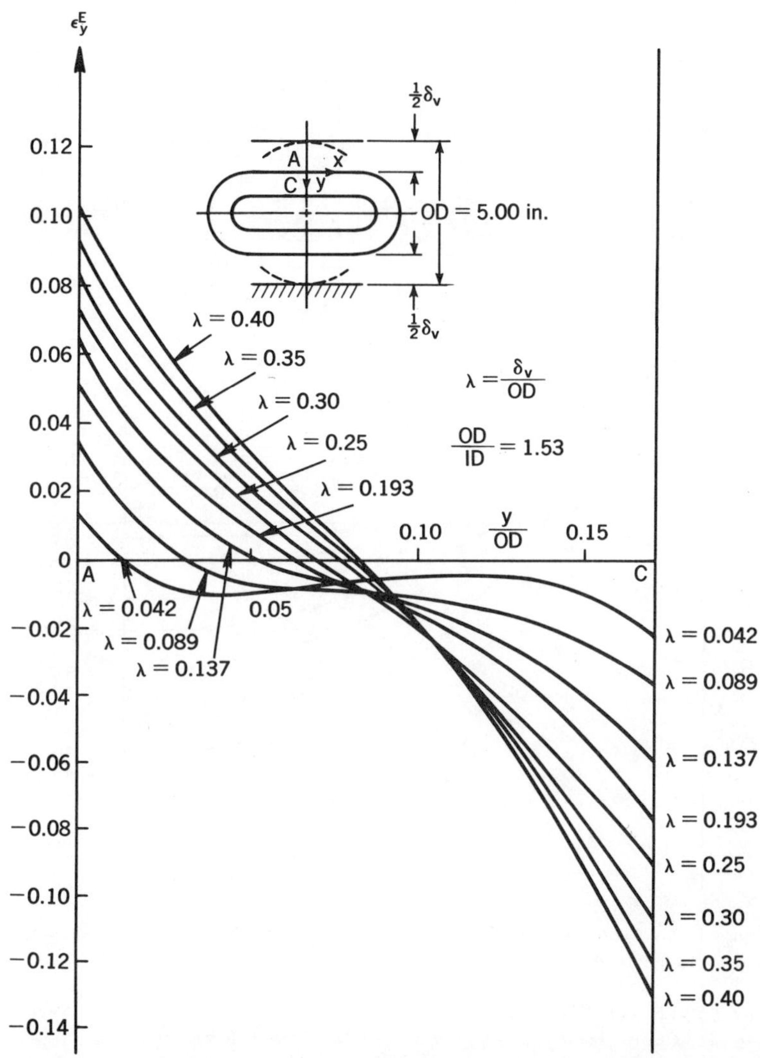

Figure 19.10. Eulerian vertical strains ϵ_y^E on the vertical axis of a circular ring subjected to several levels of diametral compression (moire method).

partial derivatives in the Eulerian–Cartesian system are of the form

$$\frac{\partial u}{\partial x} = \frac{p}{s} N = KN \tag{19.8}$$

where s is the shift of the pattern on itself and K is the p/s ratio. (The K values are given in Fig. 19.8 so that the shift can be computed as: shift $= 0.001/K$).

The partial-derivative patterns are ordered in much the same manner as

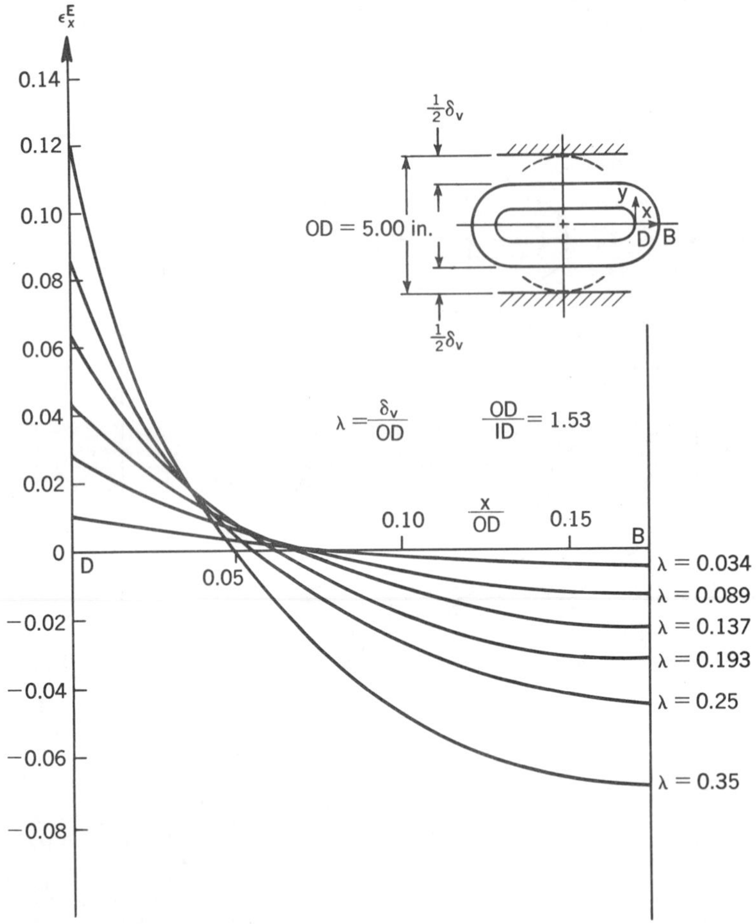

Figure 19.11. Eulerian horizontal strains ϵ_x^E on the horizontal axis of a circular ring subjected to several levels of diametral compression (moire method).

Figure 19.12. (Right) Eulerian vertical strain ϵ_y^E on the horizontal axis of a circular ring subjected to several levels of diametral compression (moire method).

the displacement patterns. In this case, the zero order is obtained at any position where a displacement fringe crosses its own image. The orders are then assigned in ascending (+) or descending (−) sequence from a knowledge of the sign at some point.

The Eulerian–Cartesian strains along the vertical axis of the ring for 8 load levels are given in Figs. 19.9 and 19.10. Figures 19.11 and 19.12 give the Eulerian–Cartesian strains along the horizontal axis. Since there is symmetry and no rotation along the axes, the Cartesian strains are principal and are obtained from one derivative ($\epsilon_x = \partial u/\partial x$, $\epsilon_y = \partial v/\partial y$).

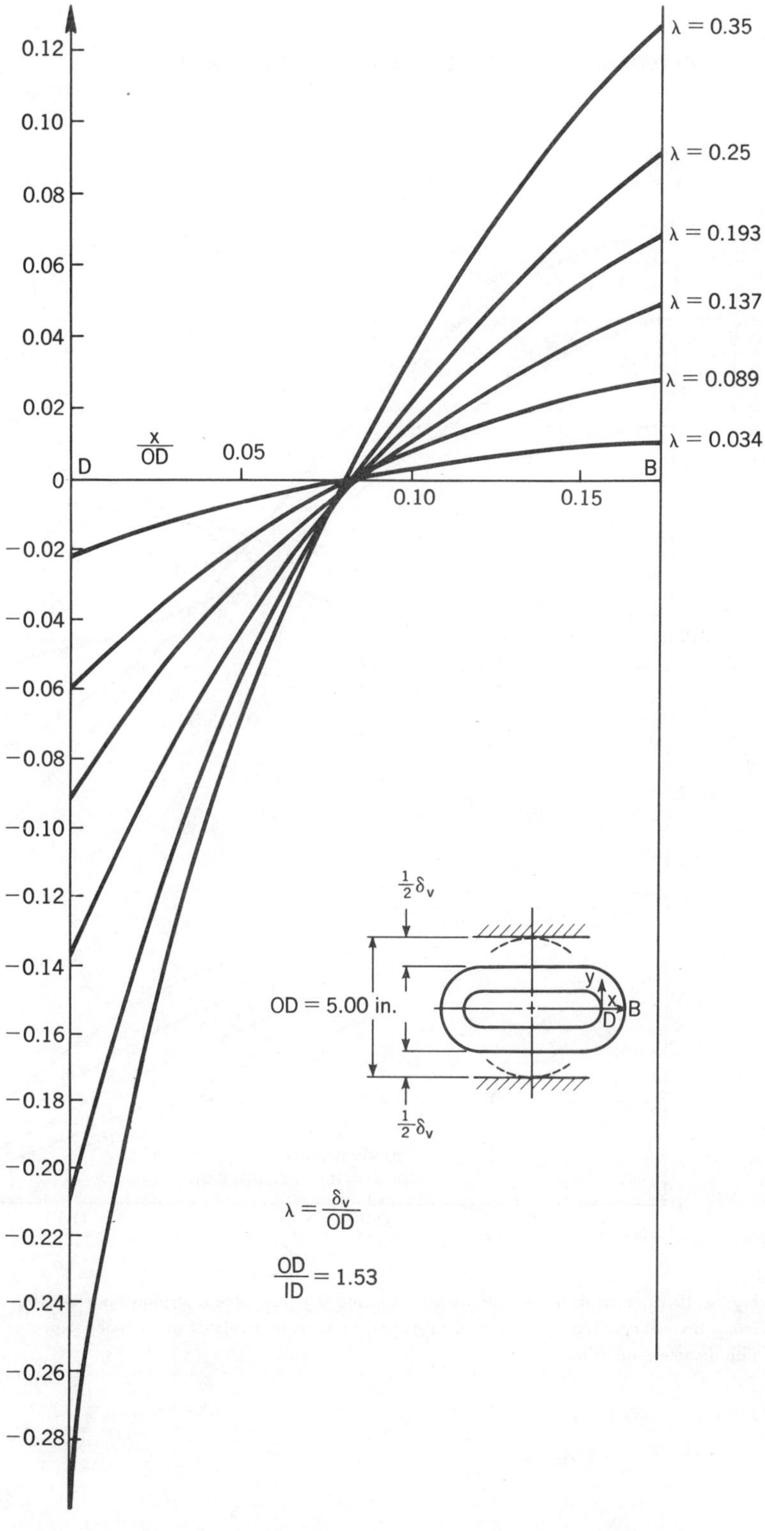

0.12
0.10
0.08
0.06
0.04
0.02
0
−0.02
−0.04
−0.06
−0.08
−0.10
−0.12
−0.14
−0.16
−0.18
−0.20
−0.22
−0.24
−0.26
−0.28
D
x/OD
0.05
0.10
0.15
B
λ = 0.35
λ = 0.25
λ = 0.193
λ = 0.137
λ = 0.089
λ = 0.034
½δv
OD = 5.00 in.
y
x
D
B
½δv
λ = δv/OD
OD/ID = 1.53

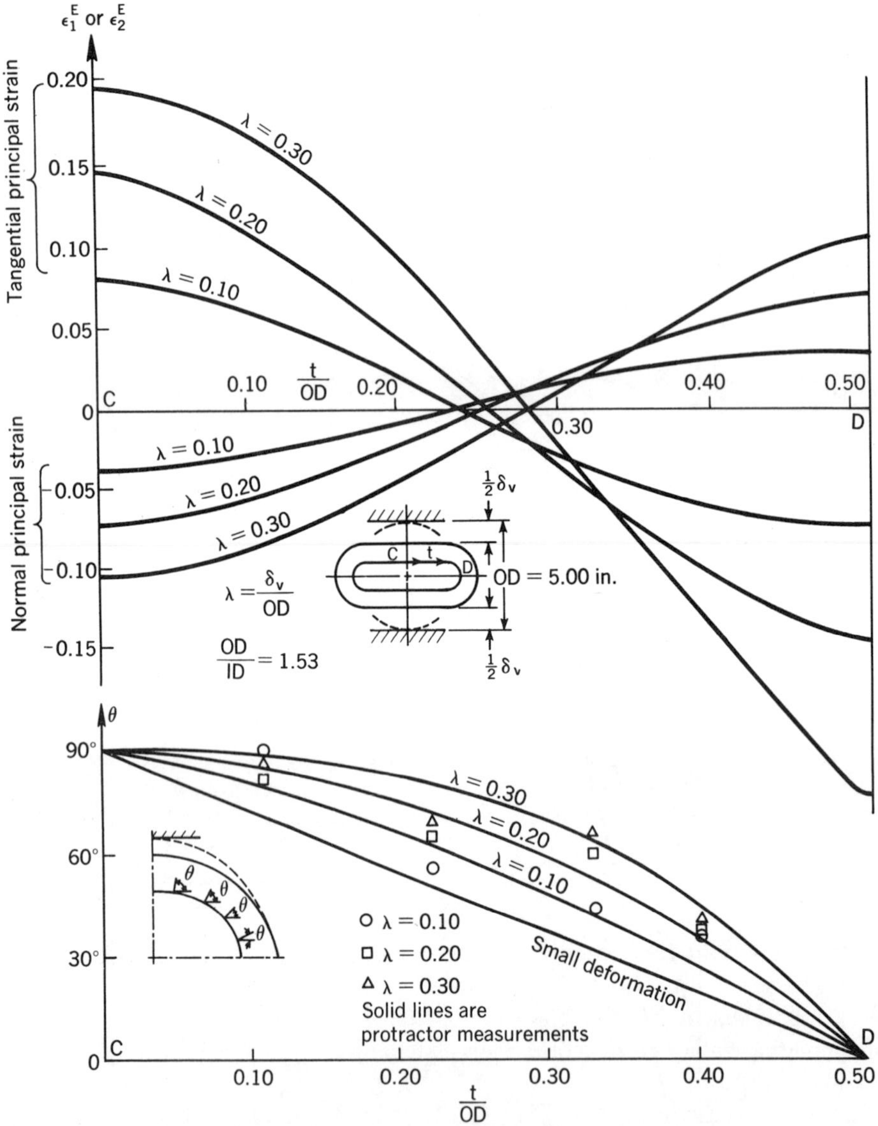

Figure 19.13. Principal Eulerian strains ϵ_1^E and ϵ_2^E and their orientation on the inner boundary of a circular ring subjected to several levels of diametral compression (moire method).

Figure 19.13 gives the Eulerian principal strains along the inner boundary of the ring. These were obtained by first computing the Eulerian–Cartesian strains ($\epsilon_x^E, \epsilon_y^E, \gamma_{xy}^E$) using Eqs. (4.8) and (4.12). The principal strains were then obtained from these after expressing them in tensorial form.

The angle of the principal strain with the horizontal direction was also computed from the moire data and compared with protractor measurements made on the free boundary. These are also shown in Fig. 19.13.

19.4.3. Photoelasticity

In the photoelastic analysis it is convenient to use another definition of strain. Two calibration strips of the same rubber sheet as used in the analysis of the ring were subjected to uniaxial stress (one to tension, the other to compression) and the strains, isochromatic-fringe orders, and stresses were determined independently. It was noted that the natural strain has a relation to the fringe orders (Fig. 19.14) which is much closer to being linear than the ones exhibited by other definitions of strains. It was decided to assume a linear relationship and to interpret the isochromatic-fringe order in the ring in terms of natural strain.[8]

Isochromatic patterns for 9 levels of load on the ring are shown in Fig. 19.15. The results are reported in the next section when compared with the results of the other methods. The strain obtained from the isochromatic fringe is

$$\bar{\epsilon}_y - \bar{\epsilon}_x = \bar{\epsilon}_1 - \bar{\epsilon}_2 = 2nF_\epsilon \tag{19.9}$$

where $\bar{\epsilon}_1$ and $\bar{\epsilon}_2$ are the natural principal strains at a point which has the fringe order n and F_ϵ is one-half the slope of the straight line in Fig. 19.14 ($F_\epsilon = 0.00242$ in./in./fringe.)

19.5. Comparisons

Figures 19.16 and 19.17 show the natural strain along the horizontal axis and inner boundary, respectively, of the ring for 3 levels of load. The isochromatic fringes have been multiplied by $2F_\epsilon$ to obtain the ordinates of the graph in Fig. 19.16. The corresponding values from moire analysis (changed from Eulerian to natural strains) are also shown.

The points on the inner boundary (Fig. 19.17) are in a state of unidimensional stress. Hence, the tangential strains can be obtained directly from the photoelastic calibration. Again, the grids and moire results were expressed in terms of natural strains for comparison.

---

[8] It must be noted here that the photoelastic effect depends on the thickness of the material. In large deformation the material will tend to change thickness appreciably under load. However, it was found that the fringe–strain relation was linear when no attempt was made to compensate for changing thickness.

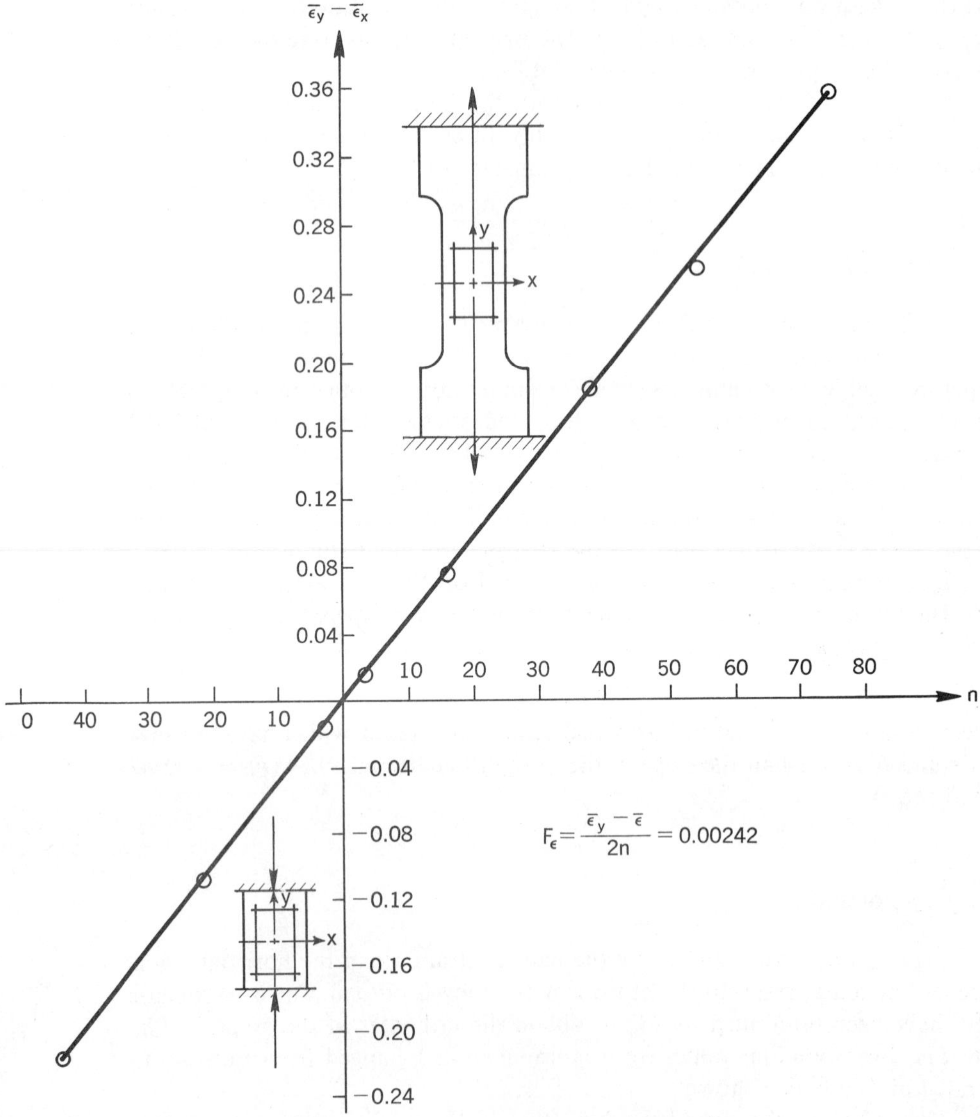

Figure 19.14. (Top) Strain-fringe curve for urethane rubber (Hysol 4485) in uniaxial tension and compression.

Figure 19.15. (Right) Isochromatics in a circular ring subjected to several levels of diametral compression (the maximum fringe orders at the inside boundary are shown).

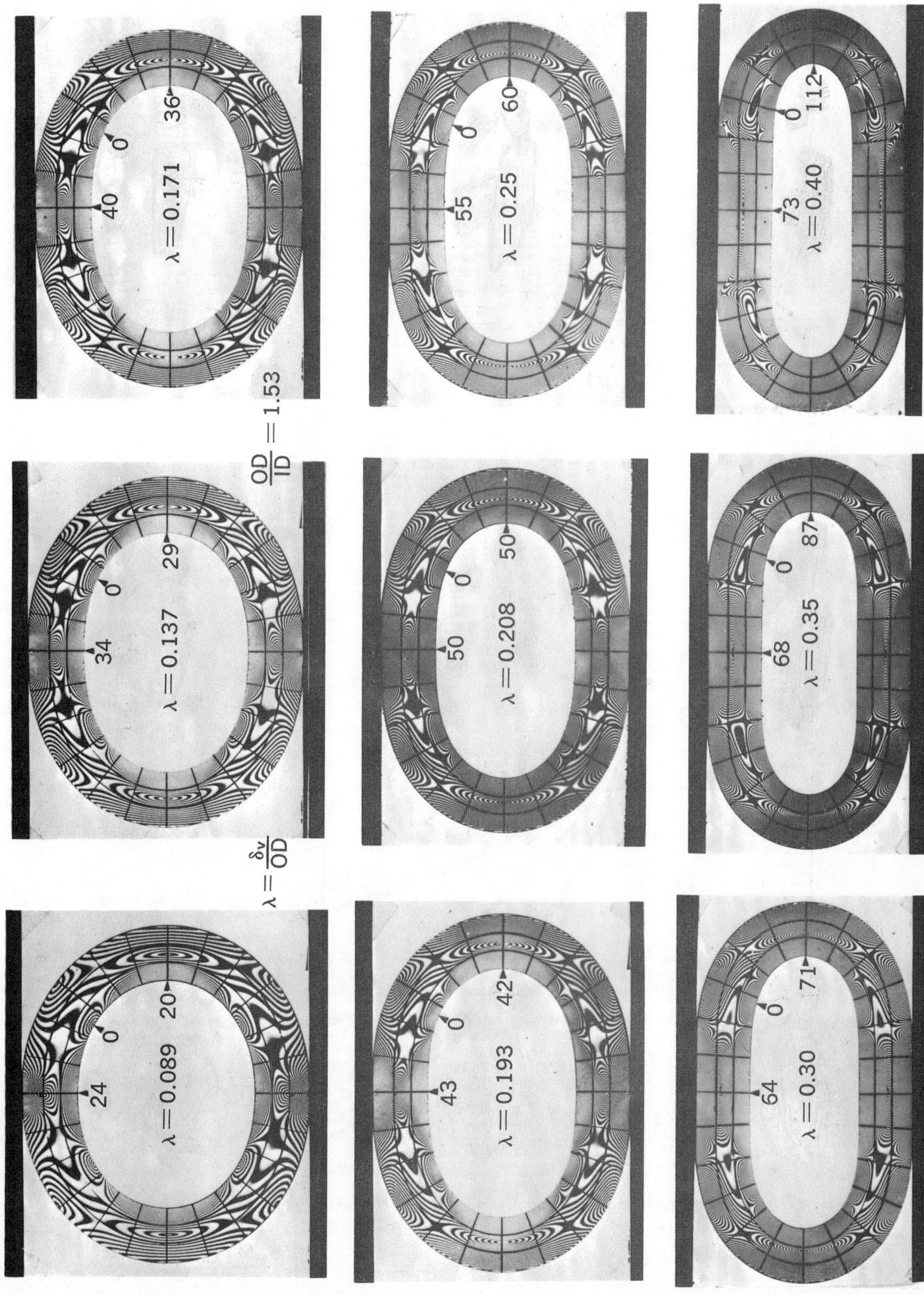

λ = 0.089
24
0
20
λ = 0.137
34
0
29
λ = 0.171
40
0
36
$\lambda = \frac{\delta_v}{OD}$
$\frac{OD}{ID} = 1.53$
λ = 0.193
43
0
42
λ = 0.208
50
0
50
λ = 0.25
55
0
60
λ = 0.30
64
0
71
λ = 0.35
68
0
87
λ = 0.40
73
0
112

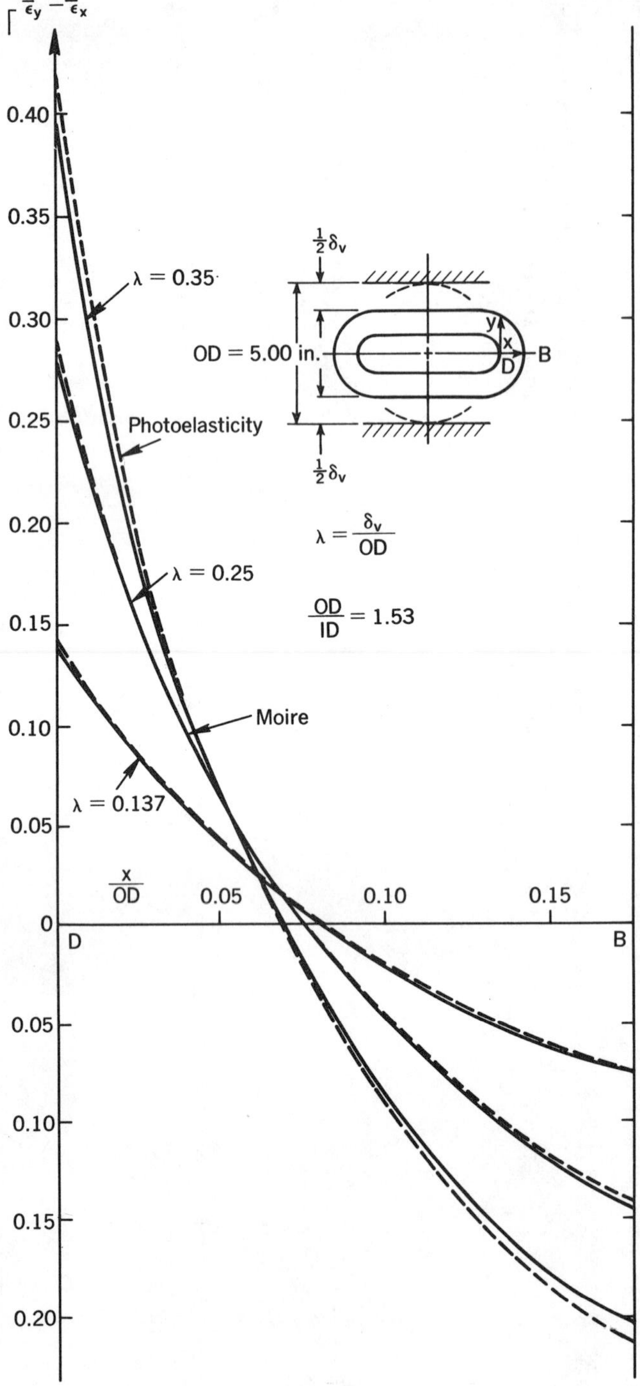

Figure 19.16. Difference of natural principal strains at the horizontal axis of a circular ring subjected to several levels of loads. Strains obtained using photoelasticity and moire.

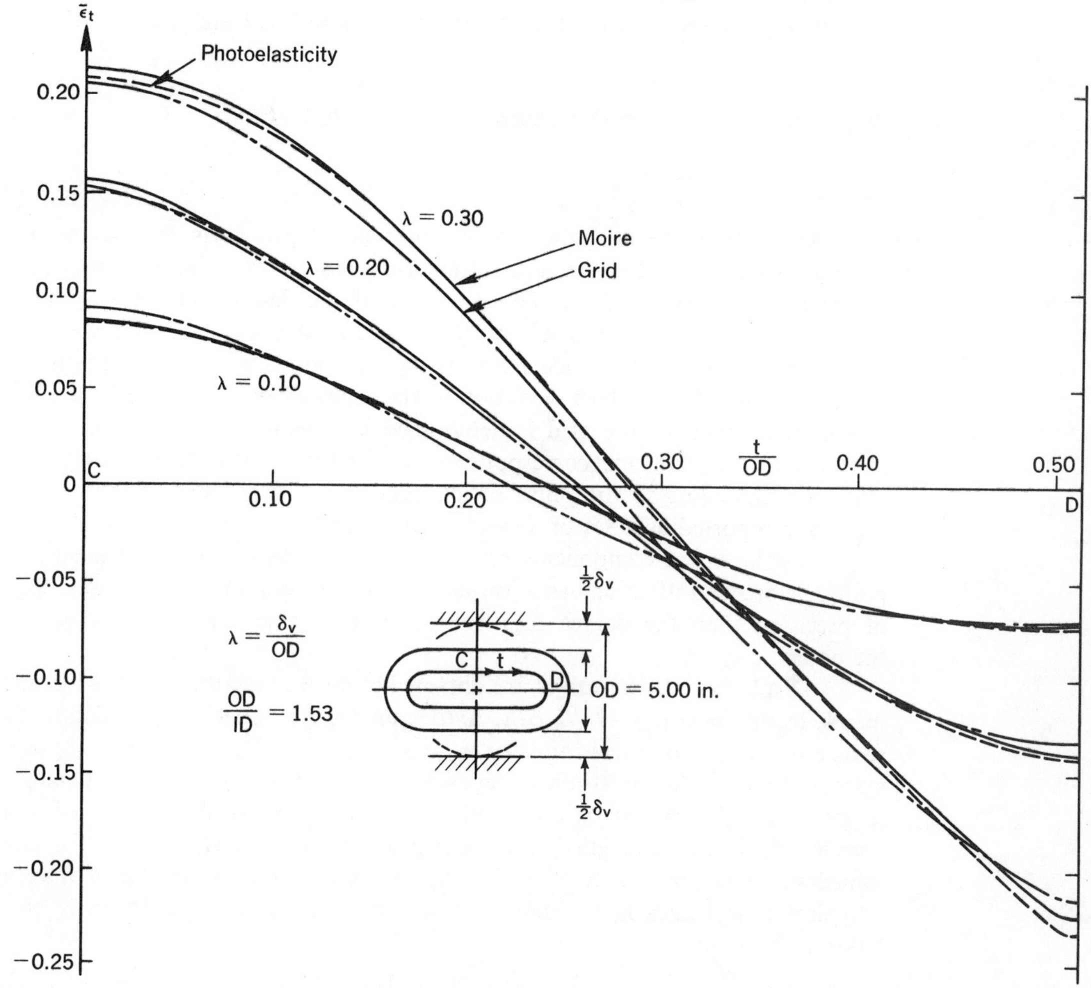

Figure 19.17. Natural strain $\bar{\epsilon}_t$ tangential to the inner boundary of a circular ring subjected to several levels of diametral compression. Strains obtained using photoelasticity, moire, and grids.

The comparisons show fair agreement, and the variations are due mostly to errors to be discussed later.

An equilibrium check on the horizontal axis can be obtained using the moire results. This requires a translation of Eulerian strains into natural strains and the computation of stresses from the strains. This has been done elsewhere.[9]

[9] A. J. Durelli, V. J. Parks, and H. Feng, "Experimental Methods of Large Strain Analysis," *Int. Jour. Non-Linear Mech.*, Vol. 2 (1967), pp. 387–404.

The area under the true-stress curve was compared with the load P for 2 levels of load. The values agreed within 4 per cent for $\lambda = 0.193$ and $\lambda = 0.35$.

19.6. Advantages and Disadvantages of Each Method

19.6.1. Grids

Grids have the advantage of giving a direct geometric representation of the body. A limitation is the level of error associated with the measurement of the geometric transformation. The precision of the determinations will depend on the sharpness of the grid marks, the recording method (photographic, etc.), and the precision of the measuring instrument. (Indices of precision are usually associated with the standard deviation of the repeated measurements.) The location of each point of the grid is determined by means of two readings on the instrument scale, the one corresponding to the zero point (or origin) and the other corresponding to the point of interest. This is true whether the measurements are reported as a set of x- and y-values or as a series of lengths "l's."

Displacement components are obtained from the difference in position of points before and after deformation and so will have a somewhat lower degree of precision than the degree of precision of the values of the coordinates of the points.

Displacement derivatives are obtained by further treatment of the measurements, in the form $(l_f - l_i)/l_i$, $\Delta l/l$, $\Delta u/\Delta x$, $\partial u/\partial x$, etc., operations which further reduce the degree of precision. For a given grid line, sharpness, and measuring system, the error in the strain depends on the magnitude of Δl. This change in length of a segment can be increased by increasing the strain or increasing the base length. The base length l, however, has to be relatively short, since the determined strain is assumed to vary linearly over its length. In the case of the ring, a typical base length is $l = 0.41$ in. and a typical standard deviation is about 0.001 in. Thus

$$\epsilon = \frac{\Delta l + 0.001}{0.41 + 0.001} \tag{19.10}$$

For a given Δl, it is seen that the strain will have a standard deviation of the order of 0.002. The overall curves presented will be more accurate due to the averaging effect of drawing a smooth curve through the points.

A second disadvantage of grids is the laborious point-by-point analysis. In the case of the ring, for the relatively small number of lines analyzed the effort was about the same for grids and moire. However, if the complete ring were analyzed, the grid method would take more time than the moire method. An attempt to reduce the time required to evaluate grid patterns has been described in Durelli, Parks, and Feng.[10]

---

[10] A. J. Durelli, V. J. Parks, and H. Feng, "Evaluation of Grid Data to Determine Displacements and Strains," *Journ. of Strain Analysis*, Vol. 2, No. 3 (July 1967), pp. 181–187.

Grids can easily be drawn to represent any desired coordinate system (polar or elliptical, for instance). In certain cases this may be an important advantage.

19.6.2. Moire

Moire has the advantage of giving a whole-field representation of displacements with high precision. The moire-of-moire patterns shown in Fig. 19.8 are approximate (as was the case for the grid analysis), in the sense that they represent the average of the partial derivative over the shifted interval. The shift in this case, however, is an order of magnitude smaller than the grid base length.

When the shift takes place normal to a boundary, the boundary value of the partial derivative has to be estimated by extrapolation from the recorded pattern. Otherwise, it should be obtained by graphical differentiation of the displacement field.

The standard deviation of the moire-grating pitch is probably less than 0.00001, and so the standard deviation in strain for a shift of 0.03 will be of the order of

$$\epsilon = \frac{0.00001}{0.03} = 0.0003 \tag{19.11}$$

This error does not include the error associated with the determination of the position of the point.

The moire grating is so dense that the fringes are the average of a large number of grating-line intersections. This averaging effect increases the precision beyond the one indicated above.

On axes of symmetry, even if the strains are large, the strains in the direction of the axes or perpendicular to them can be obtained from a single isothetic pattern.

19.6.3. Photoelasticity

Photoelasticity is also a whole-field method giving data available immediately at any point in the field. When photoelasticity is applied to large-strain analysis, a transparent birefringent material which shows a simple relation between the isochromatic fringes and some quantity of interest is required. The optical and mechanical calibrations used in the program reported here were found convenient. In some cases, other fringe–strain, fringe–stress, and stress–strain relations of this material might be used and be of special interest in some particular problem.

In comparison with grids and moire, the chief disadvantage of photoelasticity may lie in the fact that the material has to be birefringent and has to be calibrated in terms of strain.

Photoelasticity gives directly only the maximum shear strain, so that, away from free boundaries, when direct strains are required, grids and moire have an advantage.

19.6.4. Thickness Measurements

Thickness measurements using comparators have been applied to the solution of a problem described elsewhere.[11] In the case of rubber specimens, difficulty is encountered in the fact that the comparator anvil is spring-loaded and applies a force on the specimen. This force causes an extraneous deflection which decreases the precision of the measurements and which has to be corrected.

The results obtained from thickness measurements can be combined directly with those obtained from photoelasticity and both principal strains can be determined.

19.6.5. Opaque Materials

The illustration shown in this chapter uses a transparent ring. The four methods can also be applied to opaque materials (photoelasticity in the form of thin birefringent coatings), and some contributions in the field of solid-propellant grains can be found in the literature.[12] The methods are also suited to the determination of plastic strains, as shown in Chapter 18. Grids and moire can be applied directly to the surface of the specimens to be analyzed. The accuracy of photoelastic coatings may be limited by the error associated with the coating thickness when large strain gradients are present.

19.6.6. Criteria for Selecting a Method

The decision as to which method to use to solve a particular problem will depend on several points: (1) the possibility of using a birefringent material, (2) the order of magnitude of the strains to be determined, and (3) the scope of the analysis—a solution limited to free boundaries, a solution including axes of symmetry or few selected points, or a solution of the whole field.

If the analysis can be conducted with a transparent material and is limited to free boundaries, and the fringe order of isochromatics does not exceed approximately 150, photoelasticity is probably the best choice.

If the order of strains, in problems similar to the one studied in this chapter, is higher than 30 per cent, photographing grids may be the simplest method to use, mainly when the analysis is limited to selected points.

---

[11] See Footnote 3, this chapter.

[12] A. J. Durelli, "Experimental Strain and Stress Analysis of Solid Propellant Rocket Motors," Office of Naval Research, Contract 2249(06) (March 1965), Interim Rpt. No. 6, Section 2.8.

If the whole field has to be strain-analyzed and the order of magnitude of the strains is larger than 5 per cent, moire presents several advantages.

For the particular problem of the ring, it was decided that moire was the best method to use to obtain the complete strain field. Strain fields for ϵ_x, ϵ_y, γ_{xy}, ϵ_1, ϵ_2, the principal angles, and the rigid-body rotations have been computed for the case $\lambda = 0.3$, and stresses determined from strains. In what follows, a summary of this work is reported.

19.7. Complete Strain Field in a Ring Reduced in Diameter by 30 Per Cent

The two isothetic patterns over more than one-half of the surface of the specimen are shown in Fig. 19.18.

Each of the u- and v-field patterns was copied and shifted on its own image in the x- and y-directions successively to produce four moire-of-moire patterns (two from each field) corresponding to the four Cartesian derivatives of the displacement components $\partial u/\partial x$, $\partial u/\partial y$, $\partial v/\partial x$, and $\partial v/\partial y$. A representative quadrant for each of these four fields is shown in Fig. 19.19.

A grid made of (1) the inner and outer diameters of the ring, (2) two concentric circles equally spaced between the diameters, and (3) radial lines every 15°, was used to specify 28 points per quadrant of the specimen. These circles and radial lines were circular and radial in the undeformed ring, and were used as convenient coordinates to analyze the deformed specimen.

The four Cartesian derivatives of the displacement components were computed at each of the 28 points with the following equation and three similar equations (for the other three partial derivatives):

$$\frac{\partial u}{\partial x} = \frac{\text{Pitch}}{\text{Shift}} N \tag{19.12}$$

where "Pitch" is the pitch of the undeformed grating in inches (0.001 in.), "Shift" is the amount the isothetic pattern was shifted on itself, measured in inches on the actual ring size, and N is the fringe order at the required point, estimated to about one-tenth of a fringe.

Eulerian strains were determined at the 28 points using Eqs. (4.8) and (4.12).

Principal Eulerian strains were computed at the 28 points using the following equation from Sec. 4.11:

$$(1-\epsilon_1^E)^2 = \tfrac{1}{2}[(1-\epsilon_x^E)^2 + (1-\epsilon_y^E)^2 \pm \sqrt{[(1-\epsilon_x^E)^2 - (1-\epsilon_y^E)^2]^2 + 4[(1-\epsilon_x^E)(1-\epsilon_y^E)\sin\gamma_{xy}^E]^2}\,] \tag{19.13}$$

By interpolation between these 28 points, contour lines were drawn representing different levels of principal strain (Fig. 19.20). These contours are called *isotenics* and have a meaning similar to the meaning of the fringes of constant displacement (isothetics) and the fringes of constant derivative of displacement (isoparagogics) shown in Fig. 19.19.

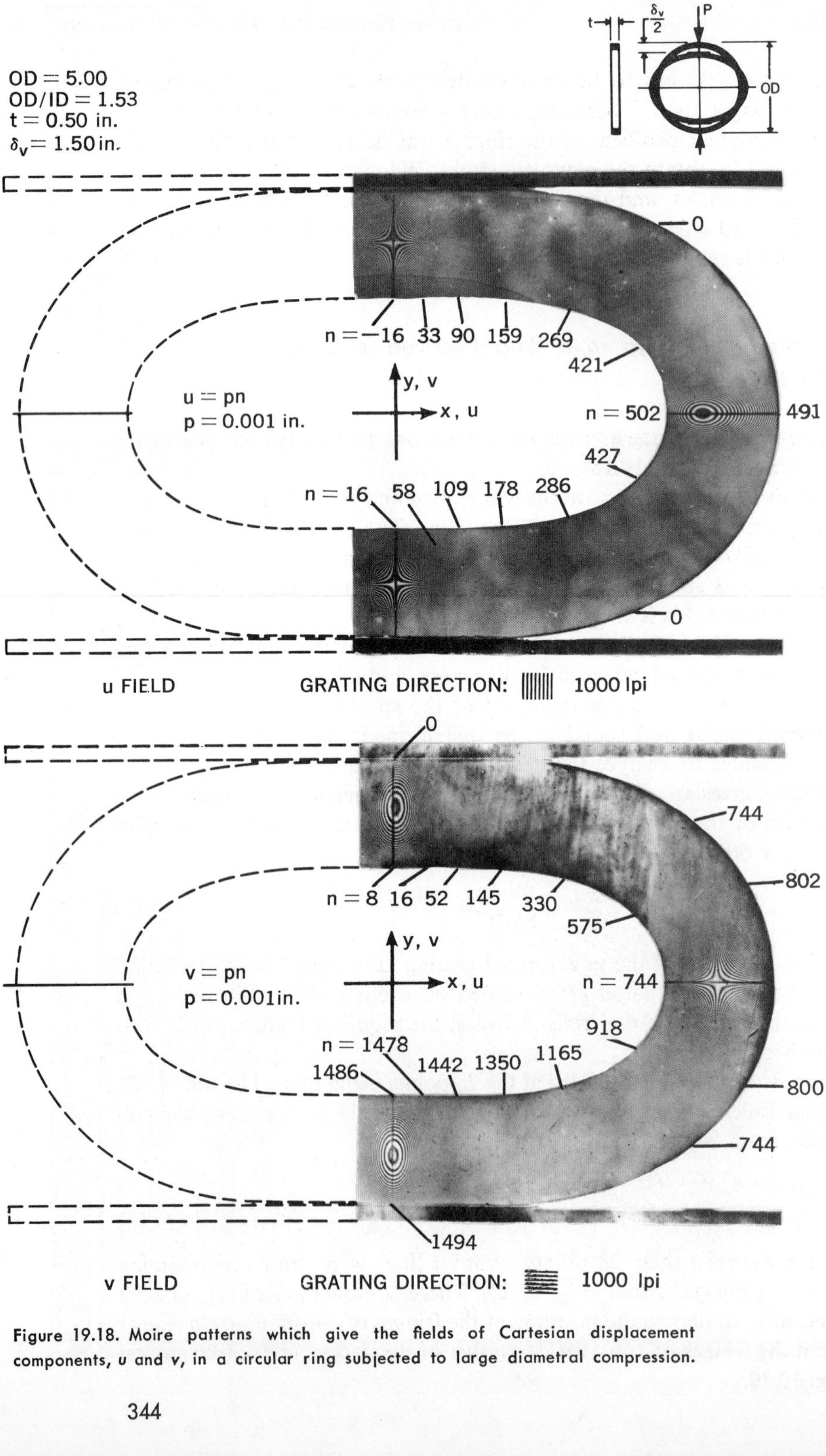

Figure 19.18. Moire patterns which give the fields of Cartesian displacement components, *u* and *v*, in a circular ring subjected to large diametral compression.

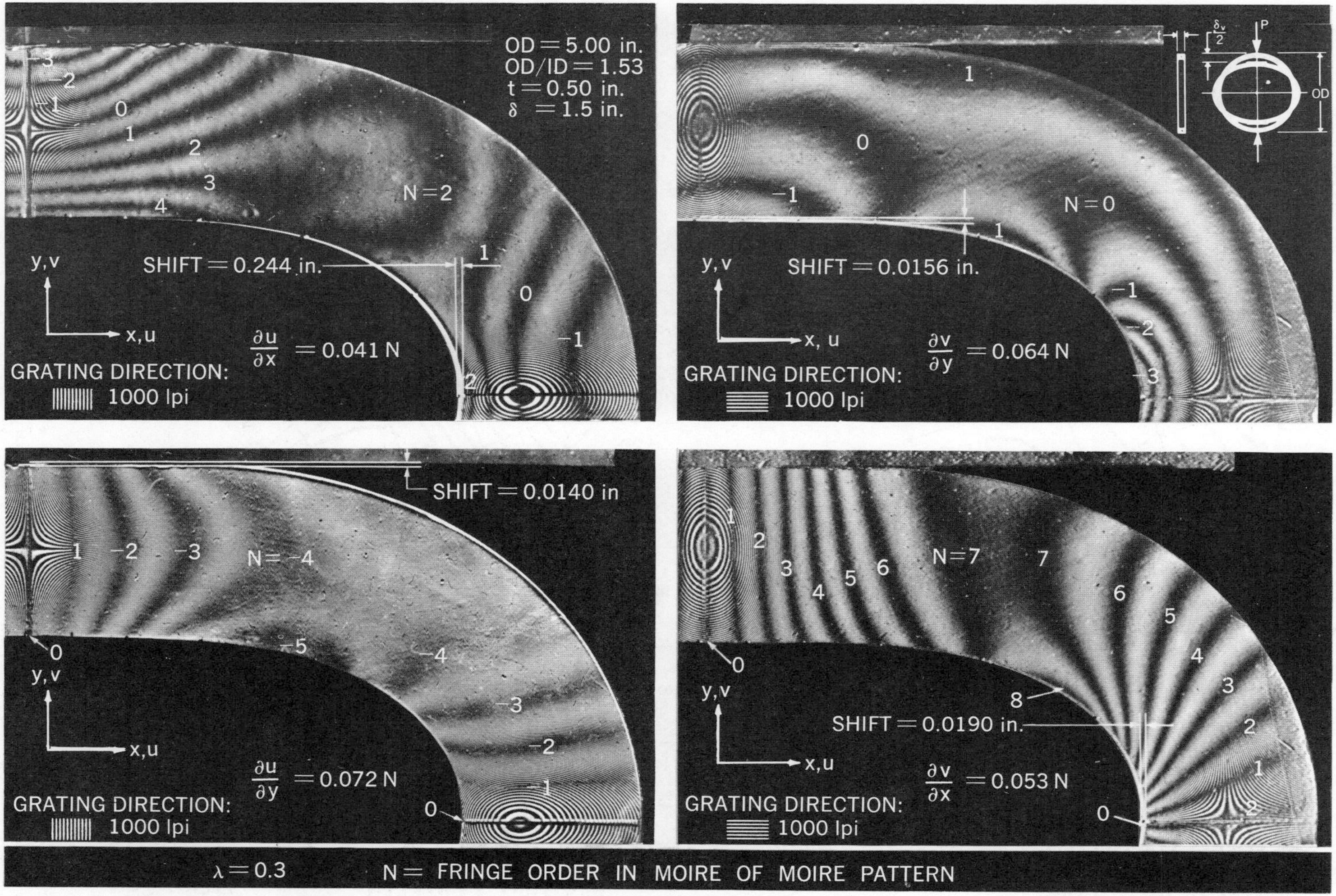

Figure 19.19. Isoparagogics in a circular ring subjected to large deformation.

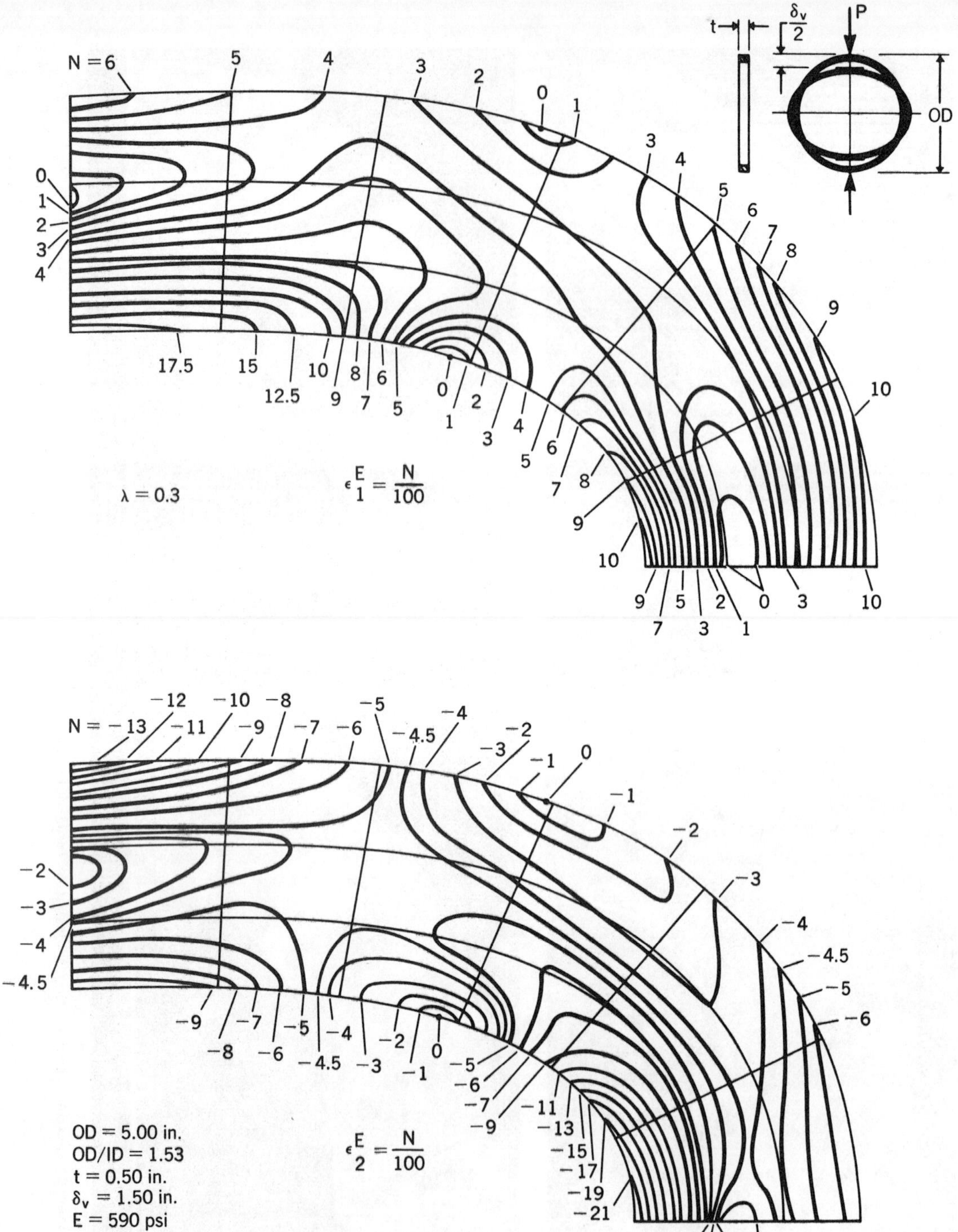

Figure 19.20. Loci of points exhibiting the same value of ϵ_1^E and ϵ_2^E (Eulerian isotenics) in a circular ring subjected to large diametral compression.

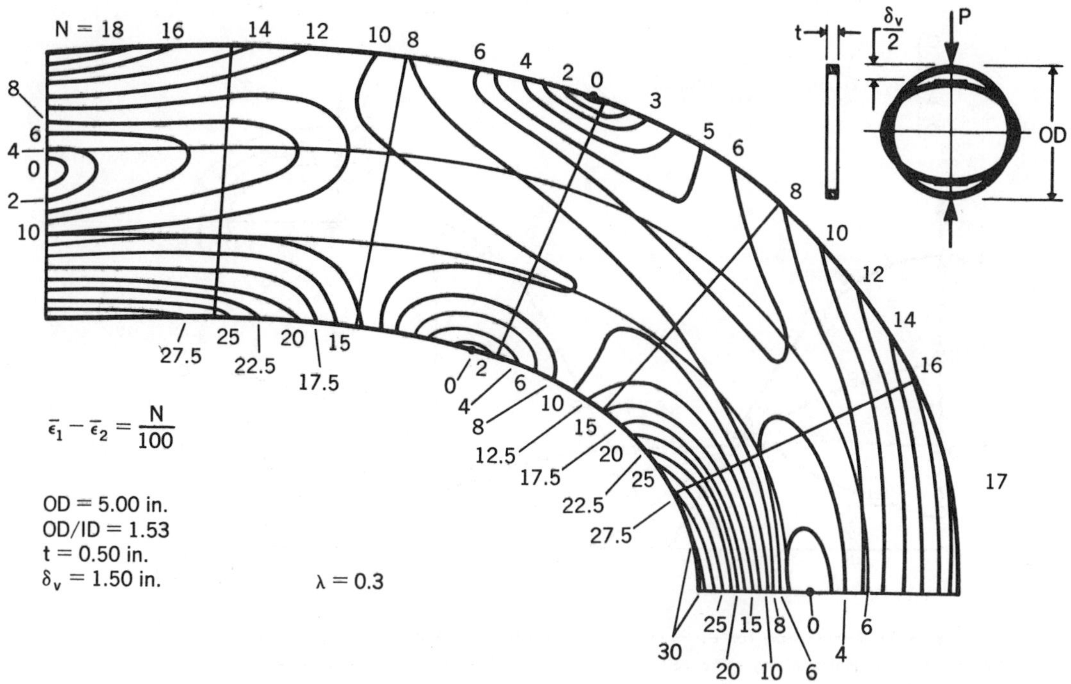

Figure 19.21. Loci of $\bar{\epsilon}_1 - \bar{\epsilon}_2$ (natural isochromatics) in a circular ring subjected to large diametral compression (obtained from moire patterns).

Figure 19.22. Isochromatics in a circular ring subjected to large diametral compression (obtained photoelastically).

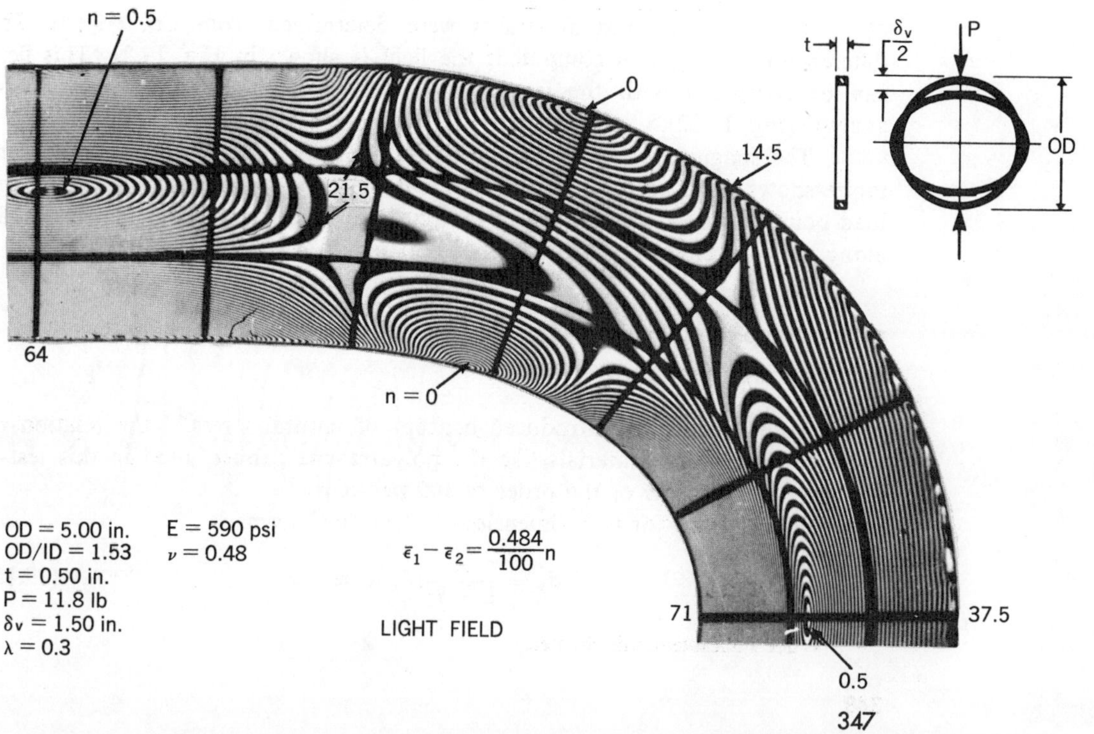

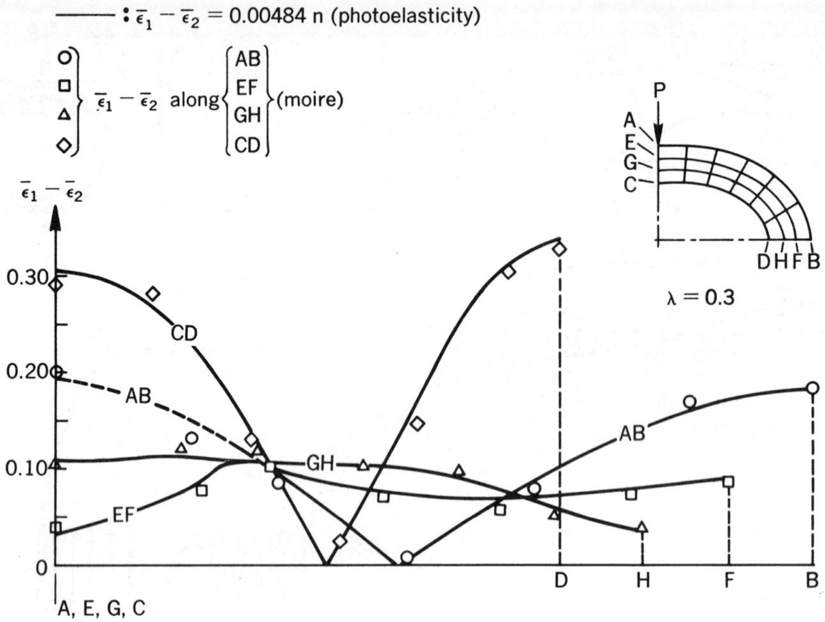

Figure 19.23. Comparison for $\bar{\epsilon}_1 - \bar{\epsilon}_2$ (natural isochromatics) obtained by photoelasticity and moire along some selected lines in a circular ring subjected to large diametral compression.

Eulerian and natural strains are related by

$$\bar{\epsilon} = \ln \frac{1}{1 - \epsilon^E} \tag{19.14}$$

The principal natural strains were determined from Eq. (19.14). The difference $(\bar{\epsilon}_1 - \bar{\epsilon}_2)$ was computed; the field is shown in Fig. 19.21. This field can be compared with the isochromatic pattern obtained from photoelastic analysis (Fig. 19.22).

The magnitude of the values of $(\bar{\epsilon}_1 - \bar{\epsilon}_2)$ in Fig. 19.21 obtained from moire shows good agreement with that obtained photoelastically, except at the load point. A direct comparison of the photoelastic and moire results was made along the four circumferential lines of the ring, and is given in Fig. 19.23.

19.8. Complete Stress Field in a Ring Reduced in Diameter by 30 Per Cent

Using the newly introduced concept of natural stress,[13] the relation of stress to strain—for materials like the polyurethane rubber, used in this test—is linear up to strains of the order of 100 per cent.

Hooke's law for two dimensions in "natural" terms is

$$\bar{\sigma}_x = \frac{E}{1 - \bar{\nu}^2}(\bar{\epsilon}_x + \bar{\nu}\bar{\epsilon}_y) \tag{19.15}$$

[13] See Footnote 6, this chapter.

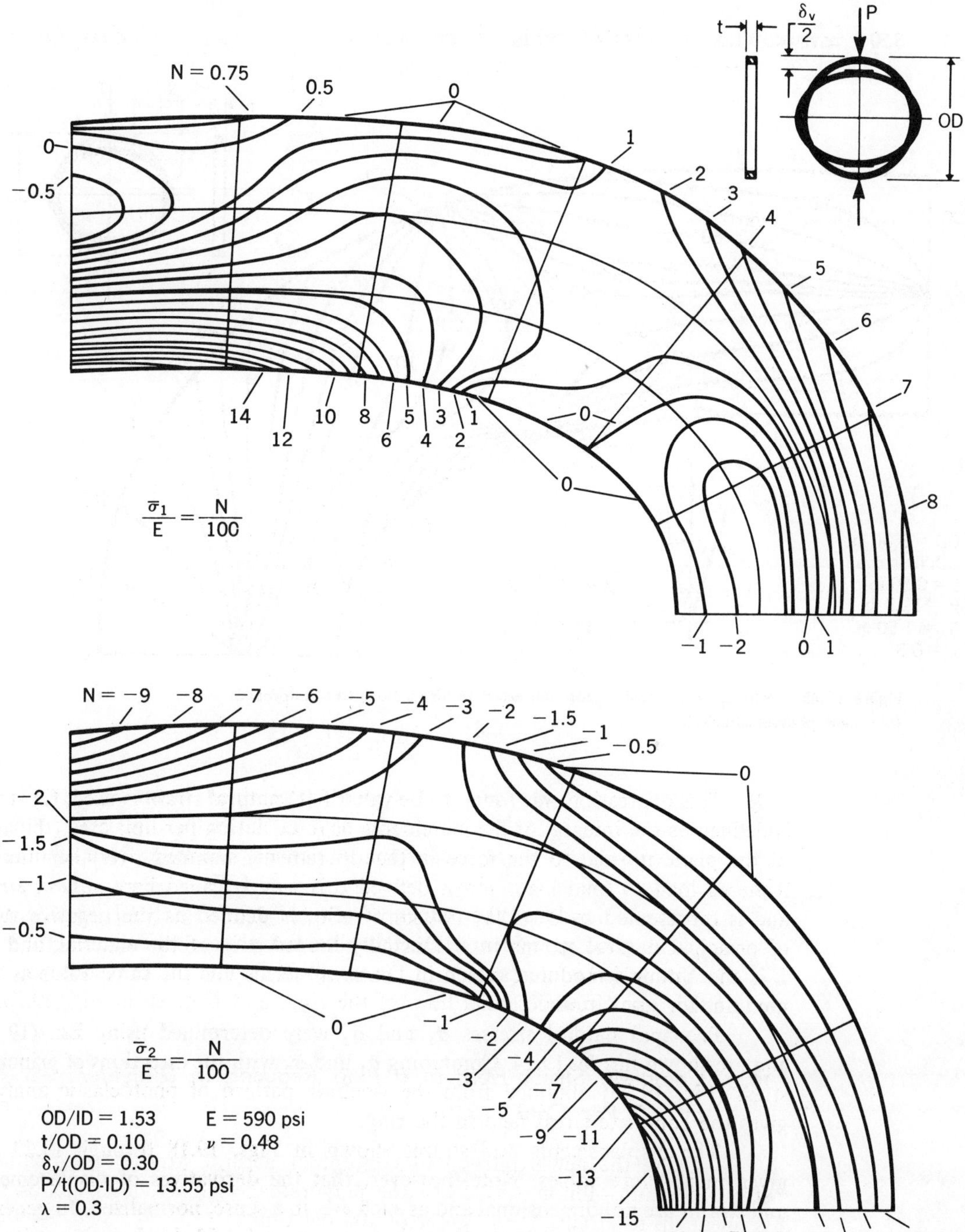

Figure 19.24. Loci of $\bar{\sigma}_1/E$ and $\bar{\sigma}_2/E$ (normalized isobars) for a circular ring subjected to large diametral compression.

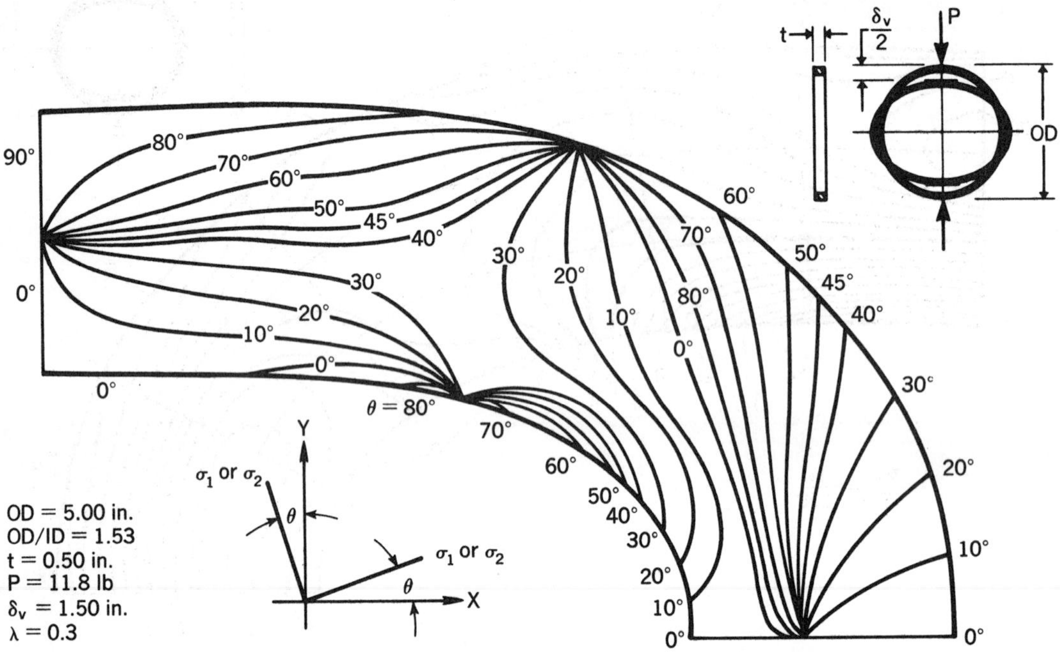

Figure 19.25. Isoclinics in a circular ring subjected to large diametral compression (obtained photoelastically).

This expression was found to be valid if (1) natural strain is used, (2) stress is defined as the integral of the increments of force acting per unit area (the area of the plane normal to the force in that increment) summed over the interval from no load to final load. Stress defined this way is called here *natural stress* and is represented by $\bar{\sigma}$. (3) $\bar{\nu}$, Poisson's ratio, is defined as the negative ratio of principal natural strains in a uniaxially loaded strip of the material, and (4) E is the Young's modulus, which in "natural" terms has the same value as the modulus for small strains.

Principal natural stresses $\bar{\sigma}_1$ and $\bar{\sigma}_2$ were determined using Eq. (19.15) and are shown in Fig. 19.24. Combining $\bar{\sigma}_1$ and $\bar{\sigma}_2$ with the directions of principal stress (Fig. 19.25) obtained from the isoclinic pattern of photoelastic analysis gives the complete stress field in the ring.

The displacements and strains shown in Figs. 19.18 through 19.23 are given in absolute values. Note, however, that the derivatives of displacements and strains are nondimensional and as such are, in a sense, normalized. In general, they apply to any thin ring with a diameter ratio of 1.53, loaded diametrically, such that the vertical displacement is 0.3 of the original diameter. The material of the ring to which they apply must also have a linear natural stress–natural-strain relation for the range of strains indicated in Figs. 19.20 and 19.21. Strictly speaking, Poisson's ratio should be matched in applying the results.

The normalized values for stress are given in the form σ/E, and have the same generality and limitations as the strains.

Thus, as in the linear first boundary-value problem, the results may be applied to different sized bodies, with different material properties. In nonlinear problems the results can be applied if the complete material behavior is matched, and the deformed geometry is matched at the required level of load.

20

DETERMINATION OF ISOCHROMATICS AND ISOPACHICS AS MOIRE OF ABSOLUTE RETARDATION PATTERNS

20.1. Introduction

The applications described in Chaps. 17, 18, and 19 required the interpretation of the moire phenomenon only as a phenomenon of light and shadow. The application described in this chapter will use the interference of light as a wave and will present the well-known isochromatic and isopachic loci from a somewhat unconventional point of view.

The common problem in photoelasticity is the determination of stresses and strains in thin plates with arbitrary shape and boundary conditions, loaded in the plane of the plate. Under certain conditions, stresses and strains in such a plate can be represented by two, two-dimensional, second-rank symmetric tensors which can be completely specified by three independent measurements. Calibration experiments show that in plates made of certain mechanically homogeneous, isotropic, linearly elastic, optically transparent materials, the velocity of light along paths perpendicular to the plane of the loaded plate can also be represented by the same type of tensor and that the three tensors are linearly related and their principal axes coincide.

The change in the velocity of light caused by loading the plate can be determined by measuring the phase shift (for a given initial plate thickness, material, and frequency) apparent in the interference pattern of one light beam with some other reference beam which is coherent with it. Hence the problem stated above can be reduced to the measuring of three quantities (two shifts corresponding to the two eigenvectors of the light velocity tensor and an angle) at each point in the plane of the plate.

Measurements usually are of (1) the magnitude of the difference of the two eigenvectors of the light velocity tensor (isochromatic interference), (2) the directions of the eigenvectors (isoclinics), or (3) the individual magnitudes of the eigenvectors (absolute retardation interference). The first two measurements can be obtained with standard polariscopes. Methods of interferometry—both classical and holographic—can provide the third measurement.

20.2. *Interferometers in Photoelasticity*

In Sec. 1.2.3 it was mentioned that moire could be considered as an interference phenomenon. This interference is sometimes called *mechanical* interference to emphasize the difference with the more commonly considered optical phenomenon of interference. In this chapter, optical interference will be studied.

Interferometers are instruments to measure optical interference and were first employed in photoelasticity by Favre[1] in 1929, for point-by-point measurement. Dose and Landwehr,[2] Post,[3] and Pirard[4] made important contributions to the development of whole-field photoelastic interferometers.

Figure 20.1 shows a whole-field interferometer set up to measure the interference between rays of light passed twice through a specimen and rays that do not pass through the specimen. Superposing the optical interference patterns from the loaded and unloaded plate gives a moire pattern of the absolute retardations. This moire effect eliminates the need of an optically flat specimen. The use of the specimen surfaces for reflection eliminates the need for the rigid set of parallel mirrors found in many photoelastic interferometers.

Absolute retardation patterns obtained from loaded photoelastic models contain, in general, two families of fringes corresponding to the magnitudes of each of the principal stresses at every point because the optical path length through the loaded models is a function of the index of refraction which can have two different principal values (depending on the polarization of the light) at each point. Nisida and Saito noted in 1964 that the same patterns could also be interpreted as containing two families of fringes corresponding to the sum (isopachics) and the difference (isochromatics) of the principal stresses if the fringes indicating

---

[1] H. Favre, "Sur une Nouvelle Methode Optique de Determination des Tensions Interieures," *Rev. Optique*, Vol. 8 (1929), pp. 193–213, 241–261, 289–307.

[2] A. Dose and R. Landwehr, "Bestimmung der Linien gleiche Hauptspannungssumme mittels Interferenzen gleicher Dicke," Ing-Arch. *21* (1953), pp. 73–86.

[3] D. Post, "A New Photoelastic Interferometer Suitable for Static and Dynamic Measurements," *Proc. SESA*, Vol. 12, No. 1, (1954), pp. 191–202.

[4] A. Pirard, "Considerations Sur la Methode du Moire en Photoelasticite," *Revue Universelle des Mines*, Tome XVI, No. 4 (1960), pp. 177–200.

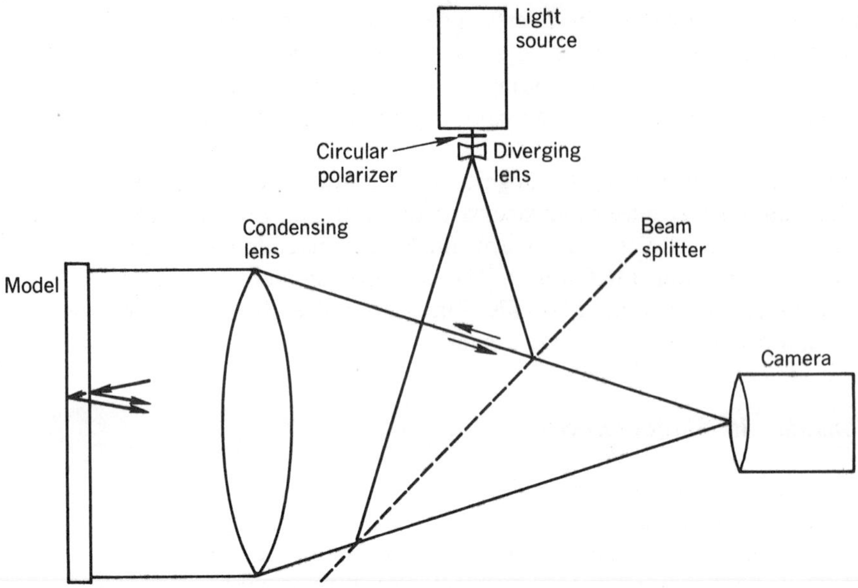

Figure 20.1. Schematic diagram of one form of extended field interferometer.

the contour of an isopachic are considered to change alternately between dark and bright whenever they cross an isochromatic fringe (meaning a switch of one half order).[5] Recently, it has also been shown that in some interference patterns the fringes indicating isochromatic contours can be expected to shift one half order instead of those following isopachics.[6] Post has shown that the fringes identified as isochromatics and isopachics in the alternative interpretation (and the fringe shift) can be understood as resulting from a moire of the fundamental absolute retardation patterns.[7] This moire effect, as pointed out in Chap. 2, can be demonstrated with two crossed gratings as shown in Fig. 20.2. The moire can be interpreted as follows: The lines of a grating can be numbered consecutively from the bottom and a function (A) of position (x, y) can be specified by assigning to it the value of the line passing nearest to a given position (x, y). A similar function (B) can be defined for a second grating. If the gratings are crossed, moire fringes appear which, when numbered consecutively, will specify the values of the function $A + B$ and $A - B$. This interpretation includes the additive moire which is usually more difficult to observe than the subtractive moire and is sometimes ignored.

[5] M. Nisida and H. Saito, "A New Interferometric Method of Two-Dimensional Stress Analysis," *Exp. Mech.*, Vol. 4, No. 12 (December 1964), pp. 360–376. See also Fig. 2.5.

[6] K. C. Thomson, "Apparatus for Seismic Modeling Optically," *Review of Scientific Instruments*, Vol. 37, No. 6 (June 1966), pp. 713–717.

[7] D. Post, "The Generic Nature of the Absolute Retardation Method of Photoelasticity," *Exp. Mech.*, Vol. 7, No. 6 (June 1967), pp. 233–241.

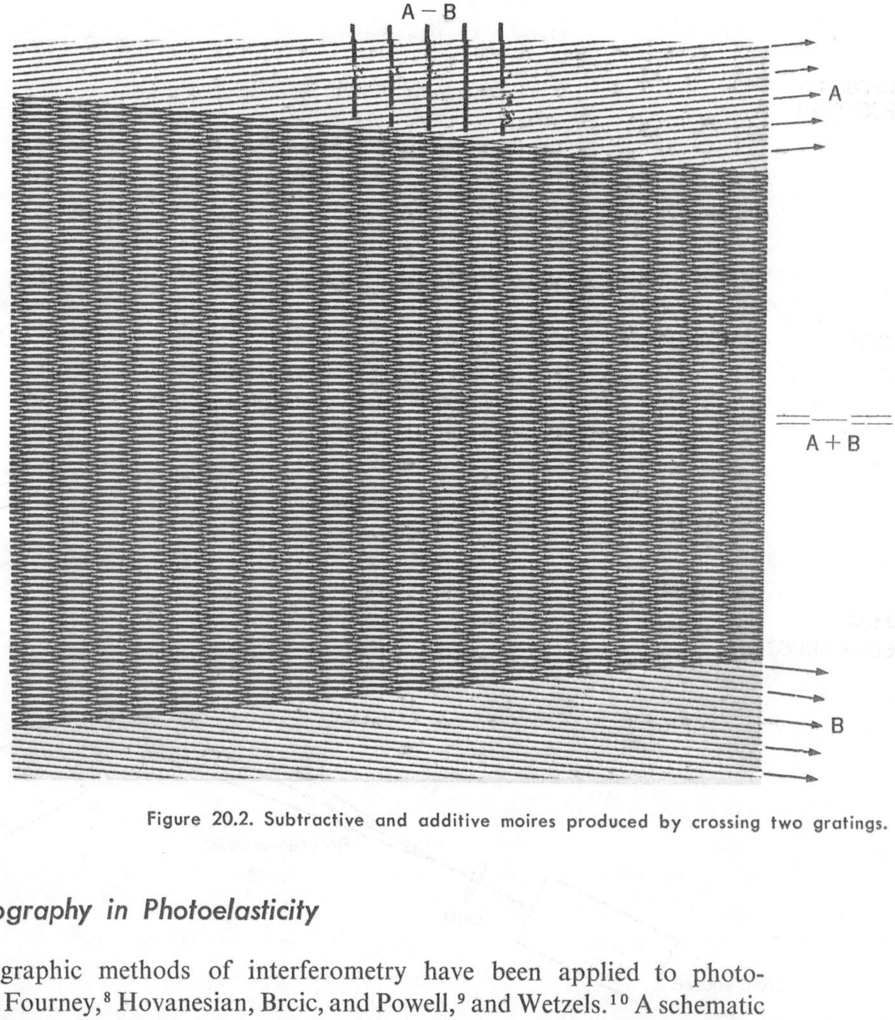

Figure 20.2. Subtractive and additive moires produced by crossing two gratings.

20.3. Holography in Photoelasticity

Holographic methods of interferometry have been applied to photoelasticity by Fourney,[8] Hovanesian, Brcic, and Powell,[9] and Wetzels.[10] A schematic diagram illustrating the main features of a holographic system is shown in Fig. 20.3. As in classical interferometers, light is divided by a beam splitter into a reference beam and a signal beam. The signal beam passes through a transparent model which has a different index of refraction than air. Mirrors are oriented so that the paths of the two beams intersect at some point beyond the model. At that point their total intensity is recorded by a photosensititive film. Coherent

[8] M. E. Fourney, "Application of Holography to Photoelasticity," *Exp. Mech.*, Vol. 8, No. 1 (January 1968), pp. 33–38.

[9] J. D. Hovanesian, V. Brcic, and R. L. Powell, "A New Experimental Stress-Optic Method: Stress-Holo-Interferometry," *Exp. Mech.*, Vol. 8, No. 8 (August 1968), pp. 262–268.

[10] W. Wetzels, "Holographie als Hilfsmittel zur Isopachenbestimmung," Optik, Band 27 (1968), pp. 271–273.

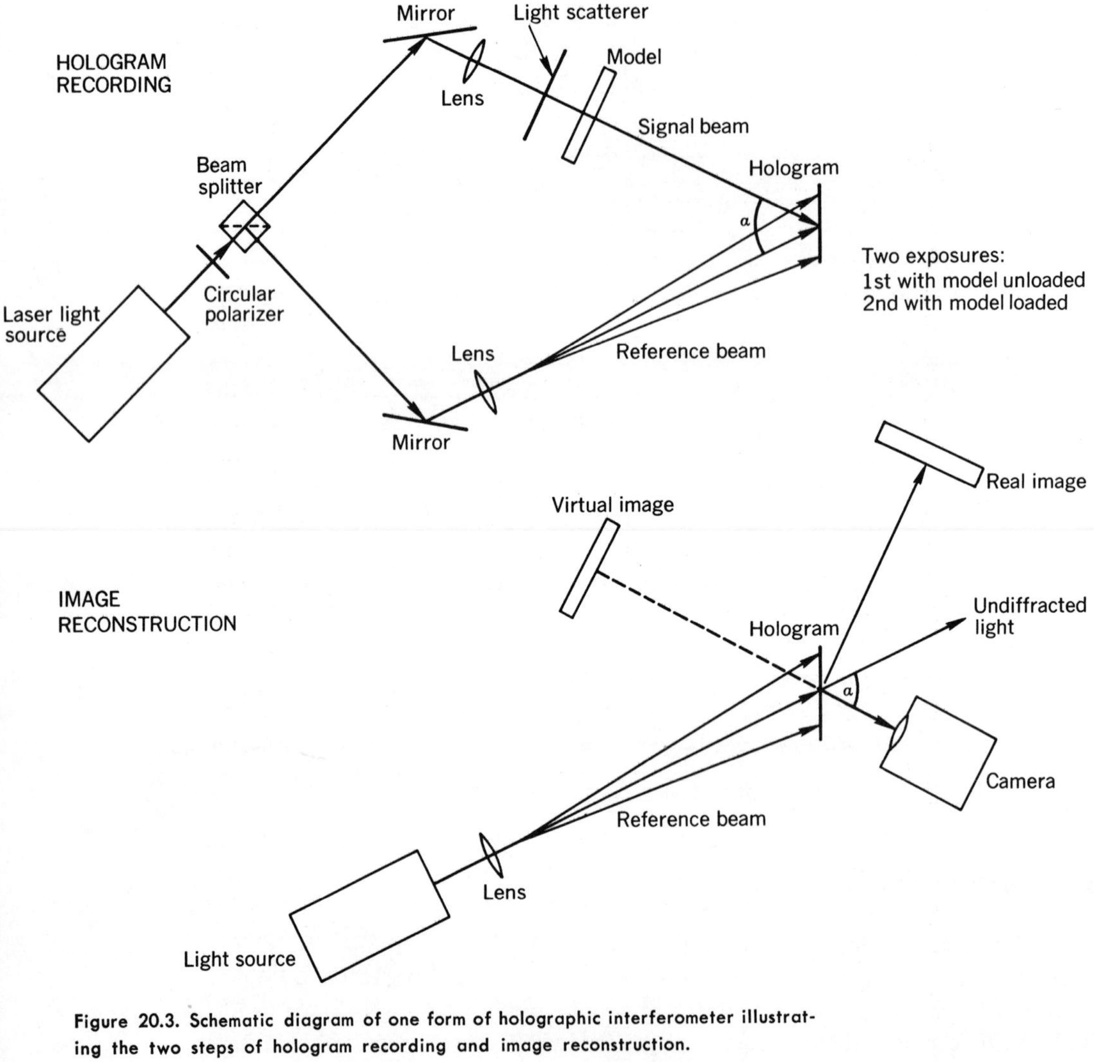

Figure 20.3. Schematic diagram of one form of holographic interferometer illustrating the two steps of hologram recording and image reconstruction.

light from a laser source is used so that the two intersecting beams will exhibit interference effects even when there are large differences in the path lengths of the two beams.

Lenses are included in holographic interferometers, as shown in Fig. 20.3, to expand the beams so that the entire model is illuminated and the reference beam diameter is large enough to illuminate an area of the recording film which is at least as large as the area of illumination resulting from the signal beam. The interference pattern of the superimposed reference beam and signal beam ("modulated" in phase by the model) which is recorded on the film is called a hologram. If a light scatterer is placed in the signal beam before the model, as shown in Fig. 20.3, the hologram acquires an additional property as described below.

When both the light scatterer and the model are removed, the hologram will consist of a family of parallel, equally spaced fringes produced by the two (approximately) plane waves that intersect at an angle α. The relation between the pitch (p) of the fringes, in inches, and the angle of separation (α), in radians, when the incident light is of wavelength 2.8×10^{-5} inch is:

$$\alpha = \frac{2.8 \times 10^{-5}}{p} \tag{20.1}$$

To obtain absolute retardation patterns by using holographic interferometry, two holograms are superimposed on the same film. One is recorded before the model is loaded and the second while the load is applied.

A second step required in holographic processes is the reconstruction of the image of the model from the hologram. As shown in Fig. 20.3, this is accomplished by modifying the apparatus so that only the reference beam illuminates the (developed) hologram. Some light is observed to pass through the hologram without being diffracted but a portion of the light is diffracted into two beams: one producing a real image, the other, a virtual image. They can be directly observed or else recorded by a camera as shown in the figure. If a light scatterer is used, as described above, the images can be viewed from many different directions. Otherwise, they can be viewed from only one direction. When images are reconstructed from holograms which are doubly exposed, as described above, they not only show the form of the object but also exhibit absolute retardation fringes (or moire patterns produced by the two families of absolute retardation fringes). Several books and papers provide more detailed information on this subject.[11,12,13]

One disadvantage of holographic interferometry has been the requirement for very high resolution of details in the hologram pattern. The effects of film grain, the limitations on the coherence of light sources, and the motion of the model, the mirrors, or even the air in the optical path during the period of exposure on hologram resolution have restricted practical applications of this method to static photoelasticity. Another disadvantage of the method is the difficulty of applying loads without producing rigid-body motions of the model between the two exposures of the hologram.

Section 20.4 describes a modified form of holographic interferometer which overcomes the difficulties noted above, but which shares with them the

---

[11] G. W. Stroke, *An Introduction to Coherent Optics and Holography* (New York: Academic Press, 1966).

[12] K. A. Stetson, and R. L. Powell, "Interferometric Hologram Evaluation and Real Time Vibration Analysis of Diffuse Objects," *Journ. Opt. Soc. Am.*, Vol. 55, No. 12 (December 1965), p. 1570.

[13] L. O. Heflinger, R. F. Wuerker, and R. E. Brooks, "Holographic Interferometry," *Journ. Appl. Phys.*, Vol. 37, No. 2 (February 1966), pp. 642–649.

advantage over classical methods that isochromatics and isopachics are produced alone rather than as a moire of other fringe patterns even with specimens that are not optically flat.

20.4. Modified Holographic Method

The modified method of holographic interferometry is shown schematically in Fig. 20.4. The light beam from a Q-switched pulse laser employing a ruby rod passes through a lens system that enlarges the beam diameter to approximately 14 inches and illuminates the model. Because the index of refraction of the model differs from that of air, light is reflected back from both of the model surfaces. The two beams of reflected light pass back through the same condensing lens and are recorded on a photographic plate at a point where the path of their beams is sufficiently separated from the path of the incident beam. Because the two beams differ only slightly in direction of propagation, their interference pattern produces a hologram of very low spatial frequency or density (typically, 10 to 300 lpi). Except for changes in magnification, the same pattern has been recorded at different positions over more than 12 inches of the reflected beam path, although portions of the incident beam are usually blocked in other positions than that shown. The patterns can also be recorded with a camera focused on the model or with a photographic plate placed on the other side of the model. Contrast in the latter recording, however, is very poor.

Figure 20.4. Schematic diagram of modified form of holographic interferometer.

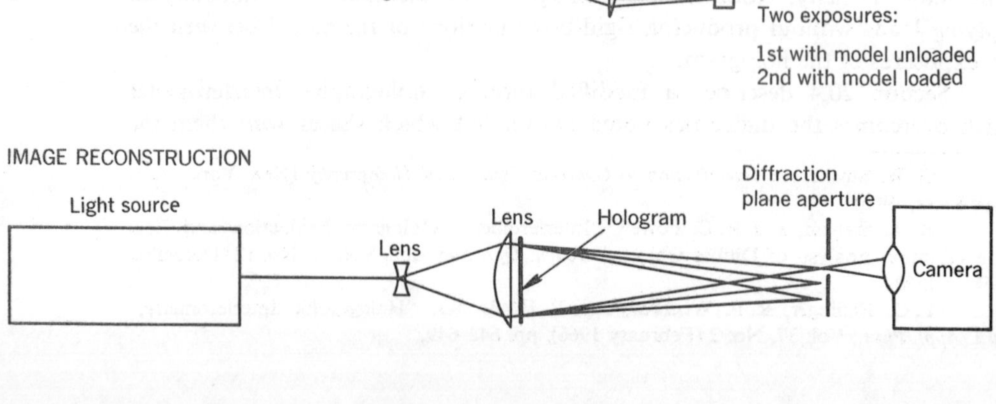

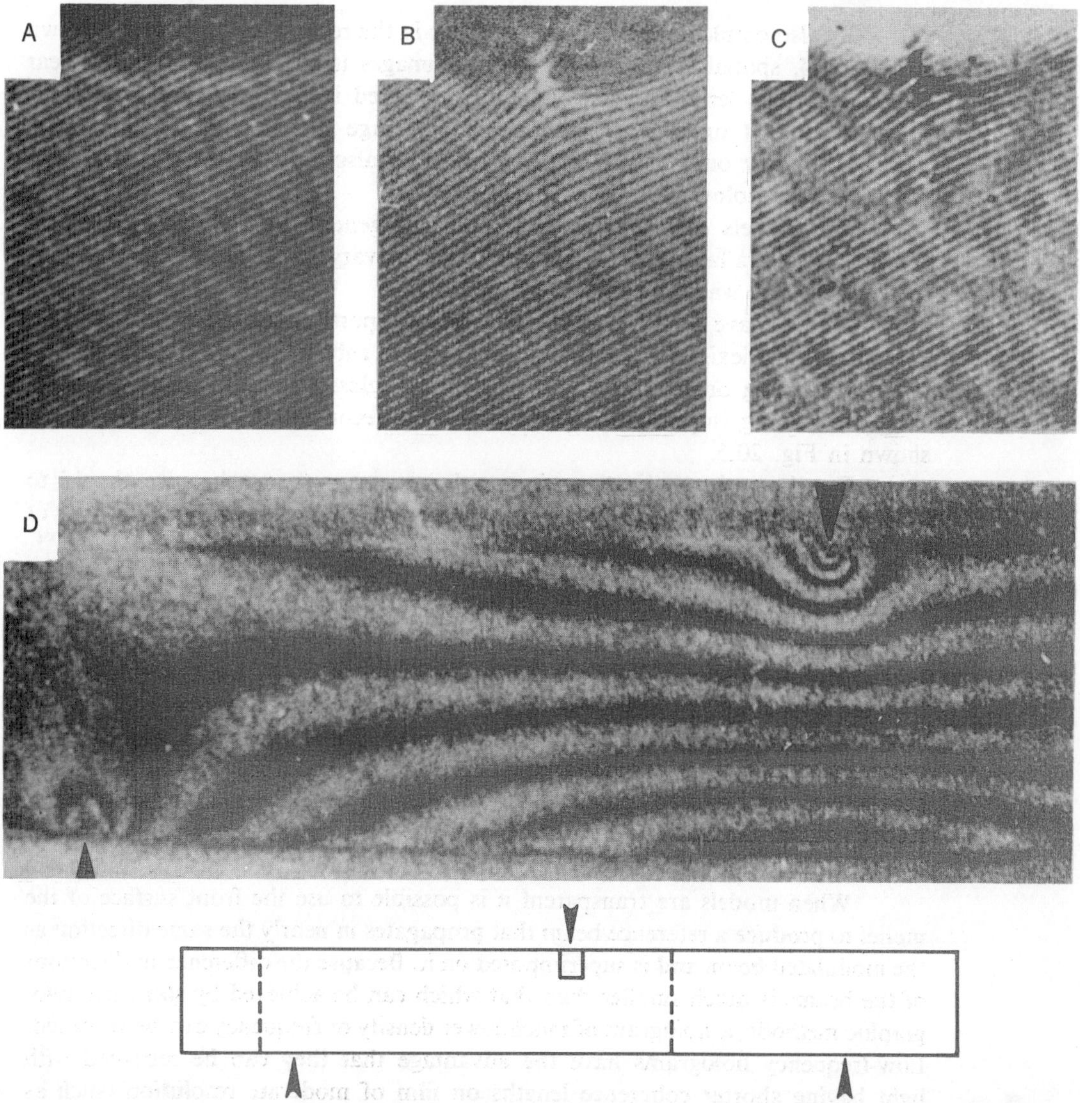

Figure 20.5. Enlarged section of holograms of (a) unloaded beam, (b) loaded beam, (c) superposition of loaded and unloaded beams, (d) isopachics obtained by filtering of the moire.

Two holograms are obtained: one corresponding to the unloaded state and the other to the loaded state of the model. They can be superimposed on the same film or recorded separately, as shown in Figs. 20.5 and 20.7. Due to the low spatial frequency of this type of hologram, the zero-order beam and the diffracted image beams produced in the reconstruction step are separated by a very small angle.

By also employing the condensing lens in the reconstruction step, as shown in Fig. 20.4, spatial separation of the two images is achieved in a region near the focus of the lens. To record the reconstructed image alone, an aperture is positioned to let only one of the diffracted image beams pass. This image is recorded directly on a photographic plate and is also photographed by a camera focused on the hologram.

The models of interest in photoelasticity generally can be prepared from commercially available stock with surfaces that vary from flatness by approximately 10 to 100 wavelengths (0.3 to 3 mils).

Models have been prepared from selected portions of commercially available sheets of Plexiglas or epoxy. Soft urethane rubber models have been prepared by casting on mercury pools or between glass plates. An example of an image containing isopachic fringes obtained by reconstruction of a hologram is shown in Fig. 20.5.

For dynamic applications it is only necessary to synchronize the 15 to 25-nanosecond flash of the *Q*-switched pulse laser with the dynamic loading of the model. This was accomplished by a method described in an earlier article.[14]

Although the setup shown in the top half of Fig. 20.4 has the form of a classic interferometer, the method described above can be called holographic because it incorporates a two-step imaging process: a hologram recording step and an image reconstruction step. In the first step a reference beam is allowed to interfere with a second beam that is modulated by (passage through) the object. Their interference pattern is recorded on a photosensitive film to produce a hologram.[15] In the second step a reference beam is diffracted by the hologram. Some of the diffracted light reconstructs a virtual image of the object that can be photographed with a camera. This second step has also been used by Brooks[16] who identifies it as a low-angle holographic interferometry.

When models are transparent it is possible to use the front surface of the model to produce a reference beam that propagates in nearly the same direction as the modulated beam and is superimposed on it. Because the difference in directions of the beams is much smaller than that which can be achieved by standard holographic methods, a hologram of much lower density or frequency can be recorded. Low-frequency holograms have the advantage that they can be recorded with light having shorter coherence lengths on film of moderate resolution (such as Kodak Contrast Process Pan or infrared film). The methods of recording high- and low-frequency holograms are compared in Fig. 20.6. Because the diffraction pattern produced by the reflection of a plane wave in a large, flat mirror is almost

---

[14] J. A. Clark, A. J. Durelli, and V. J. Parks, "Photoelastic Study of High Frequency Stress Waves Propagating in Bars and Plates," *Journ. Appl. Mech.*, Vol. 35, No. 4 (December 1968), pp. 747–753.

[15] This type of hologram has been discussed in O. Bryngdahl and A. W. Lohman, "Interferograms are Images Holograms," *Journ. Opt. Soc. Amer.*, Vol. 58, (1968), pp. 141–142.

[16] R. E. Brooks, "Low-Angle Holographic Interferometry Using Tri-X Pan Film," *Appl. Opt.*, Vol. 6 (1967), pp. 1418–1419.

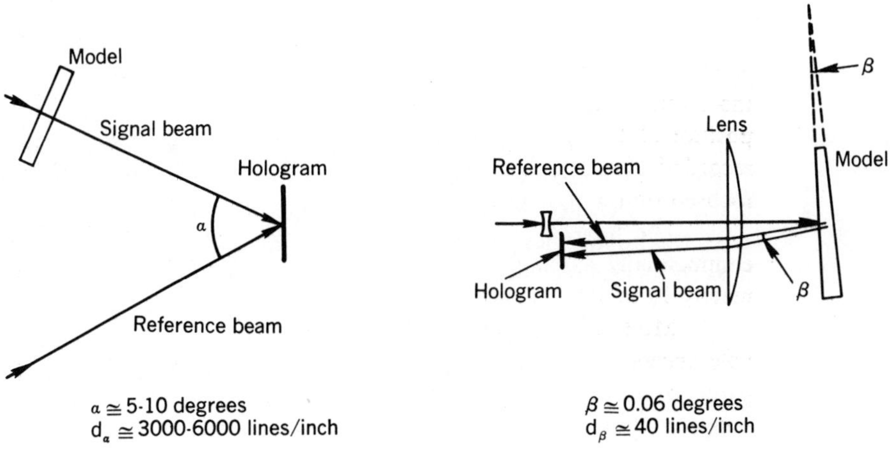

Figure 20.6. Comparison of high- and low-frequency holograms.

identical to the pattern of illumination predicted by geometric optics, the low-frequency holograms recorded by this method contain patterns that strongly resemble the original object except for the superimposed interference fringes. If a single exposure hologram were reconstructed by the method described above, the original image without the interference fringes would be obtained.

Each of the steps involved in this method, however, can be given a more familiar interpretation. The recording system corresponds to the form of classical interferometer described in the introduction except that in the case of recording the interferogram a camera focused on the model is necessary. The low-frequency hologram can be recorded without a camera. The reconstruction system is a generalization (because the aperture need not be precisely in the diffraction plane of the optical system) of a spatial filter.

The principles of operation of spatial filters are illustrated by the schematic diagram of the image reconstruction step shown in Fig. 20.4. Spatial filters were discussed in Sec. 1.8. Only a brief summary of their operation is presented here. A narrow beam of light passes through a diverging lens so that it illuminates a large portion of a condensing lens. If no obstruction is placed in the light beam after the condensing lens, it will be focused to a single point located in a plane often identified as the *diffraction* plane or *fraunhofer* plane. If a hologram with the form of a simple grating is placed in the beam close to the condensing lens, instead of a single point of light a row of evenly spaced points will appear in the diffraction plane. By placing an aperture in the diffraction plane that stops certain points and allows others to pass, modified images of the grating, which differ from the object grating in pitch, can be recorded by a camera focused on the grating as shown in Fig. 20.4. When two gratings are placed in the beam close to the condensing lens, it is possible to obtain an image that shows only their subtractive moire by allowing light from just one point to pass. An example of an application of spatial filtering to moire patterns produced by the superposition of an amplitude grating and its reflection off a curved surface, and further references concerning the method were given by Beranek in a recent article.[17]

A second modification of standard holographic methods has been used to improve the photoelastic patterns in some of the applications described in this chapter. In the recording step the pattern has been photographed with a camera focused on the model. The chief advantage of this modification is that, in some cases, the boundaries of the model are clearer in the reconstructed image than those in images reconstructed from low-frequency holograms.

Post has pointed out[18] that the superposing of absolute retardation patterns from loaded and unloaded models introduces an error not found in similar combinations such as isothetic mismatch. Since both the loaded and unloaded patterns represent specific values of retardation at a point, then if the loaded and unloaded images of a point are not superposed exactly, the distance between them may be a source of error.

Suppose a point in the deformed body is displaced a distance equal to the distance between two interference fringes in the absolute retardation pattern on the undeformed body. Then the error in the moire of the two absolute retardation patterns will be one fringe. This effect is most pronounced where an attempt is made to superpose a large field. However, any point in the field can be accurately analyzed by aligning it with its no load position. The error produced by lack of coincidence of the positions of points in the two gratings will be treated in a more general way in Sec. 20.8.

20.5. Application to Static Problems

The process of obtaining isopachics is illustrated by the sequence of photographs in Fig. 20.5. The model is a Plexiglas beam supported near both ends and subjected to a (static) centrally concentrated load. The specimen material was selected from a portion of a sheet of lucite that was slightly wedge-shaped in its thickness cross-section. The effect of the wedge shape is observed in the unloaded hologram (Fig. 20.5a) to be family of nearly parallel straight lines with a density of approximately 100 lpi. In the corresponding pattern (Fig. 20.5b) for the loaded model, some curvature of fringes near the load point can be observed. By superimposing the two patterns into a single hologram (Fig. 20.5c), a moire effect is observed similar to that described above in connection with classical interferometers. The isopachics are visible in the superimposed holograms but the advantage of using isopachics obtained by reconstructing the bar image (Fig. 20.5d) for analysis is readily apparent. Because Plexiglas exhibits negligible birefringence

[17] W. J. Beranek, "Rapid Interpretation of Moire Photographs," *Exp. Mech.*, Vol. 8, No. 6 (June 1968), pp. 249–256.

[18] Private communication.

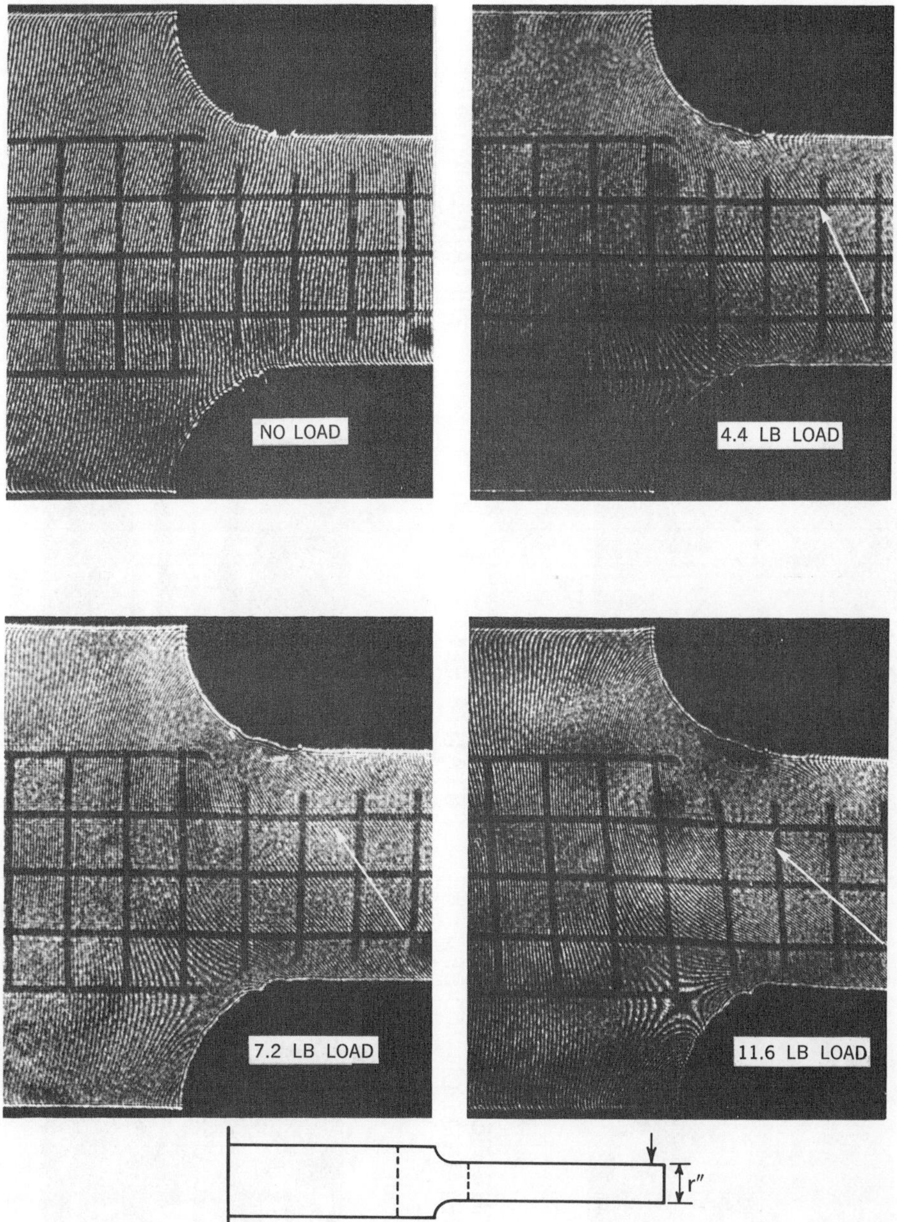

Figure 20.7. Low-frequency holograms of a cantilever beam with varying height prepared from a photoelastic material (CR-39) and loaded to various levels.

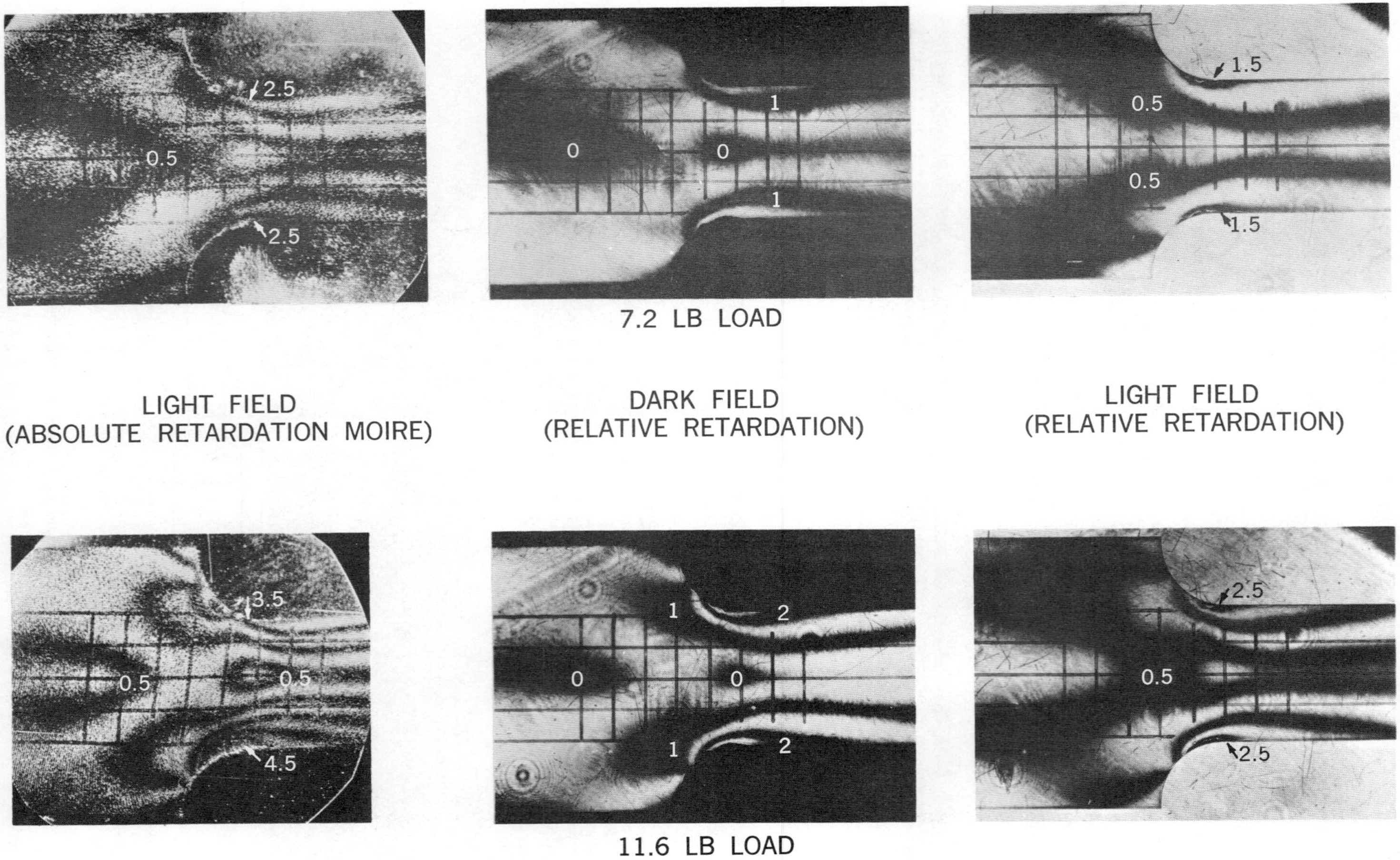

Figure 20.8. Isochromatics in a portion of a cantilever beam loaded at the free end obtained from the absolute retardation moire and by relative retardation methods.

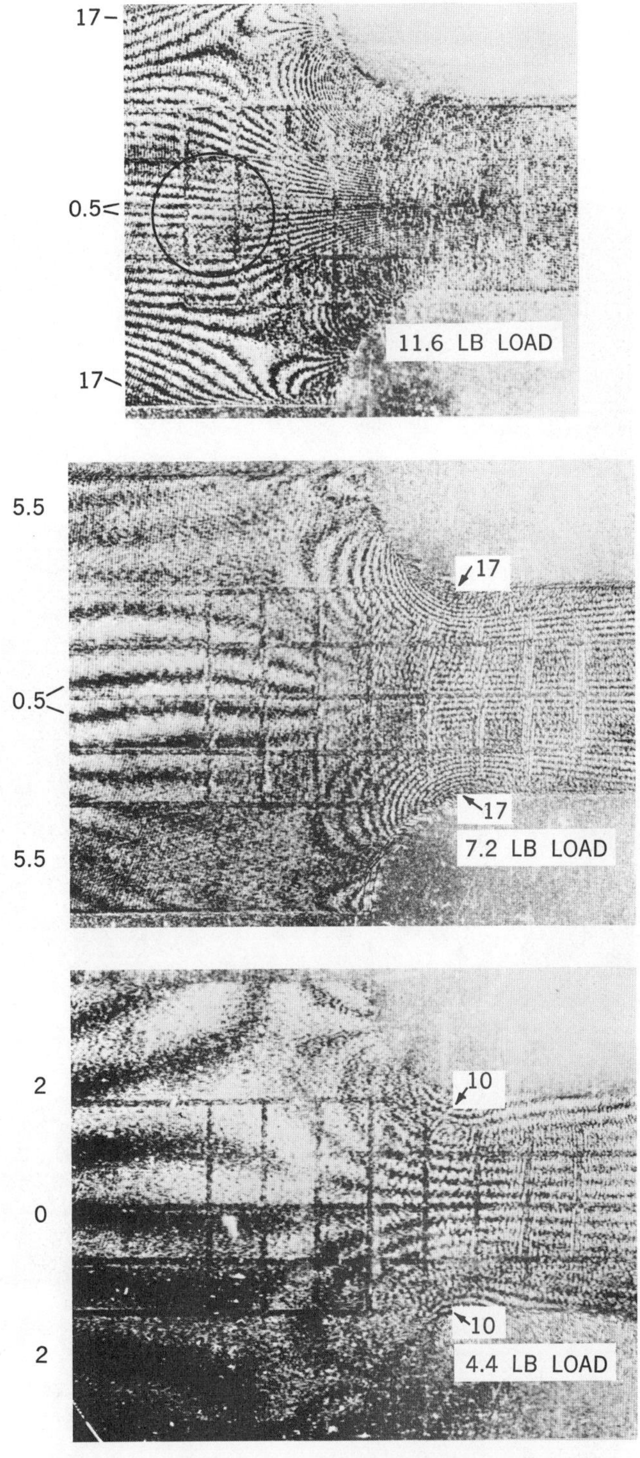

Figure 20.9. Isopachics in a portion of a cantilever beam loaded at the free end to various levels, obtained by the modified method of holographic interferometry.

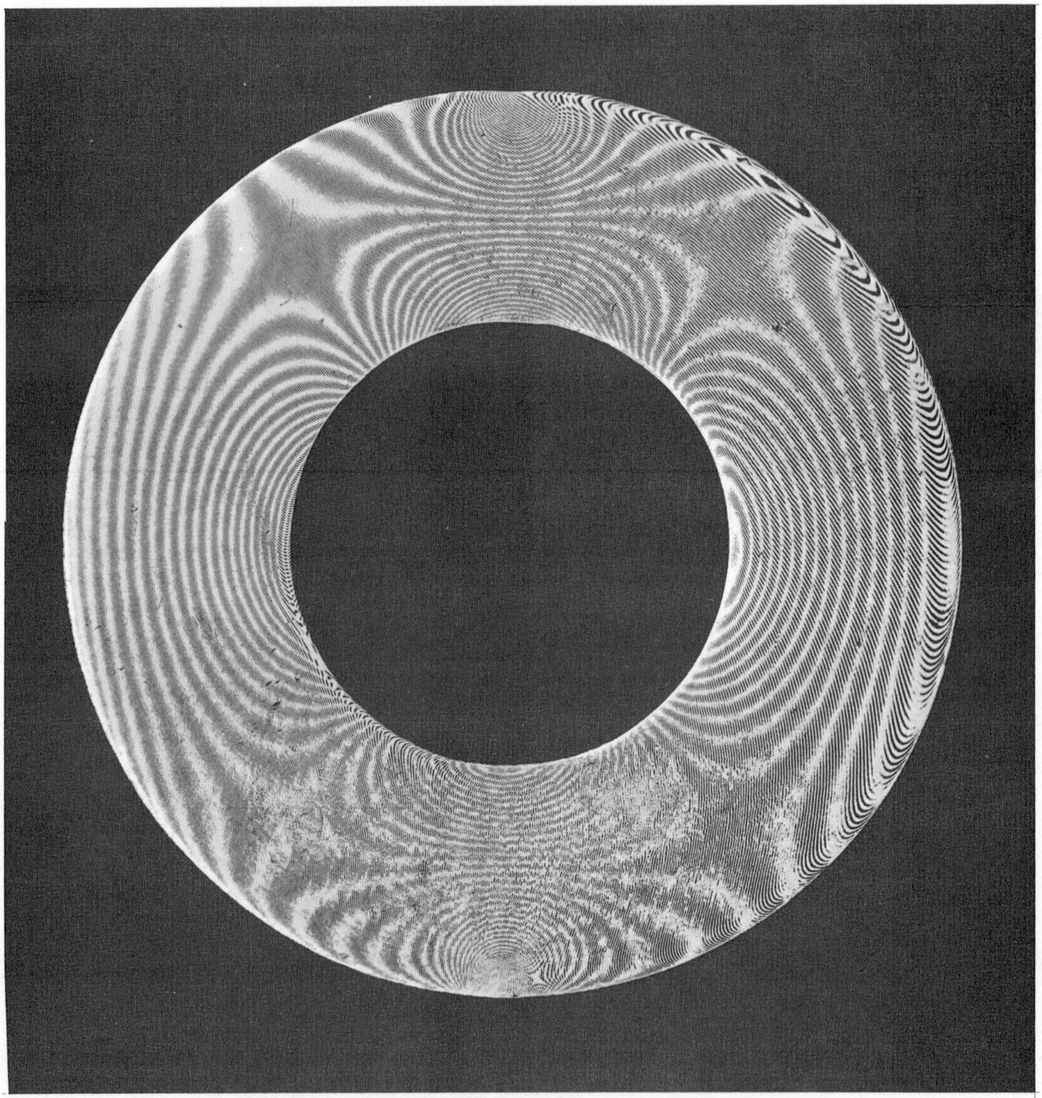

Figure 20.10. Isopachics of ring under compression before filtering.

Figure 20.11. Isopachics of ring under compression after filtering.

under the small load applied in this problem, only a single family of curves is observed.

As an example of a problem employing a material with greater photoelastic sensitivity (CR-39), patterns obtained in the analysis of a cantilever beam with varying cross-section loaded at the free end to various levels are shown in Figures 20.7, 20.8, and 20.9. Because the model was made from a sheet with a less severe thickness wedge shape, coarser fringe patterns resulted (Fig. 20.7). (The interference fringes should not be confused with the $\frac{1}{4}$-inch reference grid drawn on the specimen.) The effect of increasing load levels on portions of the beam with constant cross-section is observed as a rotation of the fringes (indicated by the arrow in Fig. 20.7). In addition, regions appear in the patterns where the fringes seem to fade. This is the result of the fringes dividing to form two families—one corresponding to each principal value of the index of refraction in the birefringent material. The two families of fringes cross each other and form a moire that is observed easily when the images reconstructed from the loaded holograms are observed (Fig. 20.8). Since the moire is a measure of the difference of the two families described above, it can be identified as an isochromatic pattern. This identification is verified by comparison with light and dark field isochromatics obtained by the usual relative retardation method under the same loading conditions. The isochromatics obtained from the absolute retardation moire should be twice as dense as those obtained by the relative retardation method because the signal beam passes through the model twice in the former method.

The isochromatics shown in Fig. 20.8 that were obtained from the absolute retardation moire were reconstructions from singly exposed holograms. When a hologram recorded at one load level is superimposed on the hologram recorded without load and a reconstruction is made, isopachic patterns are obtained, as shown in Fig. 20.9. (The isochromatics are observed in this figure only with great difficulty because they are of lower contrast than the isopachics.) The half-order shift in isopachic fringes as they cross the 0.5 isochromatic is visible in the region indicated by a circle in the figure. It is interesting to note that for this particular material the sensitivity of the method to the sum of the stresses is an order of magnitude greater than its sensitivity to the difference of the stresses. Another example of application of this method, to the determination of isopachics in a material exhibiting negligible birefringence, is shown in Figs. 20.10 and 20.11. The isopachics were obtained from a Plexiglas ring loaded statically along a diameter. Figure 20.10 shows both the isopachics and the isobathics (as fine fringes). In Figure 20.11 the isobathics have been eliminated by filtering.

20.6. *Application to Dynamic Problems*

An application to a dynamic problem is shown in Fig. 20.12. Flexural stress waves were generated in a $\frac{1}{2}$-inch thick, $\frac{3}{4}$-inch wide, and $13\frac{1}{2}$-inch long soft rubber bar (Solithane 113) by the transverse vibrations of a steel cantilever beam

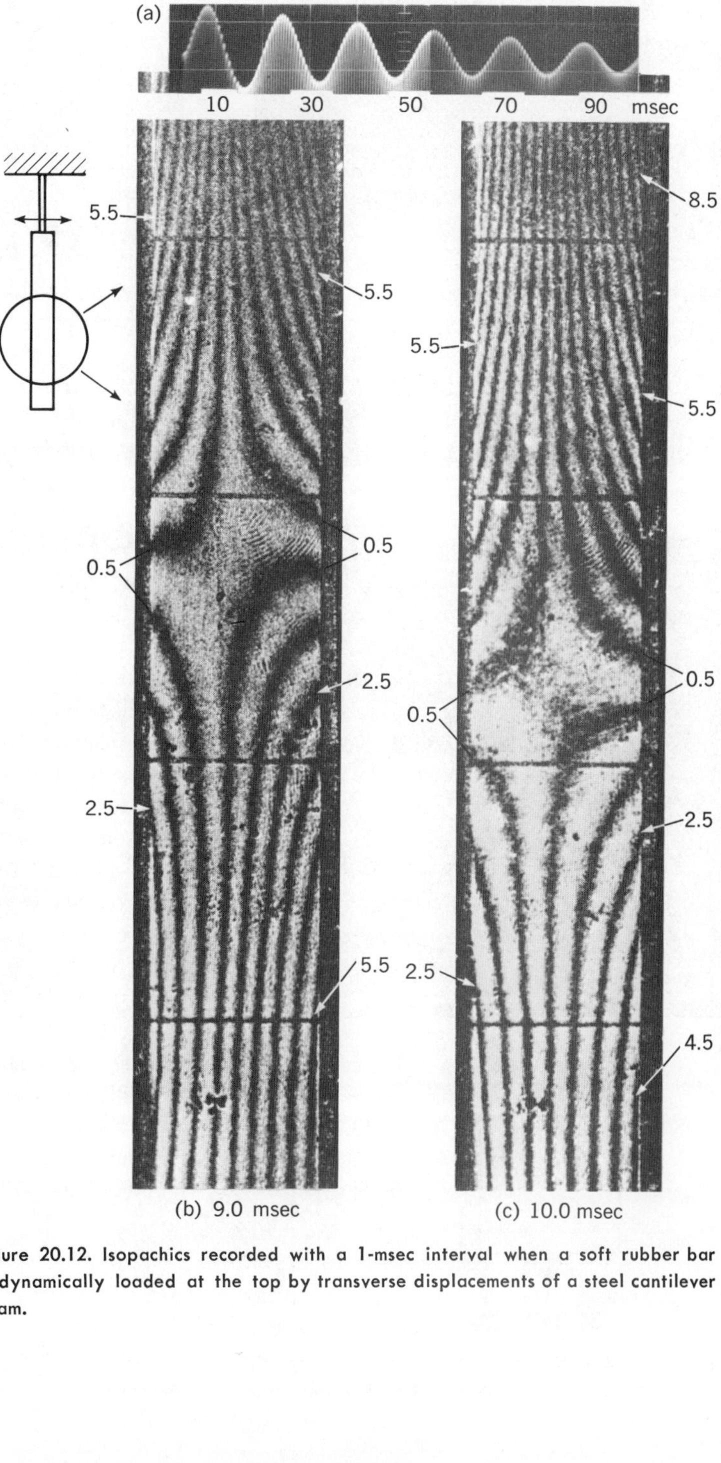

Figure 20.12. Isopachics recorded with a 1-msec interval when a soft rubber bar is dynamically loaded at the top by transverse displacements of a steel cantilever beam.

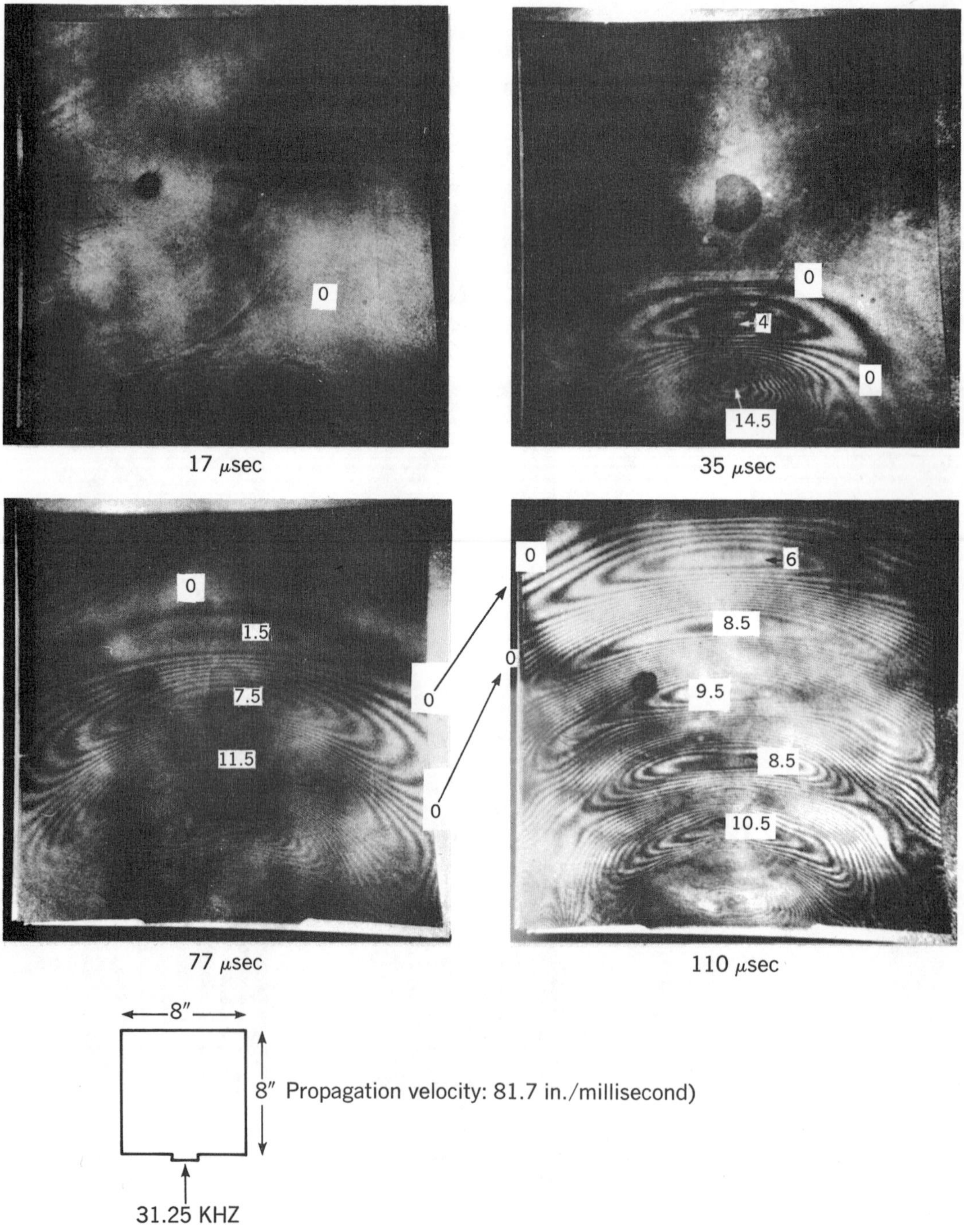

Figure 20.13. Isopachics recorded when a Homolite 100 plate is dynamically loaded at one edge by a 31.25-kHz toneburst from a piezoelectric transducer.

cemented to the top of the rubber bar. When excited by the mechanical impact, the cantilever beam vibrated at a frequency of 62.5 Hertz, as shown in Fig. 20.12a. In this case, the interference patterns were recorded by a camera focused on the model. Dynamic isopachics are shown at two times separated by a 1-millisecond interval. The velocity of flexural wave propagation computed from these photographs is 0.5 in./msec. The isopachics are in agreement along the free boundaries with isochromatics obtained by the relative retardation method. Throughout the interior of the bar they agree with the hypothesis that the low-frequency transverse loading produces flexural waves in the bar.

In a second dynamic application, stress waves were generated in a plate of photoelastic material (Homolite 100). The plate was standing on a steel platform that had a hole cut in the region where a piezoelectric transducer was located that loaded the model with a 31.25-kHz toneburst. Details of the loading system are available in an earlier paper by the authors.[19] Dynamic isopachics obtained with the modified holographic interferometer are shown in Fig. 20.13 at four different times. Also, in this case, the interference patterns were recorded by a camera focused on the model. The patterns indicate the velocity of propagation of the wave (81.7 in./msec) and also exhibit a form that is consistent with the simple harmonic form of the loading.

It is interesting to note that fringe densities as high as 22 fringes per inch are observed. In order to resolve such fringes on a photograph while they are propagating at 81.7 in./msec, it is necessary to use a light source whose duration is less than the time required for a fringe to move a small fraction of the distance between fringes in order to obtain recordings with sufficient fringe contrast. The densest fringes in Fig. 20.13 move $\frac{1}{4}$ of the distance between fringes in 0.14 microsecond. Since the shortest duration, high-intensity light sources—with the exception of pulse lasers—have durations greater than 0.5 microsecond, these fringes could not have been recorded without a Q-switched laser light source.

20.7. *Spatial Derivatives of Isopachics*

A method to obtain photographically, by moire-of-moire, the spatial derivatives of the displacement components, has been explained in Sec. 7.8. The same method can be used to obtain the spatial derivatives of isopachics (loci of points exhibiting the same value of change of thickness in a plate) or the spatial derivative of the thickness of the plate. To obtain these derivatives, two photographic films of the pattern are superposed and shifted in the direction of differentiation. Figures 20.14 and 20.15 show the partial derivatives, in the x- and y-directions, respectively, of the original thickness in the beam (no load) shown in Fig. 20.7. The patterns show a marked gradient of thickness in the x-direction, and very little change in the y-direction. As could be expected the moire-of-moire

[19] See Footnote 14 this chapter.

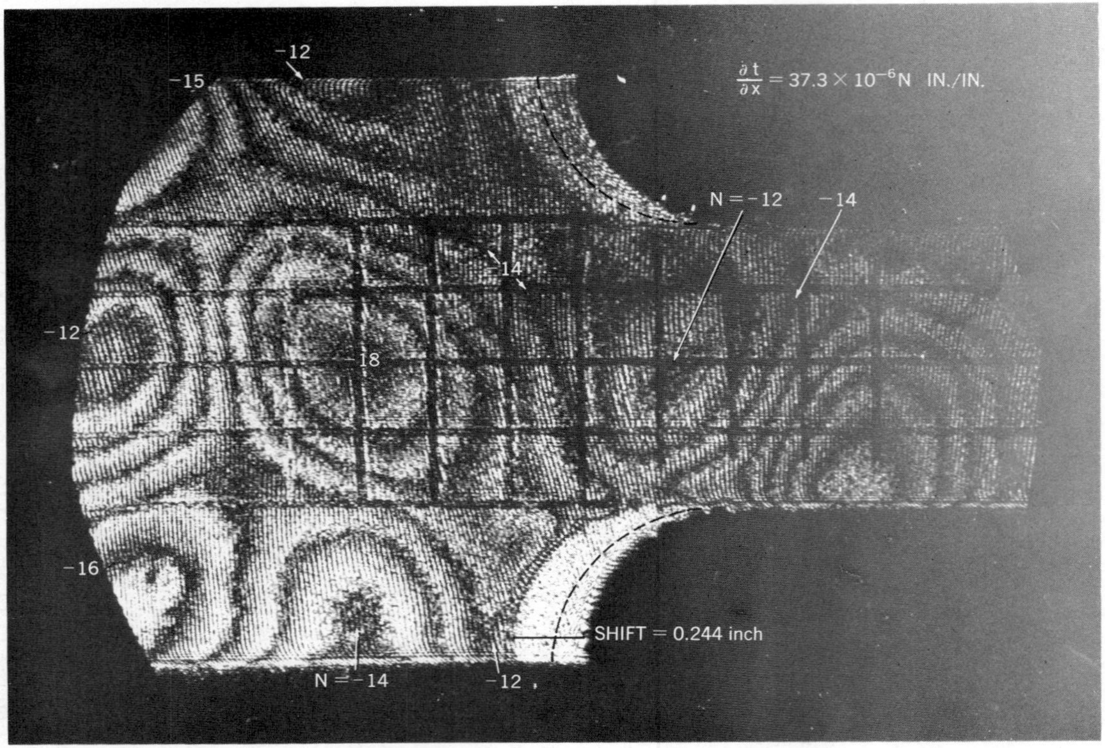

Figure 20.14. Spatial derivative $\partial t/\partial x$ of isopachics in a nonloaded beam.

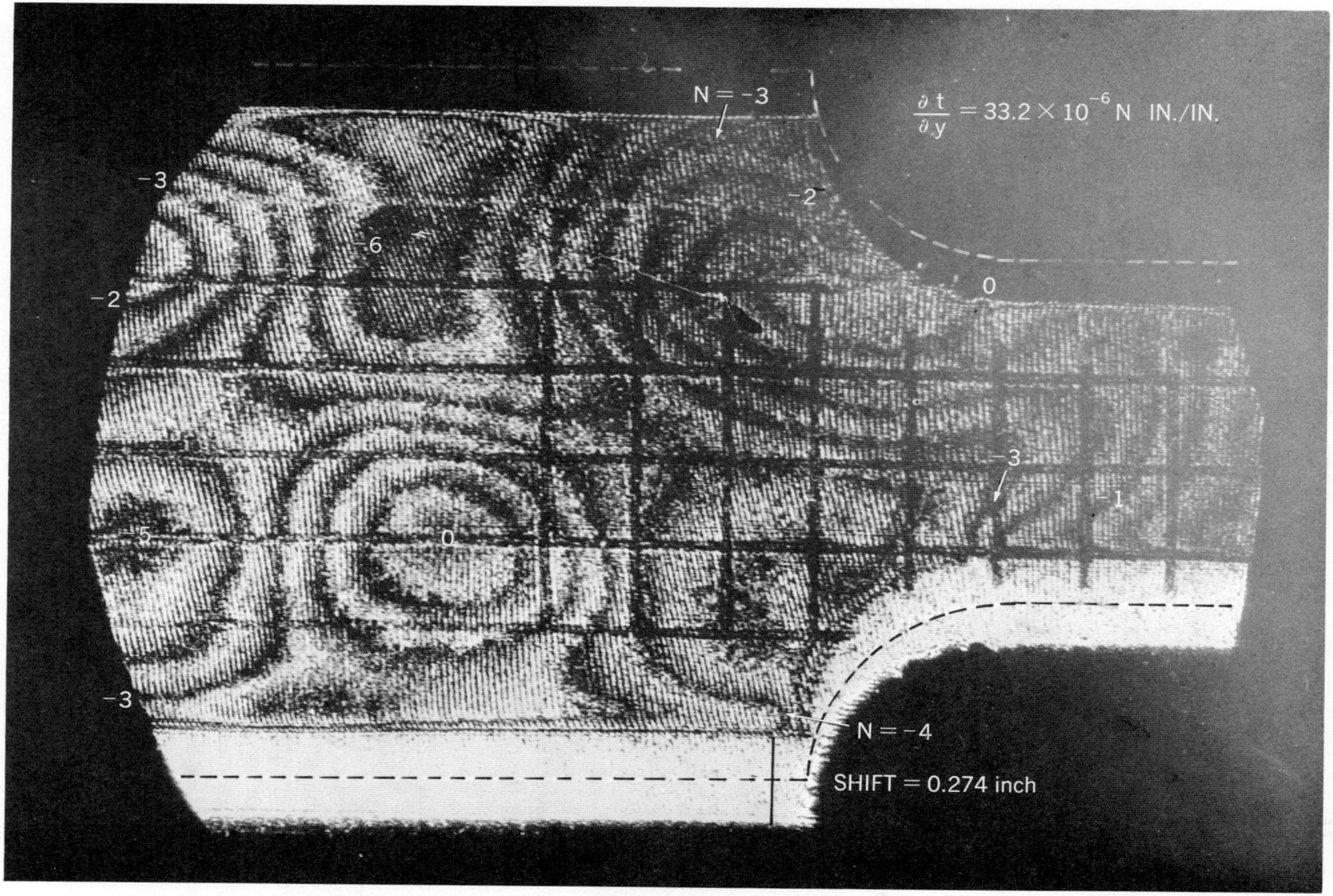

Figure 20.15. Spatial derivatives $\partial t/\partial y$ of isopachics in a nonloaded beam.

fringe of zero order in the pattern of $\partial t/\partial y$ is the locus of points of the fringes of the first-order moire, at which the tangent to those fringes is parallel to the y-axis. It can also be pointed out that in these patterns (Figs. 20.14 and 20.15) the order of the moire-of-moire fringe is equal at any point in the pattern to the number of first-order moire fringes in an interval equal to the shift.

20.8. *Lack of Coincidence of Points in Superposed Patterns*

If a moire pattern is produced to obtain the rate of change of a function by superposing patterns of the function obtained at two levels of load (as when velocities are obtained by superposing displacement patterns obtained at two different load levels), the patterns should be superposed so that the positions of the points of interest coincide in the two patterns. Since load, in general, is associated with deformation, it means that the distance between points of the body changes, and this condition of coincidence cannot be satisfied for every point of the loaded body,

On a moire pattern produced by two gratings (which can be two first-order moire) associated with two load levels, consider a region where the positions of the same physical point do not coincide in the images of the two load levels. Consider two points E and F in this region of the body. Say that the image of point E from the lower load level is superposed on the image of point F from the higher load level. The order of the fringe passing through the two superposed points is equal to the difference in order of the lines of the gratings passing through that point. The moire at the point should have been produced by images of the same point at the lower and higher levels. The fact that the image of point E was used at the lower level will introduce an error in the moire fringe value of a magnitude equal to the numer of lines between the images of points E and F in the lower level image.

This error becomes more severe, the larger the model. For large models it may be necessary to adjust and readjust the two images to produce coincidence of specific regions of the model so as to minimize the error.

This correction applies to the velocity patterns described in Chapter 10 as well as those obtained with a mismatch pattern, such as the isopachics obtained from an initially wedge-shaped plate as described above.

No correction has to be applied to the basic moire pattern of a displacement component since displacement of a point is precisely the quantity that this moire measures.

21

IN-PLANE MOIRE STRAIN ANALYSIS OF BENT PLATES

21.1. Introduction

The out-of-plane moire method developed by Ligtenberg[1] and later applied by Bradley[2] is described in Chapter 13. The method is widely used for the experimental strain and stress analysis of bent plates and requires the manufacturing of a model of the plate made with a reflecting surface which is used as a mirror to observe a coarse grating of equidistant parallel lines. When the model is loaded, the image of this grating changes its position due to the change in slope of the bent plate. By superposing the two images, the one obtained before and the one obtained after loading (for example, by double exposure on the same film), moire fringes are produced. Two such moire patterns obtained with the grating set in orthogonal directions (for example, x- and y-directions) are required for the complete solution of the problem. The two moire patterns so obtained represent contour lines of the slope of the bent plate. The moment distribution in plates can then be determined by differentiation. Since this slope is associated with an out-of-plane displacement, this method may be called an out-of-plane moire method.

---

[1] F. K. Ligtenberg, "The Moire Method—a New Experimental Method for the Determination of Moments in Small Slab Models," *Proc. Soc. Exp. Stress Analysis,* Vol. XII, No. 2 (1955), pp. 83–98.

[2] W. A. Bradley, "Laterally Loaded Thin Flat Plates," *Proc. ASCE,* Vol. 85, No. EM4, (October 1959), pp. 77–107.

21.2. Limitations of Out-of-Plane Moire Method

The following strain-curvature relationships are the basic equations that apply to plates subjected to lateral loads and small deflections.

$$\epsilon_x = -z\frac{\partial^2 w}{\partial x^2}$$

$$\epsilon_y = -z\frac{\partial^2 w}{\partial y^2} \tag{21.1}$$

$$\gamma_{xy} = -2z\frac{\partial^2 w}{\partial x \partial y}$$

The out-of-plane moire method has distinct limitations when applied to the solution of certain bending problems; for example, (1) plates with large deflections, (2) plates subjected to combined in-plane and lateral loads, and (3) plates with discontinuities, the size of which is of the order of the thickness of the plate. The following are some of the difficulties that may be encountered.

(a) Since the strain-curvature relationship (Eq. 21.1) used for the interpretation of out-of-plane moire data is based on the small deflection theory of laterally loaded plates, errors may be introduced when it is applied to plates subjected to large deflections or to combined in-plane and lateral loads.

(b) In case of plates with relatively small cut-outs, the out-of-plane moire pattern may not show the effect of the cut-out since strains near the cut-outs are no longer proportional to the curvature of the plate.

21.3. In-Plane Moire Method

It is possible to use the general in-plane moire method described in this book, for the analysis of bent plates. This general method is not subjected to the limitations listed in the previous section, but it does require a special technique.

The moire pattern for the bent plates is obtained, as is general in moire applications, by superposing a master grating on the deformed grating of the plate, but in this case the master grating has to follow the bent surface of the plate. The master grating is subjected to strain due to bending; however, the error introduced by this strain can be computed. For example, assume that the plate and master grating film are subjected to cylindrical bending. The strain in the plate and grating will be of opposite sign and will be given by

$$\epsilon = \pm\frac{h}{2\rho} \tag{21.2}$$

where h is the thickness and ρ is the radius of curvature.

The radius of curvature will be the same for plate and film, so that the unwanted bending strain of the master will be in the same ratio to the bending strain of the plate as the ratio of film thickness to the plate thickness. If the film thickness is of the order of 0.003 in., the plate should be at least 0.100 in. thick for the error to be less than 3 per cent.

To have a perfect contact between the specimen and master gratings during bending of the plate, a thin film of oil is used between them. It is assumed that this thin oil film keeps the master in contact with the specimen without transmitting any shears.

The moire pattern, obtained for the bent plate using the above procedure, can be directly photographed using a camera if the deflection of the plate is small. However, if the deflection of the plate is large, this procedure does not give a pattern readily interpretable. To overcome this difficulty, a technique, which does not require the use of a camera, is indicated below in the description of the experiment.

Two solutions are presented below: (i) a simply supported square plate subjected to a uniformly distributed load producing small deflection; (ii) a square cantilever plate subjected to a concentrated load at one corner, producing large deflections.

A theoretical solution of the stress distribution for the first problem is available.[3] For the second problem, a general solution of the displacements and moments has been given recently by Lin et al.[4]

21.4. *Experimental Work*

21.4.1. Simply Supported Square Plate under Uniform Pressure

A 1000-lines/in. grating was printed on the surface of a Plexiglas plate 6 in. × 6 in. and $\frac{1}{4}$ in. thick. The plate was placed in a loading fixture in such a way that it was simply supported along the four sides leaving a $\frac{1}{4}$-in. overhang all around. A master grating of 1000 lines-per-inch was placed in contact with the printed surface of the plate. A thin layer of paraffin oil was used between the master grating and the specimen to insure contact without producing air gaps. A column of mercury applied the uniform pressure to a sealed cell that rested against the back surface of the plate. The moire patterns produced in the loaded plate were photographed using a camera. Typical u and v moire patterns are shown in Fig. 21.1.

21.4.2. A Square Cantilever Plate Subjected to Corner Load

The case of a square cantilever plate subjected to a concentrated load at one corner was chosen to illustrate the application of in-plane moire method to

[3] S. Timoshenko and S. Woinowsky-Krieger, *Theory of Plates and Shells*, 2nd ed. (New York: McGraw-Hill, Inc., 1959).

[4] T. H. Lin, S. R. Lin, and B. Mazelsky, "Large Inextensional Deflection of Thin Cantilevered Plates," *Journ. of Applied Mechanics*, Vol. 35, No. 4 (1968), pp. 774–777.

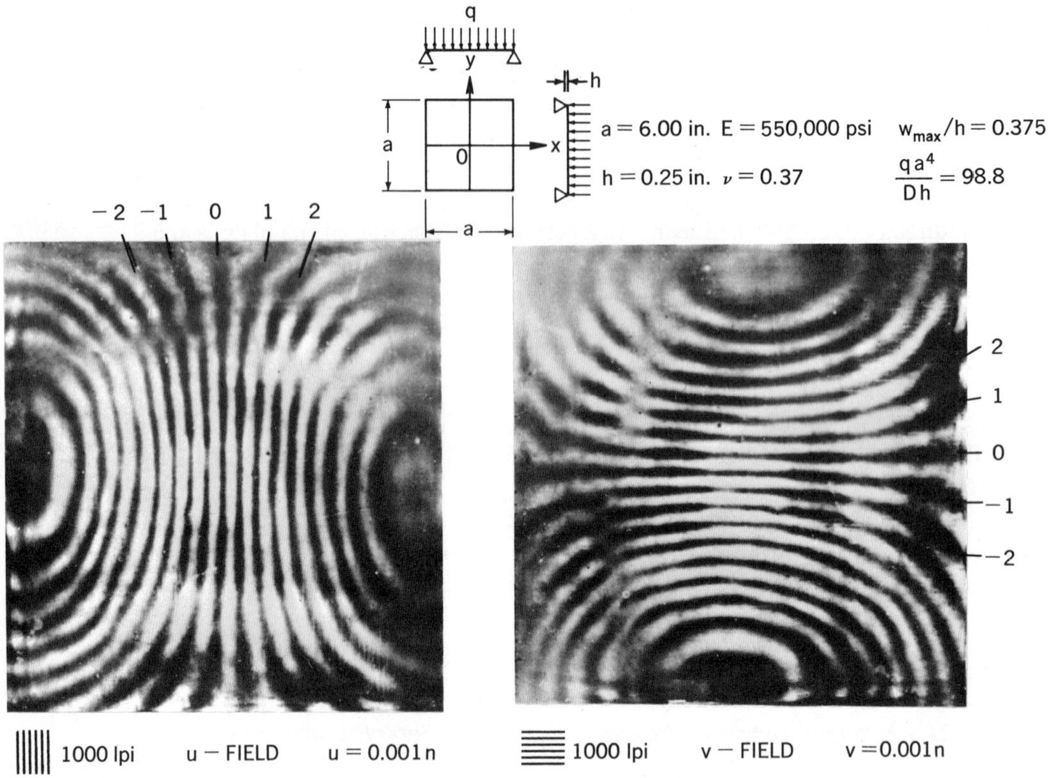

Figure 21.1. Isothetics in a largely deflected cantilever plate subjected to a corner load.

laterally loaded plates with large deflections. The size of the plate was 7.875 in. × 7.875 in. × $\frac{1}{2}$ in. thick and was made from a polyurethane rubber sheet (Hysol 4485). A 300-lines/in. cross grating was printed on the plate. The plate was mounted vertically and clamped along the bottom edge. The load was applied at one corner on the top by means of a string going over a pulley.

A 300-lines/in. one-way grating was used for the master, printed on a photographic film 0.003 in. thick. In order to record the undistorted image of the isothetics on the surface of the specimen, the photographic technique mentioned above was used.[5] A sheet of photographic film (covering the full size of the specimen) was placed on top of the plate and in contact with the master grating. A diffused white light source on the opposite side of the plate was used to expose the film, transmitting the light through the transparent specimen. The photograph

[5] A. J. Durelli, V. J. Parks, and L. Ferrer, "Photography Without Lenses," *Exp. Mech.*, Vol. 8, No. 3 (March 1968), pp. 142–144.

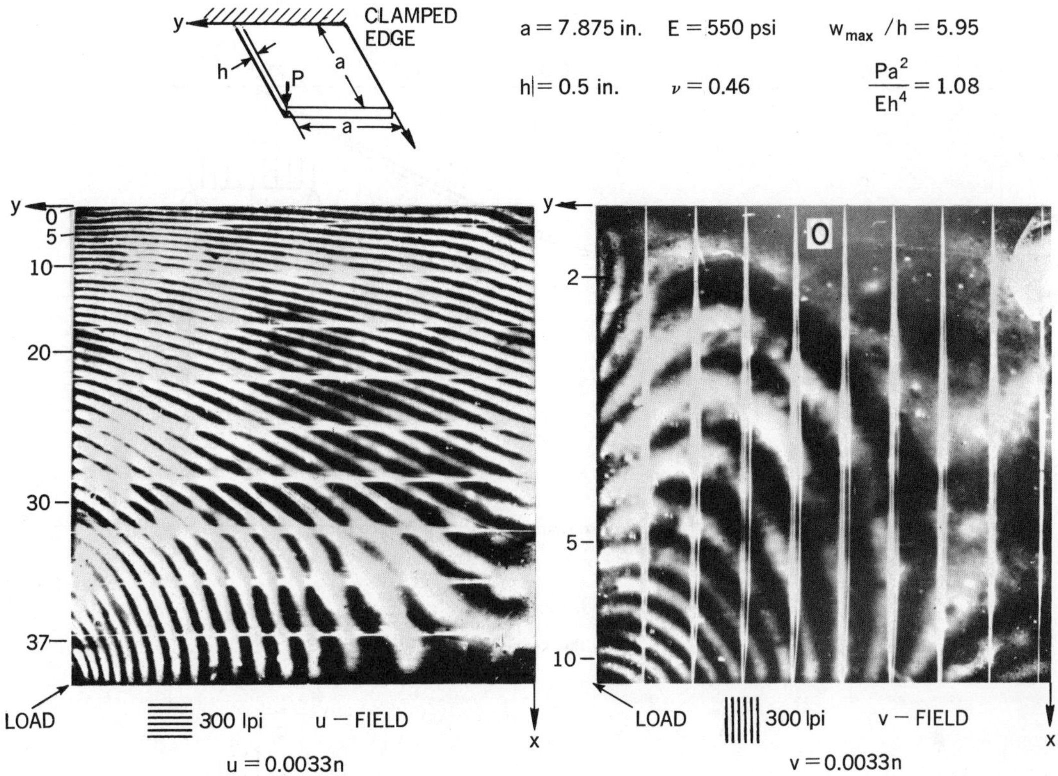

Figure 21.2. Isothetics in a largely deflected cantilever plate subjected to a corner load.

obtained by this technique gives a correct moire pattern that can be used for the analysis. A typical pattern obtained from a largely deflected plate is shown in Fig. 21.2.

In order to verify the results obtained by out-of-plane and in-plane moire methods, a test was also conducted on a square aluminum cantilever plate 9 in. × 9 in. × $\frac{1}{16}$ in. using electrical strain gages. The position of the strain gages are shown in Fig. 21.7.

21.5. Analysis and Results

Two families of moire patterns, u- and v-isothetics were recorded for each level of load applied to the plates. From these patterns the displacements u and v along any line of interest were plotted and the values of $\partial u/\partial x$, $\partial v/\partial y$, $\partial u/\partial y$,

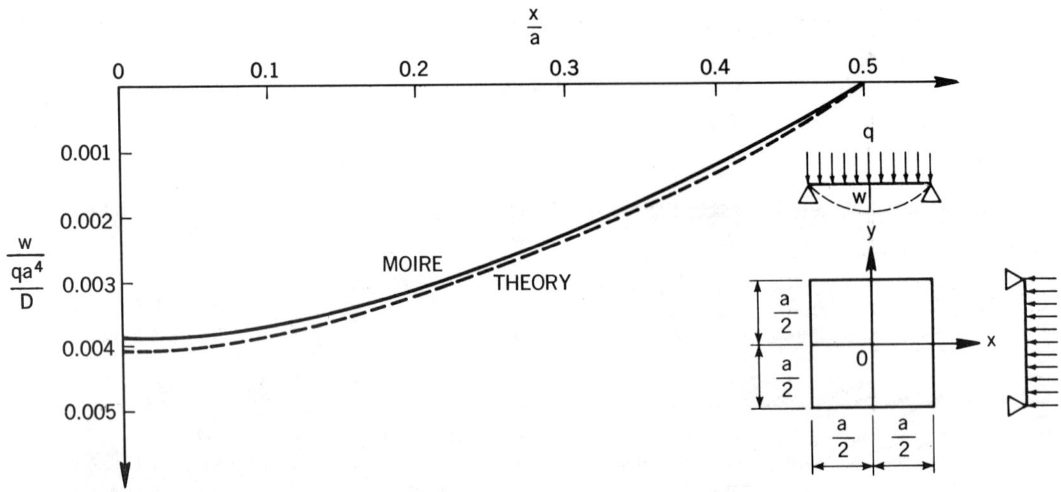

Figure 21.3. Deflection curve along the x-axis of a simply supported square plate under uniform pressure.

Figure 21.4. Cartesian strain components and along the x-axis of a simply supported square plate under uniform pressure.

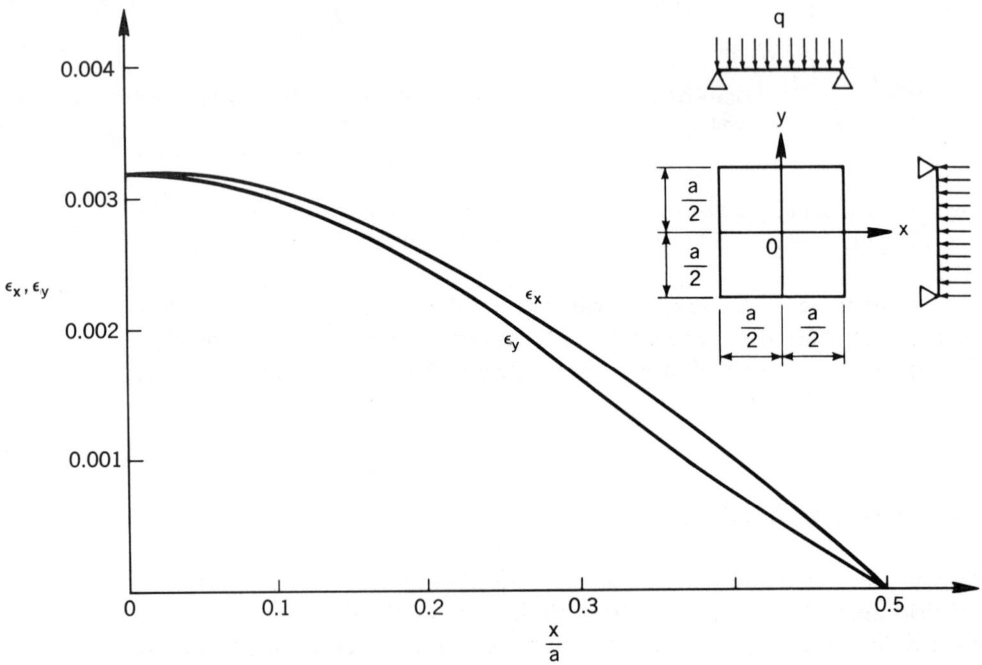

and $\partial v/\partial x$ obtained graphically. These values were then used in the strain displacement relations to determine the strains. From the strains and stress-strain relationship, stresses were determined.

Figure 21.3 shows the deflection curve for the simply supported plate obtained by integration (Eqs. 21.1). The theoretical deflection curve obtained for the same case is also included in this figure for comparison. The strain components ϵ_x and ϵ_y along the center (x-axis) of the plate are shown in Fig. 21.4. Normalized bending stresses σ_x and σ_y along the center of the plate are shown in Fig. 21.5 together with the results of the theoretical solution.

Figure 21.6 shows the nonlinear relationship between the applied load and (1) the maximum deflection and (2) the displacements at the loaded point, for the cantilever plate subjected to different levels of load. The bending moment coefficient k at different sections in the cantilever plate is shown in Fig. 21.7

Figure 21.5. Bending stresses σ_x and σ_y along the x-axis of a simply supported square plate under uniform pressure.

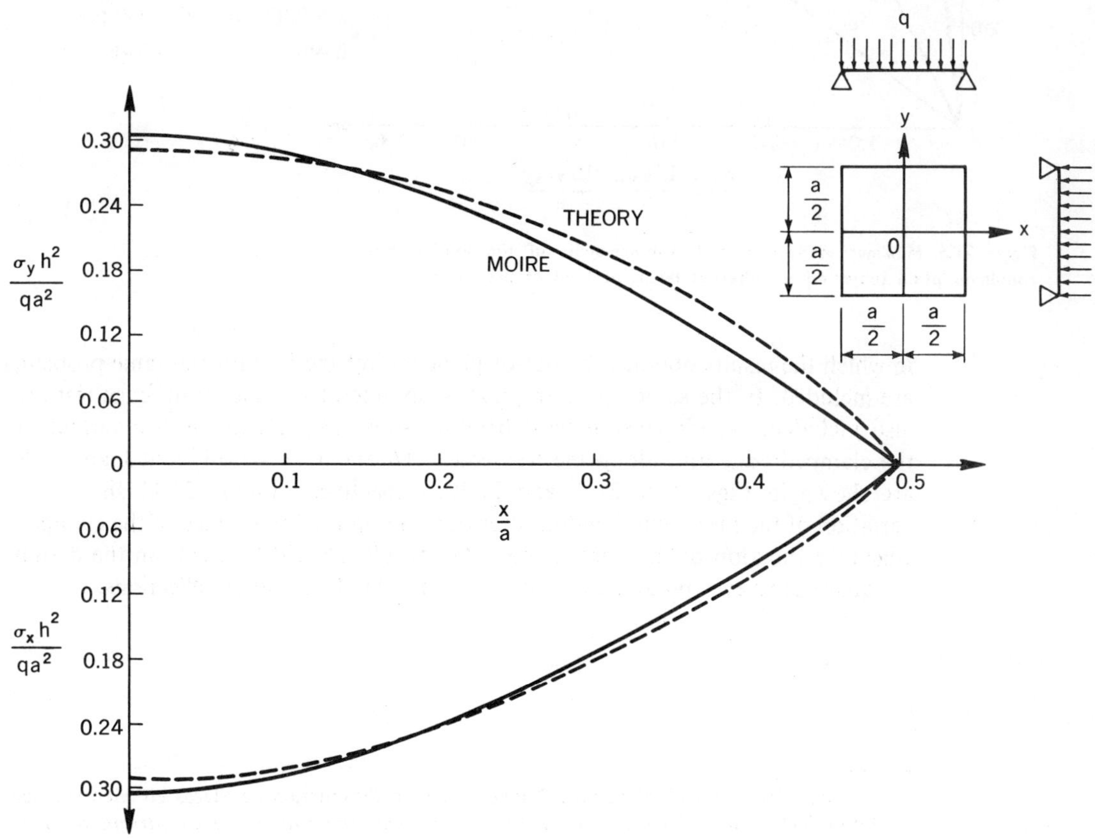

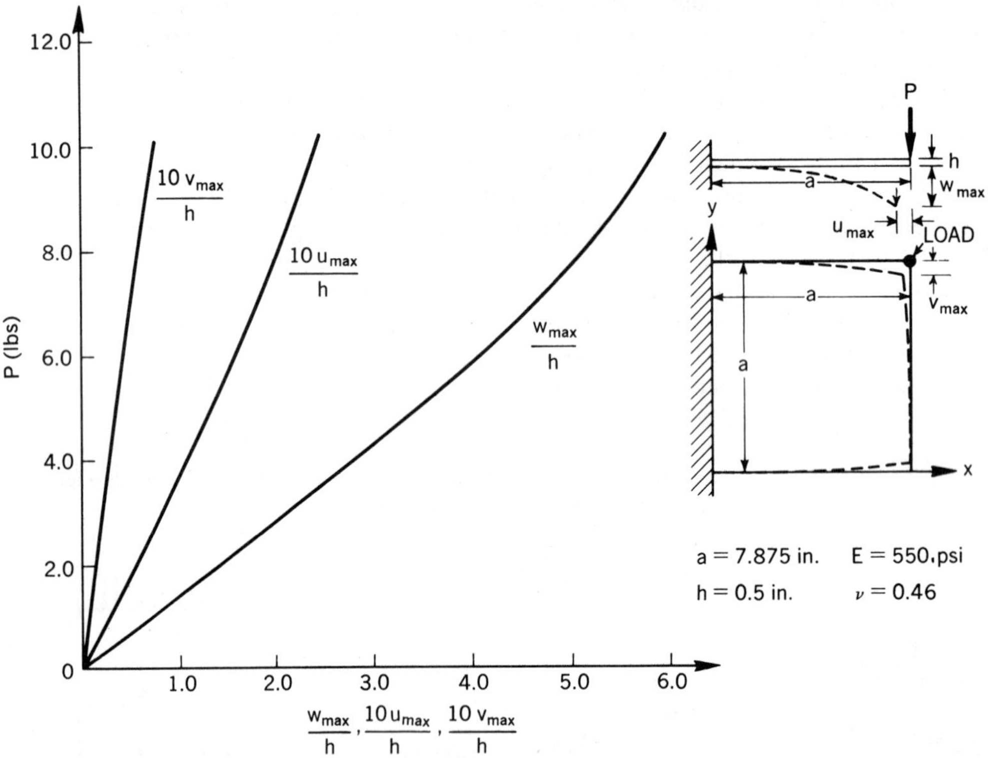

Figure 21.6. Maximum deflection and displacements at the loaded point in a cantilever plate subjected to different levels of load at the corner.

in which the results obtained by out-of-plane moire[6] method for the same problem are included. In the same figure the results obtained from the strain gage test are also included. Nondimensionalized bending stress (σ_x) along the top surface of the clamped edge and along the free edges *AB* and *CD* for different load levels are shown in Figs. 21.8, 21.9, and 21.10, respectively. Figure 21.11 shows the variation of the maximum bending stress (σ_x) along the top surface of the clamped edge as a function of the applied load. The results should be useful in the design of square cantilever plates subjected to a corner load and large deflections.

---

[6] A. Ponce, and H. Fernandez Tuneu, "Estudio Sistematico de Placas en Mensula por el Metodo del Moire," *Memorias de las IX Jornadas Sud-Americanas de Ing. Estructural*, Montevideo, Pub. No. 25 (1964), pp. 267–287.

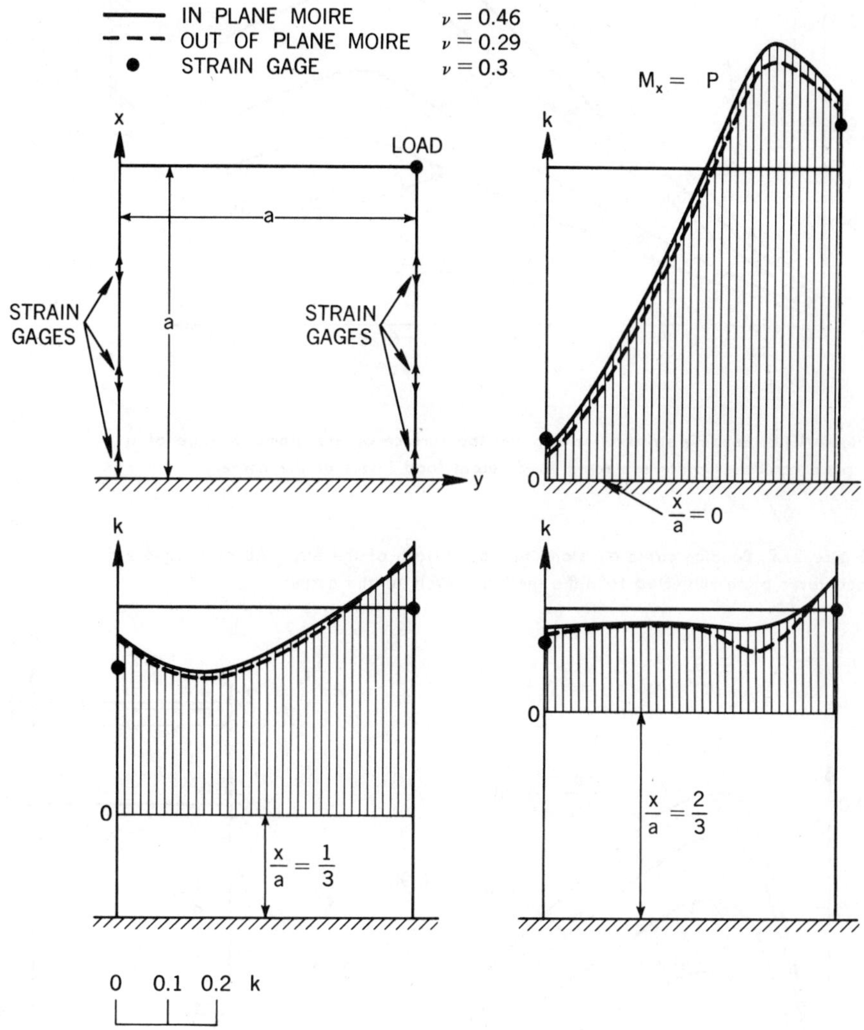

Figure 21.7. Bending moment M_x at different sections in a square cantilever plate subjected to a corner load.

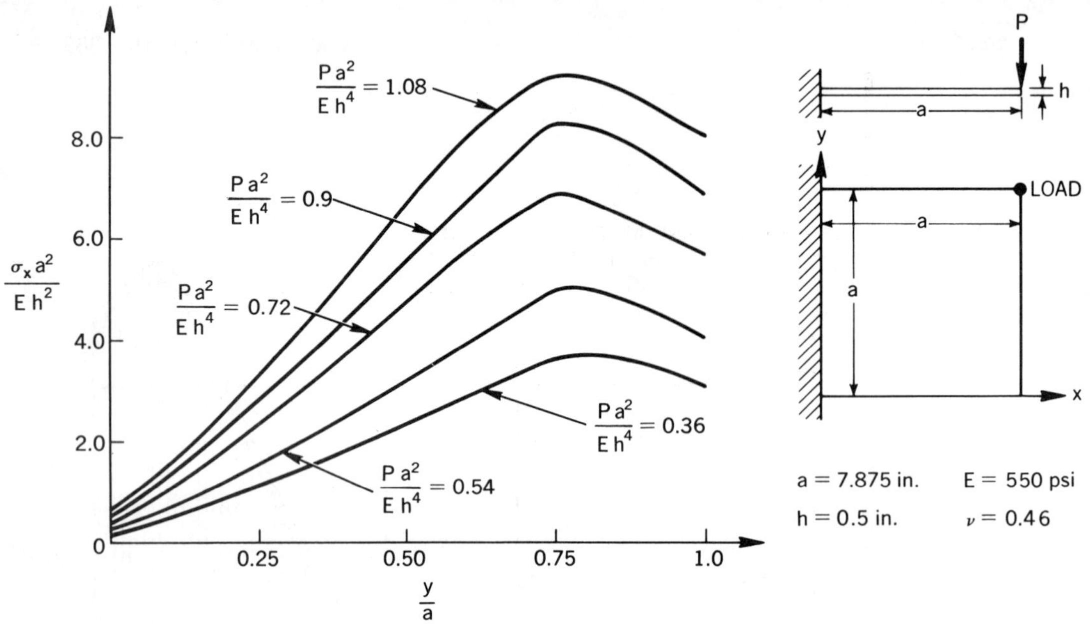

Figure 21.8. Bending stress σ_x along the top surface of the clamped edge of a square cantilever plate subjected to different load levels at the corner.

Figure 21.9. Bending stress σ_x along the top surface of the edge AB of a square cantilever plate subjected to different load levels at the corner.

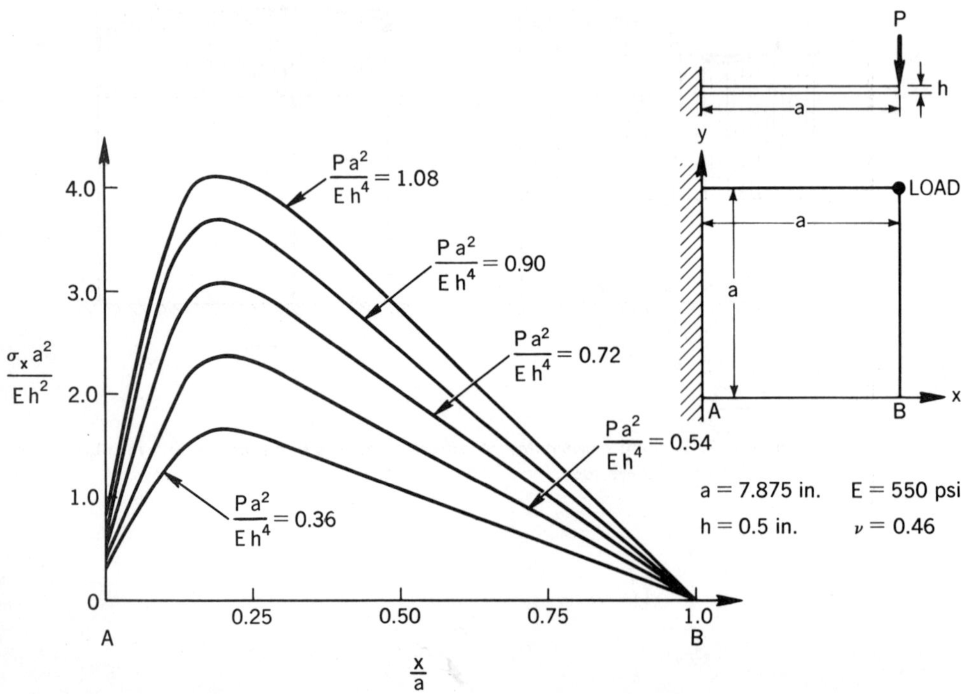

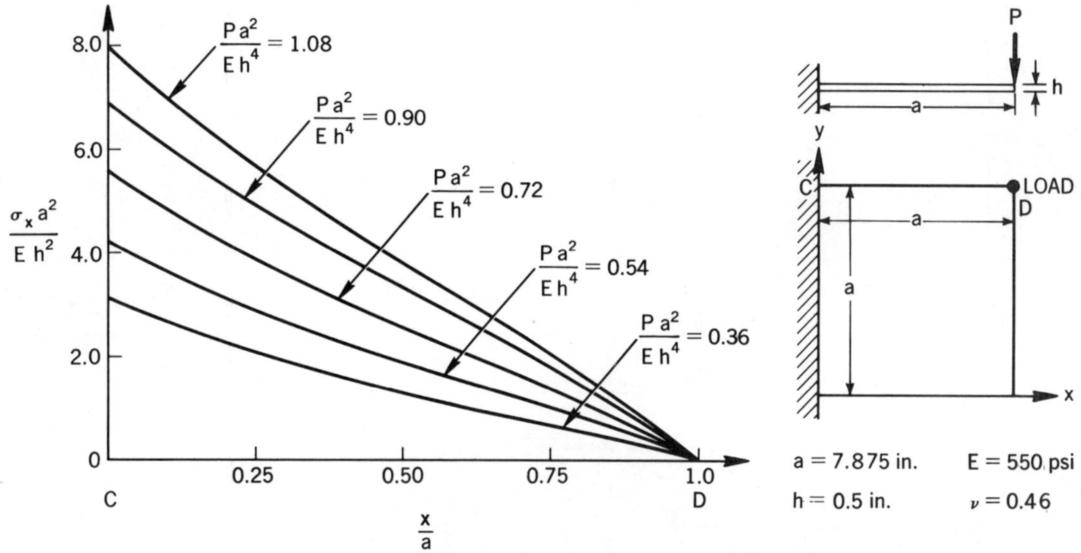

Figure 21.10. Bending stress σ_x along the top surface of the edge CD of a square cantilever plate subjected to different load levels at the corner.

Figure 21.11. Maximum bending stress σ_x along the top surface of the clamped edge in a square cantilever plate subjected to different load levels at the corner.

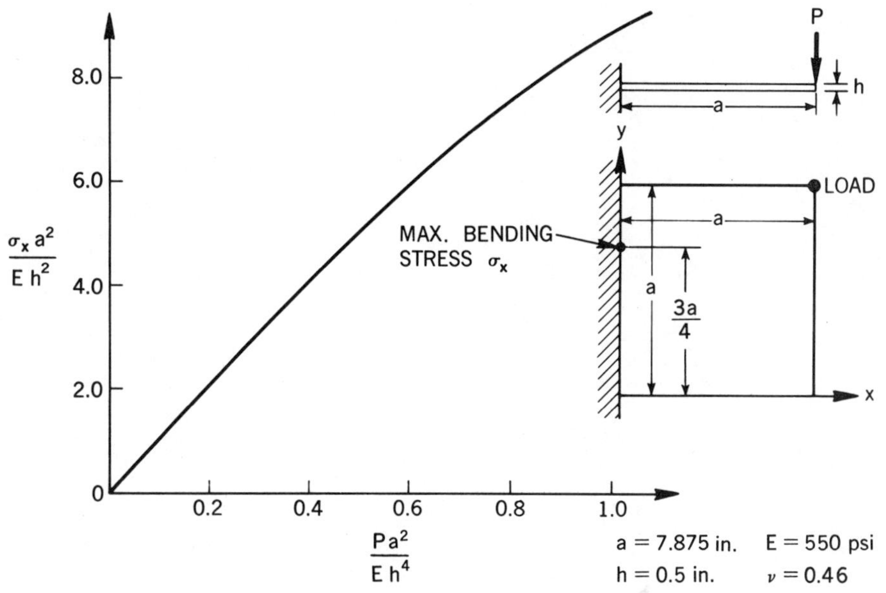

EXERCISES

Chapter 1

1.1. Translation of a Rigid Body

Obtain commercially (see Sec. 14.2), or draw, two gratings of about 50 lines/in. on transparent sheets. If a copy camera is available, copies can be made from one of the gratings shown in the text (for instance, Fig. 1.2). The film negative can be used without need of printings.

Fix one grating to a table. Put second grating on top of the first one. Shift this second grating over the fixed grating, about $\frac{1}{4}$ in. in the direction perpendicular to the grating lines. Count the number of dark-light cycles that occur throughout the field as the grating is shifted. Next, shift the grating $\frac{1}{2}$ in. at a 60° angle to the perpendicular to the direction of the grating lines and again count the number of cycles. Compare with theory $v/\cos\eta = np/\cos\eta$.

1.2. Rotation of a Rigid Body (Cartesian Grating)

Using the same two gratings prepared for Exercise 1.1, fix one grating at a table and attach the second copy to the first with a thumb tack passed near the center of both copies. Mark two points on the fixed grating, each 1 in. from the center of the thumb tack, one in a direction perpendicular and one in a direction parallel to the grating lines.

Align the two sets of grating lines and then rotate the upper grating 10°. Count the dark-light cycles that occur at each of the two points due to the rotation. Compare the number of cycles with the number of fringes that appear between

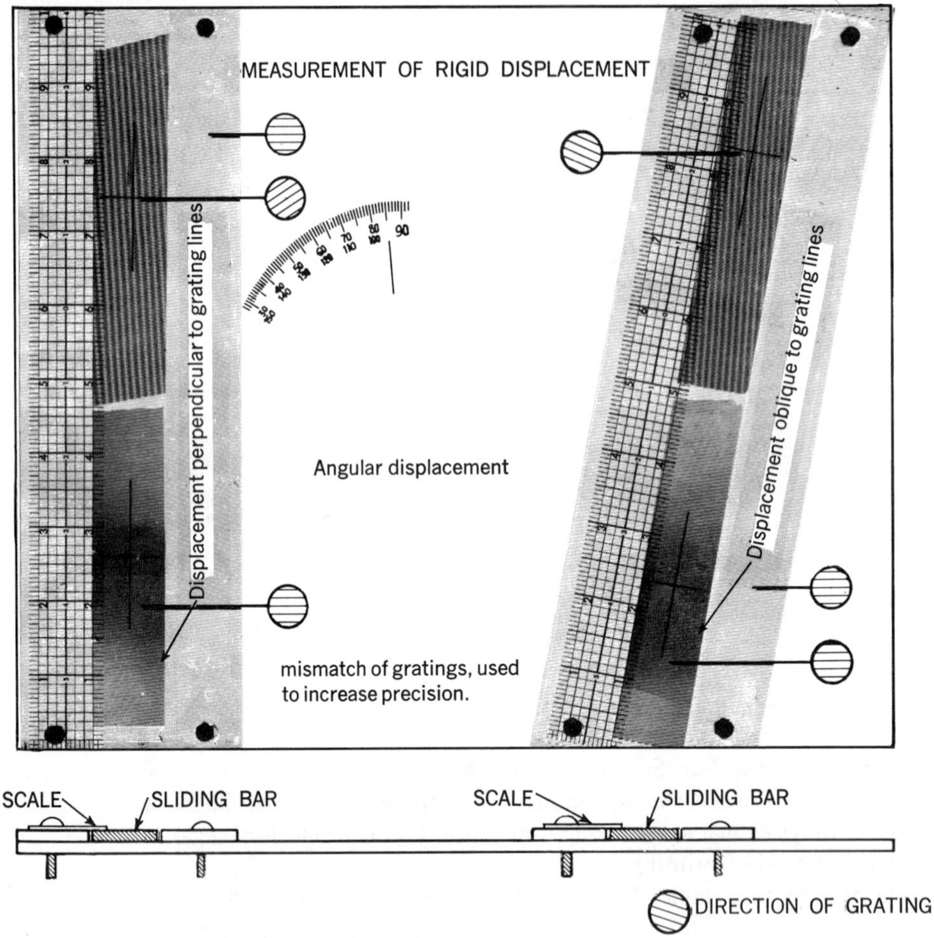

Figure E.1.1. Translucent board with coarse gratings to illustrate the measurement of rigid translation and rotation using moire.

each point and the center of the thumb tack. They should be the same. Explain the motion that occurs at each of the two points in the terms of $v = np$. Show that the motion at the two points should be equal to $(1 - \cos 10°)$ and $\sin 10°$, respectively.

Figure E.1.1 shows an arrangement of gratings and strips that has been found practical to solve Exercises 1.1 and 1.2.

1.3. Rotation of a Rigid Body (Radial Grating)

As in Exercise 1.1, obtain or prepare two gratings on transparent sheets, only this time the grating lines should be radial lines spaced about $\frac{1}{2}°$ apart.

Again fix one grating to a table. Attach the second grating with a thumb tack passed through the center of both gratings.

Rotate the upper gratings 10° and count the number of dark-light cycles that occur. Compare with theory.

1.4. Determining Elementary Patterns

Sketch in the blocks of Fig. E.1.2 the moire patterns obtained by placing a master grating with 100 vertical lines/in. on the following specimens:

a. A specimen stretched to 96 vertical lines/in.
b. A specimen compressed to 104 vertical lines/in.
c. A specimen with 100 lines/in., where the grating lines have been rotated 16° clockwise from the vertical lines.
d. A 1-in.-wide elastic beam with vertical lines is subjected to a pure moment (M) which produces fan-like grating lines that have a density of 96 lines/in. along the top of the beam and 104 lines/in. on the bottom of the beam.
e. The same specimen as above is subjected to the same moment, but now as a column with the grating lines running the length of the column.

Figure E.1.2. Sketching of moire patterns associated with blocks subjected to several loading conditions (Ex. 1.4).

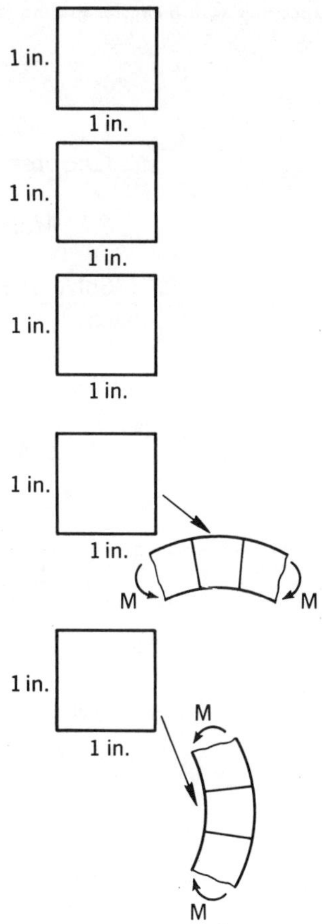

1.5. Determination of Poisson's Ratio

A grating of equispaced concentric circles was printed on a flat uniaxial tensile specimen and the specimen stretched. A circular master of the original size was placed on the deformed grating. Figure E.1.3 shows the pattern obtained. Explain the mechanism by which this particular pattern was formed and determine Poisson's ratio of the material.

1.6. Determination of Fringe Spacing

Consider an area on a specimen that is stretched 1 per cent in the x-direction, kept the same in the y-direction, and then rotated 10° counterclockwise ($\sin 10° = 0.1736$, $\cos 10° = 0.9848$, $\tan 10° = 0.1763$). If the specimen had 1,000-lines/in. gratings in the x- and y-directions, determine (1) the spacing in the x-direction of the moire fringes produced by the grating

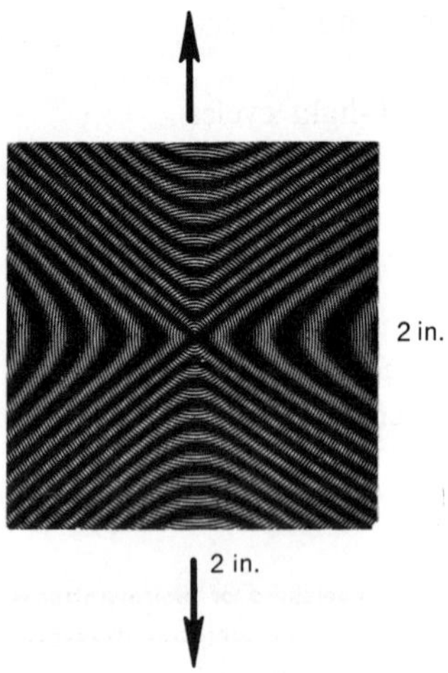

Figure E.1.3. Moire pattern on a tensile specimen with a circular grating (Ex. 1.5).

lines perpendicular to the x-axis, and (2) the spacing in the y-direction of the moire fringes produced by the grating lines perpendicular to the y-axis.

Chapter 7

7.1. Moire Analysis of a Disk

Make two photographs of the moire pattern produced by a disk of approximately 5 in. in diameter, subjected to a diametral compression, first with the grating lines horizontal and then with the grating lines vertical.

Compute strains along the horizontal and vertical axes. Determine E and ν from the computed strains, the amount of load and the known solution of stresses at the center of the disk. Compute stresses from E, ν, and from the strains along the horizontal axis and plot.

Chapter 8

8.1. Moire Analysis of a Disk

Solve the same problem described in Exercise 7.1, using the geometric approach.

Books on Moire

Guild, J. *The Interference Systems of Crossed Diffraction Gratings.* Theory of Moire Fringes. Monographs on the Physics and Chemistry of Materials. Oxford: Clarendon Press, 1956.

Guild, J. *Diffraction Gratings as Measuring Scales.* Practical Guide to the Metrological Use of Moire Fringes. London: Oxford University Press, 1960.

The following publications came to the attention of the authors after this book was being edited:

Ebbeni, Jean. *Etude Générale du Phénomene de Moirure.* Dissertation presented at the School of Applied Sciences of the Université Libre de Bruxelles. Two volumes: 1969. Part of this material has been published in issues 3, 5, and 6 of the seminar of "Analyse des Contraintes."

Theocaris, P. *Moire Fringes in Strain Analysis.* London: Pergamon Press, 1969.

INDEX

D

E

N

O

P

R

S

71 02016